Above: The celebrated Battle of Blenheim (13 August 1704) on the banks of the Danube firmly established the Duke of Marlborough's international reputation. With Prince Eugene, he devastatingly defeated Marshal Tallard's Franco-Bavarian army. (National Army Museum)

Left: The siege of Landau (12 September to 28 November 1704) was the final fruit of Marlborough's great victory at Blenheim. The operation was commanded by the Emperor of Austria and the Margrave of Baden. (National Army Museum)

Right: 'A general of the First Empire' by Edouard Detaille. An evocative if romanticized late-nineteenth century portrayal of the splendours of Napoleon's cavalry in its hey-day. Beyond the staff group the Horse Grenadiers of the Imperial Guard are forming. (Ken Trotman Arms Books)

Overleaf: The Battle of Lutzen (16 November 1632) was a notable Swedish victory over the forces of Wallenstein, but was dearly bought with the life of Gustavus Adolphus II, champion of the Protestant cause.

Atlas of
Military Strategy

David G. Chandler

THE FREE PRESS
A Division of Macmillan Publishing Co., Inc.
NEW YORK

Dedication: For Gill, patient companion on many an excursion into the military past.

The Free Press
A Division of Macmillan Publishing Co., Inc.
866 Third Avenue, New York, N.Y. 10022

Designed by David Gibbons. Edited by Tessa Rose. The publishers are pleased to acknowledge the generous assistance of the National Army Museum, Chelsea, and the Rotunda Museum of Artillery, Woolwich, in the preparation of the illustrations for this volume. Typeset by Trade Linotype Limited, Birmingham; camerawork by Photoprint Plates Limited, Raleigh; printed in Great Britain.

printing number
1 2 3 4 5 6 7 8 9 10

Title spread (overleaf): The Battle of Waterloo (18 June 1815) marked the final eclipse of Napoleon by Wellington and Blücher. The Duke is signalling his victorious troops to advance and clinch the decisive victory. (Lionel Leventhal)

ISBN 0-02-905750-7

Contents

Preface ... 6

Glossary ... 6

Key to Symbols ... 7

Introduction ... 7
The Value of Military History ... 7
General Definitions ... 8
The Principles of War ... 9
Variable Factors in Warfare ... 10
Generalship ... 11

The Classical Manoeuvres of Warfare ... 12

Strategic Legacies of the Past ... 14

PART I, 1618-1721: WARS OF RELIGION AND NATIONAL EXPANSION ... 18

Population Growth and Army Strengths, 1600-1800 ... 22

The Thirty Years War, 1618-48 ... 24
'The Father of Modern Warfare': Gustavus Adolphus and His Imitators ... 27
The Battle of Breitenfeld, 1631 ... 27
The Battle of Rocroi, 1643 ... 27
The Peace of Westphalia ... 28

The English Civil Wars, 1642-51 ... 30
The Battle of Naseby, 1645 ... 32

The Expansion of France ... 34
The Cockpit of Europe ... 37
The Campaigns of Luxembourg and William III ... 39

Masters of Manoeuvre ... 40
Turenne's Campaigns in the Vosges, 1674-75 ... 40
Marlborough's Forcing of the Lines of Brabant, 1705 ... 40
Eugene's March to Turin, 1706 ... 41

Marlborough's Generalship ... 42
The Campaign of 1704 ... 43

The Iberian Peninsula, 1702-13 ... 48

Siege Warfare ... 50
The Siege of Lille, 1708 ... 52

Islam versus Christianity ... 54
The Siege and Battle of Vienna, 1683 ... 55
The Siege and Battle of Belgrade, 1717 ... 56

Charles XII and the Great Northern War ... 58
The Battle of Narva, 1700 ... 59
The Battle of Poltava, 1709 ... 61
The Battle of the Pruth, 1711 ... 61

Colonial Strategy ... 62

The Settlements of 1713 and 1721 ... 64

PART II, 1722-1815: EUROPEAN RIVALRIES AND THE RISE AND FALL OF EMPIRES ... 66

The Austrian Succession and Seven Years War ... 69
Marshal de Saxe: Fighter and Innovator ... 69
The Battle of Fontenoy, 1745 ... 70
The Battle of Minden, 1759 ... 71

The Contribution of Frederick the Great ... 72
The Battle of Kolin, 1757 ... 73
The Battle of Rossbach, 1757 ... 74
The Battle of Leuthen, 1757 ... 75

Colonial Rivalry between the Great Powers ... 77

The British Acquisition of India, 1615-1815 ... 78
The Battle of Plassey, 1757 ... 78
The Battle of Assaye, 1803 ... 80

Combined Operations ... 82
The Siege and Capture of Quebec, 1759 ... 82
The Great Siege of Gibraltar, 1779-83 ... 82
The Invasion of Egypt, 1801 ... 82

The American War of Independence, 1775-83 ... 84
The Battle of Brandywine, 1777 ... 86
The Battle of Saratoga, 1777 ... 86
The Battle of Yorktown, 1781 ... 86

The Challenge of Revolutionary France ... 88

North Italy: 1796 and 1800 ... 90
The Campaign of 1796-97 ... 90
The Battle of Arcola, 1796 ... 92

The Campaign of 1800   93
The Battle of Marengo, 1800   93

The French in the Levant, 1798-99   96
The British Reconquest of Egypt, 1801   97

The Austerlitz Campaign, 1805   100

The Campaign of 1806   102

Napoleon's First Setbacks in Battle   106
The Battle of Eylau, 1807   106
The Battle of Aspern-Essling, 1809   107

Europe under Napoleon   108

The War in Spain and Portugal, 1807-14   110
Wellington as Chef de Guerre   111
Wellington on the Defensive:
  Torres Vedras, 1810   112
Wellington on the Offensive:
  Salamanca, 1812   112
The Siege of Badajoz, 1812   112
The Campaign and Battle of Vitoria, 1813   114

The American War of 1812   116

The French Invasion of Russia, 1812   **118**
The Battle of Borodino, 1812   121
The Battle of the Beresina, 1812   121

The Campaigns of 1813-14   **122**

The Waterloo Campaign   **124**

**PART III, 1816-78: TOWARDS
  TOTAL WAR**   **128**

The Latin-American Wars of Independence 132

**The Growth of World Empires and
  Spheres of Influence, 1816-1914**   **134**

The Sick Man of Europe   **137**

The Expansion of Tsarist Russia   **138**

The Crimean War, 1854-56   **142**
The Battle of the Alma, 1854   142
The Battle of Balaclava, 1854   145
The Battle of Inkerman, 1854   145
The Siege of Sebastopol, 1854-55   145

The Growth of the 'Second' British Empire 148
The Battle of Ferozeshah, 1845   152
The Battle of Sobraon, 1846   152
The Battle of Gujerat, 1849   153
The Expedition to Abyssinia, 1867-68   154

The Indian Mutiny, 1857-59   **156**
Causes of the Mutiny   156
Outbreak and Spread   156
The Relief of Lucknow, 1857   156
The Reconquest of North Central India   157

**The Risorgimento: The Italian Struggle for
Independence**   **158**
The Siege of Rome, 1849   159
The Battle of Solferino, 1859   161

**The Expansion of the United States of
America**   **162**
The American-Mexican War of 1846-47   164

The American Civil War, 1861-65   **168**
The Balance of Resources and Rival
Strategies   169
The Campaign of 1861   170

The Campaigns of 1862   170
The Year of Decision: 1863   171
The Campaigns of 1864   171
The Coup de Grâce: 1865   171

**The Campaign and Battle of First
  Bull Run, 1861**   **172**

**The Shenandoah Valley and Peninsular
  Campaigns of 1862**   **174**
The Peninsular Campaign, 1862   175
The Events of the Valley Campaign and
  their Effects   175

The Chancellorsville Campaign, 1863   **179**

The Gettysburg Campaign, 1863   **180**

**The Campaign and Siege of Vicksburg,
  1863**   **182**

The Victory of the North, 1863-65   **184**
Operations in the Western Theatre and
  in the Carolinas, 1863-65   184
Final Operations in the Eastern Theatre:
  Operations around Raleigh and
  Petersburg   185

The Unification of Germany under Prussia 188

**The Six Weeks (Austro-Prussian) War of
  1866**   **190**
The Battle of Königgrätz, 1866   190

The Franco-Prussian War, 1870-71   **192**
The Battle of Gravelotte-St. Privat, 1870   194
The Siege of Paris, 1870-71   195

**The Balkan Crisis of 1877-78 and the
  European Settlement**   **196**
The Russo-Turkish War, 1877-78   196
The Siege of Plevna, 1877   196

The Genius of Helmuth von Moltke   **198**

TOWARDS ARMAGEDDON   **200**

Select Bibliography   **202**

Index   **204**

Techniques, organization and weaponry

The Military Art in the Seventeenth and Early
  Eighteenth Century   33
Weapons, 1618-1720   37
Lines of Battle and Tactical Formations   65
Tactical Experiments in the Mid-Eighteenth Century   76
Artillery of the Gribeauval System   95
Small Arms Performance   95
Tactical Sequence of French Early Nineteenth
  Century Engagements   95
Napoleon as a General: His Contribution to Warfare   98
The Evolution of General Staff Organizations   127
The Development of Fortification   140
Developments in Minor Tactics   165
The Railway Age   166
The Electric Telegraph   167
Developments in Weaponry   186
  Artillery   186
  Small Arms   187
The General Staffs   193

Preface

This Atlas is designed to illustrate the salient developments in the art of land warfare between the onset of the Thirty Years War in 1618 and the ending of the Russo-Turkish struggle in 1878. The phrase 'military strategy' has been used in the widest interpretation included in *The Concise Oxford Dictionary*: namely, all matters pertaining to 'generalship, the art of war'.

Inevitably, the subjects depicted and described are a personal selection, and considerable numbers of conflict areas have necessarily been left out or merely mentioned in passing to keep the book to manageable proportions. The emphasis has been laid on the European experience and conflict, but attention has also been paid to those parts of the world that found European soldiers serving in response to the colonial and commercial ambitions of their respective countries. The US Civil War is also treated in some detail.

In most instances, the deeds of the great commanders of these 260 years have been represented by only selected examples of their military art and science. But, in the case of Napoleon and the US Civil War generals, a rather more generous allocation of space was considered necessary. As Napoleon was the greatest commander in modern history and the US Civil War was the most costly war of the nineteenth century, this imbalance is probably justifiable. Napoleon provided the great challenge to the concepts of seventeenth and eighteenth century warfare and, although his career ended in defeat and failure, he left an indelible imprint on his own and succeeding generations. Despite this, as the heir of de Turenne, de Saxe and Frederick the Great, he was in many ways the product of his predecessors' military skills and experience. Two generations later, Grant, Sherman, Lee and (in Europe) von Moltke were still applying what they believed to be his methods, ushering in the age of truly modern warfare.

Most campaigns and battles illustrated in this book have been reduced to a bare minimum of information. Three significant campaigns—those of 1704, 1806 and 1862—have been treated in greater depth as special case studies, in order to represent the highlights of the military art in their respective centuries. Care has been taken to include as much background information as possible on such matters as military methods and weaponry, so that the achievements of the great soldiers may better be understood and placed in perspective. Occasionally, for the reader's convenience, a background subject has been extended to 1914, but the campaign and battle studies end with Plevna in 1878.

No attempt, therefore, has been made to produce a comprehensive study of even the military aspects of the two and a half centuries involved. However, it is hoped that a broad appreciation of this dramatic period of military history will be obtainable from the pages that follow, and that the reader who enjoys studying maps and diagrams as well as pictures and the printed word will derive some pleasure from this treatment, however incomplete.

I would like to thank Mrs. Janet Donaldson and Mrs. Jan Ingram for stoically typing what must have been at times a rather confusing manuscript; and also Richard and Hazel Watson for producing the striking maps, diagrams and charts from my designs. Mistakes there are bound to be, and for them I assume total responsibility. I am indebted to my old friend Brigadier Peter Young, DSO, MC (then Reader in Military History at RMA Sandhurst) for first putting the idea in my head in 1966 as a shared project, and then for entrusting the entire concept to me when pressure of other affairs made it necessary for him to beat a dignified retreat. I also owe a great deal to Lionel Leventhal, David Gibbons and Tessa Rose of Arms and Armour Press for all their enthusiasm and advice at all stages of production. Last, as always, I owe much to my wife Gill and family for their assistance and, in one case at least, with the drawing of a key map.

David G. Chandler, Camberley, 1980

Glossary

Base: The locality from which an army is reinforced or supplied.

Blockade: The loose containment of a garrison or town population by hampering free access or egress.

Bridgehead: A narrow sector of captured enemy territory, often over a river.

Communications, Lines of: The roads, canals or rivers used to link an army with its base, along which flow its supplies and munitions.

Envelop: To turn the enemy's flank and thus threaten his communications.

Exterior Lines: Lines of communication that splay outwards from the front; they are often very long, posing problems of movement and supply.

Form Front to a Flank: To wheel a force outwards until its front is parallel to, rather than perpendicular to, its communications.

Forward Slope: Exposed tactical position on a hillside facing towards the foe.

Front: The area occupied by an army facing towards its opponents, or the extent of territory being actively contended for.

Grand Strategy: The formulation of national war aims, policy and alliances.

Grand Tactics: The formulation of overall tactical plans of manoeuvre or battle.

Interior Lines: Lines of communication that close inwards from the front and, therefore, tend to be short and accessible.

Investment: The tight containment of a garrison etc. by preventing access or egress.

Line of Advance: The general direction taken by an army.

Line of Operations: An army's line of march through hostile territory.

Line of Retreat: The general direction taken by a retiring army.

Lines of Circumvallation: Trenches and field-works around a besieged fortress facing away from the target to prevent relief forces breaking in.

Lines of Contravallation: Trenches and field-works around a besieged fortress facing towards the target to prevent sorties by the garrison.

Logistics: The science of moving, supplying and reinforcing armed forces.

Outflank: To turn the enemy's flank on a small, tactical scale.

Penetration, Single or Double: The forcing of an entry, or entries, through the enemy's front, thus creating a break-through.

Re-Entrant: A bulge or backward bend in a line of battle to form a trap.

Refused Flank: To draw back a flank behind the general line of the army to reduce the possibility of being outflanked.

Reverse slope: Protected tactical position on the rear side of a hill out of the enemy's view or direct fire.

Reversed Front: When an army takes up a position facing towards its own base after severing the enemy's communications.

Salient: A bulge or forward bend forced in the enemy's line of battle or front.

Strategy: The formulation and execution of plans of campaign designed to carry out the policy decisions taken at Grand Strategic level.

Strategic Consumption: The diminishing power of the offensive as an advancing army moves far from its bases and its communications become over-strained, resulting in a reduction in battle-power.

Strategic Flank: An army's flank, which, if turned, will threaten its whole front and lines of communication.

Siege: Military operations directed to achieve the capture of an invested enemy town or fortress, by force or by starvation.

Tactics: Minor military operations planned and executed to carry through the Grand Tactical scheme for the fighting of a major engagement.

Theatre of Operations: The complete area, including bases, in which armed forces are operating.

Introduction

"Every day I feel more and more in need of an atlas, as the knowledge of geography in its minutest details is essential to a true military education."—General William Tecumseh Sherman.

General Sherman's testimony from the hard campaigns of the US Civil War in the early 1860s illustrates the significance of military geography for the professional soldier. The subject is no less vital or interesting for the layman who finds satisfaction in studying the history of the near or more distant past, in which, inevitably, wars have figured prominently and tragically generation after generation.

The purpose of this Atlas is to serve the general reader as a compendium and guide to the salient points in the development of land warfare from the start of the Thirty Years War in the early seventeenth century to the end of the Balkan struggle in the third quarter of the nineteenth. This latter conflict was the harbinger of a far greater tragedy that was to engulf mankind just 36 years later; a truly titanic struggle that would be sparked off by tensions and disputes arising in no small measure from the inadequate settlement of the Balkan problem embodied in the Treaty of Berlin. The famous conference that preceded this fateful agreement also marked the emergence of Bismarck and von Moltke's German Empire to a position of predominance among the major European powers—yet another development that would lead tragically, directly or indirectly, to the massacre of several generations of European manhood, and many Americans and Asians too, in what future generations of historians may come to call 'The Great World War, 1914-45'.

It is also hoped that the maps and diagrams upon which this volume is based will aid the student towards a fuller comprehension of the many excellent monographs devoted to the examination of the military history of these turbulent 260 years. Not all of the books are adequately supported in the matter of maps, and yet for many readers there is a particular fascination about following the stages of a manoeuvre or battle with the aid of diagrammatic representations. No one would claim that coloured rectangles and boldly drawn arrows or other symbols can do more than provide broad and inadequate indications of such dramatic and terrible events involving tens of thousands of men. "War is hell," as Sherman remarked on another occasion, and stylized maps and diagrams cannot in any way convey the extent of the fear, confusion and horror of war. However, they can assist in the proper appreciation of schemes of manoeuvre, of relative distances and obstacles dividing key points, and give some idea of the successive phases in an engagement, thus helping to complement the impressions conveyed by the written word or pictorial representation. The study of maps, therefore, can play a significant part in a visual approach to study, whether for pleasure or for serious instruction.

The Value of Military History
Few fields of study enjoy a larger or more enthusiastic popular following today than military history. Every year sees hundreds of new or reissued titles pour from the presses, and few would deny that there is considerable dross and needless duplication amongst them. The range is diverse: from slim weekly part-works to massive works of scholarship. And yet much of what is published is of value, and the public appetite seems insatiable. The reason for this degree of interest deserves a little analysis.

The relevance of military history has been queried by some academic authorities, who equate its study with the encouragement of militarism. At best, such critics avow, the subject is only suitable for military men in the furtherance of their professional knowledge. There are even a number who would claim that the study of the past is valueless even for serving soldiers; that the nuclear revolution experienced since 1945 has relegated everything that happened before that date to a mere

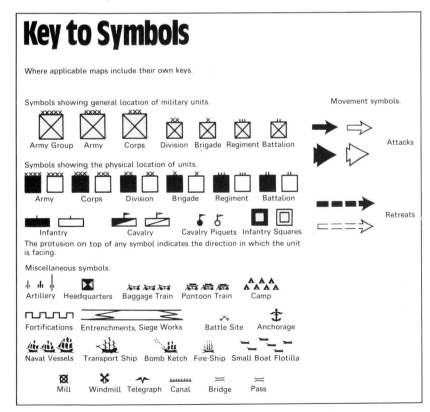

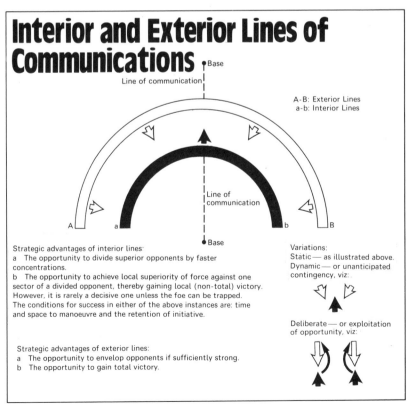

7

Introduction

archaic interest. Fortunately, there are others who hold diametrically opposite views, and avow that although weaponry and methods of communication have changed and continue to change with fearful rapidity and increase in potential, yet the essentials of the art of war remain immutable from age to age, however great the scientific developments. The essentials of generalship and man-management have not changed very dramatically over the years, and as much can be learned from the study of selected examples from the past as from more recent experience. This was self-evident to the anonymous 'Captain J.S.', author of *Military Discipline—or the Art of War*, published in London in 1689, who wrote as follows:

"It is not only Experience and Practice which maketh a Soldier worthy of his Name: but the knowledge of the manifold Accidents which arise from the variety of humane actions, which is best, and most speedily learned by reading History: for upon the variety of Chances that are there set forth, he may meditate on the Effects of other men's adventures, that their harms may be his warnings, and their happy proceedings his fortunate Directions in the Art Military."

Napoleon, equally, was under no illusions about the subject's importance: "Read and meditate upon the wars of the greatest captains. This is the only means of rightly learning the science of war." Many distinguished soldiers of modern times, including Field Marshals Wavell and Montgomery, have recorded their conviction that the study of military history—properly conducted—is of the greatest value in forming a soldier's appreciation of his profession's complexities. If a soldier is excessively bound to the examples of the past he may give too much prominence to the supposed lessons of former victories, rather than truly appreciate the even more important lessons behind failures. Thus the vaunted Prussian Army went down before Napoleon in 1806 through undue veneration for Frederick the Great's ideas, and a refusal to appreciate the ways in which warfare had changed over the intervening years. By the same token, a soldier who ignores the past may well fall into old, well-proved errors. A fuller appreciation of what befell Charles XII of Sweden in 1708-9, or Napoleon in 1812, might have helped the Wehrmacht meet the rigours of the Russian winter in 1941.

But if military history is of importance to the soldier, it is no less a fascinating and worthwhile study for the interested layman. Man's destiny has been closely associated with wars from the earliest times of recorded history, and in that fact lies much of the subject's tragedy and importance. Wars have time and again proved to be the acid test of civilizations. Sir Winston Churchill has described battles as "the principal milestones in secular history". Above all, military history from first to last is concerned with man himself: his reaction to danger, discomfort and fear, his response to extremes of stress and opportunity. The ways in which he has adjusted to these situations cannot but make for interesting reading. In this way the subject retains its appeal for old and young alike. In the 1890s, General Hamley observed that "no kind of history so fascinates mankind as the history of wars". The same could be said today, in the 1970s. This is not a morbid interest, but rather the perennial response to the appeal of adventure and drama, to the relationship between leader and led and to man's ability to rise triumphant over daunting difficulties.

No one would deny that wars are acts of regrettable human folly, causing unjustifiable distress and suffering to combatants and non-combatants alike, generation after generation. But such has been the destiny of mankind. The study of recent world events does not seem to show that man has learned to shun the scourge, although he may have learnt the need to avoid its very worst excesses in the interests of the survival of the human race, or so it must be hoped.

This present Atlas is confined to the illustration of land warfare over a period of little more than two hundred and sixty years, during which time several great leaders emerged. The pendulum of war swung twice through a broad arc: from the near-total Wars of Religion at the outset, through the period of reaction embodied in the more limited concepts of the late seventeenth and early eighteenth century, and back to the onset of near-totality as a result of the strong nationalistic and ideological passions released by the French Revolution and the technological advances of the nineteenth century. All was leading relentlessly towards the Armageddon of the first half of the twentieth century. We can but hope that the like of these great wars will never be seen again; alas, not every indicator of present-day superpowers and Third World politics is fixed at 'set fair'. The nature of man has not necessarily changed wholly for the better since 1618. However, history can offer few—if any—more dramatic periods. Our purpose has been to provide not an encyclopedic coverage, but rather to give a synoptic view of major developments in the field of land warfare.

General Definitions

To assist a reader attempting to grasp the complexities of the study of military history, it may prove useful to define the strata or levels of warfare, and also to devote a little space to a description of the so-called 'principles of war'.

Five major strata of war may be distinguished, each with its own special significance and yet inter-linked with others in varying degrees. First comes the level of *Grand Strategy*. This can be defined as the art and science of developing and employing all the resources of a nation—political, social, economic, psychological and military—to provide the greatest possible support to national aims and interests, whether in peace or war, so as to reduce the chances of frustration and defeat and increase the probability of success. As such, grand strategy is clearly the responsibility of kings, ministers and cabinets, acting with the aid of advice from their senior military advisers. The level of grand strategy, therefore, is concerned with policy. Gustavus Adolphus' decision to intervene in the Thirty Years War in 1630, or Frederick the Great's determination to seize Silesia in 1740, would come into this category.

Once these decisions have been taken, the conduct of war enters the level of *Strategy*. This comprises the art and science of employing the nation's military resources to secure the objectives of policy in the most effective and economical fashion possible by the use of force or its threat. This is the realm of senior commanders and chiefs of staff and involves the planning of campaigns in broad principle: the selection of objectives; the devising of timetables of movement and reinforcement; and the disposition of major formations in the front line, in support or in reserve. It also entails careful calculation of probable enemy resources and intentions, so that his moves or countermoves may be intelligently anticipated and guarded against. This is particularly the realm of military intelligence; the collection, collation and interpretation of information from many sources, so that an impression of what is not known can be pieced together by studying what is. Wellington referred to this as getting to know ". . . what lies on the other side of the hill". Clearly, if the foe is to be deceived and surprised, and generally thrown off balance, this aspect of strategic planning is of great significance. Napoleon was well aware of its importance, and devoted no little time and effort to building up a clear idea of his opponents, and tailored his plans to exploit their known or presumed weaknesses. If he miscalculated badly in 1812 and 1815, this should not conceal his successes in earlier years. Napoleon was probably the greatest master of strategy in modern history.

The third level of warfare is that of *Grand Tactics*, which is concerned with the planning of major operations in all their complexity in order to achieve, in the theatre of war, the declared strategic objectives and intentions. It comprises the carefully-considered placing and movement of formations in relation to each other and the enemy, and the broad planning of the stages of battle once contact has been made with his major forces. The frontier between strategy and grand tactics is often blurred, and they share many features. The achievement of speed, surprise and concentration is vital at both levels, as are the selection of correct intermediate objectives and the achievement of co-ordination, not forgetting the maintenance of an adequate reserve ready to exploit success or remedy adverse developments. Grand tactics remained very much the responsibility of army commanders in the seventeenth and eighteenth century. Gustavus Adolphus, de Turenne, Marlborough and Prince Eugene all controlled their armies in action in a highly personal and direct fashion. By the time of Frederick, however, and above all of Napoleon, the development of intermediary staffs made delegation rather more common, but commanders still often controlled the grand tactics of their battles, and many of the details as well. The same was still broadly true in the 1870s, despite the arrival of the electric telegraph.

Fourthly, we come to *Minor Tactics*: the actual methods of employing units in combat on the field of battle. These are pre-eminently the sphere of the junior leader and the rank and file, but in the period covered by this Atlas it was not uncommon for even senior commanders to become personally involved in the hurly-burly of close action. Marlborough led several cavalry charges as Captain-General, and was almost killed at Ramillies; General Bonaparte courted death at both Lodi and Arcola in 1796; Wellington was wounded at Salamanca in 1812, and had to shelter within the British squares at Waterloo; while Gustavus of Sweden was killed at Lützen in 1632; and Charles XII was shot down at the siege of Frederickshal in 1721. The dangers of the battlefield and siege-trenches were, therefore, not remote for senior officers. Of course battles have always been won or lost at tactical level; the adage 'in strategy there is no victory' makes the point, for strategy and tactics are inextricably related. As the Prussian military philosopher, Karl von Clausewitz, defined the relationship: "Tactics is the art of using troops in battle; strategy is the art of using battles to win the war." Field Marshal Earl Wavell placed a slightly different emphasis in his definition: "Tactics is the art of handling troops on the battlefield, strategy is the art of bringing forces to the battlefield in a favourable position." Both descriptions illustrate the inter-relation-ship of the two strata.

Lastly, it is necessary to mention *Logistics*. This comprises everything to do with the provision, movement and supply of armed forces, and rather than being a separate strata in its own right it is all-embracing, forming a critical part of all the other four levels of warfare. Admiral Mahan stressed its link with grand strategy when he wrote of it as ". . . the processes of industrial mobilization and the functions of a war-time economy in support of military operations". Napoleon's well-known maxim, 'an army marches on its stomach', was itself a reflection of Frederick the Great's conviction that 'the foundation of an army is the belly', and both stressed the importance of the science of logistics.

Commanders who ignored matters relating to provision and supply courted disaster. One such was Marshal Vendôme, who instructed his staff in the 1700s that "he was not to be troubled with matters concerned with the mere subsistence of the army". Napoleon, his cited view notwithstanding, was not notable for the priority he in fact accorded logistics in his planning or on campaign; some part of the cataclysms suffered in Russia, 1812, and in Spain, 1807-14, by French armies was due to this neglect. But most of the other great generals of our period paid great attention to what Wavell described as "the crux of general-ship—superior even to tactical skill". These included Gustavus Adolphus, who virtually created the first modern army; Marlborough, who paid great attention to his soldiers' welfare; Frederick the Great, who devised firm rules for the support of Prussia's armies; and of course Wellington, whose 'triple-line' system of lines of communication and depots in the Peninsular War contrasted so markedly with the far more haphazard and unsatisfactory French arrangements.

The Principles of War
These form another framework which can aid the analysis of military operations and attempts to assess the standard of generalship displayed by various commanders. It is necessary to accept at the outset, however, that it is very deceptive and even dangerous to place too much reliance on external criteria applied to so fluid and complex a subject as war, which is as much an art as a science. The so-called 'principles of war' should never be regarded as immutable laws governing military behaviour, but rather as guide-lines towards the ingredients that, in the correct combination and quantity, can lead to success. The achievement of a true

Karl von Clausewitz (1780-1831), Prussian soldier and military thinker, whose incomplete work, On War, *provided a philosophy which had great influence throughout the world during the nineteenth century.*

balance between the various 'principles' is of great importance, for all too often they can appear to be at variance with one another: thus the requirements of 'concentration of force' can clash with those of 'economy of effort', and similar conflicts of priority can occur between other principles.

Nevertheless, these criteria do provide useful yardsticks for assessing and comparing the performances of past commanders, and they can be applied to any period of military history. As guides to study, therefore, the principles can be of some assistance.

The Prussian, von Clausewitz, writing in 1818, was the first military philosopher to attempt a written formulation of principles. However, a study of the writings of earlier commanders shows that they held a number of convictions in common, many of which continued to be reflected through the years down to the present day training doctrines of modern armies.

Ten major 'principles' can be distinguished: a number are obvious in their relevance and significance, but all bear definition and illustration.

1. *Selection and Maintenance of Aim.* A clear intention and objective is of basic importance to any operation of war. The aim should be clear and simple, and closely adhered to once selected. Obviously, the selection of aims needs to be carefully based upon available means. Thus Charles XII's decision to plunge Sweden into an all-out war with Russia in the early eighteenth century was unrealistic in view of his country's limited resources, which in the end proved incapable of withstanding the strains involved.

2. *Maintenance of Morale.* This is a perpetual concern for all commanders and always has been. The rank and file need faith in their cause, their leader and their ability to win, and this confidence must be carefully inculcated and fostered. The example of General Bonaparte in 1796—faced by a demoralized and near-desperate first command on the

Introduction

Ligurian coast—brings out the point. He succeeded in inspiring the troops to make an immense effort, which reversed the situation.

3. *Offensive Action*. Wars and campaigns can only be won by positive action. Periods on the defensive inevitably occur—while absorbing enemy attacks, or preparing an attack—but the importance of gaining the initiative has always been central to the achievement of success, providing offensive action is not undertaken in a completely foolhardy fashion. Therefore, Marlborough's decision to attack Villeroi at Ramillies was strongly justified; but Charles XII's attack on Peter the Great at Poltava was questionable, given the disadvantage of the known odds.

4. *Security*. This connotes the positioning and organization of a force so that it cannot be taken at a disadvantage by the enemy. An ideal example is Wellington's position in the Lines of Torres Vedras in 1810 which was designed to protect the British Army's vital base of Lisbon.

5. *Surprise*. One of the best ways of securing the initiative and gaining psychological dominance over the enemy. It can produce results out of all proportion to the effort involved, and can help overcome a situation based on unfavourable odds. To succeed, this plan requires careful attention to secrecy, concealment and deception as well as originality. In execution it needs both speed and boldness. Charles XII's attack at Narva, 1700, during a blizzard, which concealed his weak numbers and completely surprised the Russians, is a good example.

6. *Concentration of Force*. This principle suggests the need to mass sufficiently superior forces against the enemy's position at the correct place and moment to ensure success. Obviously, superior strength applied in the correct fashion can be a major factor in achieving success, and this was always a major aim of the Great Captains of the past. Marlborough's massing of cavalry opposite Tallard's weakened centre at Blenheim (see p. 44) was the vital preliminary to winning the battle. Similarly, the whole purpose of Frederick the Great's 'oblique attack' (as at Leuthen, see p. 75), was to bring superiority to bear on the flank of his opponent's line as a preliminary to rolling up his position. Napoleon invariably maintained a sizeable reserve, 'la masse de décision', ready to clinch his great battles.

7. *Economy of Effort*. This postulates the need to assess accurately the numbers of troops required to cover all requirements, both defensive and offensive, including the provision of a reserve ready to launch the decisive blow at the correct place and time. Equally a part of this concept is the use of small numbers of troops to distract and tie down larger enemy forces in irrelevant areas pending the major attack on the critical sector. History abounds with examples of this type of deception. Marlborough's use of Orkney's British battalions to distract Villeroi at Ramillies (see p. 46), or Frederick's employment of his advance guard at Leuthen to attract Austrian attention while the main Prussian force marched unseen towards his flank are good examples. This was also a vital part of Napoleon's strategic systems (see pp. 98 and 99): to use a small force to distract the enemy's attention from major developments. But it is equally dangerous to use too few troops for an attack or manoeuvre if the result is to run an unacceptable degree of risk and thus court defeat. Similarly, to use too many troops can lead to fatal results; as, for example, with the Allies at Austerlitz in 1805 (see p. 100). Therefore, the calculation has to be carefully made: false economy or over-insurance can be equally fatal.

8. *Flexibility*. The ability to adjust a plan to actual circumstances is an equally important factor in generalship. Too stubborn an adherence to a pre-conceived scheme can lead to great difficulties. Napoleon was a pre-eminent 'master of the alternative plan' and was rarely caught at a complete loss, whatever surprise circumstance faced him. But 'Flexibility' does not permit the abandonment of the main aim, only the search for a way to achieve it by another approach. To be able to judge the situation correctly—and then make any necessary adjustments to the moves in progress—calls for a commander to be in the right place to enable him to reach the correct decision speedily. Both Marlborough and Wellington were famed for their ability to appear at critical moments to rally and encourage their men; a knack that Napoleon also displayed in the years of his prime. At Borodino and Waterloo, however (see pp. 121 and 124), he took too little direct part in the control of the battles, remaining too far to the rear.

9. *Co-operation*. The necessity of close teamwork between allies, between commanders, or between arms of the service. The excellent co-operation between Marlborough and Eugene at Blenheim contrasts markedly with the lack of it between Tallard and his colleagues—the Elector of Bavaria and Marsin—at the same battle. Similarly, the loyalty of Blücher, in marching to Wellington's assistance on 18 June 1815, is another good example of this principle in action.

10. *Administration*. The scientific organization of the day to day running of armed forces and the organization of the means of support-ing them has already been largely covered in the remarks devoted to logistics. An under-administered army, however hardy and determined, is likely to run into grave difficulties; as Napoleon's forces found in both Spain and Russia. An over-administered army, on the other hand, is prone to defensive-mindedness; commanders beset by excessive red tape rarely have the chance to show originality. Such was the fate of the Prussian Army in the early 1800s.

Variable Factors in Warfare

As the late Colonel Alfred Burne demonstrated in *The Art of War on Land*, warfare is an art rather than a science. If an appreciation of the strata and principles of war can assist a comprehension of the complexity of the subject and of the elements that go into the production of a successful military plan, a study of the main strands of experience that equally go to make for success or failure in the field can prove just as helpful.

First, there are a number of quantitative factors that need to be taken into account. The strength of contending forces is perhaps the most obvious of these: the numbers of troops involved, the amount and types of weaponry available to the rival sides, the quantity of supplies, munitions and other resources each can call upon—all are relevant to the outcome. But it would be dangerous to assert that fate has always favoured the big battalions. The pages of military history show many examples of the weaker side triumphing over the numerically stronger; as at Narva in 1701 (see p. 59).

Clearly, therefore, qualitative factors are significant. The calibre and characteristics of both leaders and led, the quality of their weapons and equipment and, above all perhaps, the prevailing level of morale and determination, all play an equally important part in determining the outcome of armed conflicts. The same may also be said of the attitude of the populations from which the armies spring, and of the efficiency and quality of the national administrative machinery that supports and backs them. History is full of examples of Davids destroying Goliaths.

The outcome of wars can also be affected by a number of decidedly variable factors, which together, individually or in various combinations, go to make up what Clausewitz termed "those elements of friction and uncertainty that are the eternal concomitants of war". The variations in the types of terrain to be encountered, the fluctuations in weather situations and climatic conditions, and the ravages of diseases that may be encountered, all belong to this category. Great commanders can rise above these problems and even exploit them, but many more have found them the final straw that led to disaster. Luck, that indefinable characteristic, has a part to play in these circumstances, as Napoleon was always aware. War is the realm of the unexpected, but he was convinced that a successful soldier must give the imponderable a place in his calculations.

Land campaigns are also governed in large measure by geographical conditions. Distance poses great strains on men and matériel—whether in victory or defeat. The campaigns in Russia in 1708 and 1812, (see pp. 58 and 118) illustrate the point. Physical features—rivers, mountainous regions, forests or swamps—can impose delay on marching armies or provide sanctuaries, and chances for achieving surprise, for forces presently on the defensive. Geography and climate determine levels of fertility in different areas; factors that are of great importance when ascertaining the operational viability of rival armies. The long Peninsular War (see pp. 110-115) illustrates this point. The French relied on haphazard logistical concepts of living off the countryside, which caused immense problems for a whole succession of marshals faced by the superior administrative skill of Wellington.

All the factors in the preceding paragraphs obviously have a bearing on the fortunes of war; a combination of moral and material influences, the former being highly volatile and shifting, the latter being more static and calculable. The successful commander takes them all into account, and makes his plans and arranges his disposition so as to exploit the available advantages and minimise the patent difficulties of the current situation. But the qualities of generalship need fuller analysis and description.

Generalship
Four hundred years before Christ, the Greek philosopher Socrates defined his conception of the qualities that go to make a good general:

"The general must know how to get his men their rations and every other kind of stores needed for war. He must have imagination to originate plans, practical sense, and energy to carry them through. He must be observant, untiring, shrewd, kindly and cruel, simple and crafty; a watchman and soldier; lavish and miserly; generous yet tight-fisted; both rash and conservative. All these and other qualities, natural and acquired, he must have. He should also, as a matter of course, know his tactics; for a disorderly mob is no more an army than a heap of building materials is a house."

Two thousand, two hundred years later, Napoleon stressed the following characteristics:

"A general's principal talent consists in knowing the mentality of the soldier and in gaining his confidence." Or again: ". . . a military leader must possess as much character as intellect—the base must equal the height". In 1804 he listed three basic requirements for a successful general: concentration of force, activity of body and mind, and a firm resolve to perish gloriously. "They are the three principles of the military art that have disposed luck in my favour in all my operations," he advised Lauriston. "Death is nothing, but to live defeated is to die every day." From the experience of his campaigns, he might have added a fourth basic concept: Surprise your opponents. Napoleon also believed a successful general must possess luck, as already mentioned.

In 1939, on the eve of the Second World War, General Sir Archibald Wavell enumerated his views on the essentials of generalship in a series of lectures at Cambridge. The first quality he defined as "robustness, the ability to stand the shocks of war"; what Clausewitz earlier called 'friction'. To Wavell, the mind of a commander must be as hardy as his body: "All material of war, including the general, must have a certain solidity, a high margin over the normal breaking strain". Like Voltaire, he particularly admired Marlborough's 'serenity of soul' on days of battle, although, in fact, it was more an external than an internal calmness.

Further qualities demanded of a good general by Wavell included boldness, which he stressed was what Napoleon really meant by luck: "A bold general may be lucky, but no general can be lucky unless he is bold". Wavell illustrated the point by citing the case of Admiral Byng, who was shot for carrying out his orders unimaginatively, thereby permitting the French fleet to escape. Next, he recommended sound common sense allied to a mastery of the 'mechanism of war', by which he included the mastery of topography, movement and supply. These administrative factors, although so often neglected in learned works devoted to the strategy and tactics of campaigns, he regarded as being of the first importance. Wavell also mentioned the importance of a commander being skilled in the accurate interpretation of intelligence, and in his possessing the means for both the rapid transmission of his orders and the receipt of reports to him. Wellington's code-breakers in the Peninsular War enabled him to make the most of the mass of military intelligence he was able to secure from his 'correspondents' (or secret agents) and the guerrilla bands. He often knew enemy orders days ahead of the French recipients for whom the instructions were intended. A century

earlier Marlborough was equally blessed with useful contacts, who fed him news of French movements and intentions. Whilst on the battlefield, his system of hand-picked aides-de-camp and 'running footmen' made it possible for him to keep his finger on the pulses of a battle in all its varied sectors.

All these qualities are desirable in a general, but one more of transcendant importance demands mention. A general's knack for man-management, allied to genuine humanity and a real interest in the welfare of his men, was of even greater significance in the seventeenth and eighteenth century than in more recent times. Not all commanders were notable for this quality: Frederick the Great was more feared than loved, and the Duke of Wellington inspired more respect than affection from his men. But genuine humanity must not be confused with softness or weakness; a general often had to be tough to the point of callousness if he was to achieve success. Yet, the Duke of Marlborough, with his calm courtesy and evident concern for his rank and file's welfare, and more so Napoleon, who could inspire his men to an unprecedented degree of loyalty to his person, could make the greatest demands upon their men's resolution and endurance and still retain the human touch. Gustavus Adolphus, de Turenne, Eugene and Charles XII also had this knack in greater or lesser degrees. The importance of this attribute was partly due to the close personal proximity to their men on days of battle. However vast the social gulf that then divided leaders from led, the sharing of physical perils and discomforts created a real bond between them when a commander took pains to foster such a relationship.

Of course there is no such phenomenon as the ideal type of general. There are many different types, each suited for a special rôle. If the fighting-general is the one that will figure most prominently in the pages that follow, we must equally acknowledge the importance of the soldier-statesman, or the superb trainer; the co-ordinator of divergent Allies; the commander skilled at leading a team or at building morale. All have had their parts to play, and continue to do so in the present day. Occasionally, a 'grand chef'—a Marlborough, Wellington or Napoleon—will combine in his person a number of these traits, but such men are in every way rare and exceptional. Most commanders were fortunate if they passed muster in one single role. To expect a man to be a true master of the kaleidoscope of war amounts to setting an impossible standard, but history has, nevertheless, seen the exploits of a surprising number of great generals.

The Classical Manoeuvres of Warfare

From earliest times, the conduct of war has been based upon seven basic manoeuvres employed either singly or in combination. Warfare has always been a kaleidoscope of rapidly-shifting moves and intentions, and such material considerations as weaponry and tactics have frequently changed or been adapted. But, in the final analysis, a commander has only a restricted selection of methods to choose from. This was particularly true of the period covered by this Atlas, for much of it was dominated by a degree of formalism that only the greatest commanders of the day could escape from. The detail invariably varied to meet special circumstances, but the underlying concepts remained much the same.

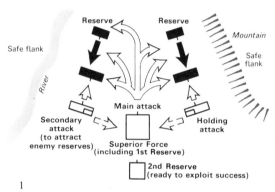

1

Penetration of the Centre is perhaps the oldest manoeuvre of all. A general combat was often resolved by the unleashing of a reserve to force the enemy's centre, and thus wreck the cohesion of his resistance. Of course skill was required in the mounting of the actual blow, and to prepare the way for it the application of deceptions and distractions was also called for. When well-employed, the method could lead to the encirclement of part of the hostile force. But, if prematurely or incorrectly used, it held certain dangers. If insufficient force were expended for the central blow, it might fail to achieve a breakthrough and invite a telling riposte. If, to mass enough force to secure superiority in the centre, a commander reduced the strength of his flanks excessively, he would run the risk of being enveloped. A past-master in the successful employment of this gambit was Marlborough, who based three of his four battle successes on the coup de grâce in the centre. In the process, however, he became predictable; at Malplaquet (see p. 47) for example, his success was dearly bought. But as at Blenheim and Ramillies, he had so thoroughly prepared his final blow by means of preliminary attacks on the French flanks that Villars was unable to withstand the attack when it came, even though it was anticipated.

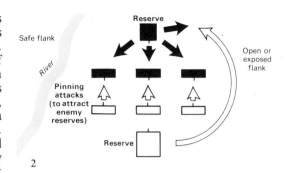

2

Envelopment of a Single Flank was a second frequently employed battle tactic. If sufficient force could be deployed against one extremity of the enemy line, resulting in its being turned, this could lead to the rolling up of the enemy line towards its centre in growing chaos. This is what Condé achieved at Rocroi, and Wellington at Assaye (see pp. 27 and 80), while a generation earlier the concept had played a part in several of Frederick the Great's major engagements. Of course, a general needed to guard against the danger of excessively weakening his centre to build up sufficient strength against the flank. The Allies fell into this error at Austerlitz (see p.100), when they denuded their centre on the Pratzen Heights in order to achieve what they hoped would be a decisive strength against Napoleon's right flank. In fact, they provided the French with the chance to deliver a fatal counter-blow, which cost the Russians and Austrians the battle.

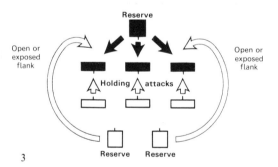

3

Envelopment of Both Flanks was more difficult to achieve, as it required massive superiority of force on the part of the army attempting what amounted to a complete encirclement of the enemy. Unless such force was available, or the foe was totally demoralized, the general attempting this manoeuvre ran the risk of over-extending his resources, thereby courting an enemy break-out attempt with superior force at one or more points. In our period, there were examples of both situations. At the Pruth (see p. 61), the Turks compelled Peter the Great to capitulate after completely surrounding his army with a huge force. However, they were less successful against Prince Eugene at Belgrade (see p. 56) where the Imperial Army, trapped between a Turkish garrison they were besieging and a huge relieving army, managed to smash through the latter's lines decisively before compelling the former to surrender. Thus the stakes were high when this manoeuvre was attempted: a case of win or lose all.

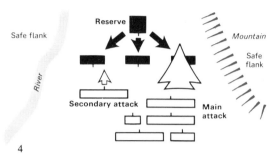

4

Attack in Oblique Order. By this manoeuvre, a general contrives to mass increasing strength against one wing of the hostile army until it cracks under the remorselessly mounting pressure, while small forces keep the remainder of the enemy line of battle in play in order to distract reserves from the key sector. The method dates back to classical antiquity, but in our period was strongly restated by Frederick the Great, whose conduct of the celebrated Battle of Leuthen in 1757 (see p. 75) was based on this manoeuvre. Subsequently, it was adopted in a varied form by Napoleon, who first attempted to employ it at Castiglione, 1796. In later years the concept formed a significant part of his system of the strategic battle (see p. 98); and at Bautzen in 1813 it was probably used to its greatest effect against the Russians and Prussians. Wellington's handling of the Battle of Salamanca in 1812 (see p. 112) saw a greatly superior force unleashed against Marmont's leading divisions as they straggled across the army's front. So, it is a method that has attracted the attention of a number of great commanders.

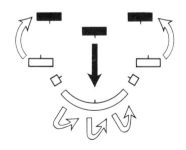

5

The Feigned Withdrawal has in historical terms often been employed to lure the enemy into an ambush or other unfavourable positions by tempting him with an apparent flight or precipitate retreat, thereby offering the prospect of an apparently easy victory. Of course, the method holds perils as well as advantages: a feigned retreat can too easily become a real one unless extreme care is exercised. This method was an established favourite of Ottoman forces, who often used it to entice Imperialist armies to abandon their formal lines of battle in the hope of looting the rich Turkish encampments. As Prince Eugene once remarked, "I fear the Turkish Army less than their camp". However, he withstood the temptation in several battles fought in the Danube region (see p. 41).

The Duke of Wellington was also an expert exponent of the method. In 1810 he used it to lure Marshal Massena through Portugal towards the secretly-prepared Lines of Torres Vedras (see p. 112), which proved both impregnable to attack and a death trap for the French Army of Portugal. Similarly, it was the appearance of a British retreat towards Ciudad Rodrigo that induced Marmont to over-extend his army near Salamanca in 1812 in an attempt to head off the Allies. This afforded Wellington with the opportunity, already mentioned above, to destroy much of the French Army.

Although not a method to be employed rashly, when properly applied by a sound general with a developed sense of timing, this manoeuvre could surprise an opponent and throw him off-balance: the psychological repercussions of finding a supposedly-dominated enemy suddenly reversing his previous behaviour can be highly unsettling.

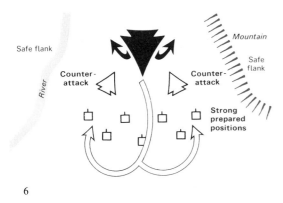

6

Attack from a Defensive Position. Napoleon described the art of war as comprising ". . . a well-reasoned and extremely circumspect defensive, followed by a rapid and audacious attack". To adopt this method, an army prepares a well-chosen defensive position or sanctuary, to which the commander proceeds to lure his foe before choosing his moment to

resume the offensive and sally out to the attack. Clearly, this manoeuvre is often associated with the one described immediately above, and, indeed, Wellington's use of his triple-line Torres Vedras position first to confound Massena and then to serve as a spring-board for a counterattack is the best example in our period. But Prince Eugene's use of the strong siege lines before Belgrade in 1717 provides a second, although less deliberate, example of the same type of system being put into operation.

On the other hand, this type of idea can lead to excessive defensive-mindedness, and may provide an excuse for the indefinite postponement of offensive action. Undue reliance on strongly-prepared lines and positions may therefore prove something of a snare and delusion.

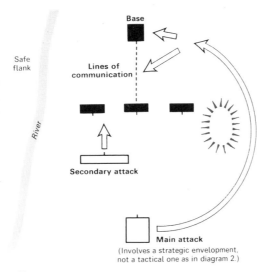

Base

Safe flank

Lines of communication

River

Secondary attack

Main attack

(Involves a strategic envelopment, not a tactical one as in diagram 2.)

7

The Indirect Approach. By this method the enemy's attention is diverted by secondary operations whilst the main body, or at least a major force, marches to envelop strategically the foe's flank or rear. If this move is successfully completed without the victim becoming aware of what is intended, the enemy's lines of communication are severed, and with the interruption of his supplies and reinforcements his position becomes critical. The only course open under normal circumstances would be for the threatened army to turn about in its tracks and try to reopen its links with its depots and bases, accepting in the process a 'reversed front' battle (one in which each army is facing towards its base). In the event of losing such an engagement, an army's situation becomes truly critical as the natural line of retreat is blocked. Obviously, speed and surprise are vital if this method is to succeed.

Many commanders have attempted to use this strategic manoeuvre, but the risks implicit in the method proved daunting to many. As Frederick the Great pointed out, the danger was that an opponent might mete out the same treatment as he was receiving. Throughout the seventeenth and eighteenth century all armies were very much tied to their convoys and lines of communications—armies could march just as far as they could carry their bread—and this made rapid moves hard to carry out. Marlborough's famous march to the Danube in 1704 (see p. 44) averaged less than ten miles a day, yet it was regarded as an immense feat in its day. Only with the advent of Napoleon's dynamic approach to the waging of campaigns, allied to possession of lighter artillery and the concept of the French Army living off the countryside (thus dispensing with many of the slow-moving convoys that continued to cripple its adversaries' freedom of action), did the method reach its full development. (For an example see the campaign of Ulm on p. 100.) This technique was employed by Napoleon some thirty times in various forms between 1796 and 1815. The chance of forcing a decisive action upon the enemy is the great advantage conferred by the method. Even if unwilling to accept battle, an army could not wholly ignore the implications of the manoeuvre. To succeed, however, it calls for considerable superiority of force and a well-developed sense of timing. An unwise commitment to the move could lead to an army being defeated in detail and, of course, too much repetition could lead to a counter-system being developed. This the Allies managed to contrive in 1813 and 1814, by keeping substantial munitions and supplies close to their fighting formations. The scheme made them less dependant on regular re-supply for a period, enabling them to ignore Napoleon's presence across their lines of communication and, indeed, to call his bluff on a number of occasions.

Such, then, were the seven gambits of the classical age of warfare. Most manoeuvres attempted by the commanders of the seventeenth and eighteenth century involved one or sometimes combinations of these basic ploys, employed with greater or lesser success. Although it is dangerous to place undue reliance on 'systems' underlying military movements, an appreciation of these concepts will often help to explain what a particular general was attempting to carry out. However, the minutiae and complexities of warfare in any age defy accurate analysis or clear classification, and only the broadest deductions can be drawn. It is this aspect that makes warfare at least as much an art as it is a science.

Strategic Legacies of the Past

The response to the challenge posed by these problems in their various manifestations stretches back into the mists of recorded history. Civilizations and empires have waxed and waned in close association with their martial fortunes, and armies have always reflected the social and economic condition of their peoples and countries. War and progress have been linked to a remarkable degree. Each new period of history has been heralded by a period of struggle and conflict, as the new power—whether for good or ill— has endeavoured to assert its superiority over its neighbours or rivals. In many cases, this conflict has ultimately led to constructive results; the fusion of what was best in the old with what was dynamic and significant in the new. Thus, the ancient Greek civilization became assimilated first by Alexander and then by Rome. On other occasions, the tide of change has submerged and all but destroyed what has gone before. Thus the Roman Empire in its turn went down before the barbarian peoples of the Eurasian heartland, although Byzantium—or Rome in the East—survived for another millenium before finally succumbing to the Ottoman Turks. Similarly, the great Mongol scourge devastated immense tracts of the world (see map, The Mongol Conquests, p. 17), including much of China, what would become known as Russia, and parts of eastern and southern Europe, in a burst of ferocious and largely unproductive energy that lasted from the thirteenth to the late fourteenth century. Far more constructive in terms of art and culture was the great Moslem wave of expansion, which swept through the Middle East, north and east Africa, and deep into southern and western Europe as well as into parts of Asia from the seventh century, and which only finally receded in the period covered by this Atlas. The middle years of this surge of religious energy saw a corresponding European effort in the series of Crusades aimed at freeing the Holy Land from the 'infidel'. These were the precursors of the spirit of adventure and search for gain that eventually led to the founding of great overseas empires by the Portuguese, the Spaniards and the Dutch; not forgetting England and France, whose rivalry will be found reflected many times in the pages that follow.

Warfare had become sophisticated by the days of Darius, Alexander, Hannibal and Caesar, as the scale of their martial achievements demonstrated. The Dark Ages in Western Europe saw a marked falling back in skill and organization, but Moslem, Saracen and Mongol armies and leaders commanded respect, as did the themas of Byzantium.

Primitive conflicts, based upon individual combat or crude horde tactics, had given way to organized armies in about 4,000 BC. Tactics remained simple, but leaders emerged capable of making the most of them. Basic manoeuvres that were designed to turn flanks caused the gradual extension of battle lines. Mobility entered the scene with chariots and cavalry, while bows and arrows greatly extended the use of missiles and firepower. The ancient empires of Persia, Macedon and Rome brought superior concepts of organization. The Roman legion and the Byzantine horse-armies, in particular, set standards that would not be surpassed, nor yet emulated, until the days of Marshal de Saxe and, above all, of Napoleon Bonaparte.

Battles became more complex, but in their form remained generally predictable, following set-patterns based on a number of variations we have already described. Spontaneous action was rare; the forming of lines of battle consuming so much time that battle may almost be said to have taken place only by mutual consent. Unless he had been trapped into a hopeless position an unwilling opponent could often break contact and make for some sanctuary where he would be virtually unassailable if his supplies could hold out. The test of superior generalship, therefore, right up to the eighteenth century, was largely based upon the ability to force action on an unwilling foe by use of speed and surprise allied to deception at a time and place of the attacker's choosing. Battles could be induced by the ravaging of the enemy's countryside or by the opening of a formal siege against one of his most important towns. Both methods were frequently resorted to in the seventeenth century, but the revulsion against the horrors of the Thirty Years War led to an increasing recourse to siege warfare during the late seventeenth and most of the eighteenth century. A pattern soon emerged— repeated many times—of a siege being prosecuted by the smaller part of an attacking army while the main body served as a covering force at some distance from the siege lines, challenging the enemy army to intervene. Many of the battles illustrated in this volume—including Narva (see p. 59), Malplaquet and Fontenoy (see pp. 47 and 70)—developed from situations of this general type. Competence at siege warfare came to rival, even surpass, battle skill as the chief professional requirement of a successful general. Associated with these skills was the art of manoeuvre—creating diversionary or interdictive threats against enemy supply lines—and stage by stage the prosecution of warfare became increasingly refined and complex.

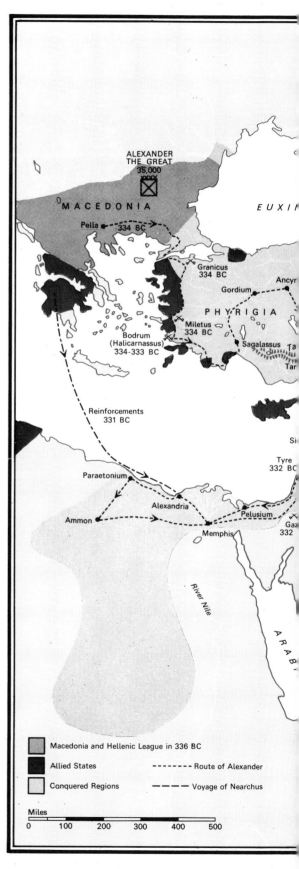

14

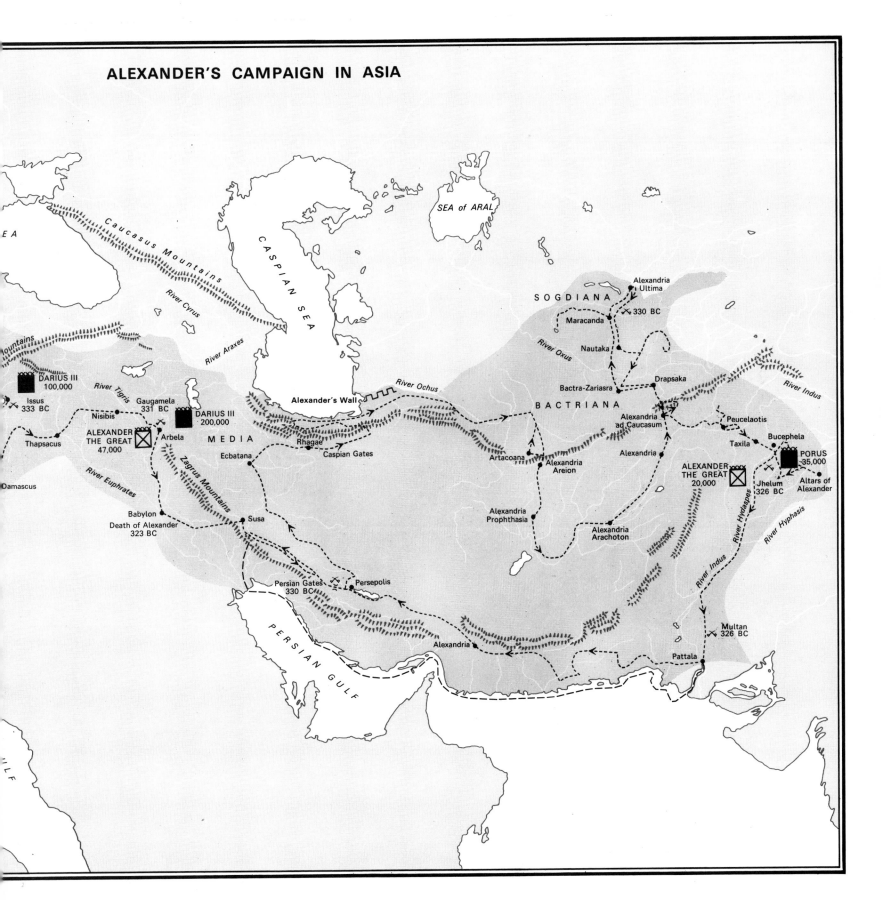

ALEXANDER'S CAMPAIGN IN ASIA

SEA of ARAL

CASPIAN SEA

Caucasus Mountains

River Cyrus

River Araxes

Alexander's Wall

River Ochus

SOGDIANA

Alexandria Ultima

Maracanda

✗ 330 BC

Nautaka

River Oxus

Bactra-Zariasra

Drapsaka

BACTRIANA

River Indus

Mountains

DARIUS III
100,000

✗ Issus
333 BC

River Tigris

Nisibis

Gaugamela
331 BC

DARIUS III
200,000

MEDIA

Rhagae

Caspian Gates

Artacoana

Alexandria
Areion

Alexandria
ad Caucasum

Alexandria

Peucelaotis

Bucephela

Taxila

PORUS
35,000

ALEXANDER
THE GREAT
47,000

Arbela

Ecbatana

ALEXANDER
THE GREAT
20,000

Jhelum
326 BC

Altars of
Alexander

Thapsacus

River Euphrates

Zagrus Mountains

Damascus

Babylon
Death of Alexander
323 BC

Susa

Alexandria
Prophthasia

Alexandria
Arachoton

River Hydaspes

River Hyphasis

River Indus

Persian Gates
330 BC

Persepolis

Alexandria

Multan
326 BC

Pattala

PERSIAN GULF

ULF

Strategic Legacies of the Past

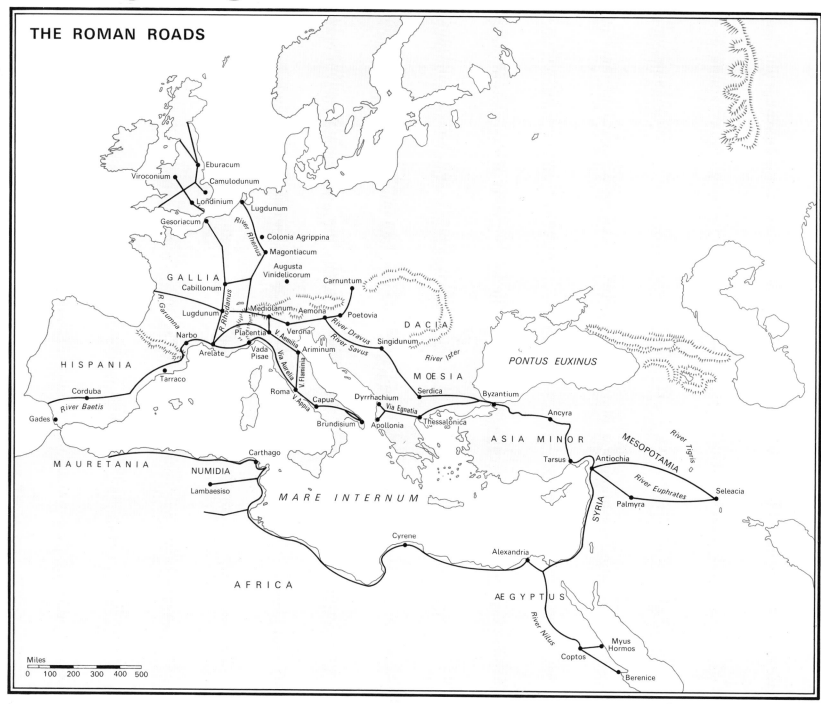

THE ROMAN ROADS

The French Revolution, and pre-eminently Napoleon, disturbed the pattern of classical warfare. Napoleon's fusion of marching, fighting and pursuing into one relentless sequence was a radical departure from the conventions of the previous two centuries. Then, armies had marched in one formation and fought in another—requiring invaluable time to convert from the one function to the other. Some of this had been foreshadowed by Frederick the Great, but it took a further generation to bring the new ideas to the fore. Napoleon exploited artillery firepower and bayonet and sabre shock action on the field of battle, and greatly reduced the reliance of his armies on formal lines of communications, convoys and pre-stocked depots and magazines; all of which tended to shackle his opponents' freedom of movement and expression. However, in the hands of a master of warfare such as Wellington the old methods still held considerable validity and power; as he demonstrated time and again against under-supplied French armies in the Peninsula and, ultimately, against the grand master of the martial arts himself during the brief campaign of 1815 in Belgium. But the future lay with adaptations of Napoleonic fluidity and blitzkrieg: armies following the French example of marching dispersed in a highly flexible network of corps

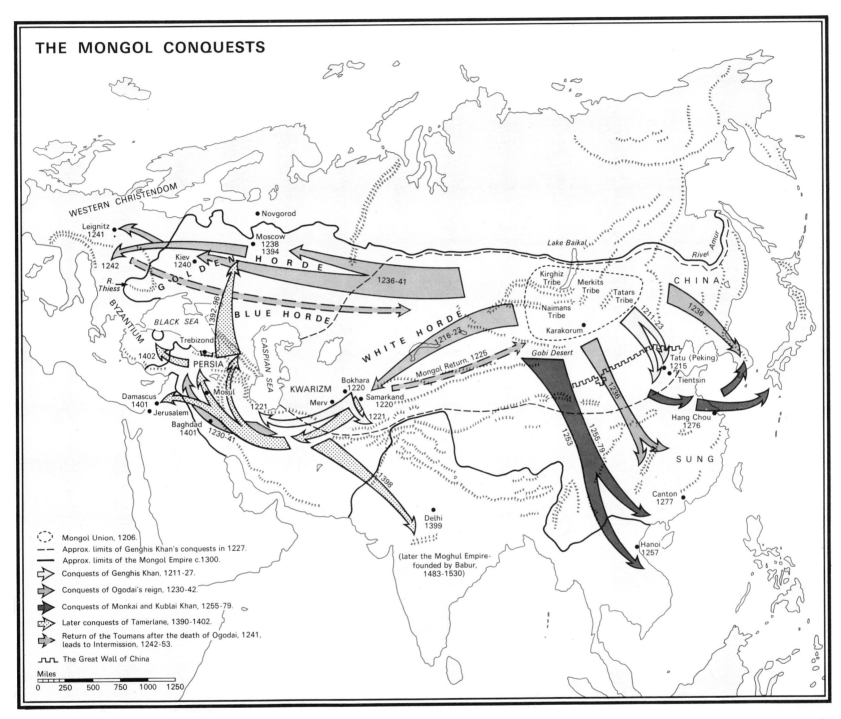

THE MONGOL CONQUESTS

WESTERN CHRISTENDOM

Leignitz 1241
• Novgorod
Moscow 1238 1394
1242
Kiev 1240
R. Thiess
GOLDEN HORDE 1236-41
BYZANTIUM
BLACK SEA
BLUE HORDE
Trebizond
1402
PERSIA
WHITE HORDE 1216-23
Damascus 1401
Mosul
CASPIAN SEA
KWARIZM
Mongol Return, 1225
Gobi Desert
Jerusalem
Merv
Bokhara 1220
Samarkand 1220
Baghdad 1401
1230-41
1221
1221
1398
Delhi 1399
(later the Moghul Empire-founded by Babur, 1483-1530)

Lake Baikal
River Amur
Kirghiz Tribe
Merkits Tribe
Tatars Tribe
Naimans Tribe
Karakorum
CHINA
1236
1211-23
Tatu (Peking) 1215
Tientsin
1236
1253
1255-79
Hang Chou 1276
SUNG
Canton 1277
Hanoi 1257

Mongol Union, 1206.
Approx. limits of Genghis Khan's conquests in 1227.
Approx. limits of the Mongol Empire c.1300.
Conquests of Genghis Khan, 1211-27.
Conquests of Ogodai's reign, 1230-42.
Conquests of Monkai and Kublai Khan, 1255-79.
Later conquests of Tamerlane, 1390-1402.
Return of the Toumans after the death of Ogodai, 1241, leads to Intermission, 1242-53.
The Great Wall of China

Miles
0   250   500   750   1000   1250

d'armée to attain maximum surprise and speed before concentrating to fight a decisive battle.

An important factor in Napoleon's success had been the development of a network of improved roads over much of Europe in the preceding generations. Not since the days of the great Roman road-builders had such manoeuvrability been possible. And as the nineteenth century progressed, the advent of railways would add a new dimension.

Much stress was placed on manoeuvres of strategic envelopment. One effect of this was to over-extend the front of operations as each side took steps to meet and counter such enemy manoeuvres by placing a blocking force in the path of each detachment. Eventually, this would lead to the stalemate conditions of continuous fronts experienced from late 1914. However, these developments—and the subsequent return to fluid warfare in the

1930s and 1940s—do not belong to the main subject matter of this volume, which is devoted to the last great flowering of the period of classical warfare when everything, or almost everything, depended on the skills and abilities of the commanders-in-chief, who were expected to lead their forces into the storm of combat as well as devise the broad movements of strategy and organize the requirements of logistics. It was the era of direct command.

Part I, 1618-1721:
Wars of Religion and National Expa

The seventeenth century was dominated by almost ceaseless wars between combinations of the major European states, many of which also faced grave internal struggles linked with religious or constitutional issues. The period saw a slow shift in the motivation of international conflicts. As the century progressed, dynastic confrontations based upon frontier adjustments with nationalistic, economic and colonial overtones replaced religious pre-occupations; although in southern Europe the intermittent Ottoman-Habsburg struggle retained a religious significance throughout.

Populations remained relatively small (see diagram, Population Fluctuations of the Nine European Powers, 1600-1800, p. 23), a fact that restricted colonial expansion as such, although the commercial aspects of imperial rivalry became increasingly important. Armies and navies, too, were small by the standards of later centuries, but there was a tendency for them to grow slightly as the governments of the day improved the machinery of administration and logistical support. The requirements of supply and transportation (Europe's communications were rudimentary in the extreme) channelled campaigning to certain key areas that became the scenes of campaign after campaign.

The same considerations tended to impose a seasonal pattern on military operations: in

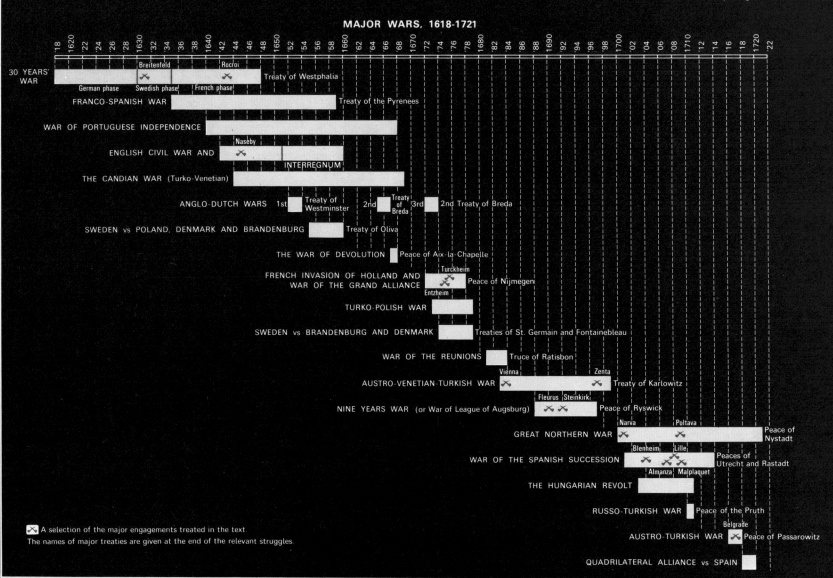

MAJOR WARS, 1618-1721

A selection of the major engagements treated in the text.
The names of major treaties are given at the end of the relevant struggles.

18

western Europe the availability of forage and the passability of mud roads and major waterways restricted campaigning in the main to the months between April and October; as a general rule the summer months in Spain and parts of southern Europe were too hot and disease-prone to permit major operations; in northern and eastern Europe winter campaigns were not unknown on account of the frozen rivers and lakes which aided movement over the mainland, despite the supply problems associated with them. As a result, the direct impact of wars on civilian populations was often relatively small and regional, although the economic repercussions were more widely felt. (Except in the case of the Thirty Years War which, it is estimated, reduced the population of central Germany by up to a third.)

The Thirty Years War, which was really a succession of wars, hastened the decline of Habsburg power in central Europe by weakening the Emperor's influence over the German princes. Austria's status declined from 'Austria Est Imperare Orbi Universo' (the proud boast of Charles V, 1558) to merely 'primus inter pares', as first Sweden and then France intervened in the struggle to further their own national interests. The military capabilities of such great soldiers as Gustavus Adolphus of Sweden, the Imperialist Tilly and the Great Duke of Condé are illustrated by the studies of Breitenfeld and Rocroi (see pp. 26 and 27).

The Spanish Habsburgs were also undergoing a serious recession of fortune. Economic weakness and an inability to control their farflung possessions (see p. 48) were aggravated by involvement in most European and colonial struggles, as the United Provinces (once a Spanish viceroyalty), England and France fought for the rich pickings of territory and commercial advantage. Portugal's war of independence, 1640-68, ended successfully, and soon the question of the Spanish succession came to dominate European diplomacy. This issue caused a number of intrigues and alliances as it became clear that the Spanish dynasty would die with the sickly Charles II, who nevertheless confounded all the experts by living until 1700.

The Ottoman Turk had also seen better days; local gains were still made at the expense of the Habsburgs until the Truce of Eisenburg in 1664, and Candia was wrested from Venice still later, but the general story was one of decline. The descent on Vienna in 1683 (see p. 55) was a flash in the pan, although the Emperor, Leopold I, had to call in the Poles to surmount the crisis. The general history of the Turkish wars against the Empire, Poland, Russia and Venice is one of Turkish withdrawal and slow eclipse, as typified by the Battle of Zenta in 1697 and the Siege of Belgrade, 1717 (see p. 56). (Both campaigns were triumphs for Prince Eugene of Savoy, who was one of the greatest soldiers of his day.) Indeed, the Habsburgs compensated for losses of prestige and influence elsewhere by their successes against the Turks.

The Peace of Westphalia in 1648 had benefited four countries in particular: namely, France, Sweden, Brandenburg and Bavaria (see p. 28). Of these, only France and Brandenburg derived lasting advantage from their gains. The history of French expansion dates from 1598 with the acquisition of Navarre under Henry IV. Cardinal Richelieu subsequently exploited the power vacuum in Germany with great skill, but it was under 'le roi soleil', Louis XIV, that the processes of French expansion were expedited following the total centralization of power in the hands of the Bourbon monarchy. An explosion of French culture and national energy was accompanied by the pushing back of the northern, eastern and southern frontiers at the expense of the Empire and Spain. Despite protestations that they were countering an international Habsburg conspiracy of encirclement, the French were the bullies of Europe for half a century. Generals such as Condé, Turenne, Catinat and Luxembourg led the best armies in Europe to numerous battlefield successes. The treaties of the Pyrenees in 1659, Aix-la-Chapelle in 1668, Nijmegen ten years later, and the episode of the Réunions, 1681, all included French acquisitions of territory (see map on p. 34) at the expense of her neighbours.

Then the decline set in; French ambitions regarding the Palatinate and parts of the Spanish Netherlands led to the so-called War of the League of Augsburg (1688-97), which faced France with the First Grand Alliance (see inset on p. 35) inspired by William III. Despite the military successes of Catinat and Luxembourg, the Peace of Ryswick saw France forced to abandon certain conquests in Lorraine and the Palatinate. Three years of intrigue over the proposed partition of the Spanish Empire followed, but all chance of a pacific settlement disappeared with Louis' acceptance of the final will of Charles II, who left the entire inheritance to Louis' grandson, Philip of Anjou. Austria (the other major claimant) was soon at war once more with her old enemy, and within a year French blunders of policy enabled an ailing William III to create the Second Grand Alliance and bring both England and the United Provinces into the war (see inset p. 35). The long War of the Spanish Succession followed, during which French military pride was humbled by the Duke of Marlborough and Prince Eugene in a quatrain of great victories and numerous sieges; the Allies also gained control of the west Mediterranean from 1704. Their decision to use this advantage to mount a major invasion of Spain in support of Charles III, the Austrian claimant, proved a costly error, for their initial successes only led to ultimate failure and disillusion.

In the end, after further indecisive campaigns in Flanders, political mismanagement of the Alliance led to France 'winning the peace' at Utrecht and Rastadt (1713-14). As a result, Philip V was confirmed in the greater part of his inheritance in exchange for colonial and commercial advantages granted to Great Britain; the Spanish Netherlands was ceded to Charles VI; Austria received the Milanese and Naples as compensation; while Sicily went to Savoy, until 1720 when it was exchanged for Sardinia. France in Europe escaped with a few border adjustments and the return of towns east of the Rhine, but the economic and colonial repercussions of sixty years of warfare were to prove disastrous during the succeeding century.

If one great theme of the seventeenth and early eighteenth century was the dynastic rivalry between the Bourbons and the Habsburgs, another was the strife between the Vasas and the Romanovs in northern and eastern Europe. The brief emergence of Sweden as a European power was confirmed by the

Wars of Religion and National Expansion

Peace of Westphalia, and despite a setback at Fehrbellin in 1675, she was still the dominant power on the southern shore of the Baltic. Unfortunately, the ambitions of Charles XII and the emergence of a relatively modernized Russia led to the Great Northern War, 1700-21. At first, the better-trained and led Swedes scored highly (see p. 58), but Charles's decision to invade Russia led predictably to catastrophe at Poltava (see p. 60) in 1709. Thereafter, Sweden as a power was in decline; her resources were hopelessly inadequate for the calls put upon them, and by the Treaty of Nystadt, 1721, most of her remaining Baltic possessions were lost.

Under Peter the Great, Russia began the painful transformation from a medieval state to a modern one. His bitterly-opposed policies of westernization bore fruit in the ultimately successful meeting of the challenge posed by Sweden. Over the same period, Russia made substantial if somewhat intermittent and haphazard territorial advances at the expense of Turkey and, above-all, Poland, which was reduced in extent and plunged into anarchy.

Another embryonic power to benefit from the eclipse of Poland and the exhaustion of the Empire was Brandenburg. At the Treaty of Westphalia the Great Elector received Eastern Pomerania and various bishoprics, and by 1688 his successors had added Magdeburg and Halberstadt. These possessions, together with East Prussia, which was wrested from Poland in 1657, established a basis for the creation of the Kingdom of Prussia in 1700. Further advantages were gained in 1713.

In conclusion, mention must be made of the maritime powers: England and the United Provinces. As a result of their lengthy struggle for independence from Spain, the Dutch were principally concerned throughout the period with considerations of Protestant support and their own national security; the war with France, 1672, redoubled Dutch anxiety on this latter score, and thenceforward the desire for a barrier of fortresses to safeguard the southern and western frontiers became a paramount issue. For the rest, the interests of the United Provinces were largely taken up with the maintenance and extension of their maritime empire in Brazil and the East Indies, created largely at the expense of Portugal. These ambitions led to serious collisions with England, resulting in three maritime Dutch Wars. However, the over-riding fear of French ambitions did much to drive the two maritime powers together, and from 1688 they acted in close co-operation in the successive Grand Alliances against Louis XIV, jointly wresting control of the Channel from France at Cap la Hogue. By 1714, however, the financial strains engendered by the long wars were taking their toll, and the United Provinces would soon be entering a long period of decline, although they gained their fortress barrier.

As for England, for the first half of the period under review, her attentions were largely absorbed by internal problems of constitutional, religious and social import; troubles that culminated in the Great Civil War (see p. 30) and the eventual emergence of a constitutional monarchy. Stuart interventions in European affairs were not blessed with much success before 1640, as the short wars with France and Spain testify. Cromwell, on the other hand, enjoyed rather more success—both against Spain and the Dutch—than either his immediate predecessors or successors. It was William III, however, who began the transformation of his new kingdom into an effective power with an important rôle to play on the Continent; he mobilized both its men and wealth to assist in the prosecution of his life-long feud against France from 1688. This process came to fruition under Queen Anne, when British armies supported by British gold and the Royal Navy made a decisive contribution to the War of the Spanish Succession; and the Peace of Utrecht, however unworthy in its timing and European repercussions, inaugurated the development of the British Empire which would be a major theme of the eighteenth century.

Armand Jean du Plessis, Cardinal Richelieu (1585-1642), the worldly ecclesiastic and soldier-statesman who masterminded France's exploitation of the Thirty Years War and centralized the power of the Bourbon monarchy.

Louis XIV of France (1638-1715) as a young man. 'Le roi soleil' had an insatiable thirst for military glory, and led his country to great successes but ultimately to the verge of bankruptcy.

Cardinal Giulio Mazarin (1602-61), the Italian guide and mentor of the youthful Louis XIV, who survived the internal revolts of the Frondes to continue Richelieu's work of strengthening the royal authority in France.

KEY STRATEGIC AREAS, 1610-1721

NORTH SEA
BALTIC SEA
St Petersburg
Gothland
River Düna
The River Gap
River Niemen
River Elbe
Plymouth
Scillies
Dover
River Rhine
River Seine
Ushant
Brest
River Oder
Pripet Marshes
River Vistula
River Dnieper
River Dönetz
River Don
River Volga
River Ural
BAY of BISCAY
River Loire
R.Garonne
Belfort Gap
Valtelline
Brenner Pass
Venice
R.Rhône
R.Po
Red Tower Pass
R.Pruth
CASPIAN SEA
Lisbon
River Tagus
River Ebro
Toulon
Iron Gate Pass
River Danube
BLACK SEA
Gibraltar
Minorca
MEDITERRANEAN SEA

Miles
0 100 200 300 400 500 600

KEY TO MILITARY AREAS

1 THE COCKPIT OF EUROPE
Significance: the fortress barriers; local waterways;
high agricultural yield; rich towns; approaches to
north German and north-east French plains.
Campaigns: 1658, 1672, 1688-97, 1701-13.
Epicentre: the Spanish Netherlands.
2 MIDDLE RHINE AND LOWER PALATINATE
Significance: approaches to Strasbourg (west),
and the Danube valley (east); Rhine valley and Black
Forest communications; Alsace.
Campaigns: 1621, 1622, 1647, 1674-75, 1688,
1704, 1707, 1713.
Epicentre: the Rhine, the Black Forest region and
Nordlingen.
3 SAVOY AND PO VALLEY
Significance: the emergence of Savoy; Habsburg and
Bourbon rivalries; good waterway communications
(Po valley); rich cities; approaches to Lyons and
Toulon (west), to Milan and Mantua (east), to

Alpine passes into southern Austria (north-east);
the Valtelline.
Campaigns: 1620, 1624, 1627, 1629-31, 1635,
1636, 1639, 1690, 1693-96, 1701-3, 1701-13.
Epicentre: Piedmont.
4 BOHEMIA AND UPPER ELBE
Significance: religious loyalties; Swedish-Habsburg
conflict; Elbe headwaters.
Campaigns: 1630-35, 1639.
Epicentres: Leipzig and Prague.
5 MIDDLE DANUBE AND TRANSYLVANIA
Significance: crucial waterways (Danube and Theiss);
Red Tower and Iron Gate passes; Limits of Ottoman
expansion; relatively high agricultural yield;
restive Hungarian population.
Campaigns: 1664, 1683, 1687, 1691, 1697, 1703-
11, 1716, 1717.
Epicentre: Belgrade.

6 THE BALTIC
Significance: Swedish expansion; Russian moderni-
zation; naval stores; Swedish need for expanded
population; control of Finland and its gulf.
Campaigns: 1657, 1658-59, 1675, 1692, 1700-21.
Epicentres: east Baltic coast and Riga.
7 AZOV-PRUTH
Significance: Russo-Turkish rivalry for control of
the Crimea; limits of expansion.
Campaigns: 1696, 1700, 1711, 1722.
Epicentres: Azov and the River Pruth.
8 VALENCIA-CATALONIA
Significance: Habsburg exploitation, particularly
of sea-support from Lisbon and Minorca.
Campaigns: 1705-14.
Epicentre: Barcelona.

KEY TO NAVAL AREAS

A ENGLISH CHANNEL
Significance: Dutch coast; trade concentrations;
invasion dangers and raids.
B GIBRALTAR STRAITS TO LISBON
Significance: entry to Mediterranean trade.
C TOULON-MINORCA
Significance: naval arsenals; trade centres
(Marseilles) and Camisards.
D CRETE-AEGEAN SEA-MOREA
Significance: Habsburg, Venetian, Turkish rivalry
—naval and trade.
E SEA OF AZOV
Significance: Russo-Turkish contest for control of
Crimean corn trade.
F EAST BALTIC
Significance: Russo-Swedish rivalry; naval stores.
G THE SOUND
Significance: Danish-Swedish rivalry for control of
entry to the Baltic; trade in naval stores.

Population Growth and Army Strengths, 160

The martial power of a country at any given period depends upon a number of variable factors relative to those of possible rivals. The economic condition, administrative skills and technological development of a nation are three quantifiable factors. Equally significant are population sizes and the proportionate military forces a country can sustain.

The average seventeenth or eighteenth century country could maintain an army of approximately one per cent of its current population without suffering severe social and economic consequences. The diagram shown provides an approximate idea of population fluctuations between 1600 and 1800. It is estimated that in 1600 the population of Europe west of the Urals totalled ninety-five million (or eighty million in 'Little Europe' excluding Poland and West Russia), and this figure grew by only five million by 1650. This represented about one fifth of the current world population, and the relative significance of Europe is partly explainable in these terms.

Of the nine representative countries examined in the diagram, three (Spain, the United Provinces and Sweden) suffered intermittent or permanent population decline over the period, a fact that was reflected in their martial power and 'great power' status. Their declines were closely linked to unsuccessful attempts to maintain, economically and socially, large armed forces through periods of protracted warfare. Those countries whose populations steadily developed, on the other hand, proved capable of maintaining sufficient troops to secure territorial and population growth—whether in Europe (France and Austria) Central Asia (Russia) or overseas (Great Britain).

The same diagram shows developments in the size of standing armies of six selected powers. No account is taken of militia or camp followers. Of the latter, in 1620 it was calculated that at least 4,000 women, boys and 'bawdy waggons' accompanied every 3,000-strong tercio in Central Europe; Turkish armies often trebled their fighting strength with slaves and attendants. An attempt has been made, however, to indicate the proportion of foreign troops included in the selected armies on the four representative dates. The use of such troops eased certain social problems associated with recruiting, but the economic strain was often aggravated. To cite two examples: in 1708 Queen Anne paid over ten per cent of the national revenue in subsidies; and all of fifty per cent of Prussia's annual budget was devoted to military expenditure under Frederick the Great. In the case of France in 1812, however, it should be pointed out that almost all her huge 'foreign' contingents were supplied by Napoleon's Allies and former foes at their own expense. Great Britain, on the other hand, did largely pay, subsist and equip her Portuguese allies in the Peninsula. No account is taken here of Hanoverian forces or of straight subsidies paid to continental powers as inducements to resist France. Thus, it only gives a generalized view.

The exclusion of Turkey and the infant USA from these tables is deliberate. Turkey, in decline from 1717 (see p. 56), maintained a standing or capiculi army of some 130,000 men, mostly Janissaries (Turkish corps d'élite), supplementing them with huge hordes of feudal levies. The United States Army, by contrast, was insignificant in size, although effective enough (with French aid) in the field. Washington's Continental Army of June 1775 started just ten rifle companies strong, and may have reached a peak of 30,000 (half being available for field operations at one time) in addition to some 4,000 French troops under Lafayette; this figure is exclusive of the ex-colonial militias. By 1789, however, this force had shrunk to exactly 718 officers and men, so suspect was the concept of a standing army in peacetime. Twenty-three years later, faced by the renewed war with Britain, this tiny cadre would be temporarily re-expanded to number some 38,000 men.

Fire, rapine, pestilence and the sword caused varying havoc in different periods (the effects of which are noted elsewhere). Economically, seventeenth century wars aggravated an already inflationary situation in Europe; the well-known spiral being often triggered off by successions of poor harvests and the generally colder weather that afflicted Europe from 1600 onwards. The mercantilist system of the period found it hard to cope with the extreme conditions. Indeed, ultimately, the Spanish economy teetered on the verge of total collapse. The price of wheat rose ten-fold between 1450 and 1750, and prices trebled between 1600 and 1750, ignoring certain record peaks in the 1620s and 1630s. In Poland between 1643 and 1681 the price of oats rose to eight times their original level; wheat prices doubled in England between 1645 and 1650; while in Italy they trebled over the same period. The need to provide and maintain armed forces both affected, and was affected by, these different stresses and strains, and population sizes inevitably reflected these severe pressures. Only those powers capable of undertaking expansion, whether in the Old or the New Worlds, were able to turn the situation to lasting advantage. Thus Spain, Sweden and the United Provinces ultimately declined, and England, France and Russia progressed in influence.

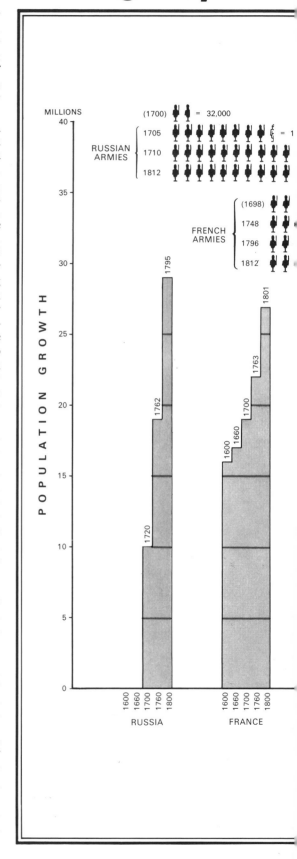

POPULATION FLUCTUATIONS OF THE NINE EUROPEAN POWERS, 1600-1800
and estimated comparative sizes of six armies in four specimen years

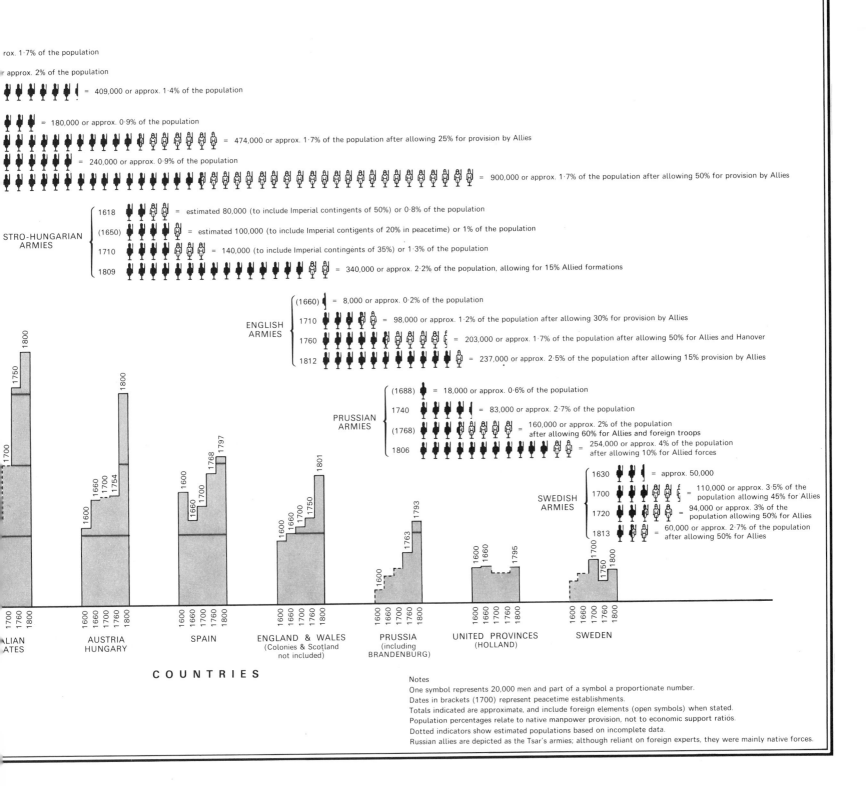

rox. 1·7% of the population

approx. 2% of the population

= 409,000 or approx. 1·4% of the population

= 180,000 or approx. 0·9% of the population

= 474,000 or approx. 1·7% of the population after allowing 25% for provision by Allies

= 240,000 or approx. 0·9% of the population

= 900,000 or approx. 1·7% of the population after allowing 50% for provision by Allies

STRO-HUNGARIAN ARMIES
- 1618 = estimated 80,000 (to include Imperial contingents of 50%) or 0·8% of the population
- (1650) = estimated 100,000 (to include Imperial contigents of 20% in peacetime) or 1% of the population
- 1710 = 140,000 (to include Imperial contingents of 35%) or 1·3% of the population
- 1809 = 340,000 or approx. 2·2% of the population, allowing for 15% Allied formations

ENGLISH ARMIES
- (1660) = 8,000 or approx. 0·2% of the population
- 1710 = 98,000 or approx. 1·2% of the population after allowing 30% for provision by Allies
- 1760 = 203,000 or approx. 1·7% of the population after allowing 50% for Allies and Hanover
- 1812 = 237,000 or approx. 2·5% of the population after allowing 15% provision by Allies

PRUSSIAN ARMIES
- (1688) = 18,000 or approx. 0·6% of the population
- 1740 = 83,000 or approx. 2·7% of the population
- (1768) = 160,000 or approx. 2% of the population after allowing 60% for Allies and foreign troops
- 1806 = 254,000 or approx. 4% of the population after allowing 10% for Allied forces

SWEDISH ARMIES
- 1630 = approx. 50,000
- 1700 = 110,000 or approx. 3·5% of the population allowing 45% for Allies
- 1720 = 94,000 or approx. 3% of the population allowing 50% for Allies
- 1813 = 60,000 or approx. 2·7% of the population after allowing 50% for Allies

COUNTRIES

ALIAN ATES

AUSTRIA HUNGARY

SPAIN

ENGLAND & WALES (Colonies & Scotland not included)

PRUSSIA (including BRANDENBURG)

UNITED PROVINCES (HOLLAND)

SWEDEN

Notes
One symbol represents 20,000 men and part of a symbol a proportionate number.
Dates in brackets (1700) represent peacetime establishments.
Totals indicated are approximate, and include foreign elements (open symbols) when stated.
Population percentages relate to native manpower provision, not to economic support ratios.
Dotted indicators show estimated populations based on incomplete data.
Russian allies are depicted as the Tsar's armies; although reliant on foreign experts, they were mainly native forces.

The Thirty Years War, 1618-48

This dire and confused struggle in Central Europe, 1618-48, was partly an internal struggle for the succession to power within the unwieldy Holy Roman Empire, partly an ideological and territorial conflict between Catholic and Protestant interests, and partly a dynastic struggle involving the royal houses of Habsburg, Vasa and Bourbon. It overlapped the Eighty Years War between the Dutch and Spanish, 1568-1648, and coincided in part with the English Civil Wars, 1642-51 (see pp. 30-32). This period also saw wars or revolutionary outbreaks involving Portugal, Catalonia, Naples and Denmark, and, of course, the ancient feud between Austria and Turkey (see p. 54)

continued in the south-east. All Europe, therefore, was in a state of intermittent turmoil. In some ways, however, the Thirty Years War was the focal point and catalyst of European unrest, and its effects were influential and dramatic in both the short and long term.

The direct cause of the German struggle can be traced to 1608 when, as a development of the long-established Reformation, the Holy Roman Empire began to split into two rival camps: those of the Protestant Union under Frederick V, Elector Palatine, and of the Catholic League led by Maximilian of Bavaria. Tension increased until flash-point was reached over the candidature to the Bohemian crown. In 1617,

the enthusiastic Catholic and pro-Jesuit Ferdinand, cousin of the aged Habsburg Emperor, Matthias, succeeded to the Bohemian throne, but was then rejected by the staunchly-Lutheran and strongly-independent Bohemian nobility, who preferred Frederick. On 21 May 1618 some imperial officials were unceremoniously flung from the celebrated window in Prague, and Bohemia was thrown into a state of rebellion. The Emperor unwillingly decided to intervene, his more extreme advisers hoping to use the revolt to restore Germany wholly to Catholicism and at the same time to strengthen Habsburg imperial power throughout the empire.

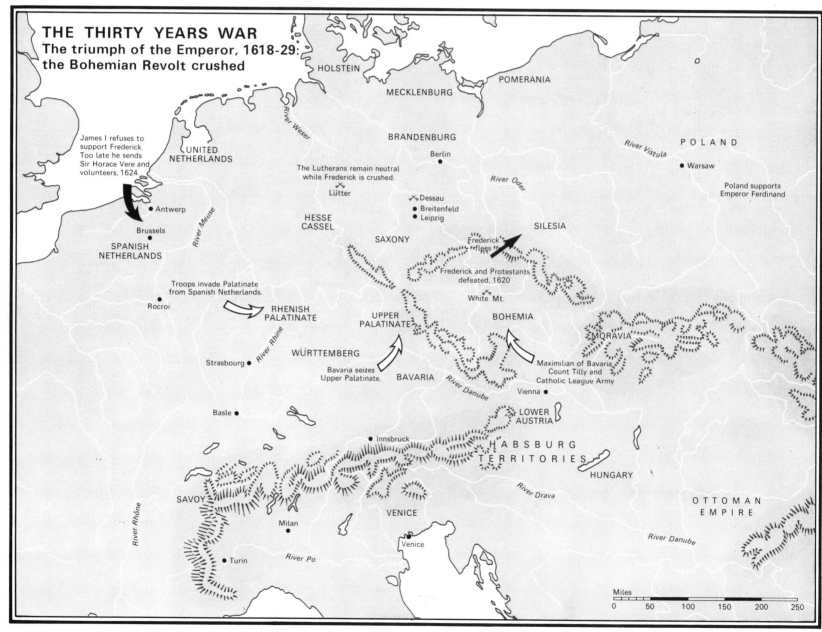

THE THIRTY YEARS WAR
The triumph of the Emperor, 1618-29: the Bohemian Revolt crushed

HOLSTEIN

POMERANIA

MECKLENBURG

BRANDENBURG

James I refuses to support Frederick. Too late he sends Sir Horace Vere and volunteers, 1624.

UNITED NETHERLANDS

Berlin

POLAND

River Vistula

Warsaw

Poland supports Emperor Ferdinand

The Lutherans remain neutral while Frederick is crushed.

Lütter

River Oder

Dessau
Breitenfeld
Leipzig

HESSE CASSEL

SAXONY

SILESIA

Frederick flees

Antwerp

Brussels

SPANISH NETHERLANDS

River Meuse

Frederick and Protestants defeated, 1620

White Mt.

Troops invade Palatinate from Spanish Netherlands.

Rocroi

RHENISH PALATINATE

UPPER PALATINATE

BOHEMIA

MORAVIA

WÜRTTEMBERG

Strasbourg

River Rhine

Bavaria seizes Upper Palatinate.

BAVARIA

River Danube

Maximilian of Bavaria, Count Tilly and Catholic League Army

Vienna

Basle

LOWER AUSTRIA

Innsbruck

HABSBURG TERRITORIES

HUNGARY

SAVOY

River Drava

OTTOMAN EMPIRE

Milan

VENICE

River Danube

Venice

Turin

River Po

River Rhône

Miles

0   50   100   150   200   250

For convenience, the war can be studied in three periods, but the full ramifications and cross-weave of clashing loyalties and interests cannot be adequately treated in the space available.

The Triumph of the Emperor, 1618-29. Ferdinand succeeded Matthias and was elected Emperor in August 1619; and Frederick V, Elector Palatine, accepted the proffered Bohemian crown later in the same month. The combined Catholic forces overwhelmed Frederick at the Battle of the White Mountain in November, 1620, and it seemed as if the Protestant cause was lost. A reign of terror began in Bohemia, but no active aid was offered by the Protestant Union or by Frederick's father-in-law, James I of England. The latter made a feeble gesture in 1624, but even when the King of Denmark intervened with an army of 30,000 in 1626 he was speedily crushed by the Bavarian general, Count Tilly, and the Imperial commander, the Catholic-convert Wallenstein. (To support the Emperor, Wallenstein had raised a quasi-independent army almost 100,000 strong which he used to further his personal interests as condottiere and war-profiteer.) Wallenstein had already defeated the Protestant adventurer from Savoy, Albert of Mansfeld, at Dessau on 25 April 1626, and the same treatment was meted out to Christian of Denmark at Lütter that August. Three years later, by the Peace of Lübeck, Denmark retired from the war. Thus, by 1629, the Emperor Ferdinand II was stronger throughout Germany than for many years, and in March he decreed the return of all ecclesiastical lands that had been absorbed by the Protestant princes. However, at the request of the Catholic League, he was induced to dismiss the unpopular Wallenstein, whose mercenaries ravaged friendly and hostile regions alike.

The Intervention of Sweden 1631-35. Although the likelihood of a Habsburg triumph appeared strong, the Protestant powers and

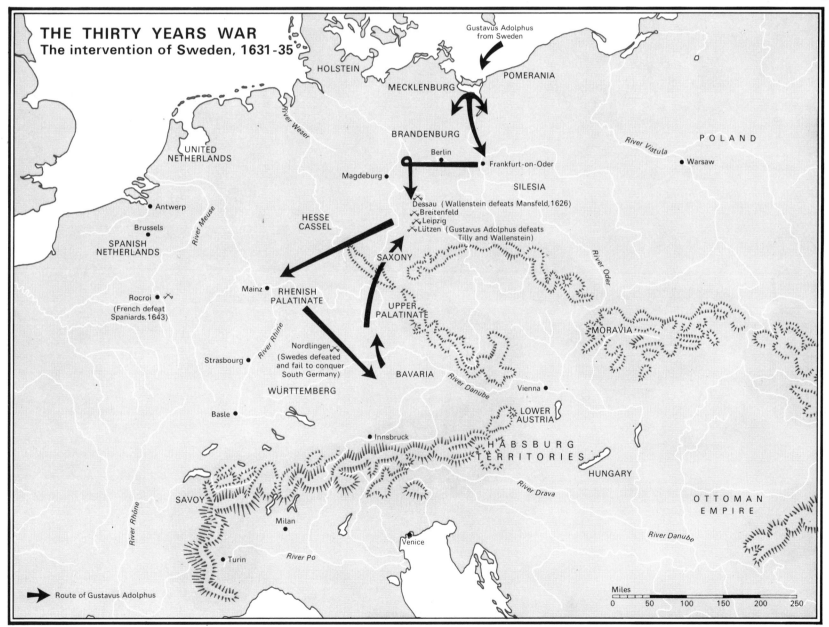

THE THIRTY YEARS WAR
The intervention of Sweden, 1631-35

Route of Gustavus Adolphus

Miles
0   50   100   150   200   250

The Thirty Years War, 1618-48

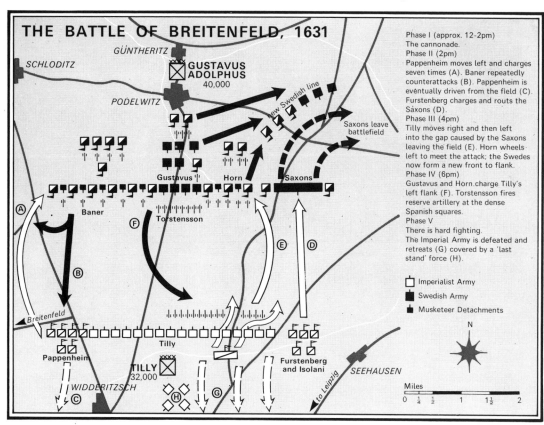

THE BATTLE OF BREITENFELD, 1631

GÜNTHERITZ

SCHLODITZ

GUSTAVUS ADOLPHUS
40,000

PODELWITZ

new Swedish line

Saxons leave battlefield

Gustavus   Horn   Saxons

Baner

Torstensson

Breitenfeld

Pappenheim

Tilly

TILLY
32,000

Furstenberg and Isolani

SEEHAUSEN

WIDDERITZSCH

to Leipzig

Phase I (approx. 12-2pm)
The cannonade.
Phase II (2pm)
Pappenheim moves left and charges seven times (A). Baner repeatedly counterattacks (B). Pappenheim is eventually driven from the field (C). Furstenberg charges and routs the Saxons (D).
Phase III (4pm)
Tilly moves right and then left into the gap caused by the Saxons leaving the field (E). Horn wheels left to meet the attack; the Swedes now form a new front to flank.
Phase IV (6pm)
Gustavus and Horn charge Tilly's left flank (F). Torstensson fires reserve artillery at the dense Spanish squares.
Phase V
There is hard fighting.
The Imperial Army is defeated and retreats (G) covered by a 'last stand' force (H).

☐ Imperialist Army
■ Swedish Army
■ Musketeer Detachments

N

Miles
0 ¼ ½ 1 1½ 2

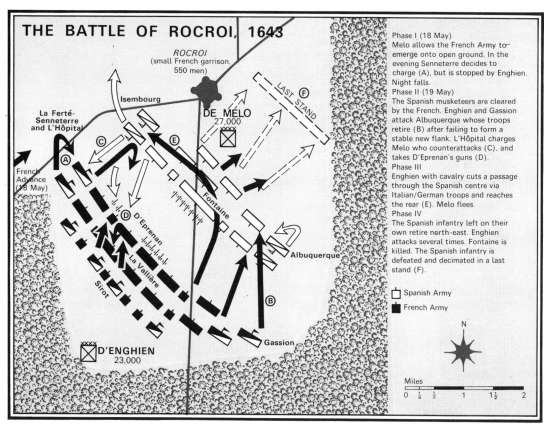

THE BATTLE OF ROCROI, 1643

ROCROI
(small French garrison, 550 men)

Isembourg

LAST STAND

DE MÉLO
27,000

La Ferté-Senneterre and L'Hôpital

French Advance (18 May)

D'Eprenan

Fontaine

La Vallière

Sirot

Albuquerque

D'ENGHIEN
23,000

Gassion

Phase I (18 May)
Melo allows the French Army to emerge onto open ground. In the evening Senneterre decides to charge (A), but is stopped by Enghien. Night falls.
Phase II (19 May)
The Spanish musketeers are cleared by the French. Enghien and Gassion attack Albuquerque whose troops retire (B) after failing to form a stable new flank. L'Hôpital charges Melo who counterattacks (C), and takes D'Eprenan's guns (D).
Phase III
Enghien with cavalry cuts a passage through the Spanish centre via Italian/German troops and reaches the rear (E). Melo flees.
Phase IV
The Spanish infantry left on their own retire north-east. Enghien attacks several times. Fontaine is killed. The Spanish infantry is defeated and decimated in a last stand (F).

☐ Spanish Army
■ French Army

N

Miles
0 ¼ ½ 1 1½ 2

secretly Cardinal Richelieu of France were determined to avert such a possibility. Accordingly, an appeal was made to Gustavus Adolphus, King of Sweden, to save the Protestant cause. 'The Lion of the North' moved into Germany with 30,000 men, hoping to gain control of the south Baltic coast for Sweden as well as to aid his co-religionaries. To this end he immediately occupied Pomerania, but failed to save beleaguered Magdeburg from its terrible fate. (In May 1631 the 20,000 inhabitants of this city-bishopric perished at the hands of Tilly's soldiery.) Gustavus then struck for Berlin and compelled the hitherto lukewarm and obstructive Electors of Brandenburg and Saxony to conclude alliances. Then, in a series of daring marches, he led his small but efficient army to defeat Tilly and Pappenheim at Breitenfeld in September 1631, before launching a winter campaign which drove Wallenstein from the field, and saw the death of Pappenheim at the fog-shrouded Battle of Lützen on 16 November 1632. It was a success for the Protestant cause, but the price was high—Gustavus lost his life. This disaster checked the tide of Protestant success; and in 1634 Bernard of Saxe-Weimar and Gustav Horn were heavily defeated at Nördlingen by a stronger Imperialist Army. In the meantime, the Emperor had decided to get rid of his reinstated but 'overmighty subject', and in February 1634 he engineered the assassination of Wallenstein.

After seventeen years of intermittent but destructive warfare, both the German princes and the Habsburg power were exhausted. The religious phase of the war now disappeared, as the unscrupulous Richelieu, who had been intriguing with both Bavaria and Sweden at the same time, scented the opportunity of making significant gains for France and declared war on Spain.

The French Incursion 1635-48. The war would drag on for thirteen more years. New French armies, which had partly absorbed Swedish concepts of warfare through the hiring of Saxe-Weimar's entire army, proved incapable of bringing off a decisive success against the Imperialist and Spanish forces, and Richelieu died before his schemes could reach maturity. However, the advent of the youthful and brilliant general, the Duc d'Enghien (later the Great Condé), on the military scene at last broke the impasse. His decisive victory at Rocroi on 19 May 1643, together with his compatriot Marshal Turenne's series of successes over the Emperor's forces, ensured French military superiority over the troops and allies of Ferdinand III, who had succeeded to the Holy Roman Empire in 1637. The scene was at last set for a general pacification, which came in January and October 1648.'

The 'Father of Modern Warfare': Gustavus Adolphus and his Imitators

The King of Sweden made the most significant contribution to the development of the art of war in the seventeenth century. From his accession in 1621, he set about creating an effective standing army. Aware that the ordinary soldier was the key to failure or success, he trained, paid, disciplined and uniformed his troops in blue, and thus instilled professional pride. Introducing a regular rank structure, he insisted that officers should be responsible for their men's welfare and made great efforts to improve administrative and supply services.

In tactical terms, he continued the work of Maurice of Nassau and carried it to a logical conclusion. Adopting the Dutch linear tactics, he demonstrated that linear formations could defeat massive tercios in the attack as well as on the defensive. His squadrons and brigades proved far more flexible than the imperialist tercios, and combined shock action with fire-power to a marked degree.

Equipping his men with shortened 8ft pikes and improved wheel-lock muskets (see p. 37) in approximately equal proportions, he introduced the firing of salvos, which he adapted to a platoon firing system in order to achieve continuous fire (see p. 65).

Lighter muskets (15 pounds each) and pre-packed paper cartridges increased mobility and rate of fire. As with his infantry, so too with his cavalry he combined missile power with mass shock impact. His troopers were trained to use both pistol and sword, but the latter was regarded as the more significant weapon. The caracole was abandoned.

Artillery was also revolutionized. First 'leather' guns and later light 3-pounder cannon were attached to every brigade to provide close fire support. Care was taken to differentiate siege guns (24-pounders) from field (12-pounders) and regimental (3-pounders) artillery. Improved design and casting made guns lighter, more reliable and, above all, more manoeuvrable. Carefully co-ordinated patterns of artillery and musketry fire were devised for battle, and all arms were trained to fight as combat groups or teams. In sum, Gustavus created the first truly modern army; one that was destined to be widely copied in France, the United Provinces and England. Between them, Maurice and Gustavus carried through what has been rightly called a 'military revolution'. Gustavus certainly deserves the title of 'Father of Modern Warfare'.

In England, Sir Thomas Fairfax and Oliver Cromwell used many Swedish concepts in building the élite New Model Army of 1645, which ultimately proved capable of defeating French and Spanish forces at the Dunes in 1658. In France, both the Great Condé and Marshal-General Henri de Turenne followed the Swedish lead. The former became especially famous for his battlefield grand tactics, the latter for his strategic skills and for his superb talents as a man-manager. Turenne's last campaign, in 1675, was probably his best (see p. 40); his use of rapid manoeuvre earned him three battle successes (Mulhouse Colmar and Turckheim) that forced France's foes to evacuate Alsace. On 27 July 1675, however, he was killed at the Battle of Sasbach. Condé succeeded to his position, and managed to drive the gifted Austrian Raimondo Montecuccoli back over the Rhine before himself retiring from active service. Thus France lost her two best generals in the same year, although Marshal Luxembourg would continue in their great tradition (see p. 39). But it was Turenne who directly inspired the Duke of Marlborough (see p. 42)—destined to be the scourge of the latter-day, and less inspired, marshals of Louis XIV. Prince Eugene of Savoy, too, would emerge as a great commander in succession to Montecuccoli. It is noteworthy that Napoleon placed both Turenne and Eugene amongst his seven greatest soldiers of all time.

The French also made great administrative contributions to warfare. Such ministers as Le Tellier and Louvois (1641-91) instituted rigid centralization and imposed close inspections of both men and material. Their supply system, based upon the frontier intendants, enabled France to keep almost 400,000 men under arms at one period. Their careful accountancy, however, never defeated the vices of corruption and speculation, and many of their regulations were flouted in practice.

The Battle of Breitenfeld, 1631

Gustavus Adolphus' great success at Breitenfeld proved to be the salvation of Protestant Germany. It also vindicated the new Swedish military methods that were to revolutionize the whole conduct of European warfare. His army comprised 192 infantry squadrons, 131 cavalry squadrons, and some 20 field-guns besides 52 light 3-pounder cannon. In overall terms, the Swedes enjoyed an advantage of some 8,000 men over the Leaguer Army of Count Tilly, who commanded 21,000 infantry, 11,000 cavalry and 30 guns.

The main stages of the unfolding struggle are given on the accompanying map. Points that

deserve reiteration include the Swedish double line of battle, each provided with a tactical reserve, and the way in which small bodies of musketeers and light guns were interspersed amongst the Swedish cavalry on the right flank. These dispositions conferred advantages of mobility and flexibility, and contrasted with both the Saxon formations on the Swedish left wing and the somewhat haphazard arrangement of Tilly's army. The latter's weak control was illustrated by Pappenheim's unauthorized attack, and the resultant attempt to carry off a double envelopment of the Swedish Army, which was virtually impossible given his numerical disadvantage. Nevertheless, at about 2pm Fürstenberg's wing proved capable of routing the Saxons, and only the flexibility of the Swedish Army enabled Horn to redress this dangerous situation; even then, for a time he was facing 23,000 opponents with only 3,500 men. Meanwhile, on the farther flank, Baner's combination of salvo fire and short, sharp charges completely foxed Pappenheim who, after repeated attacks, withdrew defeated at about 6pm.

Fortunately for Horn, Tilly required considerable time to re-order his battle line of tercios before trying to roll up the Swedish line, and Horn seized a fleeting chance to charge and scatter the Leaguer horse when they inadvertently strayed between the Swedish line and their own tercios beyond. Flung back upon their own infantry, the disarrayed cavalry held up the redeployment of their foot even longer. This won more time for the transfer of further Swedish formations to the flank in the heat of battle, and eventually they gained local superiority. Compressed and outflanked, Tilly's centre and right wings became increasingly disordered and desperate. After Tilly himself was wounded, Gustavus and Horn launched a massive counterattack against the exposed left flank, capturing the Leaguers' guns which were promptly turned upon their late masters. This forced the Leaguers to retreat in disorder, but a complete rout was averted by four tercios who fought to the last on a knoll and covered the escape of their comrades. As it was, the Leaguers lost 7,600 killed; 6,000 were captured on the field and a further 8,000 were taken prisoner afterwards. Swedish losses were put at 4,000. The emergence of Sweden as a major, albeit short-lived, European power dates from this battle.

The Battle of Rocroi, 1643

Rocroi, twelve years later, foretold the eclipse of Spanish military power. It also saw the emergence of a notable commander, aged only twenty-one, in Louis de Bourbon, Duc

1. *Gustavus Adolphus II (1594-1632), King of Sweden and champion of the cause of European Protestantism. Nicknamed 'the Lion of the North' he has also been called 'the father of modern warfare'; his reorganized Swedish Army became the model for many European armed forces.*
2. *Johan Tzerclaes, Count Tilly (1559-1632)—a Flemish mercenary who led the forces of the Catholic League with skill, but proved no match for his Swedish opponents and their modern methods.*
3. *Albrecht Eusebius Wenzel von Wallenstein, Duke of Friedland and Mecklenburg (1583-1634), the ruthless and over-ambitious Habsburg commander who almost crushed the forces of German Protestantism. He was out-matched by the Swedes, and was assassinated by a group of his own officers.*
4. *Louis II de Bourbon, 'the Great Condé' (1621-86), victor of Rocroi, whose later implication in the Fronde revolts did not prevent his re-emergence as the leading French commander (second only to Turenne) of Louis XIV's early and mid-reign.*

d'Enghien (or Condé). The French Army comprised 18 battalions, 32 squadrons and 12 guns; that of the Spaniards, 20 tercios, 7,000 cavalry and 28 guns. The latter enjoyed an initial superiority of some 4,000 men overall. D'Enghien's basic aim was to relieve the beleaguered French garrison of Rocroi, which was an important strategic fortress.

After an inconclusive evening skirmish on 18 May, in which L'Hôpital vainly strove to break through to Rocroi, the two armies ranged themselves for battle and moved into combat range early on the 19th. As the map clearly illustrates, d'Enghien made his main effort on the right, sending Gassion to execute an enveloping attack against Albuquerque, so as to compel the Spanish left to change front, and thereby present d'Enghien with the chance to unleash further forces against the newly-exposed Spanish flank. This plan succeeded, and after an hour's fighting Albuquerque's forces were routed.

Meanwhile, on the opposite flank, La Ferté-Seneterre charged home against Isembourg, who was commanding the Spanish right, only to be routed in his turn. The Spanish cavalry swept back to attack d'Eprenan's foot, and captured several of his guns which were promptly turned on the French centre. Under this pressure, General la Vallière ordered a retreat, but this was checked by Sirot—like Gassion a veteran commander who had served under Gustavus Adolphus—and the French centre recovered itself and then attacked the Italian tercios to contain their advance.

During these developments, d'Enghien swept on along the enemy centre, piercing all three lines of formations, throwing them into disorder. Isembourg attempted to counter-attack, but Sirot saw his chance and charged the Spanish lines in support of his commander-in-chief's onslaught. Deprived of cavalry support, the tercios began to withdraw to the north-east in good order. After detaching Gassion to watch for anticipated Spanish reinforcements, d'Enghien reformed his ranks and then advanced against Fontaine. The Spaniards held their fire until the last moment, then poured in a deadly volley. There followed a vicious struggle in which Fontaine was struck down. D'Enghien himself was almost killed in one confused incident, and this so infuriated the French that they fell upon the Spaniards with unmatched ferocity. By 10 o'clock the flower of the Spanish infantry was no more. Too late, Spanish reinforcements came within range of the field, but took no part.

The Spaniards lost 21,000 men (including 7,500 killed and as many more captives) to the French 4,000. This notable success inaugurated

the reign of the child-king Louis XIV, who had succeeded his father on 10 May, on a successful martial note. It also marked the passing of the age of the Spanish tercios, for a century past the best troops in Europe. From Rocroi, military predominance began to move, slowly but surely, into the hands of the French. There it would remain until the early eighteenth century.

The Peace of Westphalia

After lengthy deliberations and diplomatic manoeuvrings, the first congress of European governments at last managed to patch up a solution to some of the more outstanding issues. First, by the Treaty of Münster, 7 January 1648, Spain at last fully recognized the independence of the Protestant Netherlands—the United Provinces. This treaty, together with that of Osnabrück, formed the basis of the Peace of Westphalia which was finally signed on 24 October 1648.

One power to gain materially by the terms was, predictably, France. The House of Bourbon received rather vague rights over Alsace, and was awarded the bishoprics of Metz, Toul and Verdun and other fortresses. All in all, these represented significant strategic gains towards the achievement of the Rhine frontier. The Emperor retained Bohemia and received Upper Austria back from Bavaria, but the other participants in the war had to be bought off. Sweden received Western Pomerania and territory commanding the mouths of the Rivers Weser, Elbe and Oder. Brandenburg gained Eastern Pomerania, Minden and Magdeburg; these were significant gains for the Great Elector, forming the basis for the evolution of the powerful state of Brandenburg-Prussia in later years. Bavaria was given the Upper Palatinate. Such were the main territorial adjustments. The religious issue was resolved by according Lutherans and Calvinists equal status, and all church property transferred or seized before 1624 was confirmed to the present owners in perpetuity. Politically, the loser by these agreements was the Holy Roman Emperor. One effect of this was to strengthen the importance of the Habsburg's Austrian and Hungarian possessions. Compelled to recognize the independence of the Spanish Netherlands and the Swiss Cantons, and to concede the power to make treaties to the German Electors, the Habsburg authority over the empire became more nebulous and insubstantial than ever. Other seeds of future trouble included the likelihood of increasing conflict of interest between Bourbons and Habsburgs over the vague stipulations regarding Alsace and other

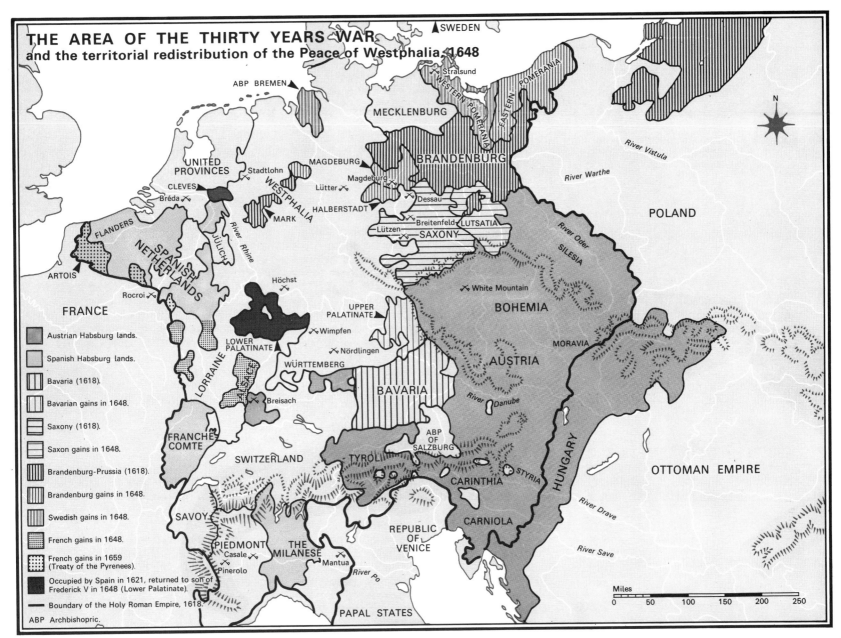

THE AREA OF THE THIRTY YEARS WAR
and the territorial redistribution of the Peace of Westphalia, 1648

SWEDEN

ABP BREMEN

Stralsund

WESTERN POMERANIA

EASTERN POMERANIA

MECKLENBURG

POLAND

River Vistula

BRANDENBURG

River Warthe

UNITED PROVINCES

Stadtlohn

MAGDEBURG

Magdeburg

CLEVES

Bréda

WESTPHALIA

Lütter

River Rhine

Dessau

River Oder

HALBERSTADT

MARK

Breitenfeld LUSATIA

SILESIA

Lützen SAXONY

JÜLICH

FLANDERS

SPANISH NETHERLANDS

White Mountain

ARTOIS

Höchst

BOHEMIA

Rocroi

FRANCE

UPPER PALATINATE

Wimpfen

MORAVIA

LOWER PALATINATE

Nördlingen

WÜRTTEMBERG

AUSTRIA

LORRAINE

River Danube

HUNGARY

Breisach

BAVARIA

OTTOMAN EMPIRE

FRANCHE COMTE

ABP OF SALZBURG

SWITZERLAND

TYROL

STYRIA

CARINTHIA

River Drave

SAVOY

CARNIOLA

PIEDMONT

Casale

THE MILANESE

REPUBLIC OF VENICE

River Save

Pinerolo

Mantua

River Po

N

Miles

0 50 100 150 200 250

PAPAL STATES

Legend:
- Austrian Habsburg lands.
- Spanish Habsburg lands.
- Bavaria (1618).
- Bavarian gains in 1648.
- Saxony (1618).
- Saxon gains in 1648.
- Brandenburg-Prussia (1618).
- Brandenburg gains in 1648.
- Swedish gains in 1648.
- French gains in 1648.
- French gains in 1659 (Treaty of the Pyrenees).
- Occupied by Spain in 1621, returned to son of Frederick V in 1648 (Lower Palatinate).
- — Boundary of the Holy Roman Empire, 1618.
- ABP Archbishopric.

frontier areas of shared interest. Louis XIV would not, in the years of his adult reign, hesitate to press every claim, and further major wars would result. Similarly, the recognition of Sweden as a major power would lead that country to ever-increasing over-extension of her limited resources as she entered fully into the vortex of Polish and East European affairs—and ultimately to the ruin and the collapse of her Baltic empire.

In military terms, the significance of the Peace would also be very important. Repugnance to the horrors of war witnessed over the thirty years led to a strong reaction. Death and displacement due to fire, rapine and the sword

is estimated to have involved a drop of forty-five per cent in the rural population, and thirty-three per cent in the population of towns, through large tracts of Germany. To be sure, many of the people returned to rebuild their ruined villages and re-establish the economic life of their regions, and some areas—such as Württemberg and Pomerania—suffered more severely proportionately than others, for example Lower Saxony. But in overall terms, what we now call Germany—recently the centre of a flowering culture which the many-sided genius of Albrecht Dürer represents—received wounds that put back her development by an estimated hundred years, with results that

Europe would come to appreciate in the nineteenth and twentieth century.

The reaction took the form of a growing recognition of the importance of international law governing the relations between nations, and a determination that future wars would be far more restricted in their means of being waged and thus in their overall effects. One strategic effect of the emotions released by this struggle and its pacification, was to launch a period of limited warfare that would extend until the late eighteenth century. The military reached a nadir in popular regard. As Descartes observed, "I cannot, in all conscience, regard the profession of soldier as a noble calling".

The English Civil Wars, 1642-51

Compared with the Thirty Years War, the three English Civil Wars were only pale imitations—with the exception of a few particular incidents. Nevertheless, the lengthy struggle between the House of Stuart and Parliament caused sufficient dislocation and misery to colour British attitudes towards constitutional and military problems down to the present day.

At its base, the struggle was about money. The problems of inflation and rising governmental costs led Charles I to exact various direct taxes—such as ship-money for the support of the Navy—without proper recourse to Parliament, which, sensing its growing power over the purse, was determined to exact constitutional concessions in return for any vote of supply. The issue then became one between Charles's claim to reign by the 'Divine Right of Kings' and his political foes' avowal that ultimate authority resided in the 'King-in-Parliament'. Religious issues soon further embittered the disputes; the King represented the established Church of the realm, but was accused by his more puritanical foes of neo-Catholic intentions for which blame was levelled at his French queen and Archbishop Laud. Rumours that his servant Strafford was preparing to bring Catholic Irish regiments into England, linked to the fiascos of the short campaigns against Scotland in 1639 and 1640—which compelled Charles to re-summon Parliament after eleven years of personal government —led to a rapid rise in tension. The Long Parliament lost no time in attacking the King's ministers, with both Strafford and Laud being sacrificed to expediency. Finally, Charles was stung into retaliation by the ever-more extreme constitutional demands.

By the spring of 1642 the country was on the verge of Civil War, and, after several months of sparring and unofficial military action by both sides, the crisis at last came on 22 August when the King formally raised his standard at Nottingham.

The early campaigns that followed were not particularly well conducted by either side. Both Charles's and Essex's armies were very amateur in composition and leadership during the first two years of the struggle, and neither party could gain a decisive advantage. The King enjoyed much support in the Midlands, Wales and the north; Parliament controlled the eastern counties, the south, much of the west (although this was rapidly lost to Generals Hopton and Goring) and, above all, the great cities and ports of London, Bristol and Hull, together with most of the fleet. Socially, all classes were split, but gradually Anglicans and Catholics rallied to the King and the more Protestant to Parliament which, from the start, enjoyed the greater economic resources that sprang from London and continuing trade through the ports.

The war was fought regionally rather than nationally. In 1642 a Royalist attempt at a concerted advance on London led to the Battle of Edgehill (23 October), which was a drawn affair, and Essex was only able to stem the tide at Turnham Green on the outskirts of the capital. Charles then fell back on Oxford, which became his capital until the end of the war.

The year 1643 saw much disjointed fighting in the north, Midlands and west with many engagements and sieges. Overall, the year ended to Charles's advantage with almost three quarters of the realm under his control. However, Parliament was able to enlist Scottish support, and the armies of the Association began to improve—including Oliver Cromwell's 'lovely company' of disciplined cavalry or 'Ironsides'. Thus, they were able to survive the death of their leader, the Parliamentarian John Pym, and to defeat the King's imported Irish forces. In 1644 Lord Leven's and Oliver Cromwell's great victory over Prince Rupert at Marston Moor, 2 July, effectively cost Charles the north, and marked the turn of the tide.

The year 1645 saw the emergence on the battlefield of Fairfax's and Cromwell's New Model Army: a disciplined, motivated, paid and uniformed force formed along Swedish lines. This formation proved more than a match for the less-efficient Royalists at the decisive Battle of Naseby (see p. 32). Although the First Civil War continued until mid-1646, Charles's cause was doomed—despite temporary local successes by Montrose in Scotland (reversed at Philiphaugh in September 1645). On 5 May the King surrendered to Parliament's Scottish allies at Southwell. Oxford fell in June.

A period of complex and shifty negotiations between King, Parliament and the New Model Army followed, with Charles trying to play his foes off against one another. Attempts to reach an agreed settlement ultimately foundered when, in 1648, the Scottish army of Hamilton invaded the north on behalf of the King, unleashing the brief 'Second Civil War'. Cromwell routed this venture at Preston in August and, thereafter, the extremists began to demand Charles's trial. This led to his execution on 30 January 1649, after which England became a Commonwealth or republic. The transition of power was far from easy. Later that year, Cromwell had to fight a bitter campaign in Ireland in the first round of the so-called 'Third Civil War'; and the massacre of Drogheda, 12 September, left a permanent scar on Anglo-Irish relations. However, Ireland was occupied, and practically all its territory 'settled' with Parliamentarian ex-soldiery and supporters. Next, in 1650, Cromwell led his army against the Scots, decisively defeating Leslie and the Covenanters at Dunbar on 3 September. However, the Scottish Stuart cause was still not lost, and in 1651 an Anglo-Scottish army, accompanied by the youthful Charles II, penetrated as far as Worcester before being finally defeated there exactly a year after Dunbar, 3 September. Charles escaped to France by way of Brighton after sundry adventures, and the Commonwealth at last seemed secure.

The following years saw increasing factional strife between Parliament, the Army, and various extreme religious sects. To avoid chaos, the Army purged Parliament, and little by little Cromwell moved towards the control of the state. In 1654 he became Lord Protector, but refused the offer of the Crown from his second Parliament in 1657. Thereafter, he ruled as virtual dictator until his death on 3 September 1658. In the meantime, he carried through an effective foreign policy: defeating the Dutch at sea; declaring war on Spain; capturing Jamaica, 1655; and, with French help, defeating the Spanish at the Battle of the Dunes in June 1658, thereby securing Dunkirk for the Commonwealth.

After Cromwell's death insecurity rapidly spread over the Commonwealth, and the final breakdown of relations between Army and Parliament led to Monk's march from Scotland on 2 January 1660 towards London. Parliament, acceding to the growing demand for a Stuart Restoration, dissolved itself in March and, two months later, after negotiations and the summoning of the Convention Parliament, King Charles II 'came into his own again' after landing at Dover on 25 May. So ended the troubled times of the Civil Wars and the Interregnum, and with them absolute monarchy, which would never be restored—although it took the Glorious Revolution of 1688 and the exclusion of James II from the throne to prove the point once and for all.

Militarily, this period is notable for the contrast between the ineffective campaigns of the earlier years—led by a few good soldiers on both sides but in the main amateurishly waged —and the emergence of the first truly professional English Army in the 'New Model' of 1645. Both Thomas Fairfax and Oliver Cromwell emerged as soldiers of genius, particularly the latter, and their army went on to earn a European reputation. Unfortunately, its involvement in politics proved a fatal complication, and created a deep-rooted

THE ENGLISH CIVIL WARS, 1642-51

Insets indicate the shifts of local loyalties and control during the period 1642-45.
The large numbers of battles and actions reveal lack of military competence.

1642

Newcastle ◄ Weapons
York
Hull
Charles
Shrewsbury • *Nottingham*
Essex
Edgehill
Bristol
Bedford
Turnham
Exeter Green
Plymouth ◄
◄ Weapons

1643

York
Weapons
Adwalton Moor *Hull*
Newcastle ◄Weapons
Winceby
Chalgrove **Manchester**
Field
Essex *Cambridge*
Gloucester *Oxford*
Bristol *London*
Stratton *Newbury*
Hopton Lansdown
Sourton
Down *Plymouth*
Weapons

1644

Berwick
Leven
Marston Moor *York*
Hull
Manchester
Shrewsbury *Newark*
Cropredy Bridge
Oxford *Cambridge*
London
Charles *Newbury*
Essex
Plymouth

1645

Newcastle
Chester
Rowton Heath
Leven **Charles**
Naseby
Hereford **Fairfax**
Bristol *Oxford* *London*
Langport
Goring

Legend

‖═▷ Royalist Force, 'Second Civil War', 1648.

━▶ Cromwell's Forces, 1648.

▨▨▷ Cromwell's March, 1649.

⟹ Charles' March, 'Third Civil War', 1650-51.

▧▧▷ Cromwell's March, 1650-51.

━▶ Royalist Forces (Insets).

▻ Parliamentary Forces (Insets).

▨ Royalist regions.

▨ Areas of Parliamentary influence.

Main map labels

Auldearn 1645
Inverness
Alford 1644
Aberdeen 1644
Inverlochy 1645
Dundee
Perth
Inverary Tippermuir 1645
Stirling
Kilsyth 1645 Edinburgh Dunbar 1650
Philiphaugh 1645 Selkirk 1645 Berwick
Derry
Belfast
Dundalk
CONNAUGHT IRISH SETTLEMENT
Athlone
Drogheda
Dublin
Rathmines 1649
ENGLISH SETTLEMENT
Limerick
Tipperary Kilkenny
Cahir
Clonmel Wexford
Waterford
Killarney
Cork
Kinsale

Carlisle
HAMILTON
Kendal
Marston Moor 1644
Sherburn
Preston 1648
Adwalton Moor 1643
CROMWELL
CHARLES II
Chester
Nantwich
Newark
Gainsborough
Winceby 1643
Shrewsbury
Naseby 1645
Worcester 1651
Edgehill 1642
Cropredy Bridge 1644
Powick Bridge
Ripple 1643
Oxford
Chalgrove Field
FAIRFAX
Colchester
Chelmsford
Gloucester
Roundway Down 1643
Pembroke
Bristol
Lansdown 1643
Newbury 1643, 44, 45
Basing
Cheriton
London
Dover
Torrington 1645
Taunton
Stratton 1643
Langport 1645
Sourton Down 1643
Exeter
Arundel
Portsmouth
Carisbrooke
Braddock Down 1643
Lostwithiel 1644
Plymouth

Newcastle
Durham
Scarborough
York
Hull
Kings Lynn
Norwich
Ely
Huntingdon
Cambridge
Northampton

Miles
0 25 50 75 100

31

The English Civil Wars, 1642-51

distrust of standing armies in peacetime that has lingered down to the present. However, the army's excellence was demonstrated by the ease with which it was disbanded in 1660, its members reassimilating themselves into civilian life with a minimum of friction or upset. The following years of Stuart neglect of the Army would only be remedied after 1689 by William III, who prepared the way for even greater achievements in the reign of Anne.

The Battle of Naseby, 1645

After losing Leicester to the Royal Army (30 May), Parliament reacted by sending the New Model Army northwards from the Thames valley (under Sir Thomas Fairfax with Oliver Cromwell as his second-in-command). A skirmish in Naseby village on 13 June led to a decision by both sides to fight a major engagement on the morrow.

The armies met between Sibertoft and Naseby. The Royalists comprised 4,000 veteran foot of the Army of Oxford, 5,000 horse and a dozen cannon. The New Model Army fielded some 7,000 foot, 6,000 horse and dragoons, and perhaps 13 guns.

The sequence of major events is shown on the accompanying diagram. By the end of the day, the Parliamentarians had inflicted 6,000 losses on the Royalists (most of them being the flower of Charles I's infantry), and sustained perhaps 1,000 themselves—a light price for a decisive victory.

The battle is notable for the contrast between Charles's rather weak leadership and the tactical acumen and drive displayed by Fairfax and Cromwell, particularly the latter. The great élan but indiscipline of Rupert's famous cavalry also contrasts with the equal fighting skill, but greater control, of Cromwell's right wing. After scattering Ireton, Rupert's men headed for the baggage trains, and only returned to the main field when it was too late to affect the issue. The Ironsides and other Parliamentary horse of the right, after defeating Langdale's wing, immediately rallied and were able to clinch the victory by falling on the exposed left flank of Astley's infantry. Similarly, Colonel Okey's dragoons, after carrying out a dismounted role amongst Sulby Hedges to defeat the Royalist right, remounted their horses to fall on Astley's right flank, thus completing the trap—a good example of flexible tactics. The New Model Army, victory gained, slightly disgraced itself by looting the Royalist waggons and molesting the womenfolk. Nevertheless, this truly professional army had marked its emergence with a fine display of fighting, thereby effectively winning the military phase of the First Civil War.

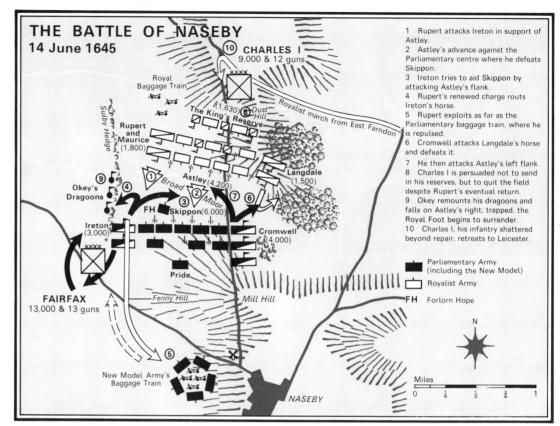

THE BATTLE OF NASEBY
14 June 1645

CHARLES I 9,000 & 12 guns

Royal Baggage Train

Royalist march from East Farndon

The King's Reserve

Dust Hill

Rupert and Maurice (1,800)

Langdale (1,500)

Sulby Hedge

Okey's Dragoons

Astley (4,200)

Broad

Moor

FH Skippon (6,000)

Ireton (3,000)

Cromwell (4,000)

Pride

FAIRFAX 13,000 & 13 guns

Fenny Hill

Mill Hill

New Model Army's Baggage Train

NASEBY

1 Rupert attacks Ireton in support of Astley.
2 Astley's advance against the Parliamentary centre where he defeats Skippon.
3 Ireton tries to aid Skippon by attacking Astley's flank.
4 Rupert's renewed charge routs Ireton's horse.
5 Rupert exploits as far as the Parliamentary baggage train, where he is repulsed.
6 Cromwell attacks Langdale's horse and defeats it.
7 He then attacks Astley's left flank.
8 Charles I is persuaded not to send in his reserves, but to quit the field despite Rupert's eventual return.
9 Okey remounts his dragoons and falls on Astley's right; trapped, the Royal Foot begins to surrender.
10 Charles I, his infantry shattered beyond repair, retreats to Leicester.

▮ Parliamentary Army (including the New Model)

▯ Royalist Army

FH Forlorn Hope

N

Miles
0   ¼   ½   ¾   1

1. Baron Thomas Fairfax (1612-71)—the creator, with Oliver Cromwell, of the New Model Army which accomplished the defeat of Charles I in the Great Civil War. 'Black Tom' was widely regarded as a model soldier and commander. (DAG)

2. Oliver Cromwell, Lord Protector (1599-1658), a country-gentleman who in middle life emerged as a great soldier and statesman. Adopting Swedish practices, he remodelled the Parliamentarian cavalry, and later led armies to victories over the Scots and Irish. (DAG)

3. Rupert, Prince Palatine of the Rhine (1619-82) was a nephew of Charles I. A dashing cavalry commander, he played a leading role in the Great Civil War, and after the Restoration commanded Charles II's fleet in several actions. He is seen here as a young man. (DAG)

4. Charles I, King of England (1600-49), whose conduct of the Great Civil War—despite occasional flashes of talent—ended in total defeat for his cause and ultimately led him to the scaffold. Well-intentioned but stubborn and crafty, he was outmanoeuvred by his foes.

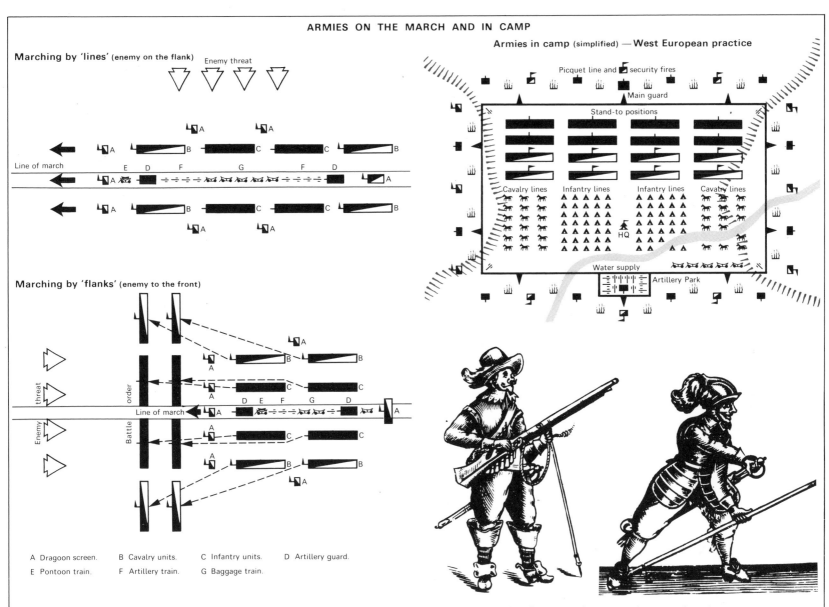

Marching by 'lines' (enemy on the flank)

Enemy threat

Line of march

E D F G F D

Marching by 'flanks' (enemy to the front)

Enemy threat

Battle order

Line of march

D E F G D

A Dragoon screen.   B Cavalry units.   C Infantry units.   D Artillery guard.

E Pontoon train.   F Artillery train.   G Baggage train.

Armies in camp (simplified) — West European practice

Picquet line and security fires

Main guard

Stand-to positions

Cavalry lines   Infantry lines   Infantry lines   Cavalry lines

HQ

Water supply

Artillery Park

A mid-seventeenth century pikeman and musketeer.

THE MILITARY ART IN THE SEVENTEENTH AND EARLY EIGHTEENTH CENTURY

Armed forces grew considerably in size over this period. During the Thirty Years War, field armies were rarely larger than 40-60,000 strong, but several topped 100,000 early in the next century; in 1705, France fielded a total of 400,000 men.

One basic consideration determining size was the problem of subsistence. The availability of fodder for the many horses and food for the men tended to channel armies to the more fertile areas. When it was deemed impracticable (through fear of mass desertions) or impolitic to feed armed forces wholly at the expense of the local peasantry (attitudes that developed with the revulsion widely felt for any form of military excess after the wars of religion), armies became increasingly dependent on pre-stocked depots and slow convoys or river-boats, supplementing these sources of supply with carefully-staged 'grand forages' in the war area. One result was to restrict

the operational range of any army to the distance it could carry its bread. However, reliance on civil contractors for bulk-supplies often led to further difficulties and delays.

The need to convoy thousands of waggons and the weighty, tandem-drawn cannon over the rudimentary roads of Europe considerably affected the speed and scale of warfare. With the average middle-sized field-gun weighing some three tons, trail inclusive, and requiring eight to ten horses or oxen controlled by civilian drivers to draw it, few armies averaged much more than five miles a day; a daily march of ten miles was exceptional. Armies habitually left their siege guns and bulkier supplies behind at the outset of a campaign, calling them up when required. However, this could lead to complicated covering operations (as before the siege of Lille, see p. 52) designed to ward off the foe's attempts at

strategic interdiction.

Armies on the march in a theatre of war moved in a number of columns ready to take up battle positions if they met an opponent. Guns and munition waggons were routed down the best available road, usually in the centre of the array; the horse and foot moving along subsidiary tracks or cross-country. If the anticipated danger lay on a flank, an army marched by 'lines': a simple turn to left or right by every unit would form the battle line. If the foe was thought to be ahead, more columns would be required (as at Marlborough's advance on the River Nebel early on 13 August 1704). Despite such precautions it still took a number of hours to prepare an army for battle; the individual regiments adopted different formations for marching and fighting, and orders took a considerable time to be passed down to the respective commanders.

Armies in Camp again took up positions as nearly resembling the battle line as was practicable. Advance parties commanded by the Quartermaster-General rode ahead to select suitable sites; the main criteria being the availability of fresh water and green forage for the horses, and the general security of the area. Such activities could lead to major engagements: as at Ramillies, 1706, see p. 46).

The areas for tents and horse lines were carefully pegged out; the artillery and munitions habitually being placed in a special encampment some way apart to reduce the fire risk. Every formation's rallying-point was designated by planting its colours, and a comprehensive series of cavalry patrols and guard posts were arranged to ensure security. If the camp was to be occupied for some time, a surrounding earthwork with redoubts would be constructed around the perimeter.

The Expansion of France

The ravaging that Germany sustained during the Thirty Years War, particularly in respect of the weakening of Habsburg authority, created a power vacuum in Central Europe that France was not slow to exploit. The processes of French expansion, already inaugurated by Henry IV and Cardinal Richelieu, came to a head under Louis XIV, who by 1660 had effectively centralized power. The continuous work of Le Tellier and, later, Louvois led to the creation of the most effective European army, whilst Colbert's reforms were rebuilding the French fleet. With the means of waging wars ready to hand, Louis XIV did not hesitate for long before launching France into a series of lengthy wars in pursuit of various territorial claims; most of them at the expense of her Austrian and Spanish Habsburg neighbours, and of various German princes with possessions west of the Rhine.

French aims were disguised as defensive responses to a supposed Habsburg policy of encirclement, but were based upon Louis XIV's ambitions to acquire the 'natural frontiers' of France (Rhine, Alps and Pyrenees), his desire to gain a share in the long-anticipated dissolution and subsequent partition of the Spanish European Empire, and his growing sense of 'la gloire'. European reaction to French aggression took the form of a series of alliances of varying efficacy, the soul of which, from 1672, was William of Orange, who was Louis' most implacable opponent. William was determined to preserve a realistic balance of power.

The circumstances created by the various hostile power-groupings—and the large size of her armed forces—involved France in fighting campaigns on a number of fronts around her borders. Throughout the wars, however, if we exclude the relatively small-scale struggles in the colonies (see p. 62), France enjoyed the advantages conferred by 'interior lines'. The main campaigns were fought out on three fronts: the north-east, the eastern and the region of the French Alps. The large number of fortresses built or repaired by Marshal Vauban (the pré-carré) in these areas reflects their military importance at this time.

The following is an outline history of French aggrandisement from 1648 to 1714.

1648: Peace of Westphalia. France received Metz, Verdun and Toul; full control of Upper and Lower Alsace, with certain reservations; possession of Moyenvic, Breisach and Pinerolo; garrisoning rights at Philippsburg.

1659: Treaty of the Pyrenees. French possession of Roussillon, Artois and certain Flemish towns, and Montmedy in Luxembourg, all confirmed; Spain drops claim to Alsace.

1668: Treaty of Aix-la-Chapelle. France

THE EXPANSION OF FRANCE, 1600-1714

Permanent acquisitions under Henry IV.

Permanent acquisitions under Louis XIII.

Permanent acquisitions under Louis XIV.

Permanent acquisitions under Louis XV.

Fortresses built or improved by Vauban.

Miles
0 20 40 60 80 100

SOUTH-EAST FRONTIER

SOUTH-WEST FRONTIER

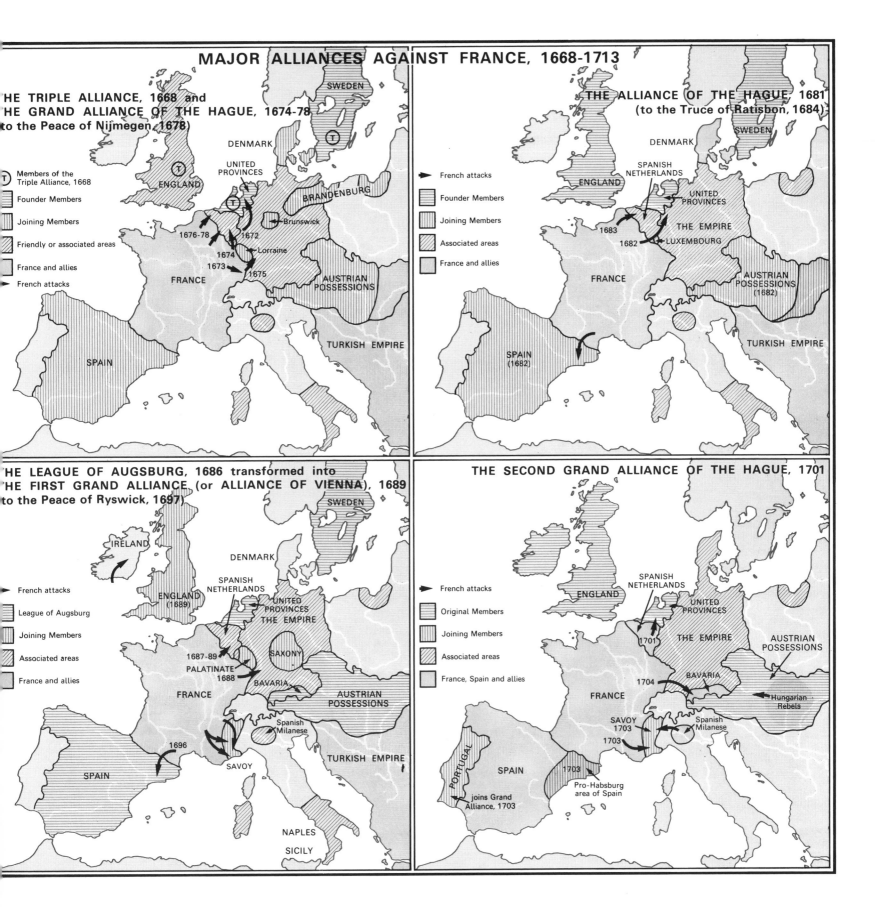

MAJOR ALLIANCES AGAINST FRANCE, 1668-1713

THE TRIPLE ALLIANCE, 1668 and THE GRAND ALLIANCE OF THE HAGUE, 1674-78 (to the Peace of Nijmegen, 1678)

SWEDEN

DENMARK

UNITED PROVINCES

ENGLAND

BRANDENBURG

T — Members of the Triple Alliance, 1668

Founder Members

Joining Members

Friendly or associated areas

France and allies

French attacks

Brunswick

1676-78 · 1672

1674 — Lorraine

1673 · 1675

FRANCE

AUSTRIAN POSSESSIONS

SPAIN

TURKISH EMPIRE

THE ALLIANCE OF THE HAGUE, 1681 (to the Truce of Ratisbon, 1684)

DENMARK · SWEDEN

SPANISH NETHERLANDS

ENGLAND

UNITED PROVINCES

THE EMPIRE

French attacks

Founder Members

Joining Members

Associated areas

France and allies

1683

1682 — LUXEMBOURG

FRANCE

AUSTRIAN POSSESSIONS (1682)

SPAIN (1682)

TURKISH EMPIRE

THE LEAGUE OF AUGSBURG, 1686 transformed into THE FIRST GRAND ALLIANCE (or ALLIANCE OF VIENNA), 1689 (to the Peace of Ryswick, 1697)

SWEDEN

IRELAND

DENMARK

SPANISH NETHERLANDS

UNITED PROVINCES
THE EMPIRE

ENGLAND (1689)

French attacks

League of Augsburg

Joining Members

Associated areas

France and allies

SAXONY

1687-89

PALATINATE 1688

BAVARIA

FRANCE

AUSTRIAN POSSESSIONS

Spanish Milanese

1696

SAVOY

SPAIN

TURKISH EMPIRE

NAPLES

SICILY

THE SECOND GRAND ALLIANCE OF THE HAGUE, 1701

SPANISH NETHERLANDS

ENGLAND

UNITED PROVINCES

THE EMPIRE

AUSTRIAN POSSESSIONS

French attacks

Original Members

Joining Members

Associated areas

France, Spain and allies

1701

BAVARIA

1704

Hungarian Rebels

FRANCE

SAVOY 1703 · Spanish Milanese

1703

PORTUGAL

SPAIN · 1703

Pro-Habsburg area of Spain

joins Grand Alliance, 1703

35

The Expansion of France

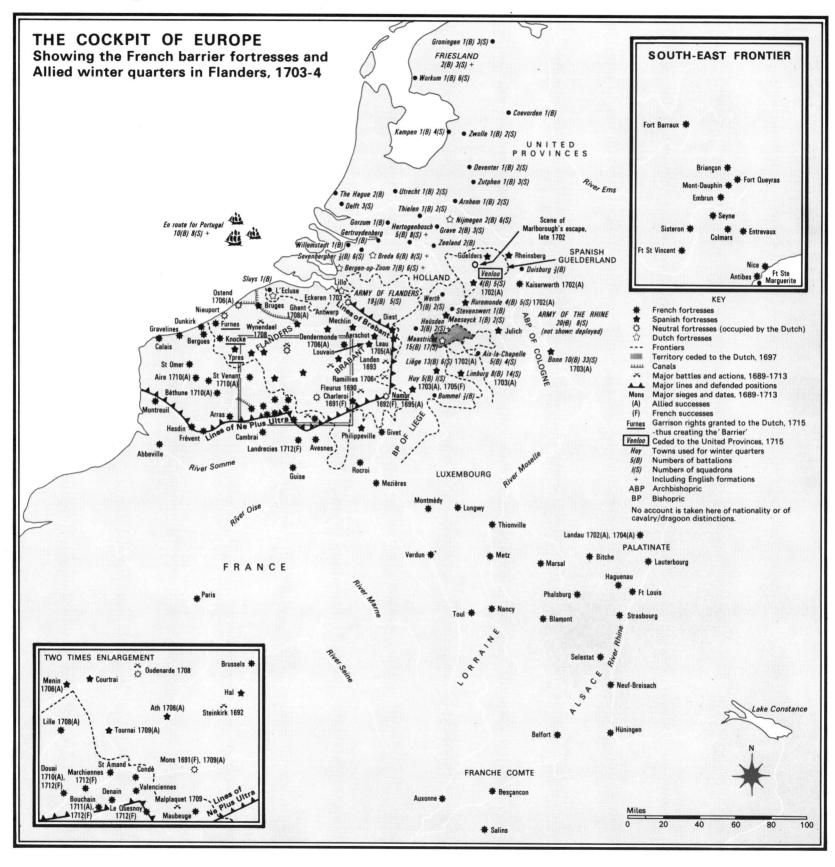

THE COCKPIT OF EUROPE
Showing the French barrier fortresses and Allied winter quarters in Flanders, 1703-4

SOUTH-EAST FRONTIER

Fort Barraux
Briançon
Mont-Dauphin — Fort Queyras
Embrun
Seyne
Sisteron — Entrevaux
Colmars
Ft St Vincent
Nice
Antibes — Ft Ste Marguerite

KEY
- ✳ French fortresses
- ★ Spanish fortresses
- ✩ Neutral fortresses (occupied by the Dutch)
- ☆ Dutch fortresses
- - - - Frontiers
- ▨ Territory ceded to the Dutch, 1697
- ⊥⊥⊥ Canals
- ✕ Major battles and actions, 1689-1713
- ▲▲ Major lines and defended positions
- Mons Major sieges and dates, 1689-1713
- (A) Allied successes
- (F) French successes
- Furnes Garrison rights granted to the Dutch, 1715 -thus creating the 'Barrier'
- Venloo Ceded to the United Provinces, 1715
- Huy Towns used for winter quarters
- 5(B) Numbers of battalions
- I(S) Numbers of squadrons
- + Including English formations
- ABP Archbishopric
- BP Bishopric

No account is taken here of nationality or of cavalry/dragoon distinctions.

UNITED PROVINCES

Groningen 1(B) 3(S)
FRIESLAND 2(B) 3(S) +
Workum 1(B) 6(S)
Coevorden 1(B)
Kampen 1(B) 4(S)
Zwolle 1(B) 2(S)
Deventer 1(B) 2(S)
Zutphen 1(B) 3(S)
The Hague 2(B)
Utrecht 1(B) 2(S)
Arnhem 1(B) 2(S)
Delft 3(S)
Thielen 1(B) 2(S)
River Ems

En route for Portugal 10(B) 8(S) +
Gorzum 1(B)
Hertogenbosch 5(B) 8(S) +
Nijmegen 2(B) 6(S)
Scene of Marlborough's escape, late 1702
Gertruydenberg 1(B)
Grave 2(B) 3(S)
Willemstadt 1(B)
Zeeland 2(B)
SPANISH GUELDERLAND
Sevenberghe ½(B)
Breda 6(B) 8(S) +
Guelders
Rheinsberg
Bergen-op-Zoom 7(B) 6(S) +
HOLLAND
Venloo 4(B) 5(S) 1702(A)
Duisburg ½(B)
Sluys 1(B)
Lillo
ARMY OF FLANDERS 19½(B) 5(S)
Werth 1(B) 2(S)
Kaiserwerth 1702(A)
Ostend 1706(A)
L'Ecluse
Eckeren 1703
Lines of Brabant
Ruremonde 4(B) 5(S) 1702(A)
Nieuport
Bruges
Ghent 1708(A)
Antwerp
Mechlin
Diest
Stevenswert 1(B)
ARMY OF THE RHINE 20(B) 8(S) (not shown: deployed)
Dunkirk
Furnes
Wynendael 1708
Maeseyck 1(B) 2(S)
Heúsden 3(B) 2(S) +
Gravelines
Bergues
Knocke
Dendermonde 1706(A)
Aerschot
Leau 1705(A)
Julich
ABP OF COLOGNE
Calais
Ypres
Louvain
Landen 1693
Maastricht 15(B) 17(S)
Aix-la-Chapelle 5(B) 4(S)
St Omer
St Venant 1710(A)
Ramillies 1706(A)
Liège 13(B) 6(S) 1702(A)
Bonn 10(B) 33(S) 1703(A)
Aire 1710(A)
Fleurus 1690
Huy 5(B) I(S) 1703(A), 1705(F)
Limbург 8(B) 14(S) 1703(A)
Béthune 1710(A)
Charleroi 1691(F)
Bummel ½(B)
Montreuil
Namur 1692(F) 1695(A)
Arras
Lines of Ne Plus Ultra
Philippeville
Givet
BP OF LIEGE
Hesdin
Cambrai
Frévent
Landrecies 1712(F)
Avesnes
Abbeville
Guise
Rocroi
River Somme
Mézières
LUXEMBOURG
River Moselle

Montmédy
Longwy
Thionville
Verdun
Metz
Landau 1702(A), 1704(A)
PALATINATE
Bitche
Lauterbourg
Marsal
Haguenau
River Oise
Phalsburg
Ft Louis
Paris
Toul
Nancy
Blamont
Strasbourg
FRANCE
River Marne
River Seine
LORRAINE
Selestat
River Rhine
ALSACE
Neuf-Breisach
Hüningen
Belfort
Lake Constance
FRANCHE COMTE
Auxonne
Besançon
Salins
N

TWO TIMES ENLARGEMENT
Oudenarde 1708
Brussels
Menin 1706(A)
Courtrai
Hal
Ath 1706(A)
Steinkirk 1692
Lille 1708(A)
Tournai 1709(A)
Mons 1691(F), 1709(A)
Douai 1710(A), 1712(F)
Marchiennes 1712(F)
St Amand
Condé
Denain
Valenciennes
Bouchain 1711(A) 1712(F)
Le Quesnoy 1712(F)
Malplaquet 1709
Maubeuge
Lines of Ne Plus Ultra

Miles
0 20 40 60 80 100

36

gains important areas and fortresses in the north-east, including Lille, Douai, Courtrai, Tournai, Ath and Charleroi, but agrees to return Franche-Comté (conquered by Condé in February).

1678: Peace of Nijmegen. France regains Franche-Comté, but restores Maastricht to the United Provinces. Returns Charleroi, Courtrai, Oudenarde and Ghent to Spain in exchange for St. Omer, Ypres, Cambrai, Maubeuge and Valenciennes, and gives up rights to garrison Philippsburg in return for Breisach and Freiburg. French occupation of Lorraine continues.

1684: Truce of Ratisbon. France retains Luxembourg, Strasbourg and Oudenarde and is confirmed in all acquisitions prior to 1 August 1681. This agreement places Louis XIV at the summit of his fortunes.

1697: Peace of Ryswick. France restores most of Lorraine; abandons claim to the Palatinate (ravaged in 1689), but retains Strasbourg. Pinerolo and Casale had already been given over to Savoy by the Treaty of Turin in 1696.

1713-14: Peace of Utrecht and Rastadt. France cedes Hudson Bay, Acadia, Newfoundland and St. Kitts (see p. 64) to Great Britain; restores Menin, Tournai, Furnes and Ypres to the Austrian Netherlands, and Nice to the Duke of Savoy. France debars Philip V from the French succession.

Thus, French fortunes declined considerably between 1688 and 1713 owing to the scale and determination of the opposition to Louis' seemingly insatiable ambition. Even so, her territorial gains since the beginning of the seventeenth century remained impressive.

The Cockpit of Europe
Many of the hardest campaigns of the period were fought out in the Spanish Netherlands and associated regions. This area became a centre of military activity for a number of reasons. Geographically, it forms a bridge between the north German Plain to the east and the relatively flat areas of northern France to the west. Operations of war were also assisted by the high agricultural prosperity of the area and the number of rivers, supplemented by canals. These advantages made the movement and maintenance of armies easier than in the adjoining region to the south—the foothills of the Central European Highlands. The prosperity of the Netherlands was a further inducement, offering the prospect of valuable booty for the troops and bargaining counters for the generals and diplomats.

WEAPONS, 1618-1720

Firearms developed rapidly during the seventeenth century. The bulky snaphance musket in common use during the Thirty Years War was so heavy that the musketeer required a forked rest to support the barrel, and was also prone to as many as eight misfires in every dozen attempted discharges. It was eventually replaced by the lighter matchlock, but this weapon still had many disadvantages and was hardly more reliable than its predecessor. The reloading drill required eighteen separate evolutions; it is estimated that it took seven minutes to fire two shots. The wheel-lock musket as widely employed in Sweden and Austria was in some respects an improve-ment, but its maximum range was still in the order of 250 yards and accuracy of even a rudimentary type was only possible at less than 100 yards.

The development of the flintlock fusil in France and flinte in Germany was an important step. Its firing system, a spark caused by a flint held by the hammer striking the steel pan, was far less prone to misfire; as many as eight shots in a dozen could now be counted on, and reloading techniques had been simplified to permit one or even two shots per minute. A persistent disadvantage was the retention of the wooden ramrod, which tended to break in the heat of action.

By the 1680s the use of paper cartridges containing ball and powder had replaced the old reliance on chargers and powder-horn. These facilitated speedier reloading as mentioned above.

Bayonets went through three main stages of development. The early plug variety fitted into the barrel, and this precluded fire-action when fitted. The later ring bayonet (c. 1670) was an improvement, but tended to slide up the barrel. Finally, the lengthy pike was made completely super-fluous by the development in about 1680 of a socket-bayonet (some authorities attribute this to Vauban), which fitted firmly to the musket-barrel without obstructing firing, although reloading with ramrod could be difficult when the bayonet was fitted. In this way 'the Queen of Weapons' passed from the battlefield, although shorter pole-arms were still carried by many officers and sergeants.

Artillery pieces were massive in bulk and limited in effect. The barrels were con-structed by several methods, boring becoming practicable by the end of the century, and several designs were followed: the 'old', 'Spanish', 'new invention' etc., each of which produced special character-istics of employment. The use of howitzers and mortars was becoming more scientific as the science of gunnery developed.

INFANTRY FIREARMS

	Length	Calibre	Weight	Range	Rate of Fire
Snaphance matchlock (c.1620)	5ft. 2in.		?20lbs. (needing a forked rest)	150 yards	1 round per 10 mins.
'Mousquet Ordinaire' matchlock (c.1648)	5ft.	20 balls to lb.	15lbs. (some versions 18lb)	250 yards	2 rounds per 7 mins. (but prone to misfire)
Swedish flintlock (c.1690)	5ft.	.67in.	12lbs. (inc. bayonet)	250 yards	1 round per min.
William III flintlock (c.1696)	5ft.	.85in.	11lbs. 8oz (inc. bayonet)	250 yards	1-2 rounds per min.
French 'Fusil Ordinaire' (c.1700)	5ft.	.68in.	12lbs. (inc. bayonet)	250 yards	1-2 rounds per min.
English 'Brown Bess' (c.1720)	5ft. 2in.	.75in.	11lbs. 13oz (inc. bayonet)		2 rounds per min.

All these firearms were muzzle-loading and prone to misfire.

ARTILLERY

Representative Cannon, c.1688	Weight of barrel	Point-blank range	Max. range
3pdr. 'Minion'	800lbs.	300 paces	1,500 yards
6pdr. 'Saker'	1,500lbs.	350 paces	2,500 yards
8pdr. 'Demiculverin'	2,000lbs.	400 paces	3,500 yards
12pdr. Spanish 'Quart-canon'	3,400lbs.	450 paces	3,750 yards
16pdr. Culverin or 'Demi-canon de France'	4,500lbs.	800 paces	4,000 yards
24pdr. Demi-canon d'Espagne	5,600lbs.	900 paces	4,500 yards

Optimum range for most cannon was 800 yards or less, and few were fired over more than a mile. The rate of fire for medium-sized artillery was 1 round per minute; 1 round per 5 minutes for heavy artillery; and up to 4 rounds per minute for light artillery.

Howitzers, c.1700	Weight	Length of barrel	Bore	Effective range (approx.)
English harbitzer	1,500lbs.	4ft.	10in.	800 yards
Dutch howitzer (Firing shells at an angle of 15°)	900lbs.	3ft. 6in.	6in.	800 yards

Mortars: used for sieges only. The barrel was placed on a bed of solid timbers, and many types had a fixed elevation of 45°. They were capable of firing larger projectiles (usually bombs) than other types of artillery, and were usually conveyed in carts.

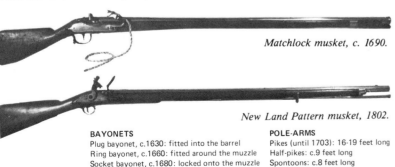

Matchlock musket, c. 1690.

New Land Pattern musket, 1802.

BAYONETS
Plug bayonet, c.1630: fitted into the barrel
Ring bayonet, c.1660: fitted around the muzzle
Socket bayonet, c.1680: locked onto the muzzle

POLE-ARMS
Pikes (until 1703): 16-19 feet long
Half-pikes: c.9 feet long
Spontoons: c.8 feet long

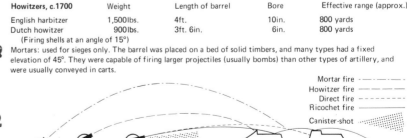

Mortar fire ·—·—·—·
Howitzer fire - - - - - -
Direct fire – – – – –
Ricochet fire ————
Canister-shot ∷∷∷∷

The main limitations of all artillery pieces were their bulk, weight and difficulty to transport. More mobile versions did exist, such as Gustavus Adolphus' leather-guns (1630), and Coehoorn's small mortars (1670), but their destructive power and range were limited.

The Expansion of France

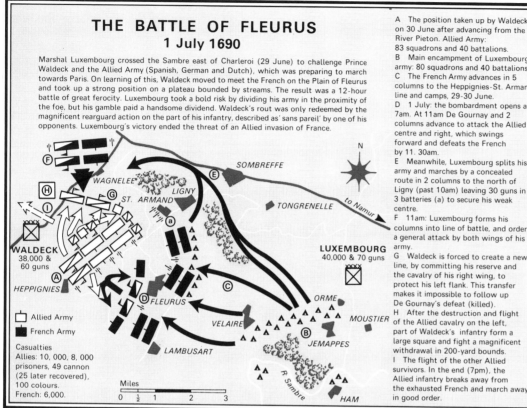

THE BATTLE OF FLEURUS
1 July 1690

Marshal Luxembourg crossed the Sambre east of Charleroi (29 June) to challenge Prince Waldeck and the Allied Army (Spanish, German and Dutch), which was preparing to march towards Paris. On learning of this, Waldeck moved to meet the French on the Plain of Fleurus and took up a strong position on a plateau bounded by streams. The result was a 12-hour battle of great ferocity. Luxembourg took a bold risk by dividing his army in the proximity of the foe, but his gamble paid a handsome dividend. Waldeck's rout was only redeemed by the magnificent rearguard action on the part of his infantry, described as 'sans pareil' by one of his opponents. Luxembourg's victory ended the threat of an Allied invasion of France.

A The position taken up by Waldeck on 30 June after advancing from the River Pieton. Allied Army: 83 squadrons and 40 battalions.
B Main encampment of Luxembourg's army: 80 squadrons and 40 battalions.
C The French Army advances in 5 columns to the Heppignies-St. Armand line and camps, 29-30 June.
D 1 July: the bombardment opens at 7am. At 11am De Gournay and 2 columns advance to attack the Allied centre and right, which swings forward and defeats the French by 11.30am.
E Meanwhile, Luxembourg splits his army and marches by a concealed route in 2 columns to the north of Ligny (past 10am) leaving 30 guns in 3 batteries (a) to secure his weak centre.
F 11am: Luxembourg forms his columns into line of battle, and orders a general attack by both wings of his army.
G Waldeck is forced to create a new line, by committing his reserve and the cavalry of his right wing, to protect his left flank. This transfer makes it impossible to follow up De Gournay's defeat (killed).
H After the destruction and flight of the Allied cavalry on the left, part of Waldeck's infantry form a large square and fight a magnificent withdrawal in 200-yard bounds.
I The flight of the other Allied survivors. In the end (7pm), the Allied infantry breaks away from the exhausted French and march away in good order.

Allied Army
French Army

Casualties
Allies: 10,000, 8,000 prisoners, 49 cannon (25 later recovered), 100 colours.
French: 6,000.

WALDECK 38,000 & 60 guns
LUXEMBOURG 40,000 & 70 guns

SOMBREFFE · WAGNELEE · LIGNY · ST. ARMAND · TONGRENELLE · to Namur · HEPPIGNIES · FLEURUS · VELAIRE · ORME · MOUSTIER · LAMBUSART · JEMAPPES · R. Sambre · HAM

Miles
0 ½ 1 2 3

THE BATTLE OF STEINKIRK
3 August 1692

HERINNES · Bois de Triou · HAL · to Brussels · ALLIED CAMP · WILLIAM III 63,000 · ENGHIEN · LUXEMBOURG 57,000 · HOVES · Plaine St. Martin · River Senne · STEINKIRK · R. Braine · to Soignies · FRENCH CAMP

A The Allied advance by night (2-3 August) over very broken ground to surprise the French camp.
B Gen. Mackay's attack with the 8 British battalions of the advance guard.
C Mackay routs 2 lines of French infantry including the Swiss Guards.
D A flank attack by the French Gendarmerie checks Mackay's advance; his withdrawal is covered by Overkirk's 2 Dutch battalions.
E The advance of the French under Marshal Boufflers.
F The Allied withdrawal from midday onwards.

Allied Army
French Army

Miles
0 ½ 1 1½ 2 2½

The political organization of the area also made it of the greatest significance. For France, the French-speaking Walloons of the southern Spanish Netherlands afforded an excuse for intervention and partial assimilation; whilst the strategic security of the French northern frontiers called for a measure of domination of the area. For the Dutch, the dangers inherent in French ambitions (particularly after 1672) made them highly sensitive to the situation prevailing in the Spanish Netherlands. Their anxiety for national security eventually crystalized into a desire for a barrier of fortresses and a buffer zone with which to face the imposing fortifications built-up by Vauban along the French frontiers. The Emperor was concerned in the fortunes of the various princelings and ecclesiastics whose small states stood on the west bank of the Rhine, faced as they were by the territorial ambitions of 'le roi soleil' and his ministers. Spanish interest in the area was self-evident, as the Netherlands formed an important part of their European possessions, which until 1601 had also comprised Holland. Lastly, English interest in the area was assured by London's commercial rivalry with Antwerp (until 1713) and by the strategic preoccupation with the control of the mouth of the River Scheldt; the security of the Thames estuary was widely considered to require a friendly neutral power to possess this important waterway. The Netherlands also provided the most convenient venue for English military intervention in the affairs of the Continent, the main point of access from 1688 being the Hague.

The number of fortified towns and the suitability of the countryside for the construction of semi-permanent lines, utilizing the many waterways to provide flooded areas, tended to favour armies adopting the defensive and led to the prosecution of siege wars rather than campaigns of bold manoeuvre. Many battles fought up to 1697 were associated with sieges in one way or another, e.g. Steinkirk. However, Marlborough's three great battles between 1706 and 1709 were inspired by more dramatic intentions. Nevertheless, he was compelled to conduct almost thirty major sieges in the area during the War of the Spanish Succession; the most famous of which were those of Lille in 1708 (see p. 52) and Bouchain in 1711. On two occasions he also forced defended lines (see p. 40).

The great river lines of the Meuse, the Scheldt and the Sambre became the focal point of many major military operations in war after war. Lille, Mons, Namur and Ghent were among the most important fortresses dominating the area, which inevitably became the focal point of campaign after campaign.

The Campaigns of Luxembourg and William III

The ablest French commander after Condé and Turenne was François Henri de Montmorency-Bouteville, Duc de Luxembourg (1628-95), who came to the fore after the retirement of the former and the death of the latter, both of which events took place in 1675. Marshal Luxembourg was popular with his men and highly competent, if utterly ruthless, as a field commander; characteristics he well displayed during the campaigns of the Nine Years War (or League of Augsburg) fought out in the Spanish Netherlands. As a strategist, he was not wholly convinced that the defensive was the stronger form of war, and repeatedly (though politely) challenged Louis XIV's demands that he should remain wholly on the defensive, avoiding major battles. His skill brought satisfactory results. As a tactician he was both bold and cool: he excelled in adjusting his forces to meet particular situations, as at Steinkirk, and in devising tactical envelopments, as at Fleurus and Landen. Luxembourg was nicknamed 'le tapissier de Notre Dame' in recognition of the large numbers of captured enemy colours and standards he sent back to the French capital to adorn the cathedral. "Marshal Luxembourg was a worthy disciple of Turenne", wrote Prince Frederick of Prussia in 1876; "The

Battle of Fleurus was a masterpiece of the art of manoeuvring, that of Neerwinden [Landen] quite possesses the characteristic of Napoleon's battles." (Luxembourg's main victories are illustrated in this chapter.)

His chief adversary was William III, Prince of Orange and King of England (1650-1701). William's considerable abilities as a statesman and constitutional monarch were not matched by equal gifts as a general. Successes in Ireland were not repeated on the Continent, as Luxembourg's victories testify, although in 1695 William regained Namur, which had been lost to the French shortly before Steinkirk. Despite the failure of most of his offensive

movements, William was skilled at extricating his forces from lost battles to 'live and fight another day'. His unfortunate record as a field commander in no way dimmed his determination to continue his lifelong crusade against France, which he never forgave for the ruthless invasion of the United Provinces in 1672. He also saw himself as the leader of Protestant Europe, but was prepared to ally with Catholic powers in the prosecution of his feud with the House of Bourbon, and in his determination to safeguard the balance of power. William also played an important part in training-up the English Army to Continental standards; under Marlborough, it would achieve marvels.

1. Henri de la Tour d'Auvergne, Vicomte de Turenne (1611-75). The greatest French commander of the first half of Louis XIV's reign, Napoleon included his name amongst the seven great soldiers of history.
2. William III, King of England (1650-1702). As Prince of Orange he inspired the defence of the United Provinces against the French in 1672, and all his life remained the inveterate opponent of Louis XIV's expansionist ambitions, seeking to achieve a European balance of power. (Ken Trotman Arms Books)

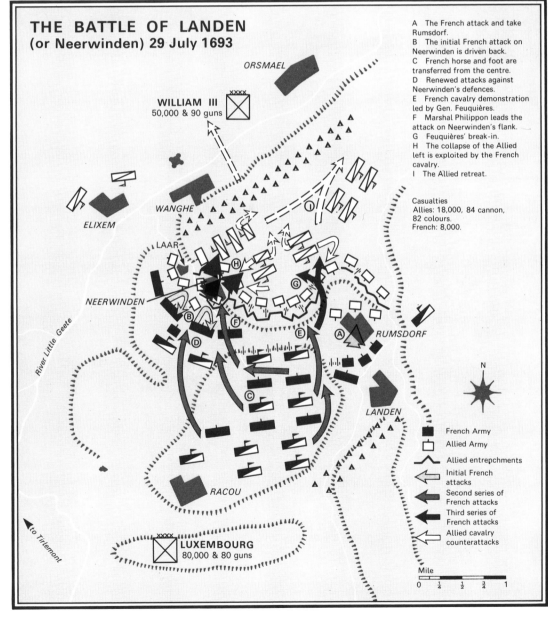

THE BATTLE OF LANDEN
(or Neerwinden) 29 July 1693

ORSMAEL

WILLIAM III
50,000 & 90 guns

WANGHE

ELIXEM

LAAR

NEERWINDEN

RUMSDORF

River Little Geete

to Tirlemont

RACOU

LANDEN

LUXEMBOURG
80,000 & 80 guns

A The French attack and take Rumsdorf.
B The initial French attack on Neerwinden is driven back.
C French horse and foot are transferred from the centre.
D Renewed attacks against Neerwinden's defences.
E French cavalry demonstration led by Gen. Feuquières.
F Marshal Philippon leads the attack on Neerwinden's flank.
G Feuquières' break-in.
H The collapse of the Allied left is exploited by the French cavalry.
I The Allied retreat.

Casualties
Allies: 18,000, 84 cannon, 82 colours.
French: 8,000.

N

Mile
0   ¼   ½   ¾   1

French Army
Allied Army
Allied entrenchments
Initial French attacks
Second series of French attacks
Third series of French attacks
Allied cavalry counterattacks

39

Masters of Manoeuvre

If the Netherlands theatre of war tended to be dominated by the large number of fortresses in the area, other regions and sectors favoured wars of movement. Turenne, Marlborough and Eugene all proved themselves masters of manoeuvre.

Turenne's Campaigns in the Vosges, 1674-75
Marshal de Turenne was entrusted with the secondary front of the Rhine while the main French campaign of the year was launched in the Netherlands. Seriously out-numbered, he had to check and repulse Bournonville's invasion of Alsace. His methods were typical of the man but untypical of the period in their imagination and sheer audacity. Using fast marches to confuse his opponent, Turenne snatched a success at Entzheim, but the outcome was tactically indecisive as the French had to retire on Haguenau to rest and refit their tired army; strategically, however, the battle blunted Bournonville's initiative and he gratefully entered winter quarters near Colmar. No soldier on either side anticipated any further activity until the next spring.

Turenne, despite his sixty-four years, planned otherwise. He saw that a surprise attack on the Imperialist Army might gain great advantages—although winter campaigns were virtually unheard of in western Europe. In great secrecy, Turenne gathered a 33,000-strong army. On 29 November, using the Vosges mountains and a moving cavalry screen to conceal his move, he led them towards the Belfort Gap. Catching his opponents in quarters, he gained a quick success at Mulhouse on 29 December and, pressing on, turned the foe's right flank entrenched between Colmar and Turckheim, 9 January 1675. By mid-January the Imperialists were re-crossing to the right bank of the Rhine. These campaigns were typical of Turenne's generalship: the use of surprise, fast movement, and small-scale, quick victories to unsettle the enemy and dominate the situation. As Napoleon remarked, Turenne's "audacity grew with years and experience".

Marlborough's Forcing of the Lines of Brabant, 1705
Marlborough's skill at large-scale marches is typified par excellence by his campaign of 1704, which changed the whole outlook of the War of the Spanish Succession. On a smaller scale, his operations in Brabant the following year formed a masterpiece of 'conventional' manoeuvre as understood by most of his contemporaries. Attacks on prepared lines could result in insupportable losses unless the defenders were fooled concerning the point of attack. By use of consummate bluff, rapid

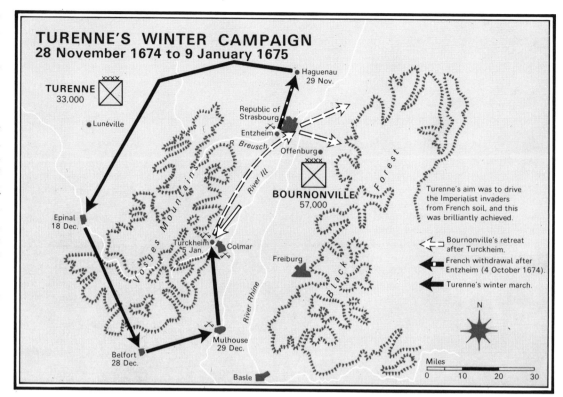

TURENNE'S WINTER CAMPAIGN
28 November 1674 to 9 January 1675

TURENNE 33,000

Lunéville

Haguenau 29 Nov.

Republic of Strasbourg

Entzheim

Offenburg

R Breusch

River Ill

BOURNONVILLE 57,000

Vosges Mountains

Black Forest

Epinal 18 Dec.

Turckheim 5 Jan. · Colmar

Freiburg

River Rhine

Belfort 28 Dec.

Mulhouse 29 Dec.

Basle

Turenne's aim was to drive the Imperialist invaders from French soil, and this was brilliantly achieved.

Bournonville's retreat after Turckheim.

French withdrawal after Entzheim (4 October 1674).

Turenne's winter march.

N

Miles
0   10   20   30

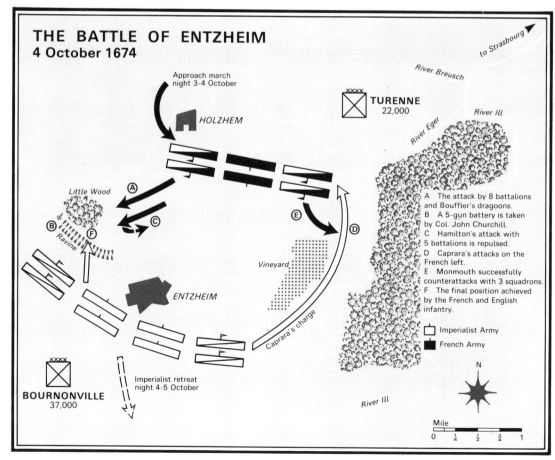

THE BATTLE OF ENTZHEIM
4 October 1674

to Strasbourg

River Breusch

Approach march night 3-4 October

HOLZHEM

TURENNE 22,000

River Eger   River Ill

Little Wood

A

B   F   C

Ravine

E

D

Vineyard

Caprara's charge

ENTZHEIM

BOURNONVILLE 37,000

Imperialist retreat night 4-5 October

River Ill

A The attack by 8 battalions and Bouffler's dragoons.
B A 5-gun battery is taken by Col. John Churchill.
C Hamilton's attack with 5 battalions is repulsed.
D Caprara's attacks on the French left.
E Monmouth successfully counterattacks with 3 squadrons.
F The final position achieved by the French and English infantry.

☐ Imperialist Army
■ French Army

N

Mile
0   ¼   ½   ¾   1

movement by night, and the willing exertions of his troops, Marlborough tricked Marshal Villeroi and induced him to abandon the complete Lines of Brabant and retire behind the River Dyle. In fact, this outcome was a disappointment for Marlborough, who had hoped to force a major engagement on his opponent whilst he was in retreat. In the event, however, the combination of Count Caraman's gallantry and Dutch obstruction robbed him of the opportunity. The passage of the lines of Ne Plus Ultra in 1711 was an even more celebrated achievement of the same general type, leading to the capture of Bouchain.

Eugene's March to Turin, 1706

The Italian theatre saw much activity, most of it centring on Savoy, Piedmont and Nice. To prevent the French successfully completing the siege of Turin (commenced 14 May), Prince Eugene prepared to march over 200 miles from the east bank of the River Adige. On receiving news of the Battle of Ramillies (see p. 46), Louis XIV recalled Marshal Vendôme to Flanders. His impending replacement in north Italy by the less competent Duc d'Orléans gave Eugene his opportunity to slip past the considerable French forces holding the central reaches of the River Adige. To do this necessitated the infringement of Venetian neutrality by crossing the river near Rovigo.

The French were completely fooled by this move, and Eugene's advance up the Po valley was virtually unopposed, the French generals everywhere falling back towards the west. Imperial troops suffered greatly, despite the use of night marches to avoid the heat of the day, but by early September they had crossed the Po to the south of Turin and successfully united with Victor Amadeus's small force which had been harassing the French siege lines from without. Subsequently, Prince Eugene pressed his advantage over the largely demoralized foe to force the decisive Battle of Turin on 7 September. Badly beaten with 9,000 casualties, the French fled homewards, leaving their garrisons in the Milanese completely isolated. At this point, Eugene realized that "Italy is ours and its conquest will not be costly". Early the next year the French agreed to surrender their posts in Italy in return for free withdrawal to France. Thus, the period of Austrian hegemony over north Italy was inaugurated. Victor Amadeus was also thenceforward secure in the kingdom of Savoy.

Eugene's march lasted sixty days and covered some 225 miles. It bears comparison with Marlborough's march to the Danube and shows audacity and breadth of vision overcoming the limitations of eighteenth century warfare.

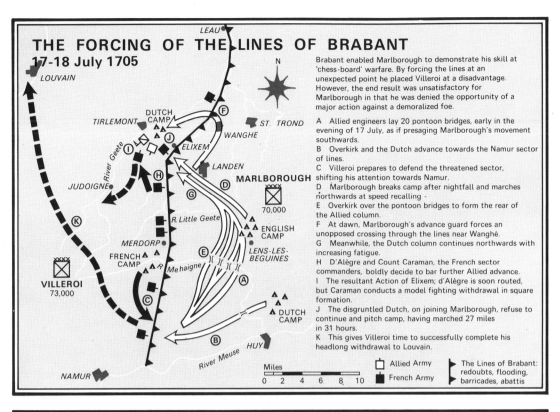

THE FORCING OF THE LINES OF BRABANT
17-18 July 1705

Brabant enabled Marlborough to demonstrate his skill at 'chess-board' warfare. By forcing the lines at an unexpected point he placed Villeroi at a disadvantage. However, the end result was unsatisfactory for Marlborough in that he was denied the opportunity of a major action against a demoralized foe.

A Allied engineers lay 20 pontoon bridges, early in the evening of 17 July, as if presaging Marlborough's movement southwards.
B Overkirk and the Dutch advance towards the Namur sector of lines.
C Villeroi prepares to defend the threatened sector, shifting his attention towards Namur.
D Marlborough breaks camp after nightfall and marches northwards at speed recalling –
E Overkirk over the pontoon bridges to form the rear of the Allied column.
F At dawn, Marlborough's advance guard forces an unopposed crossing through the lines near Wanghé.
G Meanwhile, the Dutch column continues northwards with increasing fatigue.
H D'Alègre and Count Caraman, the French sector commanders, boldly decide to bar further Allied advance.
I The resultant Action of Elixem; d'Alègre is soon routed, but Caraman conducts a model fighting withdrawal in square formation.
J The disgruntled Dutch, on joining Marlborough, refuse to continue and pitch camp, having marched 27 miles in 31 hours.
K This gives Villeroi time to successfully complete his headlong withdrawal to Louvain.

Miles
0 2 4 6 8 10

☐ Allied Army
■ French Army
► The Lines of Brabant: redoubts, flooding, barricades, abattis

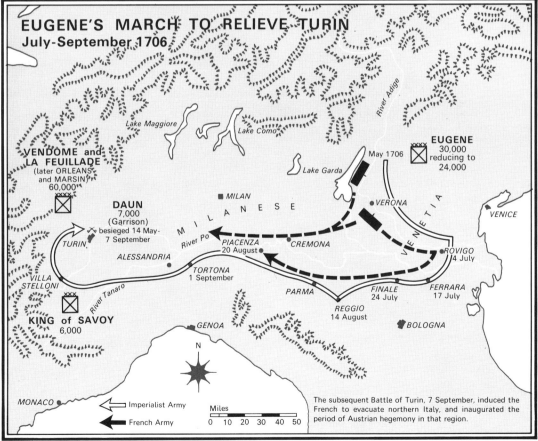

EUGENE'S MARCH TO RELIEVE TURIN
July-September 1706

Imperialist Army
French Army

Miles
0 10 20 30 40 50

The subsequent Battle of Turin, 7 September, induced the French to evacuate northern Italy, and inaugurated the period of Austrian hegemony in that region.

41

Marlborough's Generalship

John Churchill, first Duke of Marlborough (1650-1722), was probably the greatest soldier of his day; his only possible rivals for the title being his friend and colleague Prince Eugene (see p. 55) and the ill-fated Charles XII of Sweden (see p. 58). His early career was not particularly spectacular, but as a relatively junior officer he served under Turenne, Waldeck and William III, and built up a solid basis of military experience, which included service as a marine (1672). Distrusted as a 'trimmer' by William III (Churchill had played a leading role in the deposition of James II, his former patron), he only emerged as a leading soldier after the accession of Queen Anne in 1701; however, earlier he had been trusted with the negotiation of the treaties that eventually burgeoned into the Second Grand Alliance against France.

Throughout his period of influence, Marlborough played the difficult double-role of soldier-diplomat. The problem of coalition warfare—conflicting interests of member-states, irregular coordination of effort, etc.— took up much of his time, but his famous charm and patience enabled him to hold the Alliance together through good times and bad, although sometimes at the expense of military considerations. The obstructions often placed in his way by the wary Dutch were perhaps his greatest burden in this respect.

As a strategist, Marlborough believed that the only way to a real victory in the War of the Spanish Succession lay through the Netherlands and thence towards Paris; nine of his ten campaigns were fought in the Flanders area. He was also aware of the value of using sea-power to mount distractions in the Mediterranean and along the French Atlantic coasts. But during the first campaigns the French enjoyed the initiative, and in 1704 Marlborough was forced to concentrate on saving Vienna from the Franco-Hungarian threat. His grasp of the realities of coalition strategy is nowhere better illustrated than in his conduct of operations in that year. A second triumph came in 1707 when he persuaded Charles XII of Sweden not to become involved in the war with France and Spain.

The Battle of Blenheim changed the aspect of the War of the Spanish Succession, and thereafter the Allies were basically on the offensive. It proved hard, however, to get Allied agreement on the best means of achieving victory and Marlborough never had the authority to impose a common policy. Consequently, years of brilliant achievement in Flanders—as in 1706 when Ramillies led to the conquest of the Spanish Netherlands—were balanced by years of frustration. But, little by

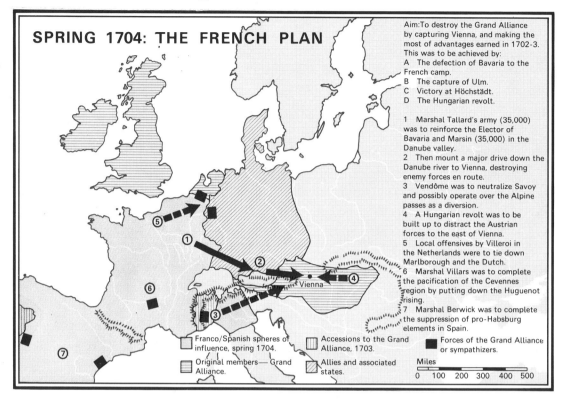

SPRING 1704: THE FRENCH PLAN

Aim: To destroy the Grand Alliance by capturing Vienna, and making the most of advantages earned in 1702-3. This was to be achieved by:
A The defection of Bavaria to the French camp.
B The capture of Ulm.
C Victory at Höchstädt.
D The Hungarian revolt.

1 Marshal Tallard's army (35,000) was to reinforce the Elector of Bavaria and Marsin (35,000) in the Danube valley.
2 Then mount a major drive down the Danube river to Vienna, destroying enemy forces en route.
3 Vendôme was to neutralize Savoy and possibly operate over the Alpine passes as a diversion.
4 A Hungarian revolt was to be built up to distract the Austrian forces to the east of Vienna.
5 Local offensives by Villeroi in the Netherlands were to tie down Marlborough and the Dutch.
6 Marshal Villars was to complete the pacification of the Cevennes region by putting down the Huguenot rising.
7 Marshal Berwick was to complete the suppression of pro-Habsburg elements in Spain.

☐ Franco/Spanish spheres of influence, spring 1704.
☐ Original members—Grand Alliance.
▨ Accessions to the Grand Alliance, 1703.
▨ Allies and associated states.
■ Forces of the Grand Alliance or sympathizers.

Miles
0 100 200 300 400 500

little, the Allies reduced the number of French-held fortresses until, by 1711, they were operating on French soil. By this time, Marlborough was fast falling from favour in England. In some ways his tact as a statesman anxious to maintain the cohesion of his discordant alliance conflicted with his common sense as a soldier. In 1705, for instance, he did nothing to prevent the escalation of the Mediterranean diversion into a major effort for the direct conquest of Spain. This creation of a costly and ultimately unsuccessful second front —largely at Austria's bidding—proved the greatest strategic error committed by the Grand Alliance. Then again, after Oudenarde in 1708 (see pp. 46 and 47), he quietly dropped his imaginative scheme for an invasion of France along the coast and up the Seine via Abbeville, ignoring the fortress barriers and relying on supply by the fleet, in favour of the difficult siege of Lille which his Allies preferred (see p. 52). In both cases, the instincts of Marlborough the diplomat triumphed over those of Marlborough the soldier.

In the realm of grand tactics and tactics he had no peer. His skill as a military administrator and logistician underwrote all his successes and made him genuinely beloved by his officers and men. This mutual trust made possible great exertions; whether of a sustained nature, as on the march to the Danube, or of a shorter variety, as in the forced overnight

marches which preceded both Blenheim and Oudenarde, or the passage of the French lines in 1705 and 1711. Marlborough was also a master of deception, and could excel at the exploitation of success. Blenheim was the prelude for carrying the war to Landau on the Rhine; Ramillies opened the way for the whirlwind conquest of the greater part of the Spanish Netherlands; Oudenarde, the most dangerous battle he ever undertook, led to the capture of Lille and the regaining of Ghent and Bruges; Malplaquet, although less spectacular in its immediate effects, ensured the capture of Mons and prepared the way for the advances made in 1710. Marlborough was ahead of his time in his realization of what use could be made of a large-scale victory. At the same time he was a master of every aspect of 'conventional' warfare, an expert at both siege and manoeuvre.

On the battlefield, his eye for ground (as at Ramillies and Oudenarde) combined with his ability to keep firm control of the overall situation (intervening personally at every crisis point) and his use of massed cavalry to clinch the successes gained by the infantry were important features. Above all he could weld multi-national armies into a team, and aided by such generals as Eugene and Overkirk he was more than a match for the lesser French commanders who succeeded Luxembourg. However, his tactics became possibly a shade

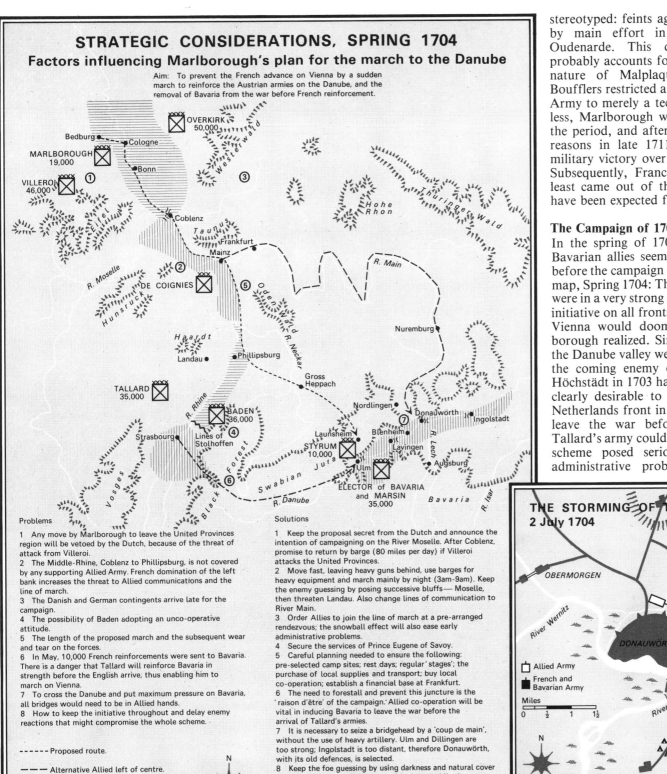

STRATEGIC CONSIDERATIONS, SPRING 1704
Factors influencing Marlborough's plan for the march to the Danube

Aim: To prevent the French advance on Vienna by a sudden march to reinforce the Austrian armies on the Danube, and the removal of Bavaria from the war before French reinforcement.

OVERKIRK 50,000

Bedburg — Cologne

MARLBOROUGH 19,000 — Bonn

VILLEROI 46,000 ①

Westerwald

③

Coblenz

Eifel

R. Moselle

Taunus

Frankfurt

Mainz

DE COIGNIES ②

Hunsruck

⑤

Odenwald R. Neckar

R. Main

Hohe Rhon

Thuringer Wald

Haardt

Landau

Phillipsburg

Nuremburg

TALLARD 35,000

R. Rhine

Gross Heppach

Nordlingen

Donauwörth

Ingolstadt

BADEN 36,000

Strasbourg

Lines of Stolhoffen ④

Launsheim

Blenheim

STYRUM 10,000

Lavingen

Black Forest

Swabian Jura

Ulm

Augsburg

⑥

⑦

Vosges

R. Danube

ELECTOR of BAVARIA and MARSIN 35,000

Bavaria

R. Lech

R. Isar

Problems

1 Any move by Marlborough to leave the United Provinces region will be vetoed by the Dutch, because of the threat of attack from Villeroi.
2 The Middle-Rhine, Coblenz to Phillipsburg, is not covered by any supporting Allied Army. French domination of the left bank increases the threat to Allied communications and the line of march.
3 The Danish and German contingents arrive late for the campaign.
4 The possibility of Baden adopting an unco-operative attitude.
5 The length of the proposed march and the subsequent wear and tear on the forces.
6 In May, 10,000 French reinforcements were sent to Bavaria. There is a danger that Tallard will reinforce Bavaria in strength before the English arrive, thus enabling him to march on Vienna.
7 To cross the Danube and put maximum pressure on Bavaria, all bridges would need to be in Allied hands.
8 How to keep the initiative throughout and delay enemy reactions that might compromise the whole scheme.

- - - - - Proposed route.

— — — Alternative Allied left of centre.

⌧ Major bridges.

▢ Area covered by Allied Armies.

▥ Area dominated by French and Bavarians.

Solutions

1 Keep the proposal secret from the Dutch and announce the intention of campaigning on the River Moselle. After Coblenz, promise to return by barge (80 miles per day) if Villeroi attacks the United Provinces.
2 Move fast, leaving heavy guns behind, use barges for heavy equipment and march mainly by night (3am-9am). Keep the enemy guessing by posing successive bluffs— Moselle, then threaten Landau. Also change lines of communication to River Main.
3 Order Allies to join the line of march at a pre-arranged rendezvous; the snowball effect will also ease early administrative problems.
4 Secure the services of Prince Eugene of Savoy.
5 Careful planning needed to ensure the following: pre-selected camp sites; rest days; regular 'stages'; the purchase of local supplies and transport; buy local co-operation; establish a financial base at Frankfurt.
6 The need to forestall and prevent this juncture is the 'raison d'être of the campaign.' Allied co-operation will be vital in inducing Bavaria to leave the war before the arrival of Tallard's armies.
7 It is necessary to seize a bridgehead by a 'coup de main', without the use of heavy artillery. Ulm and Dillingen are too strong; Ingolstadt is too distant, therefore Donauwörth, with its old defences, is selected.
8 Keep the foe guessing by using darkness and natural cover (e.g. Swabian Jura). Exploit possible delays while the French commanders refer to Versailles (messages take 6 days there and back) for new orders.

Miles
0 20 40 60 80 100

stereotyped: feints against the flanks followed by main effort in the centre, except at Oudenarde. This degree of predictability probably accounts for the unreasonably costly nature of Malplaquet, where Villars and Boufflers restricted a massively-superior Allied Army to merely a technical victory. Nevertheless, Marlborough was the greatest soldier of the period, and after his disgrace for political reasons in late 1711, all chance of a final military victory over France and Spain faded. Subsequently, France 'won the peace' or at least came out of the war better than might have been expected from her military record.

The Campaign of 1704

In the spring of 1704, the French and their Bavarian allies seemed set to capture Vienna before the campaign ended. As is shown on the map, Spring 1704: The French Plan, the French were in a very strong strategic position, with the initiative on all fronts except at sea. The fall of Vienna would doom the Alliance, as Marlborough realized. Since the Imperial forces in the Danube valley were not sufficient to thwart the coming enemy offensive (their defeat at Höchstädt in 1703 had lowered morale), it was clearly desirable to reinforce them from the Netherlands front in order to force Bavaria to leave the war before the full strength of Tallard's army could reach the Danube. Such a scheme posed serious politico-military and administrative problems, but Marlborough

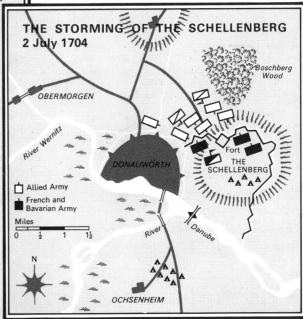

THE STORMING OF THE SCHELLENBERG
2 July 1704

Boschberg Wood

OBERMORGEN

River Wernitz

DONAUWÖRTH

Fort THE SCHELLENBERG

▢ Allied Army

◼ French and Bavarian Army

Miles
0 ½ 1 1½

River Danube

N

OCHSENHEIM

Marlborough's Generalship

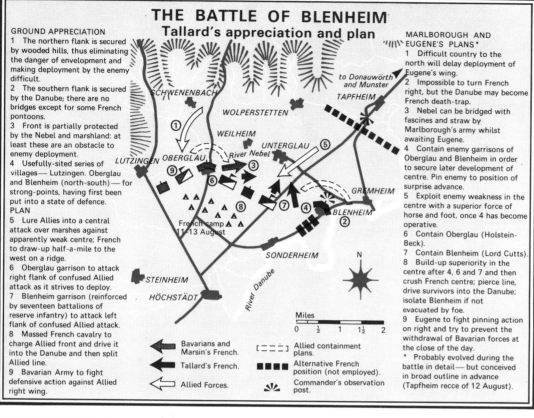

THE BATTLE OF BLENHEIM
Tallard's appreciation and plan

GROUND APPRECIATION
1 The northern flank is secured by wooded hills, thus eliminating the danger of envelopment and making deployment by the enemy difficult.
2 The southern flank is secured by the Danube; there are no bridges except for some French pontoons.
3 Front is partially protected by the Nebel and marshland: at least these are an obstacle to enemy deployment.
4 Usefully-sited series of villages— Lutzingen, Oberglau and Blenheim (north-south) — for strong-points, having first been put into a state of defence.
PLAN
5 Lure Allies into a central attack over marshes against apparently weak centre; French to draw-up half-a-mile to the west on a ridge.
6 Oberglau garrison to attack right flank of confused Allied attack as it strives to deploy.
7 Blenheim garrison (reinforced by seventeen battalions of reserve infantry) to attack left flank of confused Allied attack.
8 Massed French cavalry to charge Allied front and drive it into the Danube and then split Allied line.
9 Bavarian Army to fight defensive action against Allied right wing.

MARLBOROUGH AND EUGENE'S PLANS*
1 Difficult country to the north will delay deployment of Eugene's wing.
2 Impossible to turn French right, but the Danube may become French death-trap.
3 Nebel can be bridged with fascines and straw by Marlborough's army whilst awaiting Eugene.
4 Contain enemy garrisons of Oberglau and Blenheim in order to secure later development of centre. Pin enemy to position of surprise advance.
5 Exploit enemy weakness in the centre with a superior force of horse and foot, once 4 has become operative.
6 Contain Oberglau (Holstein-Beck).
7 Contain Blenheim (Lord Cutts).
8 Build-up superiority in the centre after 4, 6 and 7 and then crush French centre; pierce line, drive survivors into the Danube; isolate Blenheim if not evacuated by foe.
9 Eugene to fight pinning action on right and try to prevent the withdrawal of Bavarian forces at the close of the day.
* Probably evolved during the battle in detail— but conceived in broad outline in advance (Tapfheim recce of 12 August).

Miles
0   ½   1   1½   2

← Bavarians and Marsin's French.
← Tallard's French.
⇦ Allied Forces.
⌐ ┐ Allied containment plans.
■ ■ ■ Alternative French position (not employed).
☼ Commander's observation post.

THE BATTLE OF BLENHEIM
13 August 1704: French and Allied dispositions

18 Bns EUGENE SCHWENENBACH
92 Sqns WOLPERSTETTEN
Holstein-Beck MARLBOROUGH 52,000 & 60 guns
67 Sqns
12 Bns 72 Sqns 28 Bns
River Nebel WEILHEIM to Donauwörth
MARSIN and ELECTOR Churchill
16 Bns 14 Bns UNTERGLAU 14 Sqns
LUTZINGEN OBERGLAU Mill Cutts 20 Bns
N 9 Bns Mill
9 Bns 64 Sqns 12 Sqns
11 Bns 7 Bns BLENHEIM
TALLARD 56,000 & 90 guns
SONDERHEIM
River Danube

☐ Allied Army.
■ French and Bavarian Army.

Mile
0   ¼   ½   ¾   1

proved capable of solving them all (see map Strategic Considerations, Spring 1704).

In the event, the march to the Danube was brilliantly successful. Utilizing geographical factors wherever possible (e.g. river transport, the cover provided by the Kraichgau Hills, the Black Forest and the Swabian Jura), Marlborough fooled both his Allies and his foes as to his ultimate intentions, and successfully marched his steadily-growing army (20,000-45,000) to rendezvous with the Margrave of Baden. Brilliant deception and, above all, care for administrative detail—marches in the cool of the early morning, spare shoes and saddlery stockpiled at Heidelberg, regular rest days, etc—prevented enemy intervention and also minimized wear and tear. By these means the main column covered some 250 miles in five weeks—for that period an amazing rate of advance—and arrived at Launsheim perfectly fit and ready to fight. Leaving Prince Eugene of Savoy to observe Tallard from the Lines of Stollhofen, Marlborough took the cantankerous and obstructive Baden with him to seize a bridgehead over the Danube into the heartland of Bavaria. It was not possible to attack the main Bavarian positions at Ulm or Lavingen as they were too strong, and Marlborough had left most of his heavier guns in the Netherlands to assist his march. So the Allies successfully stormed the Schellenberg Heights and took Dönauwörth and its bridges on 2 July, after surprising the local commander by a forced march from Amerdingen and attacking immediately on arrival without a pause. Losses were heavy on both sides: 6,000 Allies (ten per cent) against 5,000 Bavarians (fifty per cent of garrison).

Good fortune now began to desert the Allies. Heartened by news that Tallard was at last on his way through the Black Forest to help him, the Elector of Bavaria refused to either fight or make terms; instead, he contained his men and the French troops already in Bavaria within impregnable positions, whilst the Allies burnt an estimated 300 villages in southern Bavaria.

With Tallard's arrival at Augsburg to join the Elector, after covering 200 miles in 36 days (compare Marlborough's performance), the strategic situation swung in favour of the Franco-Bavarian cause. By moving their superior army north of the Danube, they proceeded to threaten Marlborough's lines of communication running through Dönauwörth towards Nordlingen and Nüremburg. The Allies might even be trapped south of the Danube.

Eugene, meantime, had been retiring before Tallard's forces. His delaying actions partly account for the slow overall rate of the French

44

advance from Strasbourg, as did the need for the French to escort a huge convoy of 8,000 waggons through the Black Forest passes. However, Tallard's administrative arrangements were so poor that almost a third of his men were left by the wayside as stragglers, and nearly half his cavalry was smitten with glanders before he even entered Bavaria. This contrasts most markedly with Marlborough's record over a far longer march.

To avert the threat to their communications, Marlborough decided to rejoin Eugene north of the Danube by forced marches, but first he had to get rid of Baden. So, he sent him off with 15,000 men to besiege Ingolstadt, on the pretence that the rest of the Allied forces would merely cover the siege. In fact, Marlborough had more dramatic intentions. After a close reconnaissance of the Franco-Bavarian position near Blenheim on 12 August, Marlborough and Eugene decided to risk everything by forcing a major battle on an unsuspecting Tallard, who fully expected the Allies to withdraw towards Nordlingen.

Tallard anticipated no battle because of his strong position and his overall numerical superiority: 56,000 men and 90 guns against 52,000 and only 60 guns. He was surprised when the rising mists of 13 August revealed the Allied Army moving to attack; however, delays on the Allied right gave the French time to

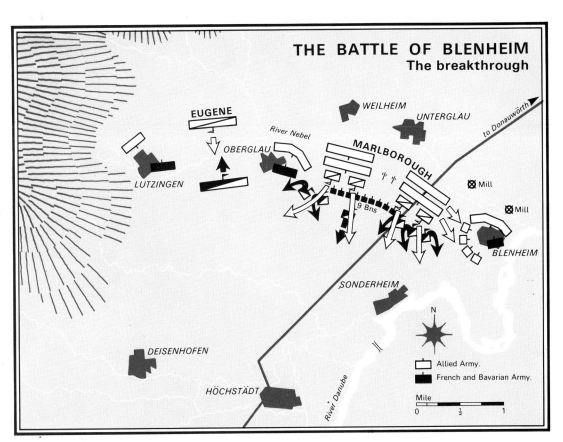

THE BATTLE OF BLENHEIM
The breakthrough

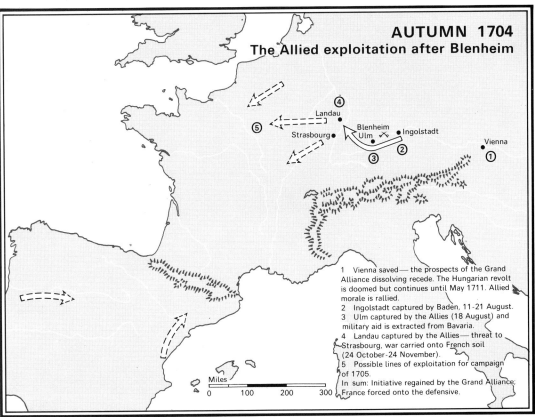

AUTUMN 1704
The Allied exploitation after Blenheim

1 Vienna saved— the prospects of the Grand Alliance dissolving recede. The Hungarian revolt is doomed but continues until May 1711. Allied morale is rallied.
2 Ingolstadt captured by Baden, 11-21 August.
3 Ulm captured by the Allies (18 August) and military aid is extracted from Bavaria.
4 Landau captured by the Allies— threat to Strasbourg, war carried onto French soil (24 October-24 November).
5 Possible lines of exploitation for campaign of 1705.
In sum: Initiative regained by the Grand Alliance; France forced onto the defensive.

Chronology of the 1704 Campaign

April: Marlborough decides to intervene in the Danube theatre.
10 May: Tallard completes transfer of 10,000 reinforcements to Bavaria.
20 May: Marlborough's force of 20,000 leaves Bedburg.
28 May: Marlborough crosses to east bank of the Rhine at Coblenz. Joined by 5,000 Hanoverians. End of the 'Moselle Bluff'.
31 May: River Main crossed; further 20,000 Allies (mainly Danes) join the column.
7 June: Column swings towards Danube. End of 'Phillipsburg/ Strasbourg Bluff'.
13 June: Council of War with Baden and Eugene at Gross Heppach. Future moves decided.
21 June: Marlborough's column joins Baden's Imperialist Army (about 50,000) at Launsheim.
2 July: Marlborough and Baden storm the Schellenberg Heights and take Dönauwörth.
1 July: Tallard at last moves over Rhine from Strasbourg; shadowed by Eugene.
July: Allies impotently ravage Bavaria, but Elector refuses to fight or negotiate.
5 August: Tallard (about 40,000) joins the Elector of Bavaria at Augsburg.
7 August: Allied emergency Council of War; Eugene joins conference.
9 August: Baden sent off with 15,000 men to besiege Ingolstadt; French begin to cross to north bank of Danube with the intention of threatening Dönauwörth and line of communication.
10-11 August: Marlborough conducts forced marches to rejoin Eugene's force north of the Danube.
12 August: Allies reconnoitre Franco-Bavarian camp near Blenheim.
13 August: The Battle of Blenheim. Complete defeat and capture of Tallard.
21 August: Ingolstadt surrenders.
11 September: Ulm captured by Allies; Bavaria occupied.
28 November: After following survivors of the French to the Rhine, Allies take Landau.

Marlborough's Generalship

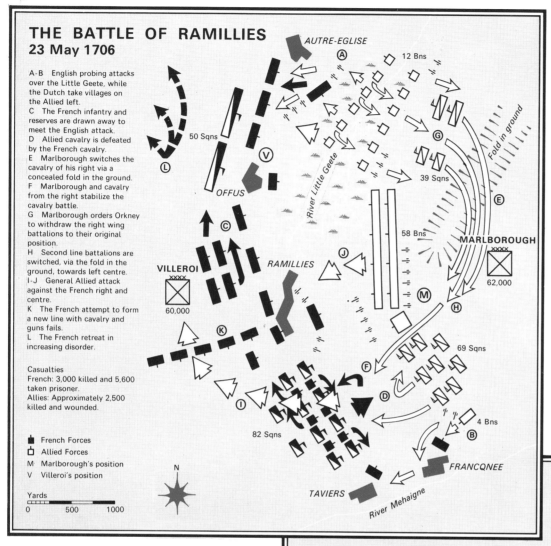

THE BATTLE OF RAMILLIES
23 May 1706

A-B English probing attacks over the Little Geete, while the Dutch take villages on the Allied left.
C The French infantry and reserves are drawn away to meet the English attack.
D Allied cavalry is defeated by the French cavalry.
E Marlborough switches the cavalry of his right via a concealed fold in the ground.
F Marlborough and cavalry from the right then stabilize the cavalry battle.
G Marlborough orders Orkney to withdraw the right wing battalions to their original position.
H Second line battalions are switched, via the fold in the ground, towards left centre.
I-J General Allied attack against the French right and centre.
K The French attempt to form a new line with cavalry and guns fails.
L The French retreat in increasing disorder.

Casualties
French: 3,000 killed and 5,600 taken prisoner.
Allies: Approximately 2,500 killed and wounded.

■ French Forces
□ Allied Forces
M Marlborough's position
V Villeroi's position

Yards
0   500   1000

routed Tallard's 68 squadrons and 9 battalions, many of whom were drowned in the Danube near Sonderheim trying to escape. On this the Elector retreated for Höchstädt, leaving the encircled garrison of Blenheim to their fate. By

John Churchill, Duke of Marlborough (1650-1722), one of England's greatest soldiers, whose skilled conduct of the War of the Spanish Succession established the British Army's reputation as the foremost in Europe. Political problems led to his dismissal in 1711.

prepare their plans, and Tallard became confident of victory.

The action began at about 12.30pm. The French fought wholly on the defensive, confident that their advantages of position and numbers would tell. However, Marlborough had the measure of the situation. By 3.30pm he had contained the garrisons of Blenheim and Oberglau; and induced the Blenheim sector commander to cram 27 battalions of French troops into the village quite uselessly. Meanwhile, Eugene fought a grim action against the Bavarians on the Allied right wing at a numerical disadvantage of two to one. Even this did not prevent him from lending aid to Marlborough in the centre at a critical moment. After deploying a superior number of squadrons and supporting battalions over the Nebel, by 4.30pm Marlborough was ready to pierce Tallard's unsupported centre. By 5.30pm the Allied 80 squadrons and 23 battalions had

THE BATTLE OF OUDENARDE
11 July 1708 (midday to 5pm)

This battle was a daring undertaking by Marlborough. It involved passing his whole army over a major river in the proximity of the foe. However his calculated gamble, based on his personal knowledge of the enemy commanders, brought off a notable success.

A The original French position (am 11 July) covering the Gavre bridge and the anticipated Allied line of approach.
B General Biron's advance guard (Swiss battalions) advances to watch Oudenarde, makes contact with Cadogan and the Allied advance guard and is routed.
C The French Army moves to a second position near Huysse; Burgundy's wing remains static for the rest of the day in the vicinity of the village.
D Vendôme commits his wing of the army, but—
E Marlborough progressively reinforces Cadogan with Rantzau,....
F Natzmer and Argyle whilst Lottum marches up to the bridgehead. The result is a long action along the River Diepenbeck with fluctuating fortunes.
G The site of the Allied pontoon bridges (completed at noon).
H Royegem Mill—the site of French headquarters.

■ French Forces
□ Allied Forces

Miles
0   ½   1   1½   2

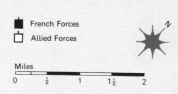

46

9pm these troops had been bluffed into surrender. Tallard was himself made prisoner.

For a loss of 12,500 casualties (23 per cent), the Allies had inflicted a major defeat on the French, who lost 21,000 battle casualties besides 14,000 prisoners of war and a further 5,000 deserters: in all, 70 per cent of their effective strength.

Note how Marlborough made use of superior tactical skill to overcome serious natural hazards, and how Tallard failed to see the superior advantages that the position near Tapfheim might have conferred. His flanks would have been equally secure, and his front only half as broad.

By the end of the campaign the Allies had saved Vienna from all peril, regained the initiative on the Danube and Upper Rhine fronts, and conquered Bavaria. Thus the campaign of 1704 entirely changed the face of the War of the Spanish Succession, and in the process made the reputation of John Churchill, first Duke of Marlborough. His ability to make use of geographical possibilities, at both the strategic and tactical level, was one major feature of his genius.

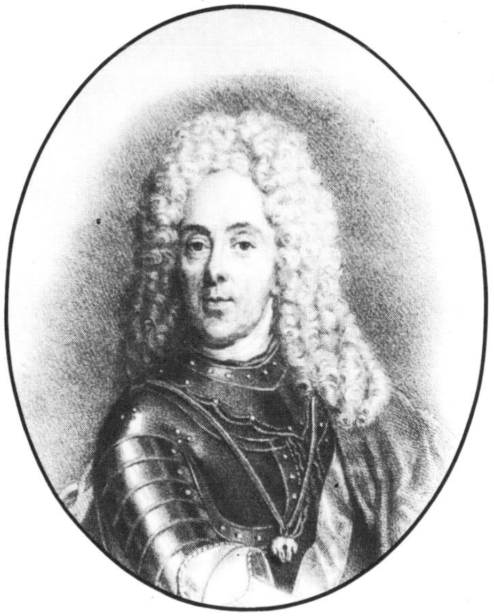

Eugene, Prince of Savoy-Carignan (1663-1736). As 'Twin-Captain' with Marlborough, he shared in three of the Duke's great victories, and his own successes at Turin (1706) and Belgrade (1717) earned him Napoleon's accolade.

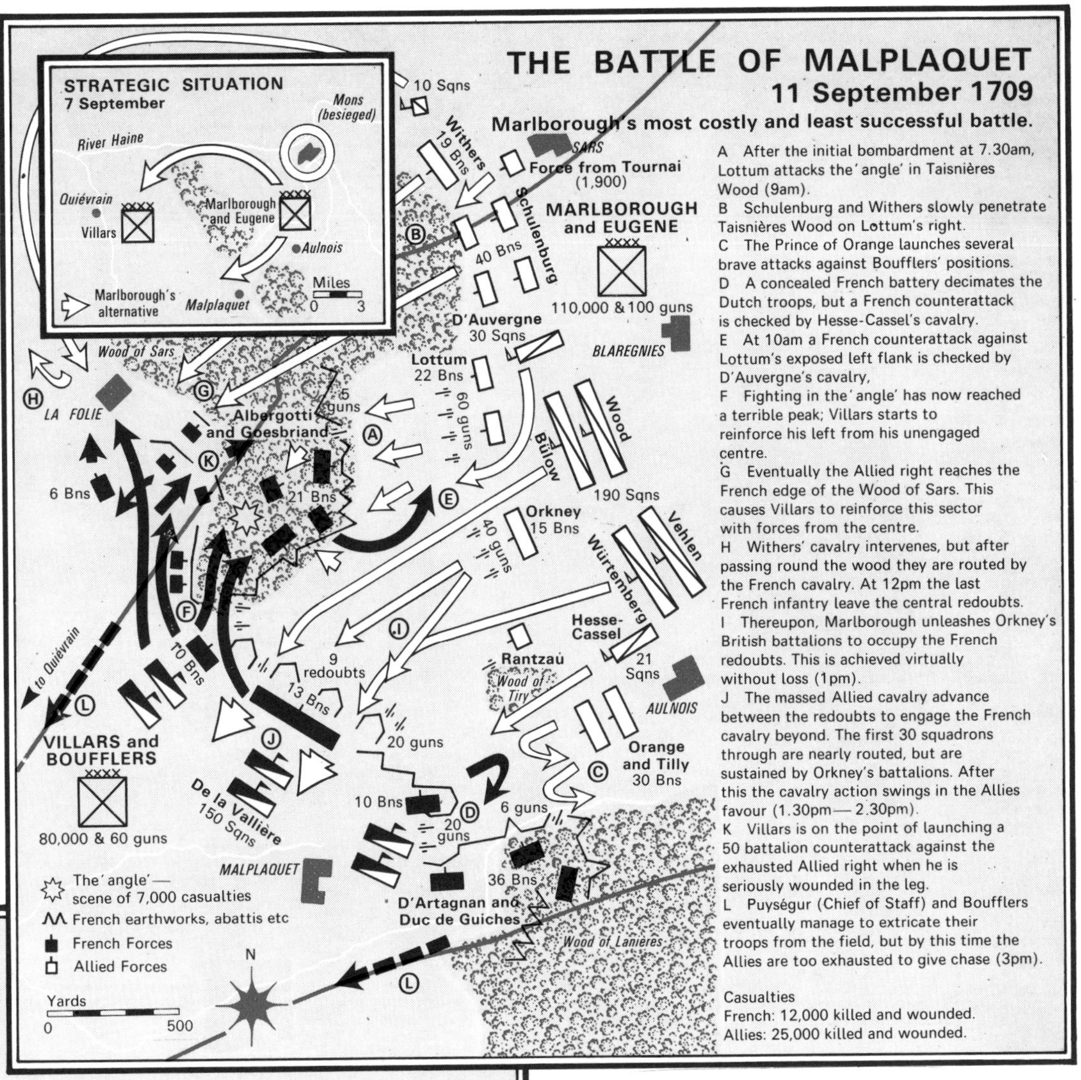

THE BATTLE OF MALPLAQUET
11 September 1709
Marlborough's most costly and least successful battle.
STRATEGIC SITUATION
7 September
Mons (besieged)
River Haine
Quiévrain
Villars
Marlborough and Eugene
Aulnois
Marlborough's alternative
Malplaquet
Miles
0
3
10 Sqns
Withers 19 Bns
SARS
Force from Tournai (1,900)
Schulenburg
MARLBOROUGH and EUGENE
40 Bns
110,000 & 100 guns
D'Auvergne 30 Sqns
BLAREGNIES
Lottum 22 Bns
Wood of Sars
LA FOLIE
5 guns
Albergotti and Goesbriand
60 guns
Wood
Bülow
6 Bns
21 Bns
190 Sqns
Orkney 15 Bns
40 guns
Vehlen
Würtemberg
Hesse-Cassel
to Quiévrain
9 redoubts
Rantzau
Wood of Tiry
21 Sqns
70 Bns
13 Bns
AULNOIS
20 guns
VILLARS and BOUFFLERS
80,000 & 60 guns
De la Vallière 150 Sqns
Orange and Tilly 30 Bns
10 Bns
6 guns
20 guns
MALPLAQUET
36 Bns
D'Artagnan and Duc de Guiches
Wood of Lanières
The 'angle'— scene of 7,000 casualties
French earthworks, abattis etc
French Forces
Allied Forces
N
Yards
0
500
A After the initial bombardment at 7.30am, Lottum attacks the 'angle' in Taisnières Wood (9am).
B Schulenburg and Withers slowly penetrate Taisnières Wood on Lottum's right.
C The Prince of Orange launches several brave attacks against Boufflers' positions.
D A concealed French battery decimates the Dutch troops, but a French counterattack is checked by Hesse-Cassel's cavalry.
E At 10am a French counterattack against Lottum's exposed left flank is checked by D'Auvergne's cavalry.
F Fighting in the 'angle' has now reached a terrible peak; Villars starts to reinforce his left from his unengaged centre.
G Eventually the Allied right reaches the French edge of the Wood of Sars. This causes Villars to reinforce this sector with forces from the centre.
H Withers' cavalry intervenes, but after passing round the wood they are routed by the French cavalry. At 12pm the last French infantry leave the central redoubts.
I Thereupon, Marlborough unleashes Orkney's British battalions to occupy the French redoubts. This is achieved virtually without loss (1pm).
J The massed Allied cavalry advance between the redoubts to engage the French cavalry beyond. The first 30 squadrons through are nearly routed, but are sustained by Orkney's battalions. After this the cavalry action swings in the Allies favour (1.30pm— 2.30pm).
K Villars is on the point of launching a 50 battalion counterattack against the exhausted Allied right when he is seriously wounded in the leg.
L Puységur (Chief of Staff) and Boufflers eventually manage to extricate their troops from the field, but by this time the Allies are too exhausted to give chase (3pm).
Casualties
French: 12,000 killed and wounded.
Allies: 25,000 killed and wounded.

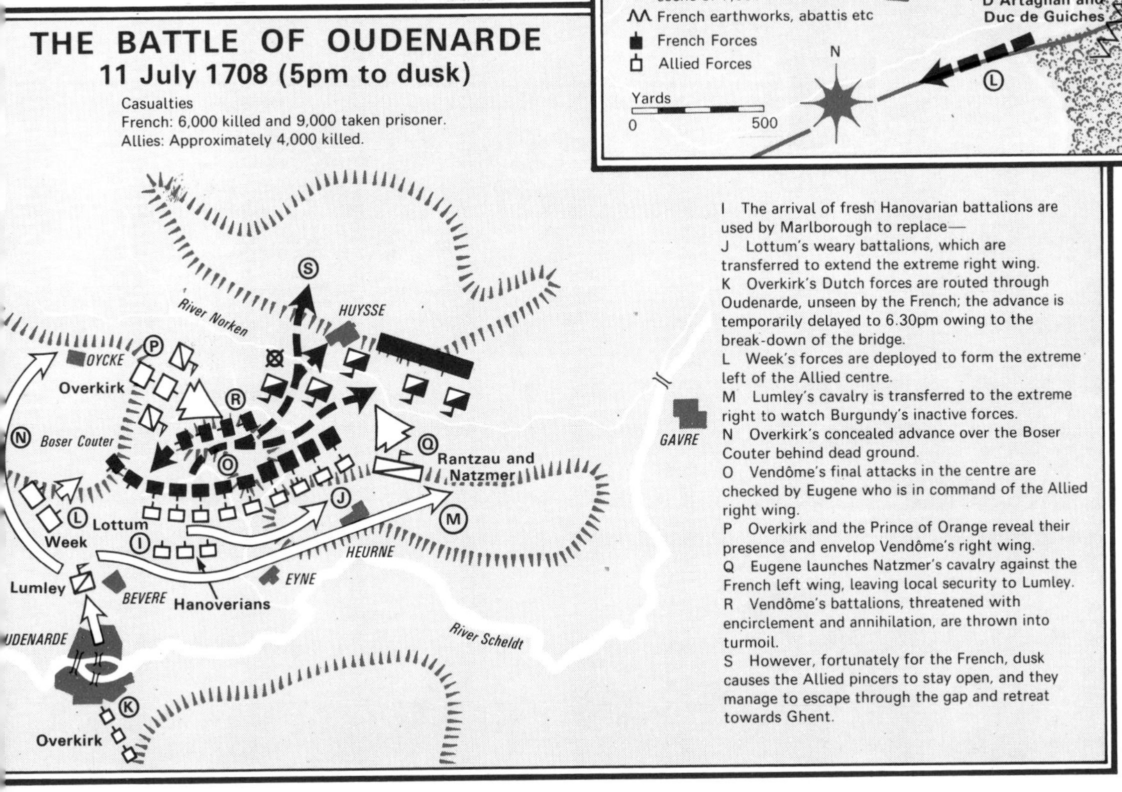

THE BATTLE OF OUDENARDE
11 July 1708 (5pm to dusk)
Casualties
French: 6,000 killed and 9,000 taken prisoner.
Allies: Approximately 4,000 killed.
River Norken
HUYSSE
OYCKE
Overkirk
Boser Couter
Rantzau and Natzmer
GAVRE
Lottum
Week
HEURNE
Lumley
EYNE
BEVERE
Hanoverians
River Scheldt
UDENARDE
Overkirk
I The arrival of fresh Hanoverian battalions are used by Marlborough to replace—
J Lottum's weary battalions, which are transferred to extend the extreme right wing.
K Overkirk's Dutch forces are routed through Oudenarde, unseen by the French; the advance is temporarily delayed to 6.30pm owing to the break-down of the bridge.
L Week's forces are deployed to form the extreme left of the Allied centre.
M Lumley's cavalry is transferred to the extreme right to watch Burgundy's inactive forces.
N Overkirk's concealed advance over the Boser Couter behind dead ground.
O Vendôme's final attacks in the centre are checked by Eugene who is in command of the Allied right wing.
P Overkirk and the Prince of Orange reveal their presence and envelop Vendôme's right wing.
Q Eugene launches Natzmer's cavalry against the French left wing, leaving local security to Lumley.
R Vendôme's battalions, threatened with encirclement and annihilation, are thrown into turmoil.
S However, fortunately for the French, dusk causes the Allied pincers to stay open, and they manage to escape through the gap and retreat towards Ghent.

The Iberian Peninsula, 1702-13

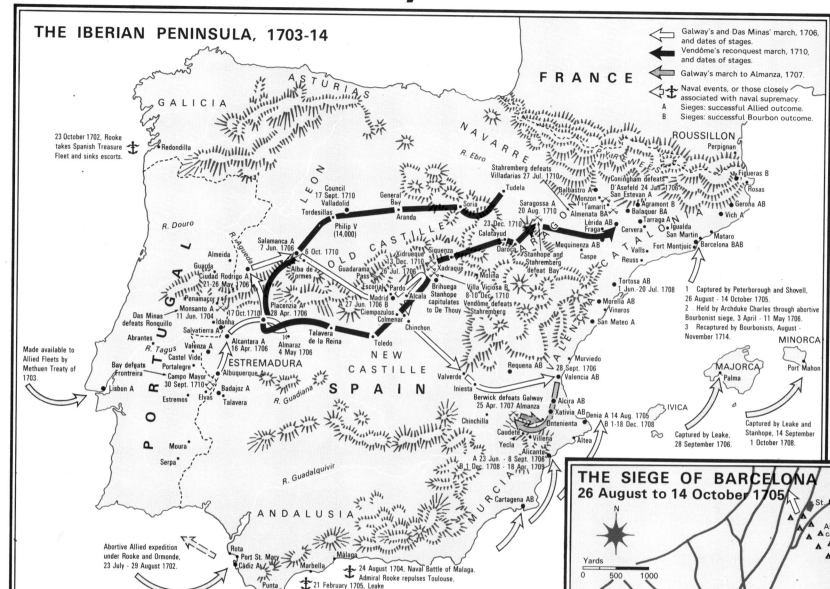

THE IBERIAN PENINSULA, 1703-14

Galway's and Das Minas' march, 1706, and dates of stages.

Vendôme's reconquest march, 1710, and dates of stages.

Galway's march to Almanza, 1707.

Naval events, or those closely associated with naval supremacy.

A Sieges: successful Allied outcome.
B Sieges: successful Bourbon outcome.

FRANCE

ROUSSILLON

23 October 1702, Rooke takes Spanish Treasure Fleet and sinks escorts.

GALICIA   ASTURIAS

NAVARRE

PYRENEES

Perpignan

Redondilla

R. Ebro

Stahremberg defeats Villadarias 27 Jul. 1710

Coningham defeats D'Asefeld 24 Jan. 1706

Figueras B

Rosas

LEON

Council 17 Sept. 1710 Valladolid Tordesillas

General Bay

Tudela

Soria

Balbastro A Monzon B San Estevan A

Agramont B Tamarit Balaguer BA

Gerona AB

Saragossa A 20 Aug. 1710

Lérida AB

Vich A

R. Douro

R. Agueda

Salamanca A 7 Jun. 1706

Philip V (14,000)

Aranda

23 Dec. 1710 Calatayud

Daroca

Mequinenza AB

Tarraga B Igualda San Martin

Cervera

Mataro

Almeida

6 Oct. 1710

Xidrueque 13 Dec. 1710

Siquenza

Stanhope and Stahremberg defeat Bay

Fraga

Fort Montjuic Barcelona BAB

Guarda Ciudad Rodrigo A 21-26 May 1706

Alba de Tormes

Guadarama Pass

Xadraque 26 Jul. 1706

Caspe

Valls Reuss

Penamaçor 17 Oct.1710

Escorial Pardo

Molina

Tortosa AB 1 Jun.-20 Jul. 1708

Monsanto A 11 Jun. 1704

Placenzia Ar 28 Apr. 1706

Madrid 27 Jun. 1706 B

Villa Viciosa B 8-10 Dec. 1710 Vendôme defeats Stahremberg

Morella AB

Idanna

Alcala

Brihuega Stanhope capitulates to De Thouy

Vinaros

Abrantes

Salvatierra A

Ciempozulos Colmenar

San Mateo A

Das Minas defeats Ronquillo

Alcantara A 16 Apr. 1706

Talavera de la Reina

Chinchon

Valenza A

Almaraz 4 May 1706

NEW CASTILLE

Murviedo 28 Sept. 1706

Castel Vide Portalegre

Campo Mayor 30 Sept. 1710

Toledo

Requena AB Valencia AB

Bay defeats Frontreira

Made available to Allied Fleets by Methuen Treaty of 1703.

Albuquerque A Badajoz A

SPAIN

Valverde Iniesta

MINORCA

MAJORCA

Port Mahon

Palma

Lisbon A

Estremos Elvas Talavera

R. Guadiana

Berwick defeats Galway 25 Apr. 1707 Almanza

Alcira AB

Xativia AB

Denia A 14 Aug. 1705 B 1-18 Dec. 1708

IVICA

Moura

Chinchilla

Ontenienta

Caudete Villena

Yecla

Altea

Captured by Leake, 28 September 1706.

Captured by Leake and Stanhope, 14 September 1 October 1708.

Serpa

R. Guadalquivir

Alicante

A 23 Jun. - 8 Sept. 1706
B 1 Dec. 1708 - 18 Apr. 1709

ANDALUSIA

MURCIA

Cartagena AB

1 Captured by Peterborough and Shovell, 26 August - 14 October 1705.
2 Held by Archduke Charles through abortive Bourbonist siege, 3 April - 11 May 1706.
3 Recaptured by Bourbonists, August - November 1714.

Abortive Allied expedition under Rooke and Ormonde, 23 July - 29 August 1702.

Rota Port St. Mary Marbella Málaga

Càdiz A

24 August 1704, Naval Battle of Malaga. Admiral Rooke repulses Toulouse.

Punta Mela

21 February 1705, Leake destroys French squadrons.

Gibraltar A

Miles
0 25 50 75 100 125 150

1 Captured by Admiral Rooke, 31 July - 6 August 1704.
2 Abortive Bourbonist siege, 24 August 1704 - 28 April 1705, — defended by Prince George.

THE SIEGE OF BARCELONA
26 August to 14 October 1705

St. Mar

N

Allied camp

Yards
0 500 1000

French attack 1697

Allied attack 1705

The Cordeliers

Jonquieres Bastion

Telles Bastion

New Bastion

Gate

Levant Bastion

Convent

St. John's Bastion

Hospital

OLD TOWN

St. Antonio Bastion

NEW TOWN

Waterport

Arsenal

King's Bastion

Harbour

Breach

1706

Mole

Montjuic Gate

The Capuchins

French attack 1706

MEDITERRANEAN SEA

Fort St. Bertran

Montjuic Fort

KEY

Route used by Prince George of Hesse-Darmstadt to surprise and take Montjuic Fo

In the seventeenth century Spain was in decline. The splendours of Charles V and Philip II proved ephemeral, and their Habsburg successors—Philip III (1598-1621), Philip IV (1621-65) and Charles II (1665-1700)— witnessed a steady erosion of national fortunes. The expulsion of the Moriscos (Christian descendants of the Moors) in 1609 was both a crime and a mistake, for it robbed Spain of perhaps her most productive element. Philip IV's attempts to create a truly centralized monarchy led to the Catalan revolt, 1640-52; an event which facilitated French intervention south of the Pyrenees and exacerbated the long Franco-Spanish War, 1635-59, which exhausted Spanish resources. Linked with this was the struggle for Portuguese independence, 1640-68, which resulted in the creation of a separate kingdom ruled by the House of Braganza, and the progressive loss of areas of the Spanish Netherlands and Franche-Comté to France over a long series of wars of French aggrandisement. The last of the Spanish Habsburgs, Charles II, was crippled throughout his life by painful diseases, and his long anticipated demise was the cause of much

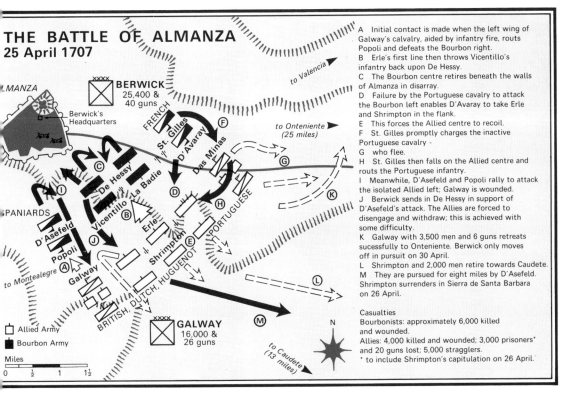

THE BATTLE OF ALMANZA
25 April 1707

BERWICK
25,400 &
40 guns

Berwick's Headquarters

FRENCH

St. Gilles

D'Avaray

Das Minas

De Hessy

La Badie

Vicentillo

D'Asefeld

Popoli

Galway

Erle

Shrimpton

HUGUENOT

PORTUGUESE

BRITISH, DUTCH, HUGUENOT

GALWAY
16,000 &
26 guns

SPANIARDS

ALMANZA

to Valencia

to Onteniente (25 miles)

to Montealegre

to Caudete (13 miles)

☐ Allied Army
■ Bourbon Army

Miles
0   ½   1   1½

A Initial contact is made when the left wing of Galway's calvalry, aided by infantry fire, routs Popoli and defeats the Bourbon right.
B Erle's first line then throws Vicentillo's infantry back upon De Hessy.
C The Bourbon centre retires beneath the walls of Almanza in disarray.
D Failure by the Portuguese cavalry to attack the Bourbon left enables D'Avaray to take Erle and Shrimpton in the flank.
E This forces the Allied centre to recoil.
F St. Gilles promptly charges the inactive Portuguese cavalry -
G who flee.
H St. Gilles then falls on the Allied centre and routs the Portuguese infantry.
I Meanwhile, D'Asefeld and Popoli rally to attack the isolated Allied left; Galway is wounded.
J Berwick sends in De Hessy in support of D'Asefeld's attack. The Allies are forced to disengage and withdraw; this is achieved with some difficulty.
K Galway with 3,500 men and 6 guns retreats sucessfully to Onteniente. Berwick only moves off in pursuit on 30 April.
L Shrimpton and 2,000 men retire towards Caudete.
M They are pursued for eight miles by D'Asefeld. Shrimpton surrenders in Sierra de Santa Barbara on 26 April.

Casualties
Bourbonists: approximately 6,000 killed and wounded.
Allies: 4,000 killed and wounded; 3,000 prisoners* and 20 guns lost; 5,000 stragglers.
* to include Shrimpton's capitulation on 26 April.

intrigue and negotiation between the European powers. Indeed, the question of the Spanish Succession came to dominate the last forty years of the seventeenth century. On his deathbed in 1700, Charles willed all his possessions to Philip of Anjou, Louis XIV's grandson, in the hope of keeping the succession intact. This action plunged Western Europe into another great war. By 1705, Portuguese, Dutch, German and English armies were marching and countermarching across the country in vain attempts to repossess it for the Austrian Habsburg claimant, the Archduke Charles. Predictably, Catalonia opted for the Archduke, and the Allies' successful capture of Barcelona on 14 October 1705, and Galway's and Das Minas' famous march to Madrid the following year seemed to bring promise of ultimate success for the Habsburg claimant; however, this proved illusory. The presence of hated foreign armies on their soil united the majority of the Spanish population around Philip V as nothing else could hope to do, and with the aid of French forces (led by the Duke of Berwick, the Duc d'Orléans and Marshal Vendôme), the tide soon began to turn. The defeat of the Earl of Galway at Almanza in 1707 began a steady decline in Allied fortunes. After a number of fluctuations Marshal Vendôme's great campaign in 1710 finally doomed the Habsburg cause. It was not until November 1714, however, that the Bourbonists regained Barcelona from the Catalan rebels. The pacification left Spain its overseas empire, but most of its European possessions were re-distributed, including the Netherlands, the Milanese, and Sicily and Naples (Kingdom of the Two Sicilies).

Campaigning in Spain presented many difficulties. The terrain was difficult: the mountainous and inhospitable Hispano-Portuguese frontier being traversable only at the two 'corridors', north and south. The Pyrenees similarly posed a serious obstacle to military movement, the main passes being at the extremities. The plain of central Spain was dominated by four great rivers: the Douro, Tagus, Guadiana and Guadalquivir. As the northern and southern coasts of Spain were infertile and mountain-dominated, there is small wonder that the Allies chose to exploit their control of the sea won at Malaga, 1704. They based their invasions on the somewhat easier east coast, close by the friendly Catalans and the Guadiana; although this decision committed them to an expensive second front which they could not satisfactorily maintain from their main logistical base of Lisbon.

Extremes of climate caused much sickness to both men and horses, particularly the intense heat of July–September which often gave rise to a virtual cessation of hostilities. Shortage of fodder, food and transport-animals was endemic, particularly as the mass of the population was strongly hostile to the Allies, especially to the 'heretic' British and Dutch contingents. Allied dissensions were also bitter: the Portuguese, British, Austrian and Catalan leaders had very different concepts of how the war should be waged. This disunity caused the fragmentation of armies and operations—a circumstance that greatly favoured the more united Bourbonists.

The inadequacy of the supply and equipment arrangements of the Allies was perhaps the most significant single cause of their ultimate failure in Spain. Flanders always received the lion's share of available resources and, consequently, the Allied armies in the Peninsula remained half-starved, ragged and far below strength for most of the war. Despite this, Britain gained possession of the naval bases of Gibraltar and Minorca, captured in 1704 and 1708 respectively, and thus emerged as both a Mediterranean and European power. Austria also gained considerable territories in Europe at the Peace of Rastadt (1714) (see map p. 64). Possibly the reduction of its unwieldy empire proved to Spain's ultimate advantage, while the new dynasty brought a fresh energy and efficiency to the conduct of Spanish affairs which would aid the country's temporary recovery during the eighteenth century and its re-emergence as an effective power.

1. Henry de Massue, Earl of Galway (1648-1720). A French Huguenot soldier who served in the English Army from 1690, he captured Madrid in the War of the Spanish Succession but was decisively defeated next year at Almanza (1707). (National Army Museum)
2. James Fitzjames, Duke of Berwick (1670-1734). Nephew of Marlborough, his Roman Catholic faith took him into Louis XIV's armies. He campaigned in Ireland, Flanders and Spain with success, and won the Battle of Almanza. (National Army Museum)

Siege Warfare

Military operations in the seventeenth century were frequently dominated by siege warfare. The period 1618-1722 saw over 120 major sieges, half of them taking place during the War of the Spanish Succession. Areas of Europe particularly prone to this form of warfare were the 'cockpit' and the North Italian Plain, where there were large numbers of fortified towns amidst fertile countryside.

Sieges were frequently protracted affairs, proving costly in both human life and materials of war (see the cited instances of Vienna, Lille and Belgrade). Disease was the greatest killer of both besieged and besiegers. Considerable progress was made in the design of fortifications over the century, the greatest names being the engineers Pagan, 1604-53 (France), Rimpler, fl 1683 (Germany), Coehoorn, 1641-1704 (United Provinces) and, above all, Marshal Vauban, 1633-1707, who is reputed to have built some 33 new fortresses and restored over 100 more along the French frontier; he also directed 53 successful sieges.

Fortifications were designed to provide the greatest possible degree of enfilade fire, mutual support and defence in depth. Defences were relatively low lying and massively constructed, the bastions being protected by glacis, deep ditches and various types of outworks. An important town would be protected by an enceinte and a separately-sited citadel.

An army planning to carry out a siege would be divided into two parts: a force for the active prosecution of the operations and another to cover them, ensuring that no relieving army could break through to sustain the garrison, or interfere with the besiegers' convoys of supplies. The first stage was the establishment of the blockade: the physical isolation of the beleaguered place. This was completed by the construction, when deemed necessary, of inward-looking 'lines of contravallation' and a corresponding series of outward-facing 'lines of circumvallation'. The besiegers established their camps and parks within the enclosed area, aided by large numbers of locally-impressed peasantry.

After a careful reconnaissance to determine the most promising sector of the defences to attack, the commanding general would order the next major step—the 'opening of the trenches'. Following the Vauban system, the engineers would mark out a large trench, the 'first parallel', some 600 yards (maximum effective cannon range) from the bastions to be attacked. Once this was well on the way to completion, indirect communication or approach trenches would be pushed forward (using saps and gabbions to protect the exposed workmen) to the proposed site for the 'second

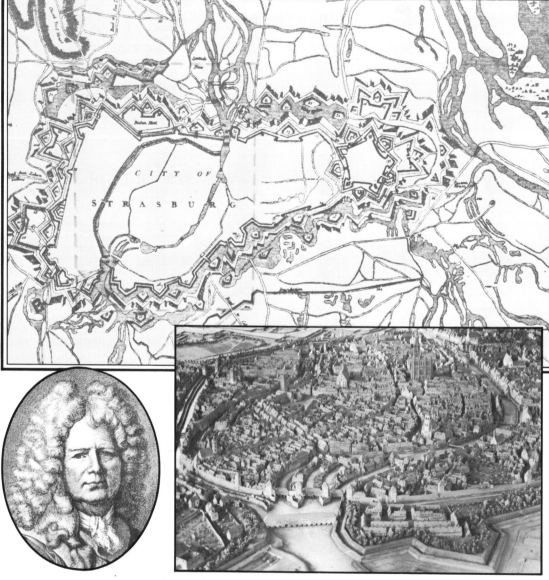

Top: Plan of Strasbourg (from The Military Library, 1802).
Above right: A model constructed in the 1720s. Note the barrage blocking the river and the incorporation of the medieval city wall.
Above left: Sebastien le Prestre de Vauban, Marshal of France (1633-1707), was the greatest military engineer of the seventeenth century. He improved the frontier fortifications of France and conducted many successful sieges. He also wrote a number of famous treatises on several subjects.
Right: A mid-seventeenth century representation of how to site batteries to breach a curtain-wall (top) or a bastion (bottom). Note the use of subsidiary cannon to neutralize neighbouring sectors of the defences, and also (in the bottom plan) the internal 'switch-line' prepared by the defenders, ready to repulse any attempted assault.

parallel' some 400 yards from the defences. Meanwhile, batteries of howitzers would endeavour to sweep the enemy's parapets and, sometimes, long-range breaching cannon would be brought into action. The whole process would then be repeated to place a 'third parallel' on the glacis itself, sap-heads being pushed forward to the very edge of the ditch and the covered way. At this stage the heaviest guns (often 24- or 33-pounders) were man-

handled forward under cover of darkness and emplaced in strongly prepared positions. The idea was to direct heavy, continuous, short-range fire against the main fortifications with a view to crumbling away a breach, filling the ditch with the debris in the process. Mining could hasten the process.

A determined garrison did not passively await its fate, but by means of surprise sorties, counter-mining and heavy fire endeavoured to

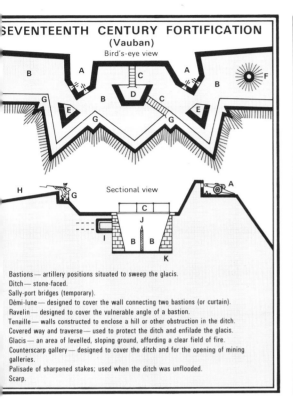

SEVENTEENTH CENTURY FORTIFICATION
(Vauban)
Bird's-eye view

labels: A, B, C, D, E, F, G, H

Sectional view

labels: H, G, A, C, J, I, B, B, K

Bastions — artillery positions situated to sweep the glacis.
Ditch — stone-faced.
Sally-port bridges (temporary).
Démi-lune — designed to cover the wall connecting two bastions (or curtain).
Ravelin — designed to cover the vulnerable angle of a bastion.
Tenaille — walls constructed to enclose a hill or other obstruction in the ditch.
Covered way and traverse — used to protect the ditch and enfilade the glacis.
Glacis — an area of levelled, sloping ground, affording a clear field of fire.
Counterscarp gallery — designed to cover the ditch and for the opening of mining galleries.
Palisade of sharpened stakes; used when the ditch was unflooded.
Scarp.

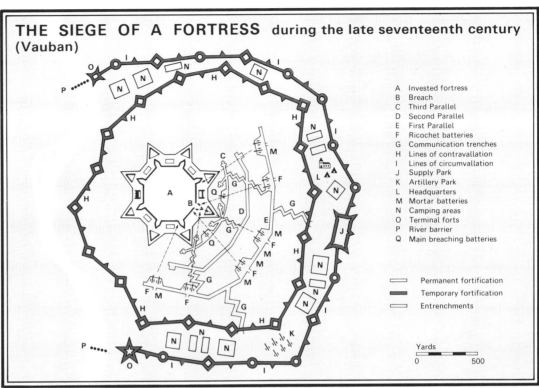

THE SIEGE OF A FORTRESS during the late seventeenth century
(Vauban)

A Invested fortress
B Breach
C Third Parallel
D Second Parallel
E First Parallel
F Ricochet batteries
G Communication trenches
H Lines of contravallation
I Lines of circumvallation
J Supply Park
K Artillery Park
L Headquarters
M Mortar batteries
N Camping areas
O Terminal forts
P River barrier
Q Main breaching batteries

☐ Permanent fortification
■ Temporary fortification
☐ Entrenchments

Yards
0   500

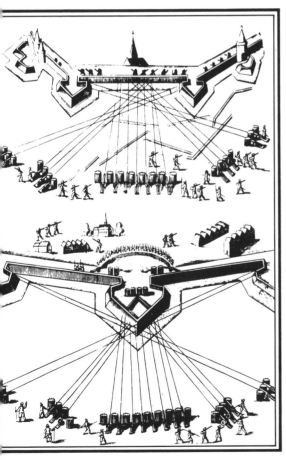

hinder the progress of the siege. Consequently, strong guards had to be continuously present in the trenches to ward off such inconveniences. Any attempts at external relief had also to be repulsed by the covering army.

Once the outlying ravelins and other works had been reduced, and a practicable breach had been made in the main defences, the possibility of a general assault had to be considered. These were rarer in the later seventeenth and eighteenth century than during the Thirty Years War, for the convention grew up that a defender was entitled to surrender on terms at the eleventh hour (or the 48th day of the siege) if there was no sign of any external relief, and if the besiegers preparations for a storming attempt were obviously complete. Indeed, harsh penalties were exactable from an unusually stubborn garrison and their hapless townsfolk in the event of a successful storm following the rejection of a formal summons to surrender. But such incidents were comparatively rare as they offended the humanitarian susceptibilities of the age, such as they were.

Vauban's timetable for the standard siege of a well-appointed and resolutely-held fortress with six bastions is included in his famous volume *Traité de l'attaque des places*:

'To invest a place, collect material, and build lines   —9 days
From the opening of the trenches to reaching the covered way   —9 days
The storm and capture of the covered way and its defences   —4 days
Descent into, and crossing of, the ditch of the démi-lune   —3 days
Mining operations, battery siting, creation of a fair breach   —4 days
Capture and exploitation of the démi-lune and its defences   —3 days
Crossing of the main ditch to two bastions   —4 days
Mining operations, and the siting of the breaching battery on the covered way   —4 days
The capture of the breach and its supporting positions   —2 days
Surrender of the town after the capitulation   —2 days
Allowance for errors, damage caused by sorties, a valorous defence   —4 days

Such was the typical progression of a siege as seen by the attacker. In most cases the attack was ultimately successful, although the details of every siege were unique and each presented special problems. In particular, the length of time taken varied enormously from a few days to many months. Some commanders—including Vauban and Marlborough—had redoubtable reputations as directors of sieges; "a town besieged by Vauban is a town taken" ran a contemporary saying about the former; as for the Duke, he "never fought a battle he did not win, nor besieged a town he did not take".

Siege Warfare

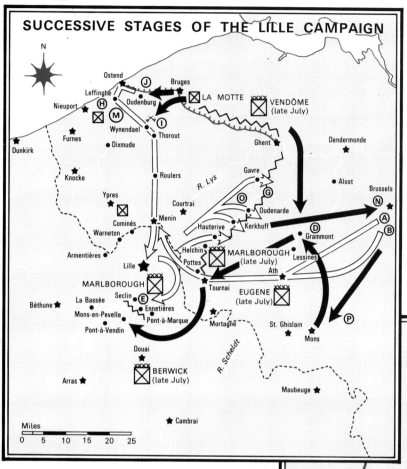

SUCCESSIVE STAGES OF THE LILLE CAMPAIGN

The Siege of Lille, 1708

The 120 days of this celebrated siege included almost every refinement of siege-craft. Defended by Marshal Boufflers and 16,000 men, it required all the skill of Marlborough and Eugene (jointly commanding approximately 100,000 men) to capture 'Vauban's masterpiece', which was the most important fortress on the French north-east frontier.

Preliminary operations, 22-25 July: The first convoy (A) moves to the Scheldt through enemy-dominated territory. 6-12 August—the 'great convoy', including 80 siege guns and 3,000 waggons, moves from Brussels (B) to Menin, covered by Eugene near Ath (50,000) and Marlborough (40,000) in the Oudenarde-Helchin-Pottes area.

The Blockade, instituted on 12 August: Lines of circum- and contravallation completed on 21 August. Northern sector of city is selected for the main attack, directed by Eugene with 50

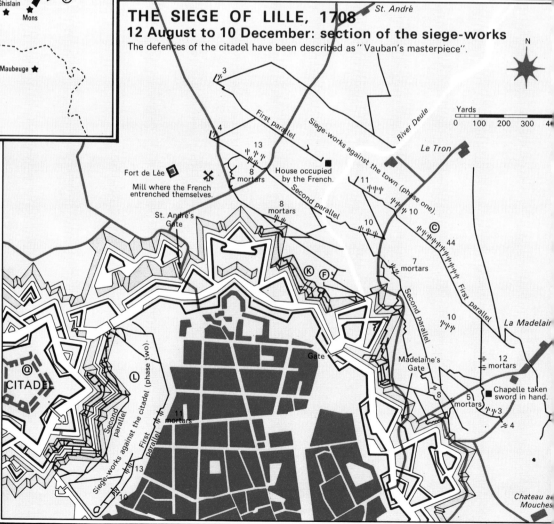

THE SIEGE OF LILLE, 1708
12 August to 10 December: section of the siege-works
The defences of the citadel have been described as "Vauban's masterpiece".

Lille was regarded as 'Vauban's masterpiece' of defensive fortification. After Oudenarde in 1708, Marshal Boufflers sustained a 120-day siege conducted by Marlborough and Eugene, but ultimately surrendered on 10 December that year. The cost was 10,000 Allied casualties. (National Army Museum)

battalions, while Marlborough carries out a covering role on the Scheldt. Meantime, Louis XIV orders Vendôme and Berwick to intervene.

The Opening of the Trenches, 22 August: 88 cannon open long-range bombardment from the completed first parallel on 27 August (C).

First Relief Attempt, 29 August: Berwick joins Burgundy and Vendôme at Grammont (joint forces, 110,000) (D). 1 September: French cross Scheldt at Tournai. 4 September: French reach Mons-en-Pêvelle. Marlborough, meanwhile, moves on inner arc to keep between Burgundy and . the siege; creates strong position around Seclin and Pont-à-Marque (E), and summons Eugene with 30,000 men to join him there. 5 September: Boufflers launches ineffectual sortie against the depleted siege lines.

Confrontation, 4-17 September: Despite their superiority of force and Louis' orders, the French dare not attack. 7 September: Eugene returns to the siege. 17 September: French withdraw.

Storming of the Counterscarp: Evening of 7 September; small gains for 3,000 casualties (F). Besieger's critical shortage of munitions relieved by arrival of convoy, 9 September. 21 September: new assaults against St. André and St. Madelaine sectors by 15,000 Allies; small gains for 1,000 casualties (including Eugene). Marlborough takes over direction of siege and finds grave administrative position: the siege is three weeks behind schedule and only four days' munitions are left. Boufflers is also very short of powder; on 28 September, Chevalier de Luxembourg tries to rush the blockade.

The Battle of the Convoys: French attempt to cut Allied lines of communication by strategic interdiction. By seizing the line of the Scheldt (G), they sever the roads from Brussels and Antwerp by mid-September. Marlborough switches to the coastal route, Ostend-Thorout-

Roulers-Menin-Lille (50 miles), but French forces at Nieuport, Ypres and Bruges interfere, opening sluices to create inland flooding near Leffinghe (H). On 25 September, Allied convoy leaves Ostend; 28 September, General Webb (11,000) beats off interception attempt by La Motte (22,000) at action of Wynendael (I); 30 September, convoy safely reaches siege lines. Furious Vendôme courts action near Oudenburg during first days of October (J), but when Marlborough (45,000) takes up the challenge on 7 October he retires to Bruges. Ostend is now almost isolated by floods, but boat flotillas maintain tenuous link, enabling the siege to continue despite shortages; the 'island' of Leffinghe fights off repeated attacks by French galleys from Dunkirk.

The capture of the city of Lille: By mid-October, Allied breaching batteries are established on counterscarp; by 20 October, several large breaches have been made (K). On 23 October Boufflers beats a parley and negotiates to surrender the city in return for a three-day truce, free withdrawal into the citadel, and free evacuation of wounded to Douai. Eugene agrees. Trenches opened against the Citadel (5,000) on Esplanade (L).

French Attempted Distraction to save Citadel, 24 October: Massacre of garrison of Leffinghe (M), who were surprised while 'drunk with joy at news from Lille. (The supply position of the Allies was still critical, but less munitions were now required.) The French try to continue strategic interdiction (decision of 3 November) but time is running out. Newly-arrived Elector of Bavaria decides to create a strategic diversion by marching on Brussels, hoping Marlborough will follow to save the city. 22 November: the French are before Brussels (N); but seven days of bombardment and assault fail to cow the garrison. Marlborough, meantime, concerts a countermove.

The recapture of the Scheldt crossings threatens the Elector's links with France. On 26 November, four Allied columns storm Gavre, through Oudenarde, Kerkhoff and Hauterive; all are captured by 27 November (O). This action restores Allied communications over the Scheldt and places the Elector's in danger. The latter abandons his guns and flees for Mons (P).

Surrender of the Citadel of Lille: Aware that there was now no hope of relief, Marshal Boufflers capitulates on 10 December and is allowed the full honours of war (Q).

This lengthy siege and varied associated operations tested every aspect of Marlborough's and Eugene's generalship. Over 15,000 Allied casualties were sustained, but the moral success ultimately achieved was of the greatest importance.

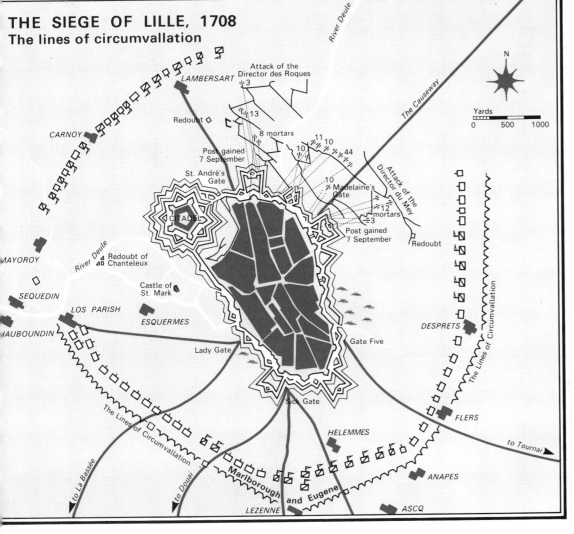

THE SIEGE OF LILLE, 1708
The lines of circumvallation

Islam versus Christianity

This time-honoured struggle gained new impetus during the seventeenth century. There were two focal areas of contention, namely the middle Danube between Vienna and Belgrade (to include lands adjoining the tributaries Drave, Raab, Save and Theiss), and the northern and western shores of the Black Sea from Azov to the River Pruth. A subsidiary area of strife (mainly naval in character) was the Aegean, centring around the Morea and Candia, which was captured by the Turks after a heroic defence in 1669.

Intermittent wars along the militargrenze or military frontiers of the Habsburg Empire took place during the first half of the seventeenth century, but Marshal Montecuccoli's great victory at St. Gotthard-on-the-Raab (1664), at the head of the forces of the Holy League created under Pope Alexander VII's auspices, led to the twenty-year Truce of Eisenburg (or Vasvar) by which Habsburg and Ottoman retained their current spheres of influence and coexisted in uneasy proximity.

Hungary was divided in two, and Transylvania was required to pay tribute to the Turk. During this lull, attention shifted northwards to Poland, where John Sobieski reversed the unfavourable trend that had brought the Turks to Lemberg. He denounced the humiliating Treaty of Buczacs, 1672, and defeated the foe at Khoczim in Bessarabia, 1673, thereby earning himself the crown of Poland, and his country a brief respite from war (1676-83). His reign (1673-96) was taken up with a crusade against the Turks in the interests of Christianity and Magyar independence. Sobieski was to prove one of the greatest men of his century, both as ruler and general. In pursuit of his cause, he was ready to ally with Russia, Venice, Persia, Sweden, France—and the Habsburg Empire.

In 1683 the supreme test came, when Sultan Muhammad IV launched the great onslaught against Vienna. The crisis was survived thanks to the intervention of Sobieski and a multi-national army, and from that time on the Porte was committed to the defensive.

Sobieski pursued the Turks and occupied Cracow. Next year, a renewed Holy League comprising Poland, Austria, Venice, Malta and, later, Russia assumed the offensive; 1686 saw the capture of Buda; and in 1687 Charles V of Lorraine won the crushing victory of Mohacs. Thereafter, the war swung out of Hungary into Croatia, Transylvania—which became an imperial field—Moldavia and Wallachia. The Venetians meantime, led by Morosini, recaptured the Morea and took Athens, 1686-87.

A new Grand Vizier, Mustafa Kuprili, checked this history of Christian triumph, retaking Nish and Belgrade, 1689-90, but in 1691 he was killed at Szlankamen by Prince Louis of Baden. The initiative restored to them, the Holy League again advanced on most fronts, a process culminating in the great Battle of Zenta in 1697 (see p. 56), which cost the Turks Serbia and Bosnia. The Peace of Karlowitz, 1699, saw Turkey renouncing all claim to Hungary and Transylvania (save the Banat of Temesvar); Kaminiec and the Western Ukraine were restored to Poland; parts of Dalmatia, most of the Peloponnese and the Aegean islands were granted to Venice; and Azov was ceded to Russia. Thus, by the end of the century the Turkish onslaught had been decisively checked, and Ottoman power in Europe was restricted to the Balkans area.

Three more struggles require mention. The Hungarian revolt of 1703-11 was the result of oppressive Habsburg rule and French intrigue, and had an important bearing on the crisis of 1704 (see p. 43), but was eventually suppressed. A new Russo-Turkish War broke out in 1710, largely due to Peter the Great's desire to extend the gains earned at Karlowitz and to the machinations of Charles XII of Sweden in exile in Bender. The result was a massive Russian débâcle at the River Pruth (see p. 61), where the Tsar was compelled to surrender with all his army. The Turkish success had been master-minded by the Swedish monarch, Russia's determined foe. The subsequent Peace of the Pruth, 1711, forced Peter to return

RETREAT OF
THE TURKISH EMPIRE,
1666-1718

Turkish possessions
Polish reconquests, 1676
Austrian territory, 1666
Vasvar Treaty limit
Austrian reconquests, 1699
Karlowitz Treaty limit
Austrian reconquests, 1718
Passarowitz Treaty limit
Venetian territory, 1686
Venetian gains
General contemporary boundary

Conquered from the Turks, 1686, but subsequently reconquered by them (1716).

Captured from Venice, 1669.
CANDIA

Miles
0 50 100 150 200 250

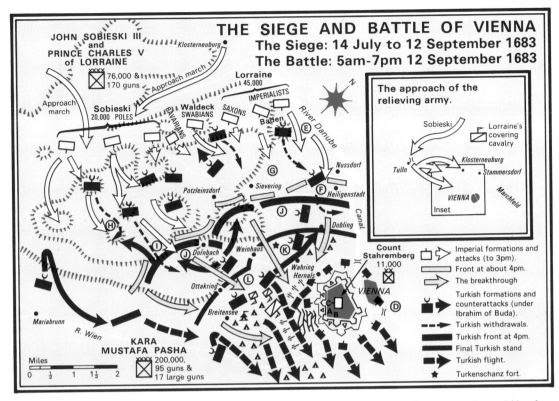

*John Sobieski III,
King of Poland
(1624-96), whose
march at the head
of a multi-national
Christian army
saved Vienna
from the Turks
in 1683, and thus
checked the tide
of Ottoman advance
in Southern Europe.*

Azov, restore Livonia and Estonia to Sweden, and to permit Charles XII a free return to his country.

The pacification was not, however, to Charles's liking, as it freed Russia from the war with the Turks, allowing her and many other enemies to fall on Sweden and its Baltic possessions. Finally, in 1716, the endemic Habsburg-Turkish wars broke out once more as the Turks attacked Venetian territory in the hope of regaining some of the ground lost at Karlowitz. The primary Turkish objective was the Morea, where they rapidly regained almost all their old losses, but the Sultan's hopes that the Emperor might be induced to remain neutral were dashed by Austria's defensive alliance with Venice in April, 1716. The Turks promptly declared war on the Emperor. Soon the fortress of Peterwardein was closely besieged by the Turks, but Prince Eugene won a great victory before the town on 5 August. Following this up, the Prince captured Temesvar—thus freeing the Banat for the first time in 164 years —and, rejecting Turkish requests for an armistice, prepared to crown his success with the capture of Belgrade. (The campaign and battle that resulted is treated overleaf.) The subsequent Peace of Passarowitz, 1718, finally excluded the Turks from Hungary, confirming the Emperor in possession of Belgrade, the Banat of Temesvar, and parts of Bosnia, Serbia and Wallachia.

Prince Eugene was by now acknowledged as one of the greatest soldiers of his age. His brilliant career (a general at twenty-two, he commanded his first army at thirty-three) and overall record during the Turkish wars was matched by his successes in Western Europe alongside Marlborough. However, he was worsted twice: at Toulon, 1707, and at Denain, 1712.

The Siege and Battle of Vienna, 1683

In the perennial struggles against the Turks, Vienna was recognized as a vital bastion of Christendom. In 1683 it became the focal point of a critical campaign which resulted in a distinct rebuff for the Ottomans.

At the end of the twenty-year truce agreed at the Peace of Eisenburg, Sultan Muhammad IV collected some 200,000 men around Belgrade, and launched them against Vienna under command of his Grand Vizier, Kara Mustafa. On the approach of this formidable host, Emperor Leopold I left the city on 7 July and retired to Passau. He entrusted Vienna's defence to Ernst von Stahremberg and a Council of five, aided by the celebrated engineer Rimpler, who was killed during the siege.

Desperate efforts were made to repair and renovate the neglected defences and to assemble a garrison of some 11,000 men (including 3,000 citizens and student militia),

but the city was barely defensible when the first Turks appeared on the 13th. The blockade began next day on the arrival of Kara Mustafa. He at once decided to open trenches against the western defences, most particularly the Lowel (A) and Burg (B) bastions and the outlying Burg Ravelin (C). With their customary energy, the Turks made rapid progress with their digging, and were not at all daunted by two half-hearted sorties.

By early August the Turkish blockade was complete; bridges linked their main encampments around Ottakring and Wahring with the detachments to the east of the canal (D), and an elaborate maze of parallels had brought the Turks to the edge of the glacis. The firing of a large mine on 12 August inaugurated three weeks of incessant fighting for possession of the Burg Ravelin, which finally passed into Turkish hands on 2 September. A large breach already gaped in the neighbouring Lowel bastion, and the possibility of a general storm caused grave concern within Vienna. Fortunately, the Burgomaster, Johann von Lienbenberg, proved capable of rallying the morale of the civilian population.

Aid for the city was not far away. During the preceding spring the emperor had been induced to make an alliance with John Sobieski, king of the Poles, and this act of foresight enabled a multi-national force of 76,000 men and 170 guns to be rapidly assembled. Of these, 21,000 men were imperial troops, 11,300 were Bavarians (led by Max Emmanuel II), the Elector of Saxony was at the head of 10,400 men and Prince George Frederick of Waldeck was in command of 9,500 Swabians. A force of 24,200 Poles, setting out from Warsaw on 20 July, made up the numbers. Overall command was entrusted to John Sobieski, seconded by the Duke of Lorraine, Charles V. Over the same period Turkish raiders penetrated as far as Amstetten which was some 100 miles west of Vienna.

While Lorraine's cavalry dominated the Marchfeld from the vicinity of Stammersdorf, the relieving army reached Tulln on 7 September and crossed to the south bank of the Danube.

55

Islam versus Christianity

Dividing into three columns, the army moved towards Vienna through the Wienerwald. Fortunately, Kara Mustafa, who was no great soldier, took few steps to meet this development, only detailing some 29,000 Turks to face the relieving army. Early on the morning of 12 September, after attending Mass at the chapel of St. Joseph on the Kahlenberg, Sobieski ordered his forces to attack the Turkish positions. At 5am the battle began.

The Duke of Lorraine, commanding the Christian left wing, set about capturing the Nussberg. (E), hoping then to penetrate the Turkish lines near Nussdorf and roll them up. After a bitter struggle these positions fell into Allied hands, and soon after midday the Turks found themselves driven back behind the Schrieberbach (F). Meantime, the Saxons had been attacking towards Sievering and Potzleinsdorf (G), and the Poles had stormed the 1,200ft.-high Gallitzinberg (H). This last sector however, also witnessed a determined Turkish counterattack (I) which delayed the development of the Christian battle plan.

Not for long, however; by late afternoon the new Turkish line based upon Dornbach and the southern bank of the Krottenbach (J) had been pierced on all sectors, most seriously in the northern area, and soon the Turks were fighting on the edge of their tented camp. A final series of onslaughts—one highlight of which was the storming of the Turkenschanz fort by the Saxon contingents (K)—caused the Turkish line to waver. At the correct moment Sobieski launched his massed 20,000 cavalry on to the open plain between Breitensee and Hernals (L), and the day was all but won.

After one last furious assault against the ruined fortifications of the Lowel-Bastion, Turkish cohesion at last broke, and very soon a mass flight towards the south and east had begun. The fourteen-hour battle had cost the Christian forces some 2,500 casualties, and the Turks no less than 15,000. The losses sustained by the garrison and militia during the sixty-day siege (during which the Turks had fired some 100,000 cannon balls into Vienna) had been in the vicinity of 5,000; no figures exist for Turkish casualties over the same period.

The saving of Vienna inaugurated another long round of wars in southern Europe, but the Turks never quite regained their ancient reputation as being masters of siege warfare.

The Siege and Battle of Belgrade, 1717

Belgrade—strategically placed at the confluence of the Danube and Save, the key to Hungary and the neighbouring Banat of Temesvar—was the scene of four dire struggles in the thirty-odd years between 1688 and 1717.

A The Turkish cavalry and a section of infantry advance towards Transylvania after passing the River Theiss.
B Turkish bridge (60 boats) and depot areas.
C On the evening of 11 September the Turkish infantry are still west of the river within incomplete fortifications.
D Eugene divides his forced march to catch the Turks into two parts by the Theiss.
E The Imperialist left envelops the Turkish right, takes the sandbank in the river and enfilades the bridge.
F Heavy attacks are launched by the Imperialist centre and right; Heister penetrates.
G The Turks flee.
H The Turkish camp is captured on 12 September.
Casualties
Turkish: 20,000 killed, 10,000 drowned.
Imperialists: Approximately 500 killed and wounded.

THE BATTLE OF ZENTA
11 September 1697

Prince Eugene surprised the Sultan Mustafa II's Army while they were crossing the Theiss, and compelled action by a forced march. He then decimated the Turkish forces on the west bank for minimal loss. Zenta is widely considered to have hastened the Turkish military decline which had begun at the Siege of Vienna in 1683.

Captured from the Turks by Max Emmanuel of Bavaria on 6 September 1688 (after a twenty-six day siege), it returned to Ottoman hands two years later after being stormed by Mustafa Kuprili on 8 October 1690. A weak Turkish garrison successfully withstood a forty-nine day siege by General Croy in 1693, and it was not until 1717 that the Imperialists were again in a position to attempt its recapture.

After his successes at Peterwardein (5 August 1716) and during the consequent reconquest of the Banat, Prince Eugene of Savoy spent the winter in the camp of Futak, where he laid plans for an expedition against Belgrade.

The next spring an imposing army of over 100,000 men and 200 guns set out for their objective, but moved on it from an unexpected direction. Instead of advancing directly over the River Save, Eugene first marched east along the north bank of the Danube, crossed the great river at Pancsoya, and then doubled back to the west. The garrison commander, Mustafa Pasha, could call on the services of 30,000 troops, and was further equipped with some 600 cannon of various calibres and 70 boats. The Imperial forces first appeared outside the city and fortress—enclosed within a tight triangle between the Danube and the Save—on 29 June 1717. Only on 22 July, however, were the lines of contra- and circumvallation completed owing to very unfavourable weather and disease that was soon ravaging the Imperial camps. These circumstances, together with the prosecution of a very active defence by the garrison (e.g. the successful sortie of 17 July), provided Grand Vizier Khalil Pasha with adequate time to set out from Adrianople at the head of a relieving army of 150,000 men (including no less than 80,000 mounted troops) and 120 guns and mortars.

Eugene clung to his positions, bombarding Belgrade incessantly—although his army had shrunk to half its original size—hopeful that he would have time enough to complete the siege before being called upon to face Khalil. However, this confidence was not borne out by events, and by 1 August the Grand Vizier's army had completely surrounded the Imperial Army: a case of the besieger being besieged. Within a short time Eugene's positions were being subjected to a terrible bombardment from two directions, while large Turkish mines were exploded under his lines of circumvallation. It was soon clear that the Turks were in no hurry to storm the positions, but were content to starve their foes into surrender.

The summer temperatures made the Imperial position barely tenable, and Prince Eugene reached the conclusion that it had to be a case of 'risk all to gain or lose all'. On 15 August he issued orders for a major attack against the Turkish lines on the morrow, stressing the need for exact compliance and absolute avoidance of looting. The attackers were to be drawn up in two lines, comprising 52 battalions, 53 grenadier companies and 180 squadrons, supported by some 60 cannon—which were to move to the flanks in two equal detachments at dusk. Spare officers and gunners were to accompany the infantry, ready to take over and operate captured Turkish artillery. Only 8 battalions were left as a trench guard, supplemented by all cavalrymen and dragoons who had lost their horses.

At 1am the cavalry of the wings began to file out of their positions covered by a dense mist, followed by the infantry at about 3am. Absolute silence and secrecy were maintained, until part of the infantry stumbled into some new Turkish trenches which gave the alarm. For some time the Imperial right wing floundered in the fog, and soon the cavalry were in serious trouble; the neighbouring infantry moved to aid their compatriots, and a gap appeared in the Imperialist front line. Very soon large numbers of Turks were pouring into this cavity, increasing the confusion. Only at 8am, as the fog rose, did Eugene become fully aware of this danger, and quickly he ordered up the second line to retrieve the situation, which was successfully accomplished.

On the Imperialist left, meanwhile, the battle opened rather later than originally planned. Severe casualties were suffered from Turkish fire from the Grand Battery. Shortly after 9am this position was stormed by Bavarian troops,

THE SIEGE AND BATTLE OF BELGRADE
The Siege: 29 June to 18 August 1717
The Battle: 16 August 1717

and soon the Turkish guns on the Bajdina heights were in action against their late masters. This caused the Turks to flee on this sector, and by 10am they were in full retreat on every part of the field. Tremendous booty fell into the hands of the Imperial forces. Two days later the city of Belgrade was surrendered to Prince Eugene in return for permission for the garrison to march out with the honours of war

to be repatriated.

The Austrians lost some 5,400 casualties in the battle (including 17 generals) besides perhaps as many as 30,000 who died of wounds and illness during the siege. The Turks suffered some 20,000 casualties in the battle, besides 5,000 killed and wounded in the siege. This defeat represented a grave blow to Turkish martial prestige.

Siege weapons of the early eighteenth century: from left to right, a 10in. mortar, a 24-pounder gun on travelling carriage; and an 8in. howitzer on travelling carriage. (These models, in the Rotunda Museum of Artillery at Woolwich, are of the Duke of Marlborough's siege train.)

Charles XII and the Great Northern War

What the great Gustavus Adolphus had built, his later successor Charles XII ended by destroying. The restless genius of this young King—he was only fifteen when he came to the throne in 1697 and but twenty-six when he undertook his fatal invasion of Russia—dominated northern and eastern Europe for almost two decades. As king he controlled both the policy and the resources of his country; as a general, he led Sweden's forces in person to great triumphs and even greater cataclysms. He was the spirit of the offensive incarnate—a master of surprise and deception, the deviser of new tactical forms—who proved more than a match for the commanders and armies of Russia, Poland and Saxony, which, together with Denmark, comprised his main foes.

The Great Northern War began in 1700 and ended in 1721 (see map). Poland and Denmark started the struggle by attacking Swedish possessions in early 1700, and under cover of these diversionary attacks Peter the Great, who ascended the uneasy Russian throne in 1689 at the age of seventeen, led an army of 40,000 towards Narva, only to be ignominiously thrashed by Charles's far smaller army.

Charles XII then turned to rend the Poles; an action that gave the Tsar time to recover from his débâcle. The Swedes, meanwhile, managed to crush their lesser opponents in a series of campaigns that ended in 1706. However, the need to garrison these conquests and their earlier possessions tied down all of 65,000 men, and the strain of this military effort proved too great a burden for tiny Sweden. Anxious in case French intrigue should induce Charles to intervene in the War of the Spanish Succession, Marlborough visited Altranstadt in 1707 and dissuaded the headstrong young monarch, but attempts to patch up the dispute with Russia (actually instigated by Peter) came to naught. Accordingly, late in 1707 Charles resumed active operations against Russia, and early in 1708 occupied Grodno. Once there, he faced a strategic choice: either to head for St. Petersburg (which was in the process of being built), or for Moscow. He opted for the latter course, and marched deep into Russia with 35,000 men, summoning General Lewenhaupt with 15,000 men and a vital supply convoy to march from Latvia to join him.

The Tsar's forces fell back before the Swedish invasion, trading space for time. As he advanced, Charles's communications lengthened, until, in July 1708, he came to a halt near Moghilev. There he awaited Lewenhaupt's convoy to renew his offensive capability. After a month's fruitless delay, Charles swung south, heading for the fertile region of the Ukraine where he hoped to find

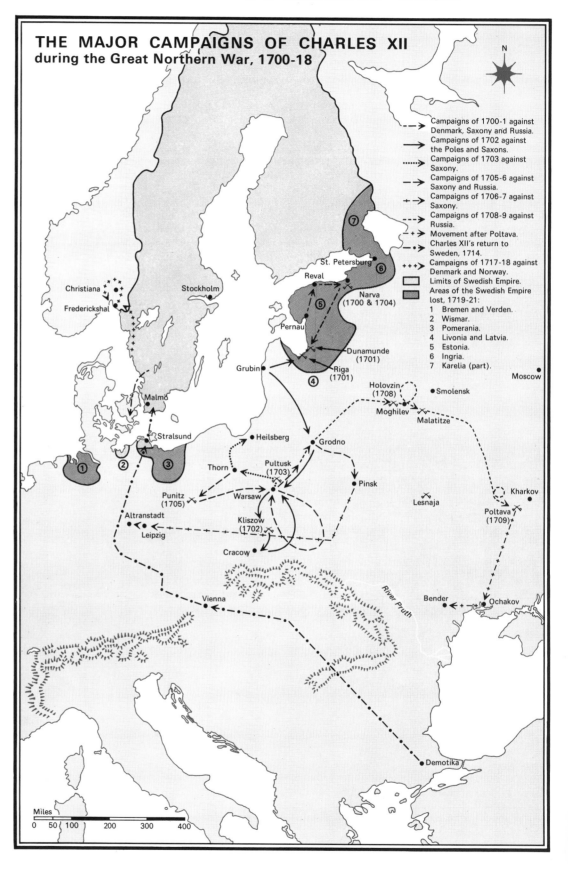

THE MAJOR CAMPAIGNS OF CHARLES XII during the Great Northern War, 1700-18

N

Campaigns of 1700-1 against Denmark, Saxony and Russia.
Campaigns of 1702 against the Poles and Saxons.
Campaigns of 1703 against Saxony.
Campaigns of 1705-6 against Saxony and Russia.
Campaigns of 1706-7 against Saxony.
Campaigns of 1708-9 against Russia.
Movement after Poltava.
Charles XII's return to Sweden, 1714.
Campaigns of 1717-18 against Denmark and Norway.
Limits of Swedish Empire.
Areas of the Swedish Empire lost, 1719-21:
1 Bremen and Verden.
2 Wismar.
3 Pomerania.
4 Livonia and Latvia.
5 Estonia.
6 Ingria.
7 Karelia (part).

Christiana
Frederickshal
Stockholm
St. Petersburg
Reval
Narva (1700 & 1704)
Pernau
Dunamunde (1701)
Grubin
Riga (1701)
Holovzin (1708)
Smolensk
Moghilev
Malatitze
Moscow
Malmö
Heilsberg
Grodno
Stralsund
Thorn
Pultusk (1703)
Pinsk
Punitz (1705)
Warsaw
Lesnaja
Kharkov
Altranstadt
Kliszow (1702)
Poltava (1709)
Leipzig
Cracow
Vienna
River Pruth
Bender
Ochakov
Demotika

Miles
0 50 100   200   300   400

both supplies and support; the latter in the form of the disaffected Cossack tribesmen led by Mazeppa. Peter, realizing that Moscow was now safe, moved south with his main army on a line parallel to the Swedes—after leading 12,000 men to intercept the dilatory Lewenhaupt. On 29 September a bitter battle at Lesnaja ended in Lewenhaupt's defeat and the loss of his precious convoy. Russian morale soared and that of the Swedes correspondingly dropped; the arrival of Lewenhaupt to join the army in person did little to ease the situation, as he was soon at loggerheads with his brother generals. The tide had turned against Charles: Mazeppa proved a broken reed, and the bitter winter of 1708-9 hastened the disintegration of the Swedish Army.

The following spring Charles tried to resume his advance east, but was headed off by mobile Russian columns. At length, Charles decided to besiege Poltava on the Vorskla. Peter at once began to mass his main army for a decisive struggle. The outcome was the Battle of Poltava on 28 June 1709, which resulted in the crushing defeat of the outnumbered Swedes.

This catastrophe effectively doomed Sweden to decline and ultimate eclipse. Charles escaped with a few followers to find uneasy sanctuary at Bender in Turkey. When relations were good he even campaigned with the Turks as an adviser, although the Peace of the Pruth in 1711 was a major setback; when they were bad, he had to fend off his hosts by dint of arms. But until 1714 he was largely separated from his country and compatriots. In that year, however, he escaped from Demotika and conducted an adventurous journey across Europe by way of Vienna to reach Stralsund and Swedish territory.

During his absence, Swedish fortunes had been ably safeguarded, so far as was possible, by Charles's generals and ministers. With the return of the King, little wiser from his experiences, a new spirit of offensive concepts once more predominated. Both Prussia and Austria were now hostile, in addition to earlier foes, and Swedish resources in men and material were just not equal to the strain placed upon them. The years 1717 and 1718 saw new Swedish campaigns in Denmark and Norway, but these represented Sweden's last initiatives. Thereafter, the rot set in rapidly, and province after province on the south and east shores of the Baltic were progressively lost. Eventually, Charles was killed at Frederikshal on 30 November 1718—traditionally by a silver bullet. With his death, Sweden's claims to martial predominance lapsed once and for all, and in 1721 the Peace of Nystadt brought the long struggle to an

THE BATTLE OF NARVA
20 November 1700

unmourned conclusion. The century of Swedish power was over.

The main beneficiary was Peter the Great. This monarch was no natural soldier or scholar, but he was an outstanding if ruthless statesman. He devoted his reign to modernizing his realm —by means of draconian reforms—dragging it from the Middle Ages into a form of modernity. He made the best use of West European practice and expertise in the process. The boyars were tamed; the streltsi (or hereditary warriors) eliminated; new forces, both on sea and land, were recruited and trained. German, French and Scottish experts laboured to good effect under Peter's supervision, and the outcome was the creation of a viable modern state. The first eight years of contest with Sweden held many defeats, but Peter could learn from his mistakes. The year 1709 saw the turn in martial fortunes. Thereafter, despite such disasters as the River Pruth capitulation in 1711 (see p. 61), which cost the Tsar his foothold on the Baltic, Russia was making progress. As Peter's long reign drew on, great progress was made in expanding the realm at the expense of its southern and eastern neighbours. Before his death in 1725, Peter was able to deploy a modern army of some 200,000 men and a fleet of over 30 capital ships.

There we have the contrast between Peter and Charles: respectively, creator and destroyer of their nations' interests. Charles's skill at battle tactics, which were based upon the maximum application of shock and force (he improved the platoon firing system, and developed cavalry attacks à l'outrance in vast wedge-shaped formations), was outweighed by his disregard for economic and logistical factors. At the height of the Great Northern War, Sweden—with a population of little over a million—was called upon to maintain an army of 110,000 men. Such a disproportionate effort placed an intolerable strain on Sweden's limited resources, even in the years of military success, and this inevitably hastened her decline from great-power status. But Swedish valour— and above all that of its unbalanced but brilliant monarch, whose personal austerity and toughness were legendary even in his own time —would be long remembered.

The Battle of Narva, 1700
Fought in a blizzard on 20 November 1700, this battle established Charles XII's reputation as a

Charles XII and the Great Northern War

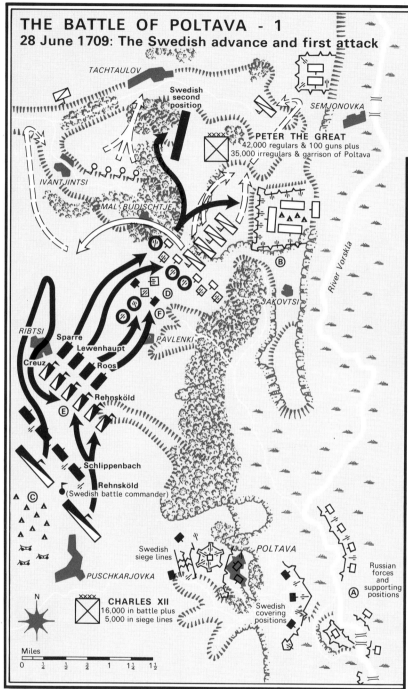

THE BATTLE OF POLTAVA - 1
28 June 1709: The Swedish advance and first attack

TACHTAULOV

Swedish second position

SEMJONOVKA

IVANTJINTSI

MAL BUDISCHTJE

PETER THE GREAT
42,000 regulars & 100 guns plus 35,000 irregulars & garrison of Poltava

B

JAKOVTSI

River Vorskla

RIBTSI

Sparre
Lewenhaupt
Creuz
Roos

PAVLENKI

D
F

Rehnsköld

E

Schlippenbach

Rehnsköld
(Swedish battle commander)

C

Swedish siege lines

POLTAVA

PUSCHKARJOVKA

Russian forces and supporting positions

Swedish covering positions

A

N

CHARLES XII
16,000 in battle plus 5,000 in siege lines

Miles
0   ¼   ½   1   1¼   1½

1. Peter the Great, Tsar of Russia (1672-1725). His modernization of Russia along Western lines laid the foundations for his country's expansion.
2. Charles XII, King of Sweden (1682-1718), led his country into bitter wars against Poland, Denmark and Russia. The victor of many battles, he was decisively defeated at Poltava (1709), and killed at Frederickshal.

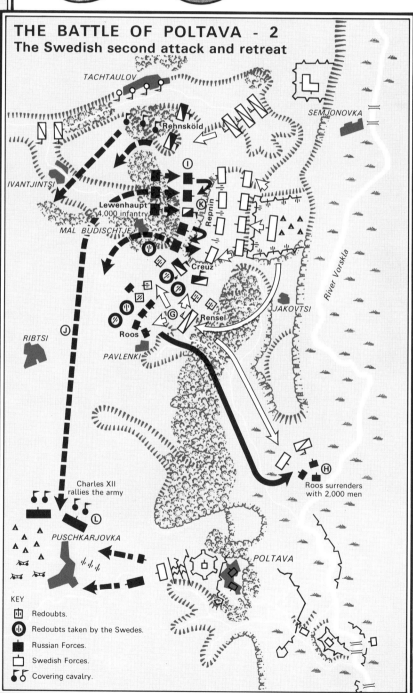

THE BATTLE OF POLTAVA - 2
The Swedish second attack and retreat

TACHTAULOV

SEMJONOVKA

Rehnsköld

IVANTJINTSI

Lewenhaupt
4,000 infantry

MAL BUDISCHTJE

I
K
Repnin

Creuz

G
Rensel

JAKOVTSI

River Vorskla

RIBTSI

J

Roos

PAVLENKI

H
Roos surrenders with 2,000 men

Charles XII rallies the army

L

PUSCHKARJOVKA

POLTAVA

KEY
- Redoubts.
- Redoubts taken by the Swedes.
- Russian Forces.
- Swedish Forces.
- Covering cavalry.

brilliant commander at the age of nineteen years. To relieve the Swedish garrison of Narva, Charles landed his army of 10,000 men nearby and immediately attacked the Russian siege lines which were manned by all of 40,000 troops. Using the driving snow to conceal his approach, Charles surprised the Russian outposts. Then, forming his cavalry into massive wedge-shaped formations, he drove into the midst of the Russian position.

Total confusion and panic gripped the Russians, who reputedly left 10,000 dead as they fled from the scene. For his part, Charles lost barely 1,000 men. Narva was thus relieved. Charles marched south to relieve Riga. By mid-June 1701 this had also been achieved.

The Battle of Poltava, 1709

This celebrated battle on the River Vorskla deep in Russia was a notable victory for Peter the Great and a crushing defeat for Swedish military and political pretensions, which thereafter began to decline.

On the day of battle, Charles XII fielded 18 battalions of infantry, some 12 squadrons of horse and a handful of guns: approximately 16,000 men when excluding a force of 5,000 besieging the town of Poltava. Against him were ranged Peter's 40,000 trained troops, almost as many irregulars, and over 100 cannon; some 30 infantry regiments and as many squadrons of cavalry making up the trained element of his army.

Charles had been besieging the town since May. Peter's relieving army came within range on 4 June, after suppressing the Ukrainian and Don Cossacks. After raising field-works opposite Poltava (A) the Tsar's main encampment (B) was built in the form of a large fortified rectangle about three and a half miles from the Swedish camp (C). The only open approach to the Russian position was defended by a series of ten redoubts (D), which were completed by the 25th. (This was the day on which Charles received the wound that effectively confined him to a litter.)

Despite the advice of his generals, Charles refused to contemplate a retreat, and instead ordered an all-out frontal attack on the Russian position, deputing the command to General Rehnsköld. Hoping that a combination of surprise and desperation might snatch a decisive victory, the Swedes drew up, (E). Advancing after midnight, they tried to rush the Russian redoubts as a preliminary to scattering the Russian cavalry to their rear and then penetrating into the main encampment.

However, lacking Charles's inspired leadership (although he accompanied his army), and bedevilled by temperamental clashes dividing its senior commanders, the Swedish Army's dawn attack proved only partially successful. Five redoubts were seized but the remainder held out, and this gave the Tsar time to alert his camp. General Roos, commanding the Swedish right, allowed his men to become tied down amongst the redoubts instead of charging through them as ordered (F). Consequently, a gap grew between his wing and the rest of the army, which made rapid progress—Creuz's

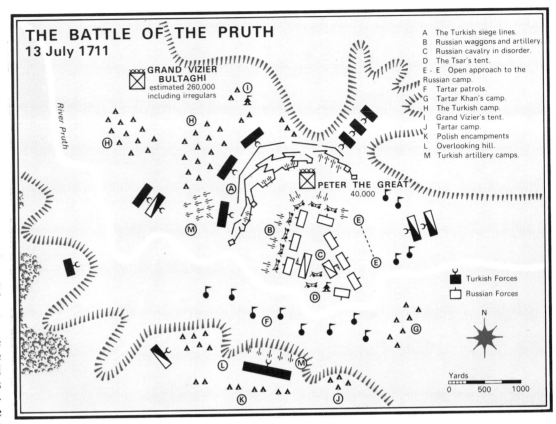

THE BATTLE OF THE PRUTH
13 July 1711

A The Turkish siege lines.
B Russian waggons and artillery.
C Russian cavalry in disorder.
D The Tsar's tent.
E - E Open approach to the Russian camp.
F Tartar patrols.
G Tartar Khan's camp.
H The Turkish camp.
I Grand Vizier's tent.
J Tartar camp.
K Polish encampments
L Overlooking hill.
M Turkish artillery camps.

GRAND VIZIER BULTAGHI estimated 260,000 including irregulars

PETER THE GREAT 40,000

River Pruth

Turkish Forces
Russian Forces

Yards
0   500   1000

cavalry scattering the Russians.

However, the division of their forces would prove the fatal Swedish flaw. General Rensel and 10,000 Russians soon isolated Roos (G), who mistook their approach for a Swedish manoeuvre until it was too late. Ultimately, he was forced to surrender after an abortive attempt to break out to the east (H).

The main Swedish army, meanwhile, had halted on the plain facing Peter's main position (I), but was now bereft of both cannon and all hope of reinforcement. Lewenhaupt had challenged the need to halt and was dangerously close to the Tsar's entrenchments. Charles ordered the dubious Rehnsköld to mount an all-out attack rather than attempt to reach Roos, but then untypically gave way to his battle-commander's view. The Swedes then began to fall back southwards (J), but at the outset, as they moved across the front of Peter's 40,000 newly-deployed troops, they were left with no alternative but to attack their numerically superior opponents (K).

Forty Russian guns decimated their ranks with case-shot as the Swedes gallantly but hopelessly tried to press home their attack. Once this had failed, the Swedish Army began a rapid flight towards its camp where it was rallied by its wounded monarch (L). Charles was borne from the field in his litter, accompanied by Mazeppa and his escort, ultimately heading for Bender. Lewenhaupt and 14,000 men capitulated. For a loss of 1,300 casualties, Peter the Great had killed 7,000 Swedes and virtually eliminated their army.

The Battle of the Pruth, 1711

Swedish preoccupations in Poland enabled Tsar Peter to reorder and retrain his armies. Finally, as has been related, he won the important Battle of Poltava in 1709. However, the deficiencies of even this reorganized Russian Army were tellingly displayed just two years later at the River Pruth on 13 July 1711. Too confident after his success over the Swedes, Peter the Great invaded Moldavia at the head of 60,000 men. Faced by a far larger Turkish Army led by the Grand Vizier Bultaghi, the Russians were pressed into a hopeless position with their backs to the River Pruth. Peter's starving troops dug entrenchments and created a waggon laager, but Bultaghi chose to negotiate rather than attack. Although Peter avoided the humiliation of a formal surrender, which must have followed had the Turks besieged him, the terms of the Treaty of the Pruth (completed on 21 July 1711) were grave enough. The key fortress of Azov in the Crimean region was returned to the Turks and a number of other major cities were dismantled.

Colonial Strategy

The far-flung colonial empires established by the Spanish and Portuguese in the sixteenth century were soon being emulated or attacked by the Dutch, French and English during the seventeenth. The most important areas of contestation became North America and South-East Asia, although other areas such as the West Indies and the slave coast of West Africa also became involved in the struggle.

Initially, the instrument of colonial expansion was often the great commercial concerns such as the rival East India Companies, English and Dutch, or the London and Plymouth Companies, both founded in 1606. World exploration was also considerably furthered by the various searches for shorter routes to the Spice Islands and Asia—by way of the north-west or north-east passages for example. All colonies were primarily regarded as sources of raw materials.

In the East, the English East India Company set up tenuous factories on the east coast of India, but all attempts to penetrate the Spice Island area were foiled by the Dutch; the so-called 'massacre' of Amboyna, 1623, restricted English traders to the Indian sub-continent, and was one cause of the hostility between England and the United Provinces, eventually leading to the three Dutch wars. The Dutch, meantime, established a monopoly of the lucrative spice trade, excluding Portugal from her former spice islands.

In the western hemisphere, rivalry centred around the eastern seaboard of North America. Here, Spaniards, French, English, Swedish and Dutch interests all came into conflict, ending with the elimination of the two last mentioned; the Dutch swallowed up the Swedish colony at the mouth of the Delaware River in 1655, only to share the same fate at the hands of the English in 1664 when New Amsterdam along the Hudson was captured and renamed New York. The Spaniards remained in control of Florida throughout the period in spite of a number of expeditions against them, and soon the main hub of activity centred around French and English possessions farther north.

The French based their possessions along their explorations of the St. Lawrence and Mississippi rivers; New France having its capital at Mont Royal, and the later acquisition, Louisiana, being ruled from New Orleans. The French proved excellent missionaries, traders, explorers and backwoodsmen, but were not particularly good as permanent settlers. They first came into conflict with the sparse English fishing settlements in Newfoundland (1583) and, after its foundation by Charles II in 1670, with the Hudson's Bay Company and its fur-trading posts.

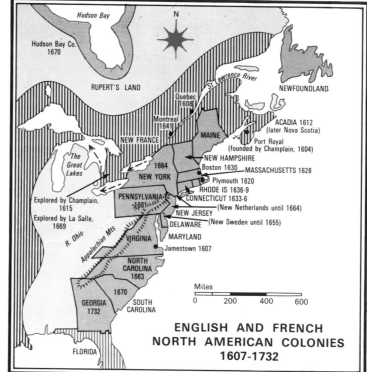

ENGLISH AND FRENCH NORTH AMERICAN COLONIES 1607-1732

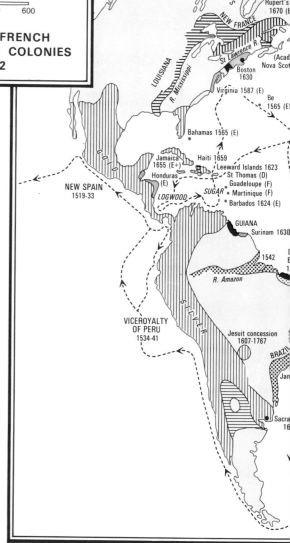

The main area of English colonization was restricted to the coastal region by the Appalachian Mountains to the west; to north and south were hostile Indian tribes, and French and Spanish settlements. The early wealth-seekers were soon replaced by the genuine settlers who for a variety of reasons desired to make their permanent homes in the New World.

The English colonies comprised three groups. Firstly, the New England Colonies: settled by religious refugees, particularly New Plymouth (1620) and Massachusetts (1628); secondly, the Middle Colonies: New York, New Jersey, Delaware and Pennsylvania (1681); and thirdly, the five 'Old' or Plantation Colonies: Virginia (1606), Maryland (1632), North and South Carolina (1663) and Georgia (1732). Their level of development was very unequal, but by 1710 most were beginning to prosper.

Military rivalry between Britain and France was relatively limited in scope. The Atlantic crossing took anything from four to seventeen weeks according to the season, and the respective military presences in North America remained small, being largely based on colonial militias of variable quality. 'King William's War'—the local name for the struggle in North America which formed part of the Nine Years War—saw the first battle of Port Royal in Acadia (Nova Scotia), which changed hands twice in as many years (1690-91) and was retained by France at the Peace of Ryswick in 1697. 'Queen Anne's War' saw three attacks on Port Royal, 1704, 1707 and 1710; the third proving successful, whereupon the town was renamed Annapolis. The French, meantime, had seized the Hudson Bay posts, but these were returned in 1713. (The effects of the Peace of Utrecht on the region are summarized on p. 64.) The eighteenth century would see a far more significant contest for control of Canada.

British island possessions included Bermuda (1609), and in the West Indies the Leeward Islands (1623), Barbados (1624) and Jamaica, which was seized from Spain by William Penn and Admiral Venables in 1655 after the repulse of their attack on Hispaniola.

A certain amount of skirmishing also took place between rival forts on the west coast of Africa, the English occupying Cape Coast Castle in 1672.

In sum, colonial considerations remained a relatively minor matter in the minds of European statesmen, but the seeds of later, larger conflicts were sown.

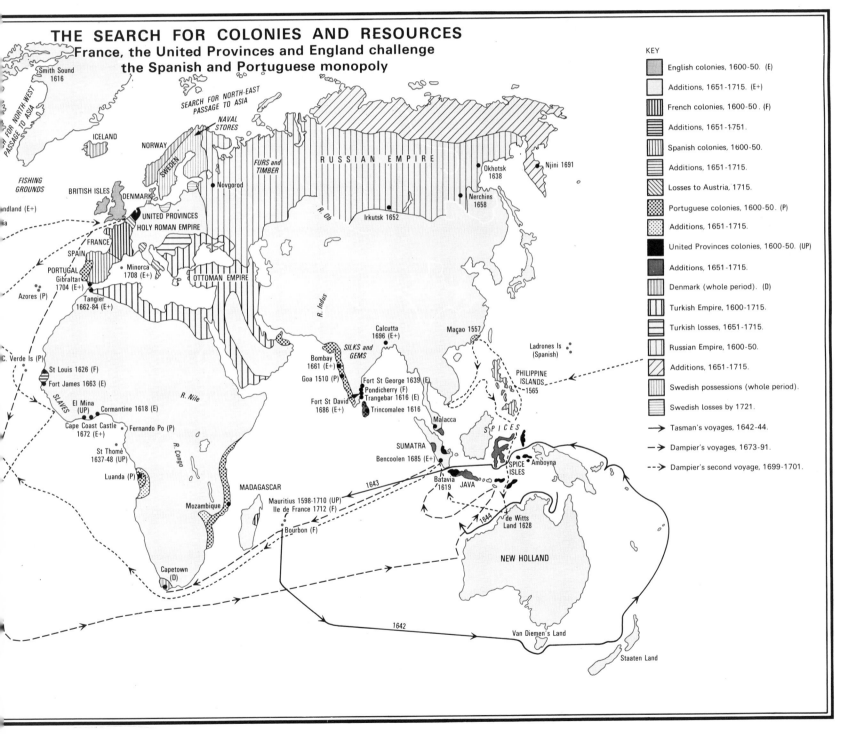

THE SEARCH FOR COLONIES AND RESOURCES
France, the United Provinces and England challenge the Spanish and Portuguese monopoly

KEY

- English colonies, 1600-50. (E)
- Additions, 1651-1715. (E+)
- French colonies, 1600-50. (F)
- Additions, 1651-1751.
- Spanish colonies, 1600-50.
- Additions, 1651-1715.
- Losses to Austria, 1715.
- Portuguese colonies, 1600-50. (P)
- Additions, 1651-1715.
- United Provinces colonies, 1600-50. (UP)
- Additions, 1651-1715.
- Denmark (whole period). (D)
- Turkish Empire, 1600-1715.
- Turkish losses, 1651-1715.
- Russian Empire, 1600-50.
- Additions, 1651-1715.
- Swedish possessions (whole period).
- Swedish losses by 1721.
- → Tasman's voyages, 1642-44.
- → Dampier's voyages, 1673-91.
- → Dampier's second voyage, 1699-1701.

The Settlements of 1713 and 1721

STRATEGIC EFFECTS OF THE TREATIES OF 1713 to 1721
The Treaties of Utrecht (1713), Rastadt (1714), Passarowitz (1718) and Nystadt (1721)

NORWAY

SWEDEN

FINLAND ← KARELIA

St Petersburg
INGRIA

ESTONIA

LIVONIA

• Moscow

DENMARK

COURLAND

• Smolensk

IRELAND

GREAT BRITAIN

UNITED PROVINCES

LITHUANIA

R U S S I A

France recognises Hanoverian succession.

Dutch Fortress Barrier established with Austrian co-operation.

Prussian monarchy confirmed; Spanish Guelderland ceded to Prussia.

EAST PRUSSIA

Scheldt estuary closed to traders.

London •

HANOVER

PRUSSIA

• Warsaw

P O L A N D

Trade restrictions equalized.

Dunkirk demilitarized.

The Hague

AUSTRIAN NETHERLANDS

SAXONY

SILESIA

VOLHYNIA

• Kiev

• Pultava

France recovers Lille.

THE EMPIRE

BOHEMIA

• Paris

For French overseas cessions see inset.

French conquests east of the Rhine returned to the Empire; in return Alsace and Strasbourg confirmed to France.

BAVARIA Elector restored.

MORAVIA

GALICIA

PODOLIA

Vienna •

AUSTRIA

Ceded to Russia, 1699. Regained by Turkey, 1711. (Peace of the Pruth)

FRANCE

SWITZERLAND

TYROL

HUNGARY

• Azov

France restores Nice to Savoy.

SAVOY

VENICE

CARINTHIA

CROATIA

Mohacs •

TRANSYLVANIA

WALLACHIA

MOLDAVIA

Sea of Azov

Independence guaranteed

SPAIN

MILAN

Zenta

Salankemen

BANAT OF TEMESVAR

Nice

GENOA

PARMA

DALMATIA

Madrid •

TUSCANY

BOSNIA

Belgrade

BLACK SEA

Philip V confirmed as King, but Spain never to be joined to France; Also trade concessions granted to Gt. Britain. (see inset)

CORSICA

PAPAL STATES

R. Danube

OTTOMAN EMPIRE

Lisbon •

PORTUGAL

Rome •

Adrianople •

Constantinople •

Gibraltar •

MINORCA

Spanish bases ceded to Great Britain.

SARDINIA (eventually exchanged for Sicily, 1720).

NAPLES

To Savoy (exchanged for Sardinia, 1720).

GREECE

ASIATIC OTTOMAN EMPIRE

SICILY

CANDIA

COLONIAL IMPLICATIONS OF UTRECHT

Hudson Bay restored to Britain.

NEWFOUNDLAND confirmed to Britain (but French fishing rights preserved).

ACADIA (NOVA SCOTIA) ceded to Britain.

Spain permits one British ship (500 tons of goods) from Porto Bello each year.

St Kitts ceded to Britain.

Spanish 'New World' possessions confirmed.

Porto Bello

FRENCH GUINEA ceded to Portugal.

Spain grants British the right to ship 4,800 slaves to Spanish America each year.

BRAZIL

British possessions.

Spanish possessions.

Portuguese possessions.

France.

STRATEGIC EFFECTS OF THE TREATIES OF UTRECHT AND RASTADT

1   Maintained the balance of power by averting the possibility of a Bourbon Empire; adjusted Westphalian provisions.
2   The neutralization of the Scheldt favours British commerce; as a result, Antwerp is ruined as an entrepôt.
3   French concessions in North America and Spanish trade agreements in favour of Gt. Britain increase the likelihood of future colonial rivalry on a larger scale.
4   The Dutch are satisfied with the Fortress Barrier as protection against France (see p. 36).
5   Gt. Britain welcomes the possession of the former Spanish Netherlands by a ' neutral' power.
6   Prussia emerges as a European power.
7   French frontiers are clearly defined.
8   British interest in the Mediterranean is confirmed by the ceding of key bases to them.
9   Franco-German hostility continues to threaten future peace.

Passarowitz ended the Turkish menace to western Europe once and for all. Nystadt implied the end of Swedish status as a major European power.

KEY

Austrian Habsburg Empire, from 1714.

Areas surrendered by Turkey to Austria. 1718 (Passarowitz).

Swedish possessions ceded in 1721 (Nystadt).

Note: text on the map shows the terms of the Treaties of Utrecht / Rastadt.

Miles
0   100   200   300   400   500

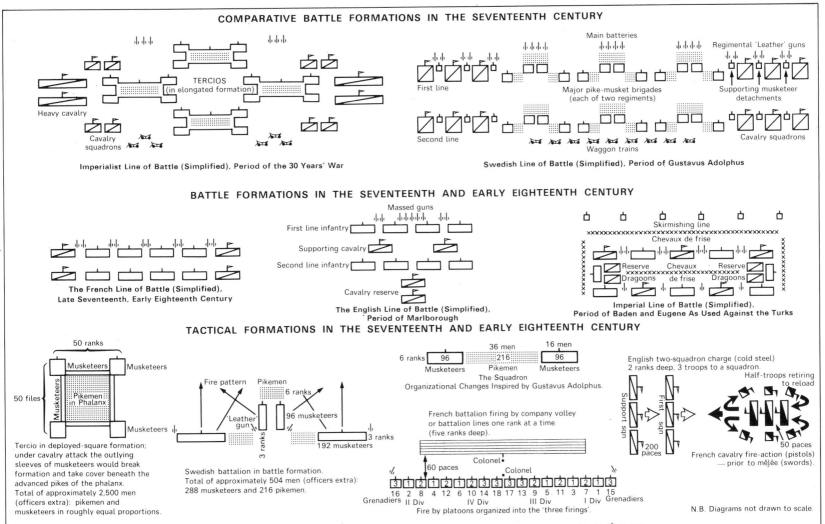

COMPARATIVE BATTLE FORMATIONS IN THE SEVENTEENTH CENTURY

TERCIOS
(in elongated formation)

Heavy cavalry

Cavalry squadrons

Imperialist Line of Battle (Simplified), Period of the 30 Years' War

Main batteries

Regimental 'Leather' guns

First line

Major pike-musket brigades (each of two regiments)

Supporting musketeer detachments

Second line

Waggon trains

Cavalry squadrons

Swedish Line of Battle (Simplified), Period of Gustavus Adolphus

BATTLE FORMATIONS IN THE SEVENTEENTH AND EARLY EIGHTEENTH CENTURY

The French Line of Battle (Simplified), Late Seventeenth, Early Eighteenth Century

Massed guns

First line infantry

Supporting cavalry

Second line infantry

Cavalry reserve

The English Line of Battle (Simplified), Period of Marlborough

Skirmishing line

Chevaux de frise

Reserve Dragoons

Chevaux de frise

Reserve Dragoons

Imperial Line of Battle (Simplified), Period of Baden and Eugene As Used Against the Turks

TACTICAL FORMATIONS IN THE SEVENTEENTH AND EARLY EIGHTEENTH CENTURY

50 ranks

Musketeers | Musketeers

50 files | Musketeers | Pikemen in Phalanx

Musketeers

Tercio in deployed-square formation; under cavalry attack the outlying sleeves of musketeers would break formation and take cover beneath the advanced pikes of the phalanx. Total of approximately 2,500 men (officers extra): pikemen and musketeers in roughly equal proportions.

Fire pattern | Pikemen

6 ranks

'Leather' gun

96 musketeers

3 ranks

192 musketeers

3 ranks

Swedish battalion in battle formation. Total of approximately 504 men (officers extra): 288 musketeers and 216 pikemen.

6 ranks | 96 Musketeers | 216 Pikemen | 96 Musketeers | 16 men | 36 men

The Squadron
Organizational Changes Inspired by Gustavus Adolphus.

French battalion firing by company volley or battalion lines one rank at a time (five ranks deep).

Colonel

60 paces

Colonel

Grenadiers | 16 2 8 4 12 6 10 14 18 17 13 9 5. 11 3 7 1 15 | Grenadiers
II Div | IV Div | III Div | I Div

Fire by platoons organized into the 'three firings'.

English two-squadron charge (cold steel) 2 ranks deep, 3 troops to a squadron.

Half-troops retiring to reload

Support sqn | First sqn

200 paces

50 paces

French cavalry fire-action (pistols) — prior to mêlée (swords).

N.B. Diagrams not drawn to scale.

LINES OF BATTLE AND TACTICAL FORMATIONS

Lines of Battle varied considerably in layout and design. Swedish practice was probably the most advanced in the first half of the seventeenth century; their lines were designed to afford maximum flexibility and fire-output associated with mutual support between horse, foot and guns. The German method in this period was less imaginative and more unwieldy. By the end of the century, Swedish concepts had been widely copied by most European powers, but local specialities were still incorporated; Queen Anne's armies habitually kept the bulk of the cavalry in reserve for the coup de grâce, or ready to cover a retreat, while Imperial armies, needing to withstand the massed rushes of their perennial Turkish opponents, made great use of chevaux de frise or portable barricades. However no formal formation larger than the brigade existed, so armies fought by lines or wings.

Tactical Formations went through considerable changes over the period, changes that were closely linked with weapon development (see p. 37). In the period of unreliable firearms and before the development of effective bayonets, all infantry formations contained a large number of pikemen, drawn up several ranks

deep for protection against cavalry. The Spanish and German tercio in full array was a massive and unwieldy affair. As we have already described (see p. 27), Gustavus Adolphus revolutionized tactical concepts by developing the earlier experiments of Maurice of Nassau. He broke down the size of infantry formations by slightly reducing the proportion of pikemen and introducing the infantry brigade. These brigades comprised two or more battalions which were drawn up in carefully-devised formations to make the most of the improved firepower of the wheel-lock musket. Gustavus also introduced regimental artillery—light pieces which could be dragged by three men or one horse—to increase the fire output, dressed his armies in blue uniforms, and trained them to march in cadenced step. Linear tactics were now appearing.

By mid-century, the improved matchlock musket was replacing the snaphance, and soon there were approximately four or five musketeers to every pikeman in the battalion. Another innovation was the introduction of grenadiers, who were originally attached to each of the twelve companies in threes or fours but gradually came to form an élite company in each

formation. Battalions still drew up five ranks deep.

The introduction of the much more reliable flintlock musket from about 1688 onwards (the changeover was virtually complete in all but Turkish armies by 1703) and the invention of the socket-bayonet (see p. 37), caused the disappearance of the last pikemen and the evolution of new tactical formations to make the most of the increased unit firepower that resulted. The French and Spaniards tended to cling to outdated formations even after the change in weapons and, consequently, their infantry was often outclassed by their British and Dutch opponents. Although the company remained as the administrative sub-unit, the platoon-firing system—introduced by William III following a Swedish idea but greatly improved in the time of Marlborough and Eugene—conferred considerable advantages in battle. The reduction to three ranks increased the battalion frontage and enabled every man to shoot; the pattern of rippling fire, platoon after platoon or 'firing' after 'firing' placed an opponent under continuous pressure with no breathing-space, while the degree of localized fire-control now possible increased the effectiveness.

Further, one third of a battalion was always reloaded ready for any emergency. Weapons and formations were still relatively clumsy, of course, but these forms represented a considerable advance on earlier practices.

Cavalry tactics and formations also saw considerable changes. The basic administrative unit was the troop of some fifty cavalrymen; the practice of grouping three of these formations into a temporary squadron for battle soon spread. Certain powers—France and Spain among them—tended to regard cavalry as instruments of sophisticated firepower, and evolved complex evolutions to enable the troopers to employ their carbines and pistols to the full, reserving the sword for the mêlée. The Swedes, however, followed by the English and the Dutch, believed in cavalry as an instrument of shock, and made the sword their chief weapon. Twin squadron 'charges' delivered at a fast trot, the troopers riding 'knee by knee', became standard British practice by Marlborough's day, and was copied by most of the members of the Grand Alliance. Under Charles XII the Swedes also adopted a wedge-shaped formation, the troopers riding 'knee behind knee'.

65

Exhausted although Europe undoubtedly was at the conclusion of the War of the Spanish Succession in 1713 and of the Great Northern War eight years later, there was to be scant time for recuperation. The Bourbon rulers of Spain—attempting to regain lost territory—engineered a succession of crises, but a series of alliances that included France and, from 1717, the Empire, successfully staved these off. In 1733 a Franco-Spanish entente threatened to upset the balance of power, but the British Prime Minister Sir Robert Walpole managed to contain the situation. Later that year a disputed succession to the throne of Poland led to another war, with France and Spain supporting one candidate, and the Emperor and Tsar the other. Walpole refused to commit Great Britain, and his efforts to restore peace prevented the spread of the war and helped bring it to an end in 1735.

However, it was not long before colonial troubles brought Great Britain into war with Spain, 1739, and this localized struggle was eventually engulfed in the next serious conflagration: the War of the Austrian Succession, 1740-48. This struggle followed the deaths of Frederick William I of Prussia and the Emperor Charles VI. Frederick II unscrupulously set out to exploit the difficult position of Charles's heir, Maria Theresa—who as a woman could not succeed to the empire— by his determination to seize Silesia. In this he was abetted by Louis XV. The conflict spread all over Europe, and had effects in India and Canada. Great Britain became involved to protect Hanover; and defeats in the Netherlands (notably Fontenoy) inflicted by de Saxe, and the Young Pretender's invasion of Scotland to mount the 1745 Rebellion posed some daunting problems. By the Treaty of Aix-la-Chapelle, 1748, Prussia received Silesia but all other conquests were exchanged by the various combatants.

Maria Theresa was resolved to regain Silesia and disillusioned with the British alliance; a feeling that was reciprocated to the full as

Austria had proved incapable Hanover or keeping the Fre Netherlands. Louis XV wa hostile towards Prussia, wh abandoned its alliance in the pre result of these stresses the fam Revolution' took place. To pr Great Britain allied with Prussi managed to secure the co-oper as well as the support of Russ Saxony. In addition, the fierce of France and Great Britain co

From these sources of conf Seven Years War, 1756-63. Som brought his country through int display of military virtuosit temporary withdrawal of Har struggle. (Prussia's desperate fi is outlined on p. 69.) Crises loom. Britain again became i struggle at William Pitt's i Ferdinand of Brunswick's vict

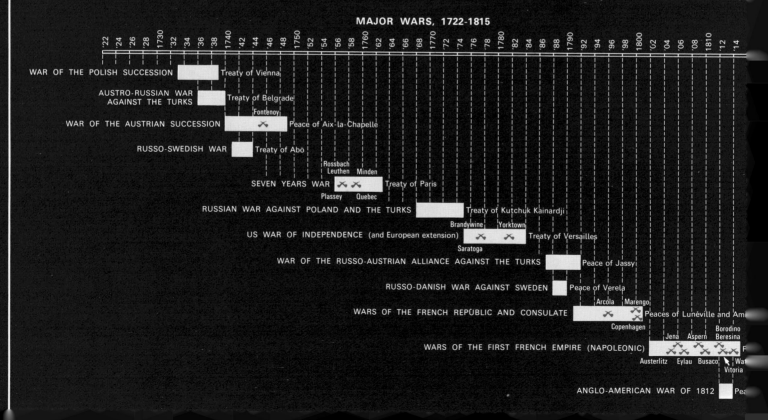

MAJOR WARS, 1722-1815

(see p. 71) repelled a French invasion of North Germany. Britain left the continental struggle in 1761, but Frederick survived once more thanks to the timely death of the Tsarina Elizabeth and the accession of Peter III who was an admirer of Prussia's monarch.

Pitt's colonial strategy and the great achievements of Wolfe and Clive in Canada and India respectively are described on pp. 82 and 78. By the Treaty of Paris, 1763, Britain made huge territorial gains for the Empire, but this was done at the expense of deserting Prussia, which was left to make its own peace.

The outstanding military aspect of this war was, however, the generalship of Frederick the Great and the fighting-power of his army. Prussian standards were emulated in many other countries, but in Prussia itself became stereotyped and unimaginatively applied.

Great Britain's supremacy overseas was not long destined to survive without receiving a telling blow. Growing tension in the American colonies eventually exploded in 1775 into the hard-fought War of Independence. After some initial setbacks, the British armies made some progress (as at Brandywine, see p. 86), but at increasing cost. As George Washington worked to improve the Continental Army so the difficulties of the British generals mounted. The surrender of Burgoyne at Saratoga (see p. 86) was the turning point of the struggle, for it persuaded France to make an alliance with the colonists. The arrival of French aid was one determining factor, but far more important was the French fleet's interference with British supply lines crossing the Atlantic.

Stage by stage the one-time colonial struggle took on European overtones. Spain declared war in 1779, and Britain became embroiled with Holland in 1780. The League of Armed Neutrality arose as a protest against British claims to the right to search neutral vessels for contraband, and involved Prussia, Russia, Sweden and Denmark. Although the stalwart defence of Gibraltar, 1779-82, was a bright episode for Britain (see p. 82) and naval success in the West Indies brought welcome respite, the war went from bad to worse. The surrender of General Cornwallis at Yorktown in 1781 proved the coup de grâce in America. George III's government was induced to

'Old Fritz'—Frederick the Great—reviewing his army. Napoleon expressed great respect for the military skills of the Prussian monarch who dominated the military scene in the mid-eighteenth century.

recognize American independence, and the Treaty of Versailles in 1783 set the seal on the fate of the 'Old Colonies'. However, British power in India, Canada and the West Indies remained intact, and although Minorca was restored to Spain, Gibraltar was retained. In southern and eastern Europe the ancient Habsburg-Turkish and Russo-Turkish struggles continued, and problems surrounding the partitions of Poland dominated diplomacy.

French satisfaction in curbing British power was bought at the high price of economic ruin. This was a major factor in producing the French Revolution, which in turn led to a devastating period of warfare. The relative moderation in the prosecution of warfare that had typified at least part of the eighteenth century now gave place to increasingly ruthless methods, as the French fought first for survival and then for the propagation of its revolutionary principles throughout Europe and parts of the Levant. Larger French armies than had been seen for many centuries marched through Europe at a speed and with a determination that first astounded and then terrified the governments of the established monarchies.

French progress was rapid after the first years (see p. 89), and with the arrival on the scene of Napoleon Bonaparte to lead the forces of his adopted country it seemed that no country could withstand his strategic methods (see p. 98). Coalition after coalition formed against him, but one after another they crumbled in the face of brilliant strategic and battlefield successes to which there seemed no answer. Once the Royal Navy had secured full mastery at sea, only Britain—secure behind the Channel—was impregnable and wholly implacable. Under the leadership of such Prime Ministers as William Pitt the Younger and later Lord Liverpool she formed the inspiration of European resistance to Napoleon's ambitions. Defeat at sea, for example, brought to naught French ambitions in Egypt and Syria, 1798-1801, while the fleet carried British troops on many an expedition.

Napoleon's attempts to tame Great Britain by all-out economic warfare rebounded against him. First, by embroiling the French ever more deeply in the Iberian Peninsula from 1807, he provided the British Army (supplied through Lisbon by the Navy) with the chance to intervene to good effect on the Continent. The great successes of Wellington made of Spain a running sore; and a selection of his achievements will be found on pp. 112 to 115.

European Rivalries

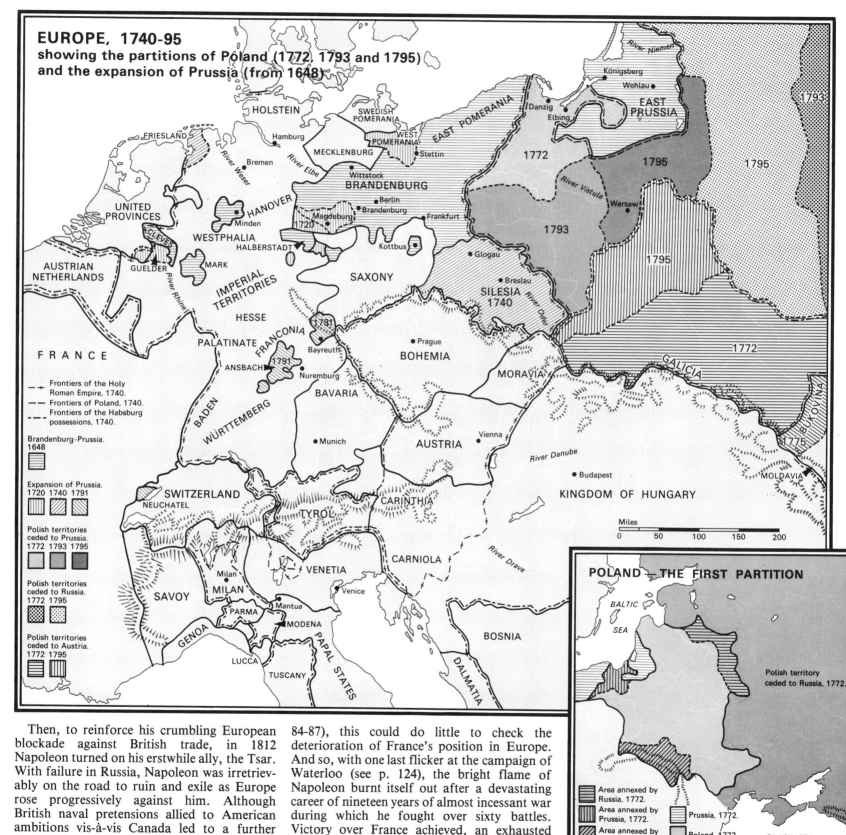

EUROPE, 1740-95
showing the partitions of Poland (1772, 1793 and 1795)
and the expansion of Prussia (from 1648)

- ┼┼ Frontiers of the Holy Roman Empire, 1740.
- ── Frontiers of Poland, 1740.
- ─·─ Frontiers of the Habsburg possessions, 1740.

Brandenburg-Prussia. 1648

Expansion of Prussia. 1720 1740 1791

Polish territories ceded to Prussia. 1772 1793 1795

Polish territories ceded to Russia. 1772 1795

Polish territories ceded to Austria. 1772 1795

HOLSTEIN · SWEDISH POMERANIA · WEST POMERANIA · EAST POMERANIA · EAST PRUSSIA · 1793 · River Niemen · Königsberg · Wehlau · Danzig · Elbing · 1772 · 1795 · 1795 · FRIESLAND · Hamburg · MECKLENBURG · Stettin · River Vistula · Warsaw · 1793 · UNITED PROVINCES · River Elbe · BRANDENBURG · Berlin · Brandenburg · Frankfurt · 1772 · HANOVER · Minden · River Weser · 1720 · Magdeburg · Kottbus · Glogau · 1795 · CLEVES · WESTPHALIA · HALBERSTADT · SAXONY · Breslau · SILESIA 1740 · River Oder · AUSTRIAN NETHERLANDS · GUELDER · MARK · IMPERIAL TERRITORIES · GALICIA · 1772 · FRANCE · River Rhine · HESSE · PALATINATE · FRANCONIA · 1791 · Bayreuth · Prague · BOHEMIA · MORAVIA · BUKOVINA · ANSBACH · 1791 · Nuremburg · BAVARIA · 1775 · BADEN · WÜRTTEMBERG · Munich · AUSTRIA · Vienna · River Danube · MOLDAVIA · SWITZERLAND · NEUCHATEL · TYROL · CARINTHIA · Budapest · KINGDOM OF HUNGARY · SAVOY · Milan · MILAN · Venice · VENETIA · CARNIOLA · River Drave · GENOA · PARMA · Mantua · MODENA · BOSNIA · LUCCA · TUSCANY · PAPAL STATES · DALMATIA

Miles
0   50   100   150   200

POLAND — THE FIRST PARTITION

BALTIC SEA

Polish territory ceded to Russia, 1772.

Area annexed by Russia, 1772.
Area annexed by Prussia, 1772.
Area annexed by Austria, 1772.

Prussia, 1772.
Poland, 1772.

BLACK SEA

Then, to reinforce his crumbling European blockade against British trade, in 1812 Napoleon turned on his erstwhile ally, the Tsar. With failure in Russia, Napoleon was irretrievably on the road to ruin and exile as Europe rose progressively against him. Although British naval pretensions allied to American ambitions vis-à-vis Canada led to a further period of conflict in the New World (see pp. 84-87), this could do little to check the deterioration of France's position in Europe. And so, with one last flicker at the campaign of Waterloo (see p. 124), the bright flame of Napoleon burnt itself out after a devastating career of nineteen years of almost incessant war during which he fought over sixty battles. Victory over France achieved, an exhausted Europe sought only a lasting peace.

The Austrian Succession and Seven Years War

The mid-eighteenth century was dominated by two great feuds: one ancient the other new. The former was the continued commercial and maritime rivalry between Great Britain and France and Spain, fought out in a thousand incidents in different corners of the world, and occasionally becoming merged in broader general conflagrations. The latter was the sudden antipathy between Austria and the state of Prussia, or rather between Maria Theresa and Frederick II.

In 1739 Britain declared war on Spain—despite the reluctance of Sir Robert Walpole—over disputed commercial claims. As a result, from that year until 1763 almost incessant colonial wars ensued, in which the main victim was France. The same proved true of the two major European wars in this period, even though the Bourbons did not instigate either of them directly.

In 1740, Frederick II seized Silesia from Maria Theresa, demonstrating Austria's military weakness at Mollwitz. This calculating and ruthless act of realpolitik inspired Bourbon France to join the war in search of her old European predominance, and her army occupied Prague. Austria's fortunes appeared at their nadir, but then unexpectedly Hungary rallied to Austria's aid, and France lost Prague; her Bavarian ally lost Munich and with it all hope of the imperial title. Britain and Sardinia further intervened in Austria's favour; King George II was present at the Battle of Dettingen in June 1743. In July of that year, Frederick unexpectedly made peace with Austria in return for the promise of Silesia, and Saxony also left the war. Maria Theresa was thus free to turn her full wrath on France and Bavaria, and in September 1743 a new anti-Bourbon coalition was signed. France seemed doomed. Suddenly, Frederick decided to re-enter the war: he invaded Bohemia and seized Prague once more, afraid to see Austria become too powerful. France was thus enabled to resume the offensive in the Netherlands, where de Saxe won brilliant successes. In the same year (1745) a Jacobite rising was instigated in Scotland. Fortunately for Britain, storms scattered the French fleet and Charles Edward Stuart's invasion was not supported. After initial successes it foundered at Culloden in April 1746. Meanwhile, in central Europe Frederick was forced out of Bohemia, but defeated Austrian armies at Hohenfriedberg and Sohr. In December 1745 Frederick again made peace with Austria, and was once more guaranteed Silesia. In north Italy, Austria also lost some ground to the Spanish Bourbons. At last, the Peace of Aix-la-Chapelle (1748) ended the war, and many possessions and conquests were

The Battle of Minden (1 August 1759) resulted in a victory for Duke Ferdinand of Brunswick's Allied Army over Marshal Contades and the French. Its effect was to save Hanover from occupation by Louis XV's forces.
(National Army Museum)

returned. But this was to be no more than a truce—although the next war only came after an astonishing diplomatic revolution.

France's decision to ally with Austria—first defensively, and then offensively (August 1756)—was extraordinary. But in September, Prussia's king, aware of deep plots against his security, suddenly invaded Saxony, and sparked off the Seven Years' War. Frederick was surrounded by foes and heavily outnumbered, but to his aid came France's inveterate opponent: Great Britain. William Pitt, Earl of Chatham, sensed a chance to destroy the French challenge in the colonial field once and for all. He devised a maritime strategy in the New World, the West Indies and India, meanwhile subsidizing Prussia with money to keep her fighting and thus distracting France. But it was Frederick's superb generalship that saved Prussia and earned him the deserved title of 'the Great'.

In 1757 his brilliant manoeuvres won two startling victories over France and Austria: Rossbach and Leuthen (see pp. 74-75). He proved less successful against Russia's armies at Zorndorf, 1758 and above all at Kunersdorf (1759), but the Tsarina's generals failed to press their advantages, and soon Allied disunion and

mutual rivalries came to Prussia's aid. Salvation came with the death of the Tsarina Elizabeth in January 1762, for Russia left the war soon after, and next year, threatened by Turkey, Austria made peace at Hubertusburg, 15 February 1763. France was forced to make peace the same year. Humiliated in Europe, the price overseas was immense. If Frederick received Silesia for the loss of 500,000 lives, Great Britain received Canada, Senegal, numbers of West Indian isles and unquestioned predominance in India—all for an expenditure of under 5,000.

Marshal de Saxe: Fighter and Innovator

Hermann-Maurice, Comte de Saxe (1696-1750), was the most distinguished commander of French armies between Marshal Villars and Napoleon Bonaparte. The latter, indeed, included his name amongst the seven greatest soldiers of all time, but a more generally accepted view would put de Saxe as first amongst notable soldiers of the second order. He proved superior to the Allied and British generals of the War of the Austrian Succession.

His life fell into two parts. First, from his earliest campaign in 1708 as a mere lad of twelve years, his career was that of a profligate

Hermann-Maurice, Comte de Saxe (1696-1750) became Marshal-General of France in recognition of his great achievements in the War of the Austrian Succession. He also wrote an important treatise on warfare.

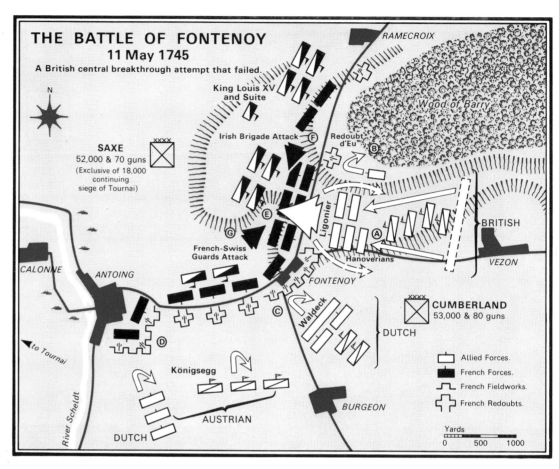

THE BATTLE OF FONTENOY
11 May 1745
A British central breakthrough attempt that failed.

SAXE
52,000 & 70 guns
(Exclusive of 18,000 continuing siege of Tournai)

King Louis XV and Suite

Irish Brigade Attack

Redoubt d'Eu

Wood of Barry

RAMECROIX

Ligonier

BRITISH

VEZON

French-Swiss Guards Attack

Hanoverians

FONTENOY

Waldeck

DUTCH

CUMBERLAND
53,000 & 80 guns

CALONNE

ANTOING

to Tournai

Königsegg

River Scheldt

DUTCH

AUSTRIAN

BURGEON

Allied Forces.
French Forces.
French Fieldworks.
French Redoubts.

Yards
0   500   1000

adventurer, soon notorious for his debauchery. At this time he saw service in Flanders during Marlborough's Wars, in Russia under Peter the Great, and above all in Hungary, 1716-17, where he fought under Prince Eugene who became his model and ideal. Finally, in 1718, he abandoned his wife and settled in France, where Louis XV appointed him Major-General in August 1720.

As the son of Frederick Augustus II, Elector of Saxony, he was deeply involved in Polish questions, and hoped to establish himself as Prince of Courland through the influence of Anna Ivanovna, but she gave him up before becoming Empress of Russia in 1730. During the War of the Polish Succession, Maurice refused the proffered command of the Saxon Army, but served on with the French who championed the claim of Stanislas, against the Saxon claimant who was in fact Maurice's brother. It was during this period that he wrote his celebrated *Rêveries* (1732). In 1734 he was promoted Lieutenant-General, and served with credit under Marshals Berwick and Belle-Isle. After the war, he was reconciled with his brother, the new King of Poland, but his hopes of Courland were dashed twice more.

The War of the Austrian Succession saw his emergence to true greatness. After aiding Broglie and de Noailles during the earlier campaigns, he was appointed a Marshal on 26 March 1744. Despite ill-health, he besieged Tournai in 1745 and won a great success at Fontenoy in May by a virtuoso display of tactical flexibility allied to the skilful choice of a position that made the maximum possible use of his available guns. Rewarded with the Governorship of Chambord, he completed the capture of Tournai on 20 June, and then

conquered much of the Austrian Netherlands. In 1746 he captured Brussels, and won the Battle of Rocoux, following this up the next year with his success at Laffeldt. His crowning achievement—and last conquest—was the capture of Maastricht in 1748. For all these services, he was appointed Marshal-General.

Marshal de Saxe's reputation rests as much on his theoretical work as on his practical record. *Mes Rêveries* were written in just thirteen nights during a severe bout of fever, but they represent the most important contribution to military thought of the pre-Frederick period. For years de Saxe had studied with Comte Folard, champion of the column as the best means of attack, but he later disagreed with his former mentor. De Saxe challenged many beliefs of his day—denying the all-importance of the great battle and challenging Vauban's reputation as an engineer of genius—but, instead, placed great reliance on uniform drill and discipline and on training for war in time of peace, insisting that the troops must be toughened by marching and exposure to the elements. He also advocated the use of large numbers of light infantry—an innovation for the day. He championed the adoption of a

formation he termed the Legion—a balanced force of all arms, with its own staff, cavalry and artillery as well as infantry, both line and light—stressing the flexibility such a formation would confer in both campaign and battle. Grassin's Legion, which fought with distinction at Fontenoy, was the prototype, but the concept did not survive de Saxe's death. However, this idea was one predecessor of the French corps d'armée, and its system (see p. 98) was destined to be the master-weapon of Napoleon half a century later. De Saxe had many enemies and critics in his lifetime, and this may account for the transient nature of his reforms and suggestions. Perhaps of greatest significance in all his work was his insistence on competent officers being truly responsible for their men. He was in many ways considerably ahead of his time.

The Battle of Fontenoy, 1745
Many eighteenth-century battles were fought in association with major sieges. Marshal de Saxe's investment of Tournai—the gateway to western Flanders—induced the Duke of Cumberland with his Allied Army of 53,000 men to intervene. But the slowness of his

approach gave de Saxe time to fortify a strong position around the village of Fontenoy, with secure flanks and a string of strong earthworks holding guns—some of them concealed—to protect his centre.

Anticipating the major Allied attack against his left, de Saxe drew up his strongest formations in dead ground north of Fontenoy, and occupied the Wood of Barry with sharpshooters, supported by his Irish Brigade. De Saxe was correct in his calculations, for Cumberland indeed decided to use his British troops massed against the northern sector, while the Dutch and Austrians attacked the Fontenoy and Antoing sectors to the south.

Cumberland's advance at 6am over the deceptively open plain in the north (A) was delayed when the hidden Redoubt d'Eu was encountered. The massed British cavalry was bombarded at the halt on the plain, while Brigadier Ingoldsby vainly tried to take the Redoubt d'Eu (B). Meanwhile, the Dutch and Austrians unsuccessfully attempted to take Fontenoy (C) and Antoing (D) respectively, suffering heavy casualties. Hoping to break the impasse, Lord Ligonier passed his British battalions through the cavalry, and at 10.30am Cumberland personally led the advance. Despite losses the British infantry moved forward in superb order, holding their fire until the last moment, when a massed volley decimated the French front line. Now compressed into a vast column, the redcoats pressed on (E) deep into the French position, but failure of a new Dutch onslaught against Fontenoy compelled them to halt to re-form. Once again the red wall moved inexorably forward, but de Saxe's gunners pounded the ranks, and the Marshal flung in his last reserves: the Irish 'Wild Geese' from the north and the Swiss Guards from the south (F and G). Cumberland ordered a retreat after midday, conceding the day to the French. Each side had lost some 7,000 casualties. The price of this defeat for the Allies was the loss of most of the Austrian Netherlands to the French.

The Battle of Minden, 1759

If Fontenoy was a setback for the reputation of British arms, Minden was a notable success, saving Hanover from the French Army of Marshal Contades. However, the British cavalry commander behaved badly.

The Duke of Brunswick had been forced to retire deep into Westphalia by the French seizure of Minden, well to his rear. As he approached the River Weser he found the French Army, at least 48,000-strong, drawn up

Field Marshal Duke Ferdinand of Brunswick Wolfenbüttel (1721-92), one of Frederick the Great's ablest lieutenants who commanded the Allied army at the notable victory of Minden.

in a strong position behind the Bastau Stream between the river and Minden marsh. With only 38,000 under command, Brunswick was faced with a difficult decision, but decided to try to lure the French out of their strong defences. First, he ordered von Wangenheim to fortify the heights near Todtenhausen due north of Minden (A). He then deliberately split his army by drawing the bulk of his men away to the south-west, at the same time sending a raiding party to threaten the French communications.

Contades obligingly fell for the bait. Late on 31 July he ordered de Broglie to neutralize von Wangenheim's position, while the rest of the French marched north from the River Bastau and its neighbouring marshes on to the open heathland beyond. Brunswick rapidly marched up his army to take up the induced challenge (B), but six British and three Hanoverian battalions under Spörcke outmarched the main body and boldly proceeded to attack the sixty-odd squadrons of the French cavalry, which were in the centre of Contades' line, without even taking the precaution of forming a square (C). Three times they were charged by successive waves of French cavalry, but each time they repelled them with well-directed volley fire. Contades vainly hoped for success on Broglie's wing, but there too the French faced disappointment, their repulse (D) being turned into near-disaster when von Holstein's squadrons charged (E) and routed the French right. Thereupon, Contades ordered a general retreat, but he would have been hard-pressed to escape with only 7,086 casualties had not Lord George Sackville on the Allied right (F) refused to charge. Nevertheless, for a loss of 2,760 casualties, Brunswick and his Allied Army had saved Hanover.

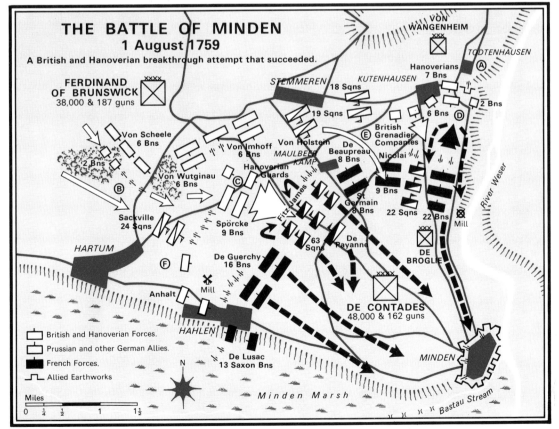

THE BATTLE OF MINDEN
1 August 1759
A British and Hanoverian breakthrough attempt that succeeded.

FERDINAND OF BRUNSWICK
38,000 & 187 guns

VON WANGENHEIM

TODTENHAUSEN

STEMMEREN
KUTENHAUSEN
Hanoverians 7 Bns (A)
18 Sqns
19 Sqns
6 Bns
2 Bns
Von Scheele 6 Bns
Von Imhoff 6 Bns
Von Holstein
MAULBEER KAMP
De Beaupreau 8 Bns
British Grenadier Companies (E)
Nicolai
Von Wutginau 6 Bns
Hanoverian Guards (C)
Fitz James
9 Bns
St Germain 8 Bns
22 Sqns
22 Bns
Mill
River Weser
Sackville 24 Sqns
Spörcke 9 Bns
63 Sqns
De Rayanne
HARTUM
DE BROGLIE
De Guerchy 16 Bns (F)
Mill
Anhalt
DE CONTADES
48,000 & 162 guns
HAHLEN
De Lusac 13 Saxon Bns
MINDEN
N
Minden Marsh
Bastau Stream

☐ British and Hanoverian Forces.
☐ Prussian and other German Allies.
■ French Forces.
⌐ Allied Earthworks.

Miles
0 ¼ ½ 1 1½

The Contribution of Frederick the Great

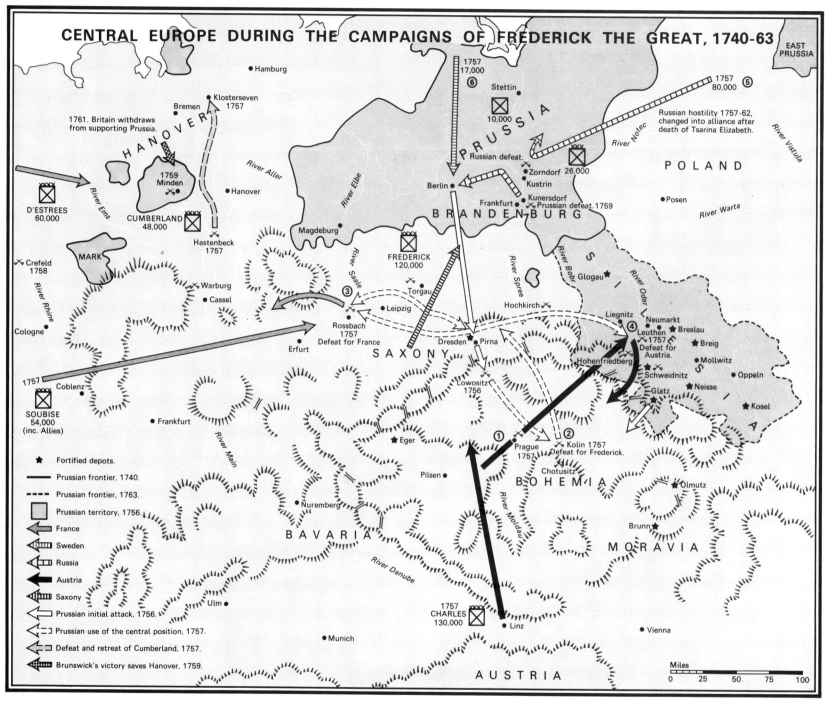

CENTRAL EUROPE DURING THE CAMPAIGNS OF FREDERICK THE GREAT, 1740-63

1761. Britain withdraws from supporting Prussia.

Russian hostility 1757-62, changed into alliance after death of Tsarina Elizabeth.

D'ESTREES 60,000

CUMBERLAND 48,000

SOUBISE 54,000 (inc. Allies)

FREDERICK 120,000

CHARLES 130,000

★ Fortified depots.
— Prussian frontier, 1740.
--- Prussian frontier, 1763.
▢ Prussian territory, 1756.
France
Sweden
Russia
Austria
Saxony
Prussian initial attack, 1756.
Prussian use of the central position, 1757.
Defeat and retreat of Cumberland, 1757.
Brunswick's victory saves Hanover, 1759.

Miles
0   25   50   75   100

If Marshal de Saxe was the luminary of the military art in the mid-1740s, he was soon to be surpassed by the genius of Frederick II, King of Prussia. Frederick was destined to have the most profound influence on the future conduct of warfare. In many ways he represents the transition from the concepts of 'limited war' that had dominated the later seventeenth and the first half of the eighteenth century towards those of 'total war' which would explode so

forcibly under the French Revolution and Napoleon. The future Emperor freely admitted that he owed much to his studies of Frederick's campaigns and battles, stating that Frederick had most of the right ideas about how to wage successful warfare but in the final analysis lacked the ideal instrument with which to carry them out.

Frederick was not originally a soldier by inclination. He spent a miserable youth and

young manhood in perpetual conflict with his brutal father, Frederick William I, the creator of the Prussian Army, who had nothing but scorn for his heir's artistic and literary inclinations. However, immediately after his accession in 1740 Frederick opportunistically seized Silesia, and from that date he was committed to a military life for most of his reign. Having sown the wind, he reaped the whirlwind with a vengeance.

As ruler, Frederick could impose his will on army and population alike; an advantage that neither Marlborough nor de Saxe had ever enjoyed. Faced by the stark alternatives of national survival or total eclipse, Prussia's king set about mobilizing all assets for the struggle. Crippling taxes, a form of conscription based on the cantonal system, reliance on considerable numbers of mercenaries, and compulsory service as officers for the aristocracy and junker-squirearchy were some of the measures Frederick relied upon to raise and maintain an army capable of defending his national interests. During his reign it developed from a strength of some 80,000 to all of 160,000, but the strains were immense. Fortunately, Frederick was a painstaking planner.

Frederick, like Napoleon after him but unlike Gustavus Adolphus, did not create a new type of army; rather, he perfected the one he inherited from his father and his adviser, the Prince of Anhalt-Dessau. Certain minor technical improvements included the adoption of the iron ramrod for the musket, the harnessing of gun teams in pairs, and an increasing reliance on the howitzer. Frederick took care to suit the capabilities of the men to available weaponry. In 1748 Frederick issued his *Secret Instructions for his Generals*, which embodied many of his ideas concerning every level of warfare, from strategy to minor tactics and logistics.

Four principles dominated his military philosophy. Foremost amongst these was an insistence on ferocious discipline: "the men must fear their officers more than the enemy". His men were to be drilled into automatons that would not think but simply obey. The aim for the infantry was to produce the maximum rate of fire possible—its psychological impact being deemed more important than its physical effect. The troops were trained to load and fire on the move, without attempting to aim accurately. Great stress was laid on manoeuvrability of formations from column of platoons to battalion in line of platoons for fire action. All this required much drill and discipline if the 'walking batteries', as his battalions became nicknamed, were to be effective in action.

Secondly, Frederick placed great emphasis on subsistence. Because of his country's limited resources, considerations of logistics virtually dictated both strategy and tactics. "Understand that the foundation of an army is the belly", he wrote, and much of the *Instructions* is devoted to the planning and provisioning of magazines, the proper protection of convoys, and the need to disrupt the foe's communications and means of subsistence. The problems of resupply meant that his armies rarely operated for more than

five days without a one-day halt, and Frederick insisted that new depots should be constructed every sixty miles as the army advanced; troops were never allowed to forage on their own for fear of mass desertions. Better roads were being developed, but communications remained a major problem.

Thirdly, Frederick preached the need for offensive warfare: "Wars are only decided by battles". This was in marked contrast to the views of such earlier commanders as de Saxe, and was in fact preparing the way for Napoleonic armageddon. But it is noteworthy that after 1759 Frederick became more concerned for the defensive, and even came to advocate massive systems of fortification and wars based upon manoeuvre as the best means to preserve his domains. This was in many ways due to the realization that his resources of manpower were far from limitless, but it was also a reversion to the concepts of war held in earlier years of the century. In essence, he was possibly more of a traditionalist than an innovator—especially in his last years.

This fact explains his fourth major principle, that of 'practicability'. Frederick was the product of the Age of Reason, a supreme rationalist. He produced excellent combined arms tactics to capitalize on available weaponry and battle-power. He took much from the study of military history. Realizing that the Austrians preferred concealed positions, he made full use of howitzers to bring fire to bear on their formations. Similarly, his so-called 'oblique order' of attack was designed to make the most of Prussia's smaller armies. Instead of attacking all along the line as was customary, Frederick used a small force to distract his opponents and used rapid manoeuvre to bring all the rest to bear against a single part of the enemy line with massive local superiority, before rolling it up from the flank.

Fighting long wars along these broad principles, Frederick proved a master at a 'strategy of survival', and his army became the pattern for many another. From first to last, as he himself described it, Frederick had been ". . . the first servant of the state". He died in 1786.

The Battle of Kolin, 1757

At the outset of the Seven Years War Frederick the Great took the initiative and advanced into Bohemia (see map Central Europe during the

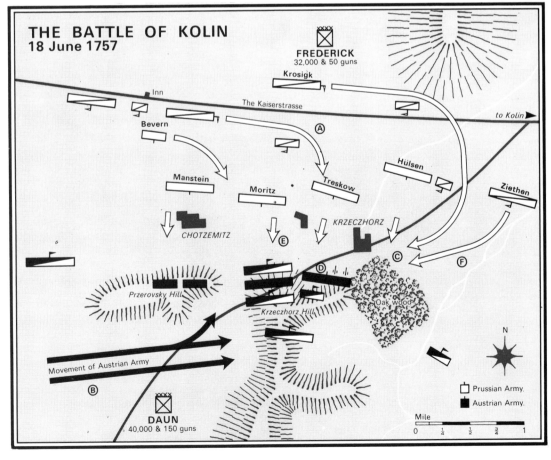

THE BATTLE OF KOLIN
18 June 1757

FREDERICK
32,000 & 50 guns

Krosigk

Inn

The Kaiserstrasse

to Kolin

Bevern

A

Hülsen

Ziethen

Manstein

Treskow

Moritz

KRZECZHORZ

CHOTZEMITZ

E

D

C

F

Przerovsky Hill

Krzeczhorz Hill

Oak wood

N

Movement of Austrian Army

B

Prussian Army.

Austrian Army.

DAUN
40,000 & 150 guns

Mile
0   ¼   ½   ¾   1

The Contribution of Frederick the Great

Campaigns of Frederick the Great, p. 72). The Austrians fought his army outside Prague on 6 May (1), but were heavily defeated, whereupon Frederick opened a formal siege of the great city. Determined to prevent the fall of Prague, the Austrians assembled 40,000 men and 150 guns under Field Marshal Daun to the east and marched to the relief of the beleaguered garrison, which was now swollen by many fugitives from the earlier battle.

Frederick, owing to the requirements of the siege, could mass only 32,000 men and 50 guns to face this new Austrian army. Marching from Prague, he discovered Daun holding a low range of hills south of the Kaiserstrasse near the town of Kolin (2), and with rash inpetuosity ordered an immediate attack without fully reconnoitring the enemy position. Although he could see nothing of its disposition and had scant information, he decided to attack the Austrian right wing.

The Prussian Army set out to march across the front of the Austrian Army, but the heat of the day slowed this manoeuvre, and the troops became so exhausted that Frederick was forced to order a halt of several hours to rest his men (A). This pause gave the watchful Daun ample time to move part of his left wing behind the concealing ridge (B) to strengthen his right wing well before the Prussian blow fell. When it came he was ready for it. Nevertheless, the spirit of the first Prussian onslaught (C) north of the Oak Wood almost took the Austrians

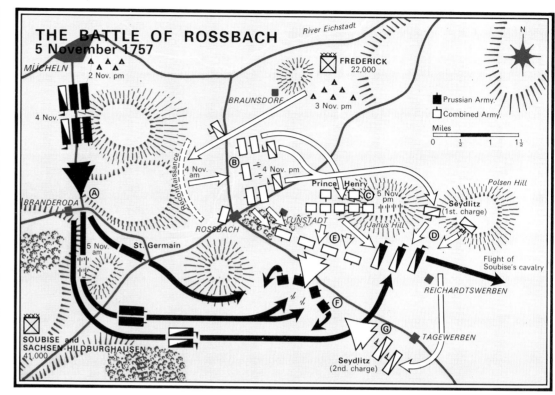

aback. General Hülsen stormed the village of Krzeczhorz with seven battalions, supported by nine more under General Treskow on his right, and captured two batteries of Austrian artillery beyond (D). The next commander in line, von Moritz, failed to advance with sufficient determination against Krzeczhorz Hill, and the impetus of the attack died away, affording Daun yet more time in which to reinforce the sector. Meanwhile, von Manstein's battalions, stung by the fire of Croatian light troops who had advanced through standing corn (E), launched an unauthorized attack against the hillside ahead of them. In this way Frederick's intended turning movement degenerated into a frontal battle of attrition, in which the Austrian superiority in both numbers and position was bound to tell. Austrian and Saxon cavalry swept down in a series of devastating attacks, and despite countercharges by Ziethen's horsemen (F) the Prussians had lost the battle. Covered by the steady withdrawal of a battalion of Prussian Guards and the Gemmingen Grenadiers, the battered Prussian Army left the field, leaving 10,000 dead and 5,000 prisoners. Daun lost an estimated 9,000 men. Frederick was forced to give up the siege of Prague, and within a month he had retreated into Saxony.

Prussia was now beset by foes and fighting for her very survival. By the year's end, however, after a stunning display of military virtuosity based strategically on the exploitation of the 'central position' (see p. 98), Frederick the Great had weathered the storm and inflicted telling defeats on two opponents and three armies within six months, two of them within a period of thirty days.

The Battle of Rossbach, 1757

After driving Austrian raiders back from the gates of Berlin, Frederick hurried west to meet the 41,000 strong Franco-Imperialist Army of Prince Sachsen-Hildburghausen and the Prince of Soubise who were currently advancing on Rossbach (3).

On 5 November the Allies marched ponderously forward in five great columns (A) planning to turn Frederick's original position (B). Frederick—with only 22,000 men—rather belatedly observed this move, and striking camp moved behind the Janus Hill, unseen by the Allies, to take up a new position (C). Believing Frederick to be in full retreat, the Allies hastened their pace, while the Prussian cavalry leader, Seydlitz, massed his squadrons behind Polsen Hill. With perfect timing, he then charged into the head of the Allied columns, routing their cavalry (D). Resisting the temptation to pursue, Seydlitz rallied his men near Tagwerben.

Meanwhile, the main Prussian force had appeared south-east of Rossbach, wheeling in line to face south-west (E). A sharp battle ensued

Frederick II, King of Prussia (1713-86), became the pre-eminent military figure in Europe. His achievements in the Seven Years' War were particularly notable, and his writings on warfare pointed the way to future developments. He was much admired by Napoleon.

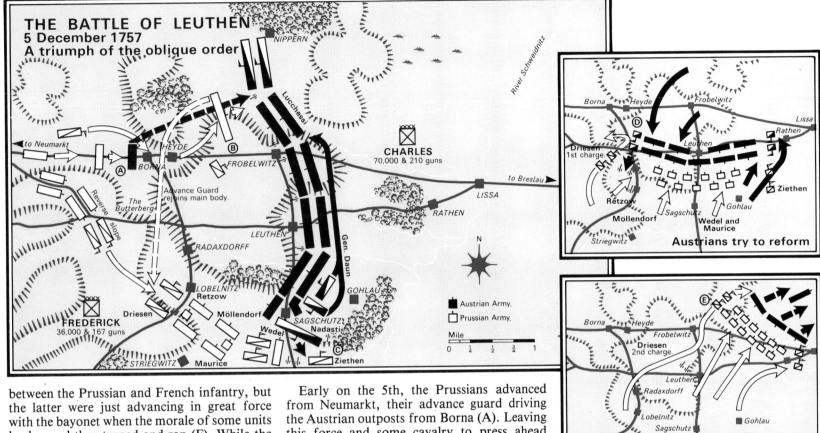

THE BATTLE OF LEUTHEN
5 December 1757
A triumph of the oblique order

to Neumarkt

NIPPERN

Lucchessi

HEYDE

BORNA

(A)

(B)

FROBELWITZ

CHARLES
70,000 & 210 guns

The Butterberg

Reverse slope

Advance Guard rejoins main body.

to Breslau

LISSA

RATHEN

LEUTHEN

Gen. Daun

RADAXDORFF

N

LOBELNITZ

Retzow

Driesen

GOHLAU

FREDERICK
36,000 & 167 guns

Möllendorf

SAGSCHUTZ

Nadasti

Wedel

STRIEGWITZ

Maurice

Ziethen

(C)

■ Austrian Army.
□ Prussian Army.

Mile
0   ¼   ½   ¾   1

Borna   Heyde   Frobelwitz   Lissa

(D)   Rathen

Driesen
1st charge.

Leuthen

Ziethen

Retzow   Sagschutz   Gohlau

Möllendorf   Wedel and
Maurice

Striegwitz

Austrians try to reform

Borna   Heyde

(E)

Frobelwitz

Driesen
2nd charge.

Leuthen

Radaxdorff

Lobelnitz

Sagschutz

Gohlau

Striegwitz

Miles
0   ½   1   1½

General Prussian advance

between the Prussian and French infantry, but the latter were just advancing in great force with the bayonet when the morale of some units broke, and they turned and ran (F). While the rest of the Allied Army hesitated in consternation, Seydlitz charged into their midst a second time—again with perfect timing (G). This proved the critical event of the battle, and the Allies were soon routed with a total loss of some 10,000 men. The Prussians had achieved this at a cost of barely 550 casualties. His western front cleared for the moment, Frederick could now turn east to face his main Austrian opponent.

The Battle of Leuthen, 1757
After its success at Kolin, 18 June 1757, the Austrian Army relieved Prague, which had been under siege since Frederick's success outside the city in early May. This setback placed Frederick on the defensive, and not until after Rossbach was he free to re-engage the Austrians.

Dashing back to Silesia with his victorious but small army (4), Frederick determined to attack Prince Charles of Lorraine's and Marshal Daun's 70,000-strong army—which was blocking his road to Breslau—with his 36,000 available troops. Despite the strength of the Austrian position, Frederick felt that deception and surprise might make up for his lack of numbers.

Early on the 5th, the Prussians advanced from Neumarkt, their advance guard driving the Austrian outposts from Borna (A). Leaving this force and some cavalry to press ahead towards the Austrian right near Frobelwitz (B) as a diversionary attack, Frederick rushed his main body behind the neighbouring ridges towards Lobelnitz. The Austrians, meantime, responded to the apparent attack on their right wing by moving Daun's cavalry reserve north, where eventually Lucchessi succeeded in repelling the small Prussian force to its front, which fell back towards the main army.

This gave Frederick time to draw up his main force between Lobelnitz and Striegwitz. Once Ziethen had met and defeated Nadasti's cavalry attack (C), he readied his formation and attacked towards Sagschutz in the 'oblique order'. With his battle line threatened with being rolled up from the south, Charles tried to re-form his army to face south, but lack of manoeuvrability led to mounting confusion in the Austrian ranks (see inset No. 1). A desperate charge by the Austrian right-wing cavalry (D) was met by General Driesen's charge and then repelled by artillery fire and Retzow's infantry and cavalry, which fell upon them from three sides. A second charge by Driesen (E) and a general Prussian advance (see inset No. 2) finally broke the Austrian resolve, and for a loss of 21,000 men they quitted the field. This victory cost Frederick 6,400

casualties, but confirmed his hold on Silesia. Together with the earlier success at Rossbach, it made 1757 a memorable year in the annals of Prussian arms.

However, even then the events of 1757 were not quite over. The impending arrival of Tsarina Elizabeth's 80,000 Russians from Poland (5) posed serious problems that required great vigilance and prompt action; although there would be no major battle until the following year (Zorndorf) and the year after that (Kunersdorf)—both of them defeats for Frederick—as the fortunes of war swung to and fro. Of more immediate consequence in 1757 was the advance of 17,000 Swedes marching towards Berlin from the Baltic (6). But the 10,000 Prussians available proved more than a match for them in the last weeks of the year. Thus Prussia, led by its dynamic ruler, survived a daunting series of threats by dint of fast marching and hard fighting. Small wonder that a generation later Napoleon would openly declare his admiration for Frederick the Great: the accolade was thoroughly deserved.

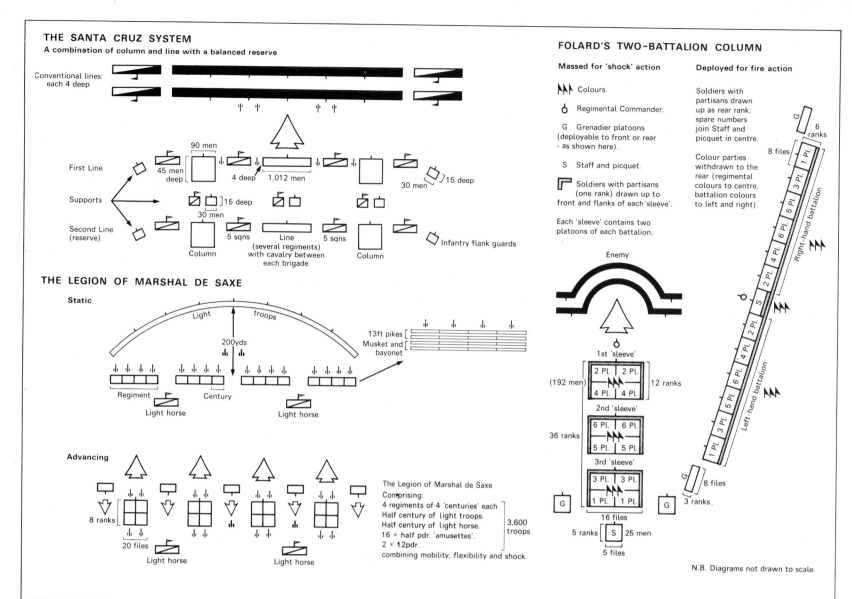

THE SANTA CRUZ SYSTEM
A combination of column and line with a balanced reserve

Conventional lines: each 4 deep

First Line
90 men
45 men deep
4 deep   1,012 men
30 men   15 deep

Supports
15 deep
30 men

Second Line (reserve)
5 sqns   Line (several regiments) with cavalry between each brigade   5 sqns   Infantry flank guards
Column   Column

THE LEGION OF MARSHAL DE SAXE

Static

Light   troops
200yds

13ft pikes
Musket and bayonet

Regiment   Century
Light horse   Light horse

Advancing

8 ranks
20 files
Light horse   Light horse

The Legion of Marshal de Saxe
Comprising:
4 regiments of 4 'centuries' each
Half century of light troops.
Half century of light horse.   3,600 troops
16 × half pdr. 'amusettes'.
2 × 12pdr.
combining mobility, flexibility and shock.

FOLARD'S TWO-BATTALION COLUMN

Massed for 'shock' action   Deployed for fire action

ΛΛΛ Colours.

⚲ Regimental Commander.

G Grenadier platoons (deployable to front or rear - as shown here).

S Staff and picquet.

⌐ Soldiers with partisans (one rank) drawn up to front and flanks of each 'sleeve'.

Each 'sleeve' contains two platoons of each battalion.

Soldiers with partisans drawn up as rear rank; spare numbers join Staff and picquet in centre.

Colour parties withdrawn to the rear (regimental colours to centre, battalion colours to left and right).

Enemy

1st 'sleeve'
2 Pl. 2 Pl.   12 ranks
(192 men)
4 Pl. 4 Pl.

2nd 'sleeve'
6 Pl. 6 Pl.
36 ranks
5 Pl. 5 Pl.

3rd 'sleeve'
3 Pl. 3 Pl.
1 Pl. 1 Pl.
16 files

G   G   G   G

5 ranks S 25 men
5 files

6 ranks   8 files   Right-hand battalion
Left-hand battalion
G 8 files 3 ranks

N.B. Diagrams not drawn to scale.

TACTICAL EXPERIMENTS IN THE MID-EIGHTEENTH CENTURY

Although many nations continued to employ the basic linear tactics and platoon-firing systems developed during the War of the Spanish Succession, a few military thinkers attempted to devise superior forms of tactical organization. These men scorned the unpractical concepts of drill-masters, and sought greater flexibility and effectiveness.

The reformers, however, fell into two main schools: those who championed firepower as the main desiderata on the battlefield, and those who believed that shock action was the critical function. This was by no means a new contention; indeed, it was almost as old as war itself, going back to the invention of missiles.

The champions of firepower had largely triumphed, in the case of infantry, during the days of Marlborough and Eugene, but shock was certainly preferred for cavalry by both these commanders and by Charles XII of Sweden. Interestingly enough, the matter remained resolved for the cavalry, and not until the late nineteenth century would

experts seriously re-advocate the mass use of 'mounted infantry', although dragoons were supposed to be capable of a mounted or dismounted role. But, in the case of the infantry, the argument raged on from generation to generation.

The Spanish Marquis de Santa Cruz, writing in the 1730s, advocated a compromise solution. Retaining the twin line of battle (see diagram), he placed some battalions in line to gain firepower and flanked them with other formations drawn up in dense columns, interspersed with forces of cavalry and backed them with small mixed bodies of supporting troops. He was particularly insistent on the retention of a sufficiently strong reserve. His ideas, although not widely used in his own day, were not far from the concept of l'ordre mixte of the French Revolution, Consulate and Empire (see p. 98).

Marshal de Saxe (see p. 69) agreed with some of Santa Cruz's views, but he doubted the importance of firepower as the pre-eminent consideration. "I have seen whole salvoes fail to kill four men", he wrote; although he was aware of the value of the delayed volley. Nevertheless, he stressed the importance of shock action, and even advocated a return to the use of a shortened version of the pike. For firepower, he placed greater faith on very light artillery pieces—his amusettes firing a half-pound ball—rather than the musket. He stressed the need for light troops out ahead serving as skirmishers, and believed in maximum tactical flexibility: line formations for static situations, regimental columns (far smaller than Santa Cruz's) for movement.

His sometime mentor, the Chevalier de Folard, 1669-1752, was for many years regarded as the proponent of infantry attack in column, l'ordre profonde (see diagram 'Folard's Column'). He sparked off a considerable contention by his outspoken preference for shock action, but in real terms he, like the others, advocated a compromise. Nevertheless, he was opposed to massed volleys, believing that shock alone could resolve battlefield situations.

The great trainer of what would become Frederick the Great's army, Leopold of Anhalt-Dessau, clung obstinately to the output of fire as the critical factor in battle. Frederick was also of this opinion, but believed that the psychological impact of blasts of fire at regular intervals delivered by moving lines of battalions was far more important than accuracy or overall effect. His goose-stepping lines, firing five or more volleys a minute, made an imposing sight on many a battlefield.

In France, Guibert and other military thinkers weighed the respective merits of the various schools of thought, and Marshal de Broglie held field experiments. The outcome—incorporated in the *Drill-Book* published in 1788—was a combination of battalions in line and battalions in column, supported by skirmishers to the fore and flanks. Great Britain, however, clung to linear concepts until after Waterloo, although from the 1790s she had begun to add light infantry units to her armies.

Colonial Rivalry between the Great Powers

Although the Peace of Utrecht had ended open fighting between British and French colonists in North America and the West Indies, friction and tension continually bedevilled the regions. French interest in developing the defences of the great rivers St. Lawrence and Mississippi was their main concern for over 25 years, but to the south the expansion of British colonies led to trouble with Spain. The building of Savannah (1733) resulted in continual friction with the Spaniards, and many minor campaigns took place—including the siege of St. Augustine —before the new British colony of Georgia felt secure.

Commercial and naval rivalry in the New World led to worsening relations between Great Britain and Spain; and from 1739 to 1741 there resulted the War of Jenkins's Ear, which was triggered off by the arrest and mutilation of a British captain (or so it was alleged) by Spanish coast guards. This was mainly a maritime struggle, but it soon expanded into a new Franco-British confrontation that coincided with, and merged into, the War of the Austrian Succession. During this conflict, and even more after the 'Diplomatic Revolution' and the outbreak of the Seven Years War (see p. 69), the colonial struggle rose to a new pitch of intensity. Not only in the traditional New World theatres but also in India, where the East India Company, backed by the Crown, took the first somewhat hesitant steps, under the prompting of such men as Robert Clive, that would lead eventually to the foundation of the British Indian Empire.

Whilst the major campaigns of the Seven Years War raged in Europe, a no less critical feud was prosecuted in North America where the old question of colonial supremacy was at stake. As first explorers and then settlers sought new areas to develop and exploit, they inevitably clashed with French 'colons' who were motivated by the same search for new markets and raw materials.

One major confrontation took place along the great River Ohio. In 1753 a force of Virginian militia, trying to establish a fort at the junction of the rivers Allegheny and Monongahela and thus pre-empt French interests in the area, was flung back. The French then established Fort Duquesne. Attempting to regain the initiative, in 1755 General Braddock led an expedition to disaster; the unsuitability of European tactics amidst dense forest became evident during the French and Indian War.

Under William Pitt the Elder, however, a consistent and effective strategy was put into operation. While British money was used to subsidize Prussia on the Continent, and thus distract French resources, British troops were sent overseas to conquer the French colonies. As early as 1748 the fortress of Louisbourg, key to the mouth of the St. Lawrence, had been captured, but at the Peace of Aix-la-Chapelle it had been returned to France. Now, in 1758, General Wolfe recaptured this key position, while General Forbes became master of Fort Duquesne on the Ohio front, renaming it Fort Pitt. As the British and American triple-pronged strategy developed, it became evident that French Canada was in peril, and in 1759 Wolfe's capture of Quebec (see p. 82) made its fall inevitable. The same year, in far-away India, Robert Clive won the significant Battle of Plassey (see p. 78), and thereby secured control of Bengal for the British East India Company. This was the first major step towards converting a series of small trading stations—Calcutta, Bombay and Madras—into the basis of a vast empire. Fruitless gains at French expense were made in the West Indies, but the 'sugar islands' were returned at the Peace of Paris in 1763.

Back in North America, the British government decided to retain a regular garrison in the Colonies. The need for this was questioned by the colonists, although the bloody Indian rising under Chief Pontiac in 1763 proved the need for the presence of troops. But friction between the colonists and the authorities steadily grew over two main issues: the attempt to restrict white expansion to a line east of the Appalachian Mountains, and the demand that the Colonies should contribute in the form of taxation towards the costs of their defence without receiving any kind of political representation. Over these issues grew the tension that ultimately led to the rebellion of 1775 and the loss by Great Britain of the 'Old Colonies'.

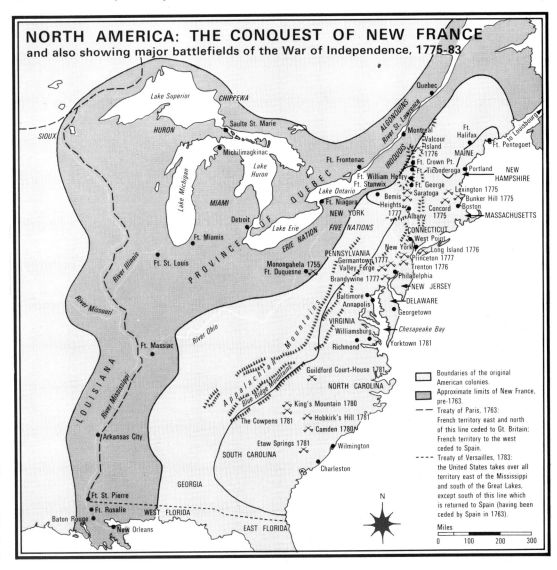

NORTH AMERICA: THE CONQUEST OF NEW FRANCE and also showing major battlefields of the War of Independence, 1775-83

The British Acquisition of India, 1615-1815

European involvement in India began in a small way with the establishment of English trading posts at Madras, 1615, Bombay, 1661, and Calcutta, 1690, while the French set up at Pondicherry (1673) and Chandernagore. With the collapse of Moghul power after 1707, both rival East India Companies became increasingly involved in intrigues and local struggles, but only when the great Frenchman Dupleix arrived on the scene did Anglo-French rivalry become intense. In 1746 the French took Madras, and the British Company at last thought seriously about defending its interests. Henry Lawrence began to organize and train a force of European and Sepoy soldiers. Madras was returned to England at the Peace in 1748.

Until 1757, both countries were more concerned with acquiring influence than with gaining territory. But Dupleix's intrigues in the Carnatic led to the French siege of a British force in Trinchinopoly. In 1751 Robert Clive endeavoured to create a diversion by seizing Arcot, which he then defended with a force of barely 500 men until the following year, when Lawrence, who had been in England, returned to drive off the French and save Trinchinopoly. Dupleix was recalled to France in 1754, and from that year, which also saw the arrival of the first British regular regiment, the 39th Foot, Madras was reasonably secure and British influence predominated in the Carnatic. Clive spent 1753-55 in Great Britain.

In 1756, Prince Siraj-ud-Daula took over Bengal and seized Calcutta. Admiral Watson and Robert Clive led an expedition from Madras, and soon brought the prince to terms. However, the outbreak of the Seven Years War persuaded Clive of the need to capture Chandernagore from the French, which would secure Calcutta once and for all. This was done in March 1757. Alarmed at the predominant British position, Siraj-ud-Daula intrigued with the French and other anti-British princes. Despite this, Clive proved his match in the resulting Battle of Plassey (1757), and the British conquest of Bengal. Meanwhile, Madras had been besieged by Lally, but held out successfully under Lawrence. Then Eyre Coote led a counter-offensive and won the Battle of Wandewash, 22 January 1760. Soon, Pondicherry also fell to the British. Thus ended direct Anglo-French confrontation in India.

The second major phase of British expansion in India lasted from 1763 to 1819, and consisted of struggles with Indian potentates rather than European rivals. Increasingly, the British aim became one of progressive conquest. A revolt in Bengal was crushed at Buxar, 1764, and in war after war, many unwillingly entered into, British power and influence spread over India.

The combination of regular British regiments with East India troops and sepoys proved superior to the native forces, although British victories were often dearly bought, and the security of their gains often challenged.

In 1779, learning of British defeats in North America, Hyder Ali invaded the Madras Presidency from Mysore, but was defeated by Eyre Coote. Soon after Hyder's death in 1782, his son Tippoo Sahib made peace in 1784. Five years later he tried to seize Tanjore, but General Cornwallis forced him to accept a new treaty. In 1799 Tippoo again plotted with the French, who were at this time in Egypt and Syria (see p. 96), and this forced Lord Harris to besiege, and take by storm, Seringapatam on 5 May. This success consolidated British control over southern India.

Another challenge was that of the Mahratta Confederacy, which took 50 years and three wars to subdue. The First Mahratta War, 1775-82, began badly for the Company with the defeat of an expedition from Bombay near Poona. However, the Governor-General, Warren Hastings, marched from Calcutta and restored British prestige, and peace ensued from 1782.

War again broke out in 1803. In this conflict General Lake won successes in the north, and Wellesley found fame at Assaye (see p. 80). By 1805 British influence had spread over other large areas, but further tough struggles followed with the warlike Sikhs and Gurkhas. In 1812 Lord Hastings had to fight the Pindari War which, from 1817, merged into the Third Mahratta War, 1817-19, during which the power of the Mahratta states was finally broken. By 1819, two-thirds of the Indian sub-continent, as well as Ceylon, was under the control of the East India Company. Thus, a series of gifted soldiers and brilliant, if often unscrupulous, administrators and rulers created a vast power-base in southern Asia which compensated in good measure for the loss of the North American colonies and inaugurated the period of the 'Second' British Empire.

The Battle of Plassey, 1757

The superiority of European troops, or European-trained local forces, over massive armies of ill-trained levies, was well demonstrated by the first of these two battles; both of which played significant parts in establishing British control over wide areas of the Indian

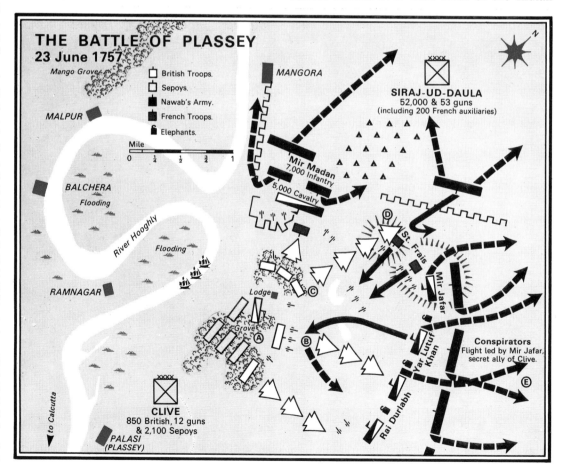

THE BATTLE OF PLASSEY
23 June 1757

Mango Grove

□ British Troops.
□ Sepoys.
■ Nawab's Army.
▤ French Troops.
🐘 Elephants.

MANGORA

SIRAJ-UD-DAULA
52,000 & 53 guns
(including 200 French auxiliaries)

MALPUR

Mile
0   ¼   ½   ¾   1

Mir Madan
7,000 Infantry

5,000 Cavalry

BALCHERA
Flooding

River Hooghly

Flooding

St. Frais

Mir Jafar

RAMNAGAR

Lodge

Grove

Yar Lutuf Khan

Conspirators
Flight led by Mir Jafar,
secret ally of Clive.

to Calcutta

CLIVE
850 British, 12 guns
& 2,100 Sepoys

Rai Durlabh

PALASI
(PLASSEY)

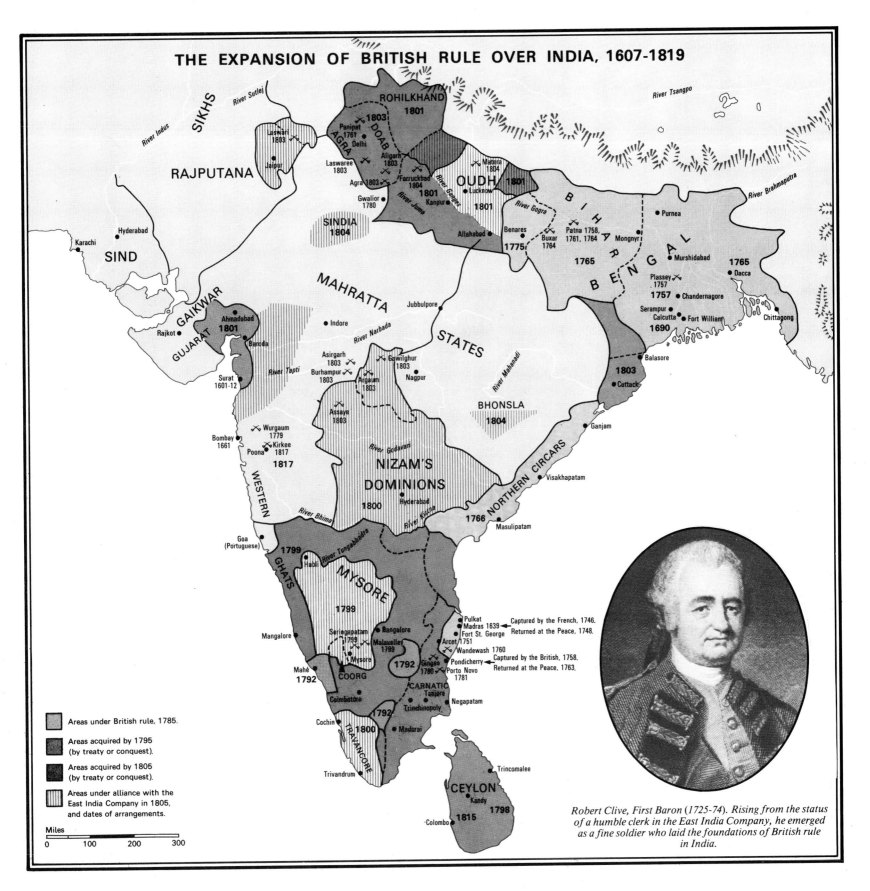

THE EXPANSION OF BRITISH RULE OVER INDIA, 1607-1819

SIKHS

River Tsangpo

River Sutlej

River Indus

ROHILKHAND
1801

1803
Panipat
1761
Delhi

DOAB

AGRA

Laswari
1803

RAJPUTANA

Jaipur

Laswaree
1803

Aligarh
1803

Mattera
1804

OUDH 1801

Agra 1803

Farruckbad
1804

Gwalior
1780

1801

River Jumna

Kanpur

1801

River Ganges

Lucknow

River Gogra

B I H A R

River Brahmaputra

Karachi

Hyderabad

SIND

SINDIA
1804

Allahabad

Benares

1775

Buxar
1764

Patna 1758,
1761, 1764

Mongnyr

1765

Purnea

Murshidabad

1765

Dacca

GAIKWAR

Ahmadabad
1801

GUJARAT

Rajkot

Baroda

M A H R A T T A

River Tapti

Indore

Jubbulpore

River Narbada

STATES

Plassey
1757

1757 Chandernagore

Serampur

Calcutta Fort William

1690

Chittagong

Surat
1601-12

Asirgarh
1803

Burhampur
1803

Gawilghur
1803

Argaum
1803

Nagpur

River Mahanadi

BHONSLA
1804

Balasore

1803

Cuttack

Bombay
1661

Wurgaum
1779

Poona Kirkee
1817

1817

Assaye
1803

River Godavari

Ganjam

WESTERN

NIZAM'S
DOMINIONS

1800

Hyderabad

NORTHERN CIRCARS

Visakhapatam

River Bhima

River Krishna

1766

Masulipatam

Goa
(Portuguese)

GHATS

1799

River Tungabhadra

Hubli

MYSORE

1799

Mangalore

Seringapatam
1799

Bangalore

Malavelley
1799

Mysore

Mahé
1792

COORG

1792

Pulkat
Madras 1639
Fort St. George

Arcot 1751

Wandewash 1760

Pondicherry

Porto Novo
1781

Gingee
1780

Captured by the French, 1746.
Returned at the Peace, 1748.

Captured by the British, 1758.
Returned at the Peace, 1763.

Coimbatore

Cochin

1800

CARNATIC

Tanjore

Trinchinopoly

Negapatam

Madurai

TRAVANCORE

Trivandrum

CEYLON

Kandy

1815

Colombo

1798

Trincomalee

Areas under British rule, 1785.

Areas acquired by 1795
(by treaty or conquest).

Areas acquired by 1805
(by treaty or conquest).

**Areas under alliance with the
East India Company in 1805,
and dates of arrangements.**

Miles

0   100   200   300

*Robert Clive, First Baron (1725-74). Rising from the status
of a humble clerk in the East India Company, he emerged
as a fine soldier who laid the foundations of British rule
in India.*

sub-continent. But the second engagement showed how formidable well-led native armies could be.

Robert Clive, with typical political cunning, prepared the way for his triumph in Bengal by exploiting the disputed succession of the Nawab, Siraj-ud-Daula, backing the rival prince, Mir Jafar, in great secrecy. Organizing a small army of 850 British troops and 2,100 Sepoys, he advanced from Calcutta to Plassey, where he encountered the 52,000-strong array of the Nawab, whose 53 guns were partly served by a French detachment under St. Fraise.

Despite his massive numerical disadvantage, Clive knew that Mir Jafar was at work undermining the loyalty of Siraj-ud-Daula's supporters. Clive's minute force took up a position amidst a mango grove (A) near the swollen River Hooghly, and at about 8.30am on 23 June a four-hour artillery duel opened. Soon, a torrential downpour put most of the Nawab's guns out of action, while Clive's better-protected thirteen cannon continued to fire and repulsed a tentative Indian cavalry attack with grape shot (B). A British officer then led an unauthorized advance to a nearby pond (C) to engage the enemy gunners beyond. Clive decided to exploit this rash action to the limit. Building up his strength, he forced the foes to his immediate front to retire, and then ordered all his men forward, storming into the Nawab's camp (D). As the astounded Indians recoiled, Mir Jafar and other conspirators took the opportunity to call for—and lead—a precipitate withdrawal (E). Siraj-ud-Daula fled the field, and was assassinated two days later. Mir Jafar became the new Nawab, and the British East India Company was virtually in control of the vast province of Bengal. The battle cost the Indians some 500 killed, and Clive just 65 casualties, but its overall political significance far outweighed its military aspects. Clive's coolness and iron nerve contrasts with the ineptitude of his enemies.

The Battle of Assaye, 1803

A generation later, Arthur Wellesley led 13,500 troops to a hard-won success over a huge army of the united princes of Scindia and Berar at Assaye, whose forces included over 20,000 famous Mahratta cavalry and 98 guns: perhaps 40,000 of their vast hordes were effective troops. The Indian position extended for seven miles behind the River Kaitna. Realizing that a frontal attack against such odds would not succeed, Wellesley decided to adopt Frederick the Great's 'oblique' attack (see p. 73). Using his cavalry to hold the enemy (A) between Kodully and Taunklee, he marched his infantry and guns to cross a ford near Warour (B) and attacked the Mahratta left, using the rivers Kaitna and Juah to secure his flanks, before the enemy could redeploy.

The Mahratta cavalry opposed Colonel Maxwell's cavalry (C), but were held in check while the infantry made their crossing. However, the Indian guns were redeployed by the European soldier-of-fortune, Pohlmann, to

Arthur Wellesley—the future Duke of Wellington—as a young man (1769-1851). He developed his military skills in India, becoming both a skilled tactician and an able administrator, attributes that would stand him in good stead in later campaigns in the Peninsula and Belgium.
Right: The Battle of Assaye (23 September 1803) established Arthur Wellesley as a skilled tactician. Noticing an unmarked ford over the River Kaitna, he outflanked the huge Mahratta army and won a decisive victory.
(National Army Museum)

face Wellesley's main force far faster than Wellesley had hoped (D). Wellesley pressed his attack despite heavy casualties, cleared the enemy guns, and the enemy infantry beyond fell back some way (E). Attacks on Assaye village itself foundered against its defences and guns (F), but when the Mahratta horsemen charged they were met and routed by Maxwell's cavalry in a spirited action (G). Maxwell then plunged into the columns of Indian infantry beyond, which also turned to flee. Thus the situation on Wellesley's right wing was redressed.

The battle was still far from won, however, for the enemy centre and reserves now took up a new position with their backs to the Juah. Wellesley did not hesitate, but used Maxwell's re-formed horsemen to watch his flank whilst he swung his handful of battalions in an arc towards Assaye (H). The Mahratta infantry broke and ran for the Juah, pursued by Maxwell (I) while Wellesley busied himself retaking some guns that the foe had turned against his rear. At 6pm the Indians fell back and the battle ended. Wellesley's army was exhausted, having sustained 1,600 casualties and inflicted 1,200. The Second Mahratta War would drag on until 1805, but after Assaye the outcome was never in doubt.

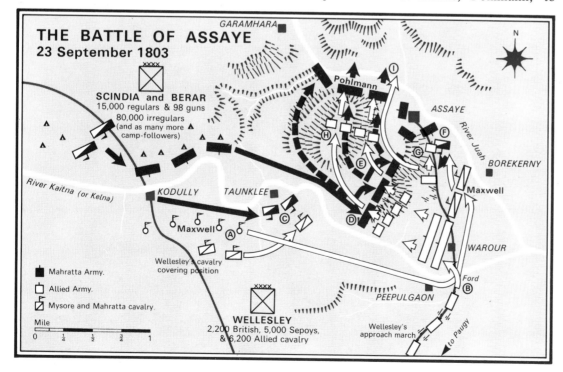

THE BATTLE OF ASSAYE
23 September 1803

SCINDIA and BERAR
15,000 regulars & 98 guns
80,000 irregulars
(and as many more camp-followers)

GARAMHARA

Pohlmann

ASSAYE

River Juah

BOREKERNY

River Kaitna (or Kelna)

KODULLY   TAUNKLEE

Maxwell

Maxwell

WAROUR

Wellesley's cavalry covering position

■ Mahratta Army.
□ Allied Army.
Mysore and Mahratta cavalry.

Ford

PEEPULGAON

Mile
0   ¼   ½   ¾   1

WELLESLEY
2,200 British, 5,000 Sepoys,
& 6,200 Allied cavalry

Wellesley's approach march

to Paugy

Combined Operations

As an island race, the British have always appreciated the significance of sea-power. When their armies fought overseas, it was the duty of the navy to transport, supply and reinforce soldiers after gaining command of the seas from their enemies. Unfortunately, it was not always easy for soldiers and sailors to co-operate wholly effectively when combined operations were the order of the day. But against such fiascos as the Isle of Rhé, 1627, the loss of Minorca, 1782, or the abortive raids on Cadiz and Ferrol in 1799, it is possible to set a list of considerable successes, of which the three described here are outstanding examples.

The Siege and Capture of Quebec, 1759

To clinch the British conquest of French Canada it was necessary to take Quebec, which was a considerable fortress and town dominating the St. Lawrence. By late May 1759, British forces were assembled at Louisbourg—namely, 22 ships of the line, a dozen frigates under Admirals Saunders and Holmes, and some 14,000 troops commanded by General James Wolfe, aged thirty-two years. On 27 June the expedition landed on the Île d'Orléans.

The French Governor-General, Vaudreuil, and the Marquis de Montcalm, commander of the forces around Quebec, mustered some 18,000 men (including Indian allies) and over 100 guns, spread between the city and the great Beauport encampments along the west bank of the river. From the first they conducted an active defence, but by early July the British had taken Point Levis and established a strong camp north of the River Montmorency. Wolfe, bedevilled by sickness and frequent disagreements with his three Brigadier-Generals, Monckton, Townsend and Murray, failed in an attempt to attack Beauport on 30 July. Changing his strategy, he sent a force under Murray and escorted by Holmes up the river past the guns of Quebec, 5 August. This threat to their sources of supply induced the French to send a 3,000-strong force up-river. Murray wrought considerable havoc before returning to the main army on 24 August.

Anxious lest the river should freeze up and compel Saunders to evacuate, Wolfe debated three plans of action. He decided to abandon the Montmorency position (31 August—2 September) and mass his forces on the east bank. By 9 September the British were masters of St. Nicolas—the aid given by Saunders's fleet being of primary importance—and Wolfe was poised for his last great effort: a direct assault against Quebec via the forbidding Heights of Abraham.

On 12 September 3,800 troops embarked on boats above Quebec and drifted silently downstream through the night. The landing at Anse de Foulon took the French completely by surprise; a feint attack against the Beauport defences by the boats of the fleet on the 12th having drawn all their attention. By 6am Wolfe's troops were on the crest of the cliffs, and advancing towards Quebec. Less than a mile from the city, Wolfe deployed his men into line—just two ranks deep to disguise their relative weakness supported by only two 6-pounder guns—while Montcalm came hastening back from Beauport. Montcalm left Quebec at 10am with 4,500 men and attacked. Wolfe and Monckton were both shot down, but the British held their fire until they were 10 yards from the French. Then, their disciplined volleys tore into the enemy. As Wolfe died, the French began to retire, and Murray set off to isolate the French from their bridges over the St. Charles River while Monckton's brigade marched on Quebec. But Townsend, who was now in command, ordered both forces to regroup. As the battle ended, Montcalm fell mortally wounded. Vaudreuil decided to slip away, and on 18 September Quebec surrendered. Further operations continued for another year, but on 9 September 1760 Canada finally became British.

The Great Siege of Gibraltar, 1779-83

If Quebec was a notable British offensive success, the holding of Gibraltar was an outstanding defensive achievement. It was one of the few successes to grace the last part of the American War of Independence, which had entered the phase of the European coalition against Britain.

Gibraltar—a British possession since 1704—was defended by General Eliott, Governor and Commander-in-Chief, who headed a garrison comprising some 5,000 British and Hanoverian troops, and 412 fortress guns. Additional support was provided by a small naval flotilla. In June 1779 a loosely co-ordinated Spanish blockade by land and sea began, but supplies continued to reach the garrison intermittently. In January 1780 the Royal Navy passed in stores and 1,000 reinforcements, a feat that was repeated in April of the next year under the very noses of the Spanish fleet. In the same month (April 1781), Spanish guns opened a heavy bombardment that was to last, on and off, for fifteen months. However, although the town suffered heavy damage the defences remained strong, and morale ran high as 'Old cock o'the Rock' Eliott maintained an iron discipline. On 27 September he launched a successful sortie against Las Lineas, spiking many enemy guns for scant loss before retiring back into Gibraltar.

The capture of Minorca in 1782 released the Duc de Crillon's French army (some 26,000 men) to reinforce Don Alvarez's 14,000 Spaniards, and the siege took on a more serious guise. A joint Franco-Spanish fleet of 40 ships was also available. On 13 September a grand assault was launched:—10 floating batteries were towed to engage the King's Bastion. A maelstrom of fire continued all night, but early on the 14th the British gained the upper hand as Captain Curtis led his 12 gunboats into action. For a loss of 2,000 men the French and Spaniards had failed, and their determination wavered. Next month Admiral Lord Howe broke the blockade and reduced the siege to a sham which nominally dragged on until 5 February 1783. Great Britain had retained possession of the key to the Mediterranean.

The Invasion of Egypt, 1801

The landing of an army on a hostile shore poses great challenges, requiring great inter-service co-operation and co-ordination. The invasion by General Abercromby's army proved a great success thanks to the planning skills of Captain Cochrane, and Major-General Sir John Moore commanding the assault troops. The attack was mounted from Marmorice Bay in Turkey, where Abercromby's 15,000 troops prepared to embark on 100 transports, and 57 Turkish vessels. Admiral Lord Keith had a fleet of five ships of the line and two frigates for the escort of this armament to Egypt.

After several weeks of practice, the invasion force set sail for its objective on 22 February 1801. Seven days later, the fleet and its convoy anchored off Aboukir Bay to the east of Alexandria, but unfortunately its final approach was known to the French. General Menou, besides holding the strong position of Aboukir Castle at one end of the bay, also placed about 6,000 men and some 40 guns amongst the dunes. Bad weather caused two postponements of the landing, giving the French even more time to prepare their positions. On the 7th the order was issued: the landing would be on the 8th.

Long before dawn the boats of the fleet began to assemble. The long weeks of practice at Marmorice now bore fruit as the three flotillas of cutters and barges carrying the troops formed up behind the cover of the inshore squadron of bomb-vessels, gunboats and sloops which were of sufficiently shallow draught to enable them to afford close support.

Keith's battle fleet lay farther out, beyond the anchored transports. Not until 9am were all boats in position, but then, despite the heavy fire of French cannon, the landing went ahead with exemplary smoothness. Within an hour of the first troops landing, Moore had 6,000 men ashore, and although French cavalry rode down to attack troops still in the boats, the central dune was stormed by Moore at the head of the 40th, 23rd and 28th Foot, who routed the French 61st demi-brigade, whilst the Coldstream and Third Guards, with the 42nd Highlanders, drove off the French cavalry on the right. For 600 casualties the British were masters of the beach, and by 4pm the whole army was ashore and a very successful operation was over. (Also see p. 97.)

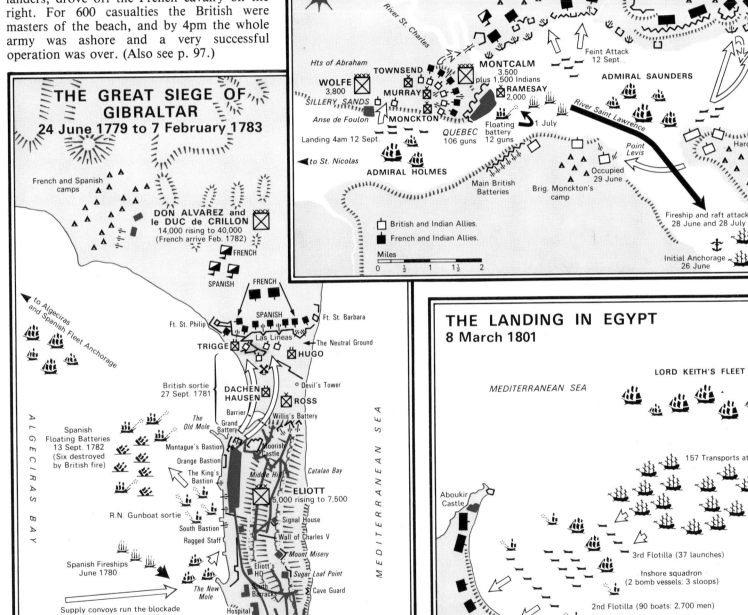

THE SIEGE AND BATTLE OF QUEBEC
June to September 1759

Falls
R. Montmorency

VAUDREUIL
FRENCH and INDIAN ALLIES
14,000 reducing to 10,800 by September

BEAUPORT

Main camp from 9 July

River St. Charles

Hts of Abraham

TOWNSEND
WOLFE 3,800
MURRAY
MONTCALM 3,500 plus 1,500 Indians

Feint Attack 12 Sept.

Attack of 30 July (fails)

ADMIRAL SAUNDERS

SILLERY SANDS
MONCKTON
RAMESAY 2,000

BRITISH and INDIAN ALLIES c. 7,500

Anse de Foulon
Landing 4am 12 Sept.

QUEBEC 106 guns

Floating battery 12 guns

1 July

River Saint Lawrence

Hardy's camp

◄ to St. Nicolas

Point Levis

ILE D'ORLEANS

ADMIRAL HOLMES

Occupied 29 June

First camp from 26 June

Main British Batteries

Brig. Monckton's camp

Fireship and raft attacks 28 June and 28 July

☐ British and Indian Allies.
■ French and Indian Allies.

Initial Anchorage 26 June

Miles
0   ½   1   1½   2

THE GREAT SIEGE OF GIBRALTAR
24 June 1779 to 7 February 1783

French and Spanish camps

DON ALVAREZ and le DUC de CRILLON
14,000 rising to 40,000 (French arrive Feb. 1782)

FRENCH

SPANISH

FRENCH

SPANISH

◄ to Algeciras and Spanish Fleet Anchorage

Ft. St. Philip
Las Lineas
Ft. St. Barbara
The Neutral Ground ►

TRIGGE
HUGO

British sortie 27 Sept. 1781

Devil's Tower

DACHENHAUSEN
ROSS

Barrier
Willis's Battery
The Old Mole
Grand Battery
Moorish Castle

Spanish Floating Batteries 13 Sept. 1782 (Six destroyed by British fire)

Montague's Bastion
Orange Bastion
The King's Bastion
Middle Hill
Catalan Bay

R.N. Gunboat sortie

ELIOTT 5,000 rising to 7,500

Signal House
Wall of Charles V
South Bastion
Ragged Staff

A L G E C I R A S   B A Y

M E D I T E R R A N E A N   S E A

Spanish Fireships June 1780

Mount Misery
Eliott's HQ
Sugar Loaf Point
The New Mole
South Barracks
Cave Guard

Supply convoys run the blockade Jan. 1780 and Apr. 1781

Hospital
Rosia Bay
Buena Vista
Little Bay
Europa Flats
Windmill Hill

Admiral Howe's Relief Fleet Oct. 1782

Europa Point

N

☐ British and Hanoverian Forces.
■ Spanish and French Forces.
▨ Areas of razed trenches.

Mile
0   ¼   ½   ¾   1

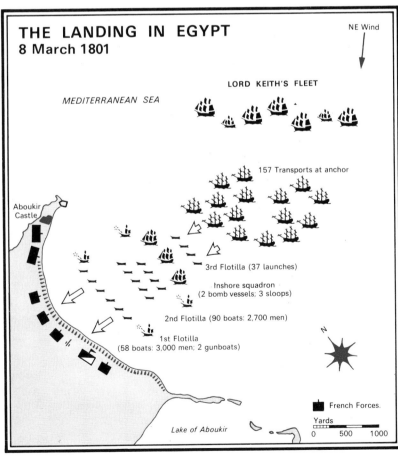

THE LANDING IN EGYPT
8 March 1801

NE Wind

LORD KEITH'S FLEET

MEDITERRANEAN SEA

157 Transports at anchor

Aboukir Castle

3rd Flotilla (37 launches)

Inshore squadron (2 bomb vessels; 3 sloops)

2nd Flotilla (90 boats; 2,700 men)

1st Flotilla (58 boats; 3,000 men; 2 gunboats)

N

■ French Forces.

Yards
0   500   1000

Lake of Aboukir

The American War of Independence, 1775-83

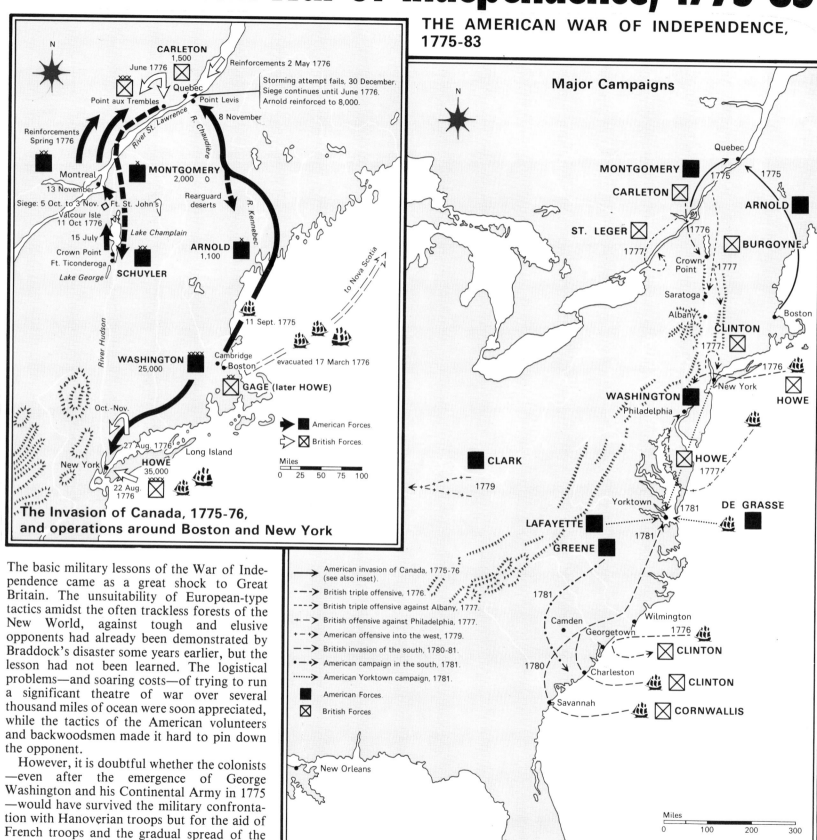

THE AMERICAN WAR OF INDEPENDENCE, 1775-83

Major Campaigns

The Invasion of Canada, 1775-76, and operations around Boston and New York

Inset map labels: N, CARLETON 1,500, June 1776, Reinforcements 2 May 1776, Quebec, Point aux Trembles, Point Levis, Storming attempt fails, 30 December. Siege continues until June 1776. Arnold reinforced to 8,000. Reinforcements Spring 1776, 8 November, River St. Lawrence, R. Chaudière, MONTGOMERY 2,000, Montreal, 13 November, Rearguard deserts, R. Kennebec, Siege: 5 Oct. to 3 Nov. Ft. St. John's, Valcour Isle 11 Oct 1776, 15 July, Lake Champlain, ARNOLD 1,100, Crown Point Ft. Ticonderoga, Lake George, SCHUYLER, to Nova Scotia, River Hudson, 11 Sept. 1775, WASHINGTON 25,000, Cambridge, Boston, evacuated 17 March 1776, GAGE (later HOWE), Oct.-Nov., New York, 27 Aug. 1776, Long Island, HOWE 35,000, 22 Aug. 1776

American Forces.
British Forces.

Miles
0 25 50 75 100

Main map labels: N, Quebec, MONTGOMERY, CARLETON, 1775, 1775, ST. LEGER, ARNOLD, 1776, 1777, BURGOYNE, 1777, Crown Point, 1777, Saratoga, Albany, Boston, CLINTON, 1777, 1776, WASHINGTON, New York, HOWE, Philadelphia, HOWE, 1777, CLARK, 1779, Yorktown, 1781, DE GRASSE, LAFAYETTE, 1781, GREENE, 1781, Camden, Wilmington, Georgetown, 1776, CLINTON, 1780, Charleston, CLINTON, Savannah, CORNWALLIS, New Orleans

Legend:
American invasion of Canada, 1775-76 (see also inset).
British triple offensive, 1776.
British triple offensive against Albany, 1777.
British offensive against Philadelphia, 1777.
American offensive into the west, 1779.
British invasion of the south, 1780-81.
American campaign in the south, 1781.
American Yorktown campaign, 1781.
American Forces.
British Forces.

Miles
0 100 200 300

The basic military lessons of the War of Independence came as a great shock to Great Britain. The unsuitability of European-type tactics amidst the often trackless forests of the New World, against tough and elusive opponents had already been demonstrated by Braddock's disaster some years earlier, but the lesson had not been learned. The logistical problems—and soaring costs—of trying to run a significant theatre of war over several thousand miles of ocean were soon appreciated, while the tactics of the American volunteers and backwoodsmen made it hard to pin down the opponent.

However, it is doubtful whether the colonists —even after the emergence of George Washington and his Continental Army in 1775 —would have survived the military confrontation with Hanoverian troops but for the aid of French troops and the gradual spread of the struggle to involve Spain as well as France in

Europe itself. Even so, this intervention might have come too late to be effective had the British government of George III and Lord North made some attempt to rally the loyalty of the mass of the colonists during the earlier years of the struggle. But as it was, even the Loyalists were scorned, ignored or exploited. Only 2,500 Loyalist volunteers were enlisted in the British ranks, and great havoc was wrought on the property of all Americans, whether friend or foe, by the Hanoverian and Hessian troops of General Gage and Lord Howe. Further, the British use of Indian allies inevitably alienated every frontiersman. Therefore, the ultimate British defeat came as a result of great political ineptitude as well as of the inability to master the military problem.

The First Continental Congress met at Philadelphia in September 1774, and this proved the unity of opposition felt towards British rule. The first military clashes came at Lexington Green on 19 April 1775, and around Concord. In May, the Second Continental Congress prepared for war, appointing Washington as commander-in-chief of the army to defend the 13 colonies on 14 June.

This mainly militia force was not at first very redoubtable, and the Americans suffered many reverses. Their attempt to invade Canada, June 1775 to July 1776, proved too complex for their resources and came to grief outside Quebec on 30 December, where General Montgomery was killed and Benedict Arnold wounded. The siege continued until June 1776, but the British commander, Carleton, held out, and by July Arnold had been forced back to Ticonderoga. General Howe's success at Brooklyn on 27 August was a further blow for the colonists, off-setting the victory they had claimed in March 1776 when Howe had evacuated Boston. But Washington's success at Trenton, 26 December, followed by another success at Princeton on 2 January 1777, rallied the rebel morale. So far neither side had gained a significant advantage.

The year 1777 saw two major British offensives against Albany and Philadelphia. But it proved impossible to co-ordinate the various columns involved, and the surrender of General Burgoyne at Saratoga (see over) came as a climax. Howe's forces won a battle at Germantown in Pennsylvania, 4 October, but the critical event of the war came the following February, when France negotiated a treaty of alliance with the colonists. That same winter Washington and his starving army narrowly survived bitter weather at Valley Forge, but the turning point in the struggle had in fact been reached.

The colonial revolt was now fast becoming a general war, and the fluctuating struggle for control of the Atlantic formed the vital background to the campaigns fought on land, where Lafayette's French corps brought Washington support and expertise. Great Britain became totally isolated, with no ally in Europe, and her

Above: The surrender of General Charles Cornwallis at Yorktown in October 1781 proved the death-knell to British rule in the thirteen American Colonies. Here he is depicted capitulating to George Washington.
Above right: Charles, Marquis Cornwallis (1738-1805) was the ablest British commander in the North American struggle. He won battles at Camden and Guilford Court-House before being forced to surrender at Yorktown. He later commanded with success in India during the Third Mysore War and also in Ireland.
Right: George Washington (1732-99), first President of the United States. His building of the Continental Army and ability to cooperate with French and German volunteers proved long-sighted, but as a field commander he was more noted for perseverence than for great talent.

resources became fatally overstretched. The British invasion of the southern states in 1780-81 made considerable progress, but turned out to be the prelude to the disaster at Yorktown which was largely due to French control of the sea approaches. Yorktown sounded the knell for British rule over the Old Colonies, and British prestige sank to a very low point that only Eliott's staunch defence of Gibraltar (see p. 82) and Rodney's naval victories did anything to restore. Thus within twenty years the brilliant successes of the Peace of Paris had given way to the humiliations reflected in the Treaty of Versailles.

The Battle of Brandywine, 1777

The British plan for the overthrow of the Revolt in 1777 has already been outlined. As Burgoyne approached Saratoga, Major-General Howe decided to call off his complementary northward advance towards the rendezvous at Albany, and instead struck south and west to threaten Philadelphia. After landing from the fleet, he found the Delaware River blocked, so the British force advanced slowly overland to seek out Washington's army. The result was the Battle of Brandywine near West Chester.

Washington (11,000 strong) was in a defensive position behind Brandywine. So, with great skill Howe executed a superb plan of envelopment; using General Knyphausen with 5,000 men (A) to pin his foe's attention to the west, Howe sent Cornwallis with 10,000 on a circuitous march over Jeffrey's Ford to outflank Washington's position (B). Taken by surprise, Washington somehow managed to extemporize a line to meet the new threat, and these units successfully covered the retreat of the bulk of the American Army. This defeat cost Washington 1,000 casualties and 11 guns. Soon after, the British occupied Philadelphia. On 4 October, however, Washington regained his prestige at Germantown. Gates' subsequent success at Saratoga further improved patriot morale.

The Battle of Saratoga, 1777

Meanwhile, only Lt. Gen. John Burgoyne's column made any real headway and, after occupying Crown Point and Ticonderoga, he pressed down the Hudson River towards his first major target, Albany. On 14 September he crossed to the west of the river by a pontoon bridge, and soon found evidence of Major General Horatio Gates' and Benedict Arnold's presence in the area. The American force, roughly 13,000 strong, had fortified the area of the Bemis Heights near the village of Saratoga.

Bedevilled by disease and shortages of supplies, Burgoyne nevertheless decided to attack. On the 19th, advancing in three columns, the British force of approximately 5,000 marched through a dense fog towards Freeman's Farm (A). At about midday the central force under Burgoyne came under heavy fire from Morgan's riflemen, while Gates on the right and Arnold on the left led up 4,000 more Americans to oppose them. The 1,500 British troops managed to hold back this superior force until Riedesel's column managed to outflank the American line (B). Whereupon, after removing the fiery Arnold from his command, Gates conceded the field.

A three week lull ensued. Burgoyne created an entrenched camp (C) while reinforcements reached the Americans. As his supply position was becoming increasingly desperate, Burgoyne decided to attack again. The result was the Battle of Bemis Heights on 7 October. Burgoyne's 1,500 men of the main attack were halted by patriot fire soon after midday, while Morgan's riflemen moved round the British right (E). The British gave ground as Poor's brigade (F) attacked their left. Then, after a pause, Benedict Arnold appeared on the scene

and brought new energy to the American side. He inspired and led two major charges; the first, against two German regiments in Burgoyne's centre, was beaten off. Undeterred, Arnold attacked the blockhouses (G) before falling wounded, but Morgan's men became masters of these posts at sunset. Burgoyne, who had lost 1,000 casualties, fell back to his camp. On 17 October the 3,000 surviving British and Hessians surrendered. This was the turning point of the war.

The Battle of Yorktown, 1781

If Saratoga proved the turning point in the American War of Independence, Cornwallis's surrender at Yorktown four years later delivered the coup de grâce to British control of the Old Colonies. Determined to recapture New York, and encouraged by the timely arrival of French naval and military aid, Washington probed for British weak points. However, Admiral de Grasse forced his hand by threatening to withdraw his fleet in mid-October, and so Washington left half his Continental Army to pin Clinton in New York while he took the rest under General Lincoln and all of Rochambeau's French forces to

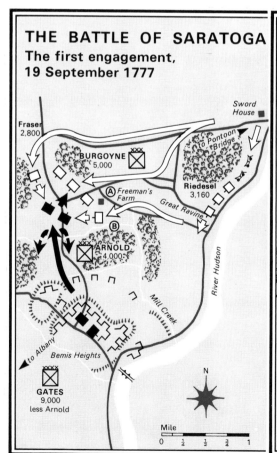

THE BATTLE OF SARATOGA
The first engagement,
19 September 1777

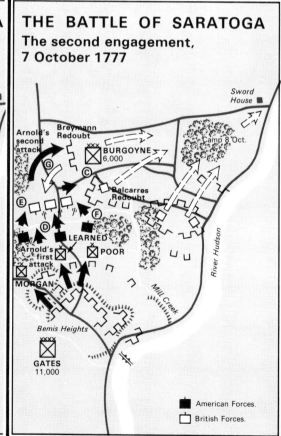

THE BATTLE OF SARATOGA
The second engagement,
7 October 1777

■ American Forces.
□ British Forces.

THE BATTLE OF BRANDYWINE
11 September 1777: A defeat for Washington

Valley Forge
Jeffrey's Ford
ⓑ SCONNELTOWN
N

Cornwallis
10,000

Main attack

Miles
0 ½ 1 1½ 2

☐ British Forces.
■ American Forces.
1 Stirling and Stephen.
2 Sullivan.
3 Wayne.
4 Green.
5 Armstrong.

River Brandywine
Jones' Ford

DILWORTH

to Chester

HOWE
15,000

PARKERVILLE
BRISTOL

Knyphausen
5,000

Diversionary attack
ⓐ Chadd's Ford

WASHINGTON
11,000

1. General William Howe, Viscount (1729-1814). Present at Quebec (1759), he won battles at Brandywine and Germantown during the American War of Independence, but resigned in 1778 over the issue of inadequate reinforcement. (National Army Museum)
2. Marie Joseph Paul Roch Yves Gilbert Motier, Marquis de Lafayette (1754-1834), placed his considerable talents at the disposal of George Washington. His later record during the early campaigns of the French Revolution was less notable.
3. General John Burgoyne (1772-92), known as 'Gentleman Johnny' to his troops, was a talented soldier, but the problems of North America compelled him to surrender at Saratoga in October 1777 after a stiff fight. (National Army Museum)

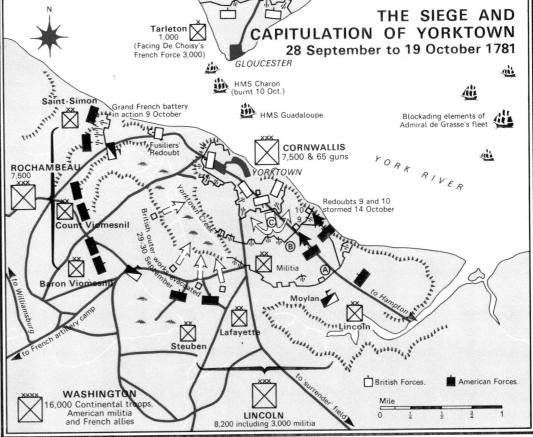

THE SIEGE AND CAPITULATION OF YORKTOWN
28 September to 19 October 1781

N

Tarleton
1,000
(Facing De Choisy's French Force 3,000)
GLOUCESTER

HMS Charon
(burnt 10 Oct.)

HMS Guadaloupe

Blockading elements of Admiral de Grasse's fleet

YORK RIVER

Saint-Simon

Grand French battery in action 9 October

Fusiliers' Redoubt

CORNWALLIS
7,500 & 65 guns

YORKTOWN

ROCHAMBEAU
7,500

Yorktown Creek

British outer works evacuated 29-30 September

Redoubts 9 and 10 stormed 14 October

Count Viomesnil

ⓒ
ⓑ
ⓐ

to Hampton

Baron Viomesnil

Militia

to Williamsburg

to French artillery camp

Moylan

Steuben

Lafayette

Lincoln

WASHINGTON
16,000 Continental troops, American militia and French allies

LINCOLN
8,200 including 3,000 militia

to surrender field

☐ British Forces. ■ American Forces.

Mile
0 ¼ ½ ¾ 1

attack Cornwallis in Virginia on the York River.

De Grasse's success at the Battle of the Capes enabled a key American convoy of munitions and siege guns to reach Jamestown, and by 28 September Cornwallis and his 8,000 men were besieged by land and sea. Progress was steady. On 8 October, heavy guns were within range, and the first parallel opened (A). On the 14th, Redoubts 9 and 10 were stormed by a combined Franco-American force, and that night the second parallel (B) was begun. A strong sortie by the British garrison on the night of 15 October (C) was repulsed, and Cornwallis's attempt to escape to Gloucester was thwarted by a gale and a dearth of shipping. The 17th saw a hurricane of shot and shell hurled at the British positions to commemorate the anniversary of the surrender of Burgoyne at Saratoga in 1777, and that night General Cornwallis, despairing of relief by land or sea, asked for terms. Almost 8,000 men, including the sick, marched out to surrender on 19 October. Thus, the military outcome of the War of Independence had been settled. Britain had, in effect, lost her American Colonies, although this was not finally acknowledged until the Treaty of Versailles on 30 November 1783.

The Challenge of Revolutionary France

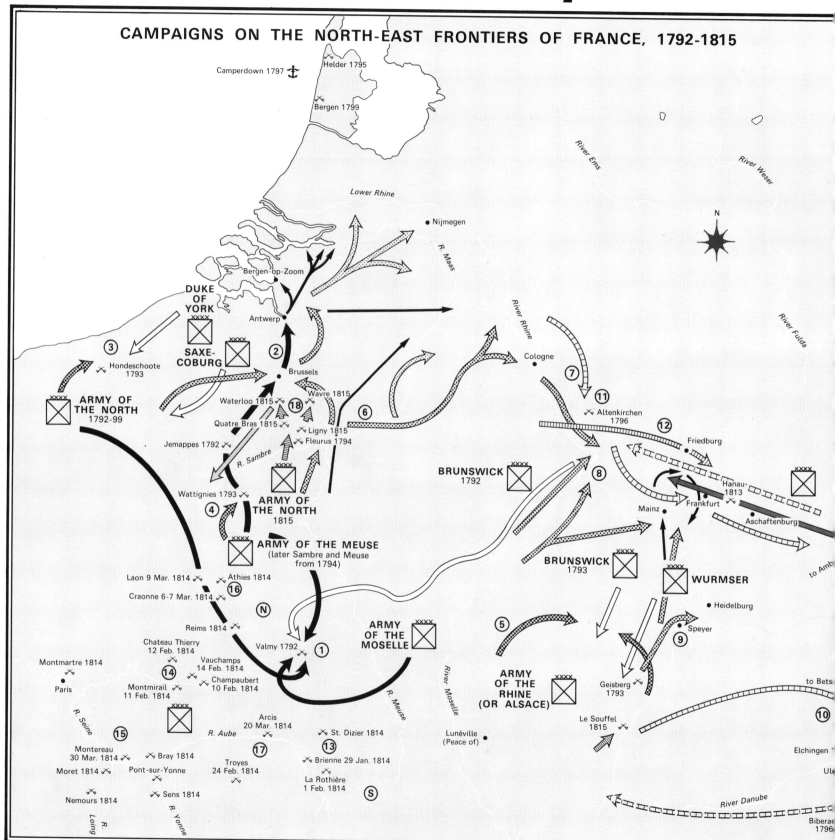

CAMPAIGNS ON THE NORTH-EAST FRONTIERS OF FRANCE, 1792-1815

Camperdown 1797

Helder 1795

Bergen 1799

Lower Rhine

River Ems

River Weser

N

Nijmegen

R. Maas

River Rhine

River Fulda

Bergen-op-Zoom

DUKE OF YORK

Antwerp

③

Hondeschoote 1793

SAXE-COBURG

②

Brussels

Cologne

⑦

⑪

ARMY OF THE NORTH 1792-99

Waterloo 1815

Wavre 1815

⑱

⑥

Altenkirchen 1796

⑫

Quatre Bras 1815

Ligny 1815

Friedburg

Jemappes 1792

Fleurus 1794

BRUNSWICK 1792

⑧

Hanau 1813

R. Sambre

ARMY OF THE NORTH 1815

Frankfurt

Wattignies 1793

Mainz

Aschaftenburg

④

to Amb

Laon 9 Mar. 1814

Athies 1814

ARMY OF THE MEUSE (later Sambre and Meuse from 1794)

BRUNSWICK 1793

WURMSER

Craonne 6-7 Mar. 1814

⑯

Heidelburg

Ⓝ

⑤

Speyer

Reims 1814

⑨

Chateau Thierry 12 Feb. 1814

Valmy 1792

①

ARMY OF THE MOSELLE

Montmartre 1814

⑭

Vauchamps 14 Feb. 1814

Paris

Montmirail 11 Feb. 1814

Champaubert 10 Feb. 1814

River Moselle

ARMY OF THE RHINE (OR ALSACE)

Geisberg 1793

to Bets

Arcis 20 Mar. 1814

R. Seine

⑮

R. Aube

St. Dizier 1814

Le Souffel 1815

⑩

Montereau 30 Mar. 1814

Bray 1814

⑰

⑬

Lunéville (Peace of)

Elchingen

Moret 1814

Pont-sur-Yonne

Troyes 24 Feb. 1814

Brienne 29 Jan. 1814

Ul

Nemours 1814

Sens 1814

La Rothière 1 Feb. 1814

Ⓢ

River Danube

R. Loing

R. Yonne

Biberac 1796

If the campaigns of Frederick the Great form something of a transitional link between the conventions of eighteenth century warfare and the blitzkrieg concepts of Napoleon, the period of the French Revolutionary Wars that preceded his emergence to senior command is of vital significance, as during these years the essential framework of France's military might was forged.

The new type of conflict that began in 1792, when the Revolutionary government challenged Austria and plunged into what became known as the War of the First Coalition, posed France with critical problems. Firstly, the emigration of many officers left the regiments badly depleted; secondly, the economic confusion of the country made it hard to mount an effective military effort; and thirdly, France found herself facing a struggle for both national and revolutionary survival. In the first campaigns, it seemed that having sown the wind France was going to reap the whirlwind.

However, a combination of enemy errors and French expedience made it possible for France to survive and emerge from the period of crisis with notable military potential. The chief architect of this transformation was Lazare Carnot, 'the Organizer of Victory'. Working on the basis of the nation in arms—the near-total mobilization of manpower and resources—the Minister of War set out to build an effective army. First, the levée en masse, once reliance on purely volunteers had proved a disappointment, made possible the mobilization of some 600,000 men. By 1798 a regular system of conscription was in force, making possible larger armies than the eighteenth century powers had ever managed to recruit. The mass of raw material provided was very inexperienced, and so tactics were adapted. The demi-brigades, as the regiments were renamed, comprised three battalions: one of experienced troops, who fought in line or in skirmishing order, and two of new troops, who were massed in dense battalion columns and trained to attack with the bayonet. These tactics were adopted largely from the armies of l'ancien régime and, indeed, the *Drill-Book* of 1788 was the basis of the 'new' system. To lead the new armies, after the failure of experiments based on the election of officers, the concept developed that only proven talent should lead to promotion. Napoleon himself was the product of this system, emerging from obscurity during the siege of Toulon in 1793 to commence his meteoric career. But other fine officers also emerged—many of them young—and the idea of the 'baton in the knapsack' became a notable ingredient of French achievement.

The supply of the large citizen armies posed

1

2

1. Lazare Nicolas Marguerite Carnot (1753-1823). A French engineer officer of outstanding ability, he was dubbed 'the Organizer of Victory' by his rebuilding of the Revolutionary armies to face a hostile Europe. (DAG)
2. Jean Victor Moreau (1763-1813) was a leading general of the French Revolutionary armies. He won the Battle of Hohenlinden in 1800, but later intrigued against Napoleon and was exiled. He was killed at Dresden whilst serving as adviser to Tsar Alexander I. (DAG)

daunting problems, but almost by chance the doctrine of living off the countryside, or making war pay for itself, was found to be practicable—at least in the more fertile areas of Europe. The consequent reduction in reliance upon waggon trains, allied to improvements in the design and reduction in the weight of artillery (once again a trend dating from Gribeauval's reforms of the 1770s), made possible a speed of marching undreamt of by earlier armies. When the inspiration of proselytizing revolutionary zeal and the supreme leadership and strategic flair of Napoleon Bonaparte were added to these factors, the result was indeed a formidable weapon.

But even before Napoleon had taken command of his first army in 1796 the French forces had made their mark. After surviving the first crises of 1792, Revolutionary France (with occasional setbacks) took to the offensive. In a series of campaigns, such able generals as Dumouriez, Jourdan and Moreau flung back their foes on the northern and eastern frontiers of France and liberated vast areas of land to the line of the River Rhine. Penetrations were made deep into Germany, and the first advances into north Italy were also undertaken. By 1795 French armies everywhere were on the offensive, and their new ideas on waging war were already well-developed. Within a few more years French troops would be the military masters of Europe, and to lead them would come one of the greatest soldiers of all time— Napoleon Bonaparte.

North Italy: 1796 and 1800

Northern Italy was the scene of two of Napoleon Bonaparte's most celebrated campaigns. Many of his military methods were first employed in the region, and it was here that he first established his reputation for military genius.

The Campaign of 1796-97

Napoleon's first Italian campaign lasted fourteen months, with a number of pauses. General Bonaparte (aged twenty-six on appointment) assumed command of the dispirited Army of Italy in early April 1796. He set about breaking inland from the barren coastal plain with his 37,600 men to reach the fertile plains of Piedmont before his force disintegrated from neglect and hunger.

This first involved defeating the Piedmontese Army of General Colli, 25,000, and the neighbouring Austrian Army, 22,000, of General Beaulieu. To achieve this, Napoleon first employed his 'inferiority strategy' (or manoeuvre of the central position, see p. 98) to divide his foes and fight each in turn with a local superiority of force. Between 11 and 17 April, the French struck north and seized the central position between the two enemy armies, fighting a number of small battles, including Montenotte, Millesimo, First and Second Dego, and Ceva. Although there were a few initial setbacks, Napoleon's plan worked well and, after being joined by Sérurier's division, his main force drove on to defeat Colli at Mondovi, 22 April. This result induced Piedmont to sign the Armistice of Cherasco on the 28th.

One opponent floored, Napoleon rushed his men back to join Massena who had been guarding the eastern flank against Beaulieu. He was too late, however, to catch the Austrians, who were already in retreat for the River Po. Napoleon allowed his foe—now outnumbered in his turn—no respite. Advancing on Alessandria and Valenza, on 6 May Napoleon put into practice his superiority strategy (or manoeuvre of envelopment, see p. 98) for the first time. While Sérurier pinned Beaulieu's attention near Valenza, the rest of the French forced-marched east screened by the Po to seize a crossing-place at Piacenza in the hope of isolating Beaulieu west of the River Adda, severing his communications with Mantua. However, the Austrians began to retreat just in time while the French, inexperienced in Napoleon's methods, also lost some vital hours. In the resulting Battle of the Bridge of Lodi on 10 May the French merely mauled Beaulieu's rearguard, although Bonaparte's hold on his troops crystallized in this otherwise insignificant action.

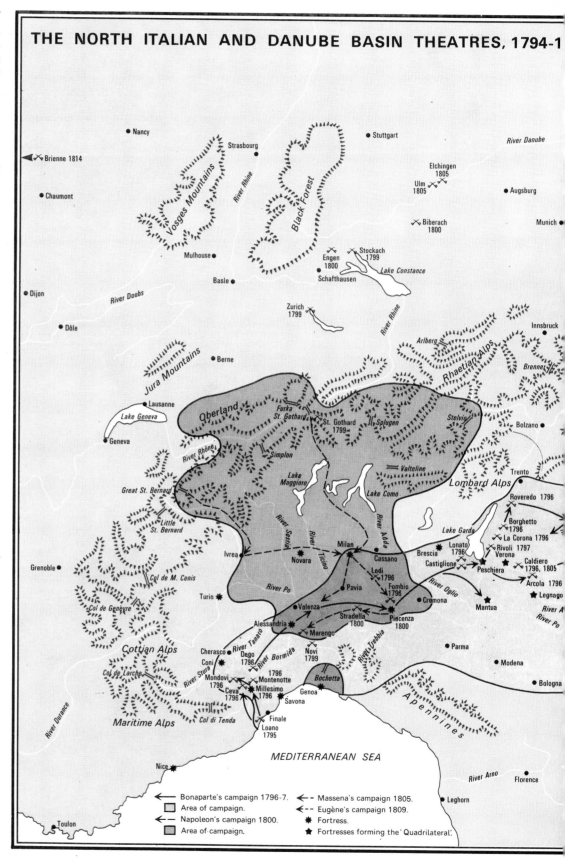

THE NORTH ITALIAN AND DANUBE BASIN THEATRES, 1794-1

Legend:
- → Bonaparte's campaign 1796-7.
- ☐ Area of campaign.
- ← Napoleon's campaign 1800.
- ☐ Area of campaign.
- ←- Massena's campaign 1805.
- ←-- Eugène's campaign 1809.
- ✳ Fortress.
- ★ Fortresses forming the 'Quadrilateral'.

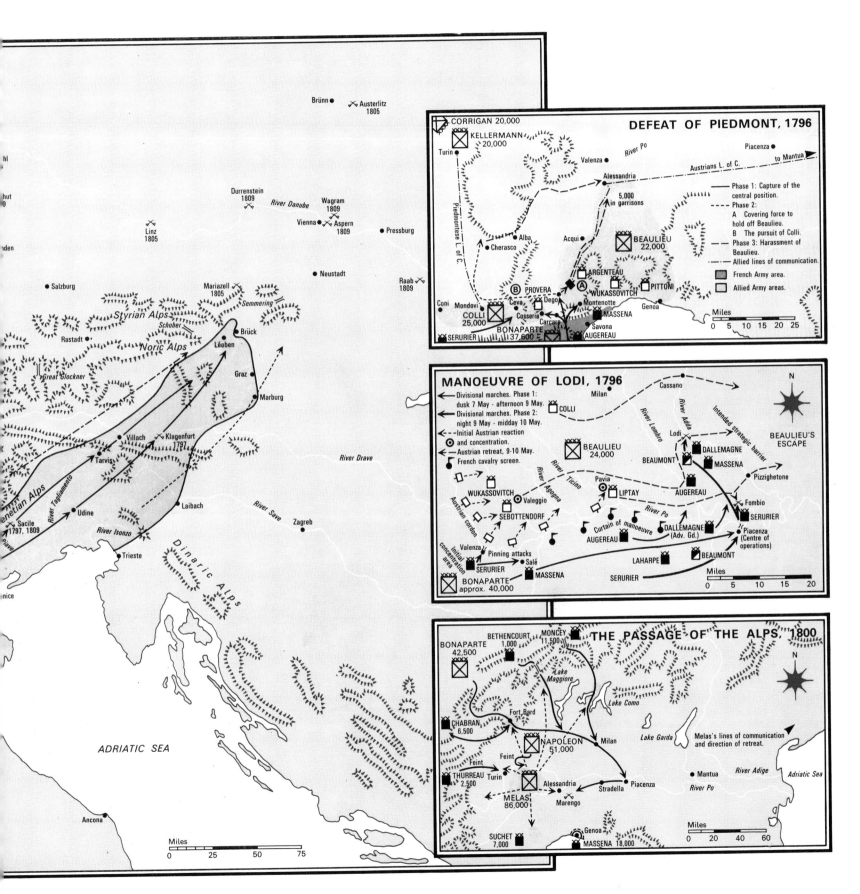

DEFEAT OF PIEDMONT, 1796

CORRIGAN 20,000
KELLERMANN 20,000
Turin
River Po
Piacenza
Valenza
Alessandria
Austrians L. of C.
to Mantua
Piedmontese L. of C.
5,000 in garrisons
Alba
Acqui
BEAULIEU 22,000
Cherasco
ARGENTEAU
(B) PROVERA
(A)
WUKASSOVITCH
PITTONI
Coni
Mondovi
Ceva
Dego
Montenotte
Genoa
COLLI 25,000
Cosseria
MASSENA
Carcare
Savona
SERURIER
BONAPARTE 37,600
AUGEREAU

Phase 1: Capture of the central position.
Phase 2:
A Covering force to hold off Beaulieu.
B The pursuit of Colli.
Phase 3: Harassment of Beaulieu.
Allied lines of communication.
French Army area.
Allied Army areas.

Miles
0 5 10 15 20 25

MANOEUVRE OF LODI, 1796

Divisional marches. Phase 1: dusk 7 May - afternoon 9 May.
Divisional marches. Phase 2: night 9 May - midday 10 May.
Initial Austrian reaction and concentration.
Austrian retreat, 9-10 May.
French cavalry screen.

Milan
Cassano
COLLI
River Lombro
River Adda
Lodi
Intended strategic barrier
N
BEAULIEU 24,000
DALLEMAGNE
BEAUMONT
MASSENA
BEAULIEU'S ESCAPE
Pavia
River Ticino
River Agogna
WUKASSOVITCH
Valeggio
LIPTAY
AUGEREAU
Pizzighettone
SEBOTTENDORF
Fombio
River Po
SERURIER
Curtain of manoeuvre
DALLEMAGNE (Adv. Gd.)
AUGEREAU
Piacenza (Centre of operations)
Austrian cordon
Valenza
Pinning attacks
LAHARPE
BEAUMONT
Initial concentration area
SERURIER
Salé
SERURIER
BONAPARTE approx. 40,000
MASSENA

Miles
0 5 10 15 20

THE PASSAGE OF THE ALPS, 1800

BETHENCOURT 1,000
MONCEY 11,500
BONAPARTE 42,500
Lake Maggiore
Lake Como
N
Fort Bard
Lake Garda
Melas's lines of communication and direction of retreat.
CHABRAN 6,500
NAPOLEON 51,000
Milan
Feint
Feint
THURREAU 2,500
Turin
Alessandria
Stradella
Piacenza
Mantua
River Adige
Adriatic Sea
River Po
MELAS 86,000
Marengo
SUCHET 7,000
Genoa
MASSENA 18,000

Miles
0 20 40 60

Brünn
Austerlitz 1805
Durrenstein 1809
River Danube
Wagram 1809
Vienna
Aspern 1809
Linz 1805
Pressburg
Neustadt
Raab 1809
Salzburg
Mariazell 1805
Semmering
Styrian Alps
Schober
Rastadt
Noric Alps
Brück
Léoben
Great Glockner
Graz
Marburg
Villach
Klagenfurt 1797
River Drave
Tarvis
Laibach
River Save
Zagreb
Venetian Alps
River Tagliamento
Udine
River Isonzo
Trieste
Sacile 1797, 1809
Dinaric Alps
Ancona
ADRIATIC SEA
Venice

Miles
0 25 50 75

Milan fell to the French on the 15th, and the Austrians were driven beyond the Mincio river line, which the French forced at Borghetto on 30 May. Some Austrians then retired to Trent; the rest entered Mantua.

The problem of Mantua was to dominate the next nine months. Set amid marshes, it was hard to besiege, while the Austrians mounted no less than four successive attempts to relieve their garrison. For practically the only time in his career Napoleon was tied to the basic conventions of classical eighteenth century siege warfare, having to divide his forces to provide a siege force and a covering army. First, General Wurmser advanced from the Alps in three columns. Napoleon was hard-pressed to win the Battles of First Lonato, 3 August, and then Castiglione, 5 August, against successive columns—the price for doing so was the temporary abandonment of the siege of Mantua. Mauled but not destroyed at Castiglione, where 28,000 French had fought Napoleon's first full-scale strategic battle (see p. 98) against 23,000 Austrians, Wurmser retired on Trent, and the French reimposed the siege.

With orders to assist General Moreau's delayed offensive against Austria from the Rhine, Bonaparte led 33,000 men up the Adige valley only to discover that Wurmser was already marching 20,000 men, covered by 14,000 left at Trent, towards Mantua by way of the Brenta valley. Napoleon boldly adopted the same route, defeated Wurmser in a 'reversed front' battle at Bassano, 8 September, and forced the Austrian Army to break up. Despite this, Wurmser fought his way into Mantua raising its garrison to 23,000 men. On 13 September the French again reimposed the siege.

Although some reinforcements were reaching him, Napoleon faced problems of an increasingly dissident population, a powerful Mantua to his rear, and renewed Austrian efforts from the Alps. The failure of Moreau's campaign enabled a third Austrian general, d'Alvintzi, to lead 28,000 through Bassano towards Verona, while a further 18,000 lured the French towards Trent. This, in turn, forced Napoleon to split his forces, and an intricate and critical period culminated in the three-day Battle of Arcola, 15-17 November.

This victory earned the French a brief respite, but in early January d'Alvintzi advanced with another 45,000 men—this time in five columns—intent on smashing the French hold on north Italy and relieving Mantua. Napoleon judged his foe's intention correctly and, ignoring the easternmost column, rushed every available man to meet the main threat at Rivoli, where he

won the celebrated battle of 14-15 January 1797. Without a break, the French troops dashed to Mantua, arriving in the nick of time to surround the fifth Austrian force. On 2 February, Wurmser at long last surrendered Mantua.

The Campaign of 1797 immediately followed, as Napoleon, freed at last from the

Napoleon Bonaparte (1769-1821) as a young general. His dynamic, opportunistic leadership achieved famous victories during the First Italian Campaign (1796-97), bringing French arms a new reputation and founding his own meteoric career. Below: directing the artillery at Lodi. (DAG)

problem of Mantua in his rear, resumed the offensive and led his 60,000 men to outwit and defeat the 50,000 Austrians of the Archduke Charles—who, although little older than Napoleon, was possibly the ablest Austrian commander. Invading Friuli, Napoleon forced one river-line after another, using Joubert at

the head of 20,000 men to threaten repeated envelopments. Massena captured part of Charles's army at Tarvis, while Napoleon pursued the main body through the eastern Alps to Leoben, barely eighty miles from Vienna. On 6 April the Austrians—in large measure fooled by Napoleon's bombast, for in fact he was very over-extended—agreed to sign an armistice, which eventually led to the Peace of Campo Formio. So ended Napoleon's formative campaign, in which he tried out and continually refined his miliary concepts.

The Battle of Arcola, 1796

This battle was attritional in nature owing to the prevailing strategic situation in late 1796, and shows Napoleon's skill at handling unfavourable circumstances. Deciding that d'Alvintzi's army of 28,700 posed the greater threat to the French siege of Mantua (see p. 92), Napoleon undertook a march down the Adige on 14 November to sever the Austrian communications east of Verona. Simultaneously, he trusted Vaubois with 10,500 men to contain Davidovitch's stronger force to the south of Trent.

Napoleon's sudden move achieved surprise, and early on the 15th the French crossed the Adige at Ronco. Sending Massena's division to Porcile, Napoleon accompanied Augereau past the marshes towards Arcola in the hope of taking the bridge there before an advance to Villanova and the subsequent severing of the vital main road to Verona. During the day, Massena held off Provera's attacks on the flank. However, Augereau only secured Arcola by dusk and, owing to the threat to Vaubois, Napoleon had to order a withdrawal south of the Adige for the night.

On the 16th, no ill news having arrived from the north, all the old ground had to be refought for against reinforced Austrian formations. Attempts to turn the enemy flank by a river crossing to the south foundered when no ford could be found, so again in the evening the French withdrew over the Adige—both sides having suffered heavily.

Early on the 17th, Napoleon gambled on Wurmser's quiescence to order up a substantial force under Vial from the siege of Mantua, with orders to pass the Adige at Legnago. Sending only part of Massena's men to face Provera, Napoleon used the rest to set an ambush in the marshes near Arcola—a ruse that secured the vital bridge. Meanwhile, Augereau found a way over the Alpone and a tough battle against d'Alvintzi's main force developed east of the river. This was clinched by the planned appearance of a small but noisy party of cavalry from the east and Vial's approach from the south. D'Alvintzi then ordered a full retreat towards Vicenza. The battle illustrates Napoleon's use of ground tactically, and his sense of strategic priorities together with his generally flexible outlook.

The Campaign of 1800

The two preceding years saw much of France's gain disappear during the campaigns of the War of the Second Coalition and during Napoleon's absence in the Levant (see p. 96). Shortly after his return to France, and his elevation to the post of First Consul in late 1799, Napoleon set about re-establishing France's martial position. A major offensive into Austria was ruled out by Moreau's obstructiveness and by the Austrians seizing the initiative in north Italy on 5 April 1800, leading to the siege of Massena within Genoa from the 22nd and the withdrawal of all other French forces over the Var.

The First Consul massed 50,000 men of the Army of Reserve around Dijon under the nominal command of Berthier. As a preliminary to destroying Melas's army and saving Genoa he planned to sweep with columns over the Alps to link with a corps from the Army of Germany in the Po valley. This was an ambitious application of the manoeuvre of envelopment, Massena's garrison forming the pinning force.

The crossing of the Alps began on 15 May, centred on the St. Bernard Pass. Such crossings were not unknown, but this one was unusually early and special measures were needed to pass the guns over the snow and ice. All went well until 19 May when Fort Bard obstructed the advance. Lannes bypassed the obstacle, four guns were rushed past by a ruse, and on 22 May the French reached Ivrea without most of their guns. Next, Napoleon marched on Milan to acquire cannon, and then set out to sever Melas's links with Mantua by seizing Piacenza. French communications were now routed over

the St. Gotthard, and between 7 and 9 June the Army passed south of the Po.

Events now intervened, for on 4 June Massena in desperation had surrendered just as Melas was about to abandon the siege. Melas headed for Alessandria to reopen his communications while Napoleon hurried on to forestall him, sending out two divisions to close possible lines of retreat towards Genoa and the Po respectively. The result was the hard-fought Battle of Marengo on 14 June.

Thereafter, Melas sued for an armistice, but not until Moreau's victory at Hohenlinden on 3 December did the Austrian government request a general pacification, which was signed at Luneville on 8 February 1801. Napoleon admitted later that 'Marengo was a lesson'.

The Battle of Marengo, 1800

The situation leading up to this famous battle has already been described. Napoleon's detachments before the battle, which he did not

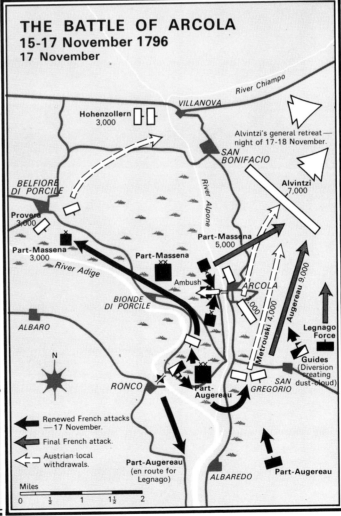

It is vital to study Arcola against its strategic background. The army of Italy was threatened from two directions at once; by Alvintzi (28,000 men) operating from Vicenza, and by Davidovitch (18,000 men) operating from Trento.

This left Vaubois with 10,500 men to face Davidovitch; while Bonaparte attempted to drive Alvintzi from the Verona region by threatening his communications via Arcola, and using the marshes to compensate for his numerical inferiority.

It was his uncertainty about Vaubois' fortunes that compelled Bonaparte to withdraw his army south of the Adige for two evenings of the 3-day action. This would enable him to march north to bolster Vaubois should the need arise. The whole episode illustrates the employment of interior lines, both strategically and tactically.

Arcola was essentially a battle of attrition. The first two days were indecisive, but the French forced the Austrians to commit more and more forces by their repeated thrusts towards Porcile and Arcola. On the third day, a combination of guile and reinforcements from Legnago enabled Bonaparte to win a considerable victory. After the battle Alvintzi gave up the offensive and retired to Bassano. Bonaparte then rushed his forces north to succour Vaubois in the nick of time, and forced Davidovitch to retire on Trento.

Casualties
French: 4,000 killed and wounded.
Austrian: 7,000 killed and wounded.

North Italy: 1796 and 1800

precisely foresee, almost cost him the day.

Phase One: the morning battle. Melas launched a totally unexpected heavy attack in three columns at 7am. Mistaking these for a feint covering an intended Austrian withdrawal from Alessandria, the First Consul did not recall Lapoype from the Po or Desaix from Novi until 9am—a near-fatal error. By midday the outnumbered and weary troops of Victor and Lannes had been forced back towards San Guiliano, and Ott's Austrian column was threatening the northern flank.

Phase Two: the afternoon battle. After a brief lull, Napoleon was forced to commit his only reserves, Monnier and the consular Guard, to hold off Ott near Castel Ceriolo. By this time his left flank was on the point of collapse before the remorseless pressure of Melas and O'Reilly. Fortunately, Desaix received his message of recall at 1pm (Lapoype only learnt of the crisis at 5pm) and reached the field ahead of Boudet's division at about 4pm. He judged

the battle lost, but ". . . there is time to win another".

Phase Three: the evening battle. By 5pm Boudet's men (5,300) had arrived behind the weary Victor. At this point, the aged Melas handed over the Austrian Army to General Zach. Desaix was advancing on the French left to challenge the massive Austrian column when an ammunition waggon exploded. To capitalize on the stunning effect, Marmont rushed up 8 guns; Kellerman charged with a few hundred cavalry on the flank; and Desaix swept forward in l'ordre mixte (see p. 98), but was shot. This attack clinched the battle, and by 6pm the Austrians were in flight. They lost some 14,000 men and 40 guns to the French 7,000 casualties. Next day Melas sued for an armistice.

Postscript. "Marengo was a lesson", Napoleon once admitted in a candid moment. To disguise the narrow avoidance of defeat and to claim the outcome as a personal success, Napoleon in later years rewrote the official

history a number of times. In the *Relation of 1803*, he made the bogus claim that his left wing was withdrawn deliberately to enable Desaix to take the Austrians in the flank as part of the standard Napoleonic battle system. In the *Relation of 1805* he added the claims that the French never lost Castel Ceriolo on the right, but used it as a pivot to lure Melas towards Desaix's 'trap' on the farther wing; and that he had set up new communications running north to the River Po to replace the threatened Alessandria-to-Voghera highway which the enemy effectively controlled by 3pm.

These 'improvements' were bogus, but Napoleon was to use them with telling effects in several of his later great battles. On 14 June 1800, however, Napoleon owed success to his subordinates and, above all, to Desaix, rather than to his own talents. Indeed, his errors almost earned him a massive defeat. But he was a master of propaganda, both on St. Helena and earlier.

THE BATTLE OF MARENGO
14 June 1800

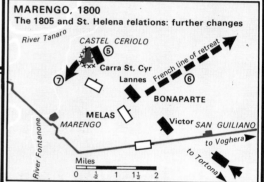

MARENGO, 1800
The 1803 relation: additions and changes

MARENGO, 1800
The 1805 and St. Helena relations: further changes

1 Deliberate, phased French retreat.
2 Practically continous occupation of Castel Ceriolo, the pivot.
3 Deliberate placing of Desaix.
4 Deletion of all mention of Monnier in favour of Carra St. Cyr.
5 Carra St. Cyr barricaded into Castel Ceriolo.
6 New French line of retreat.
7 Envelopment of Austrian left by French right.

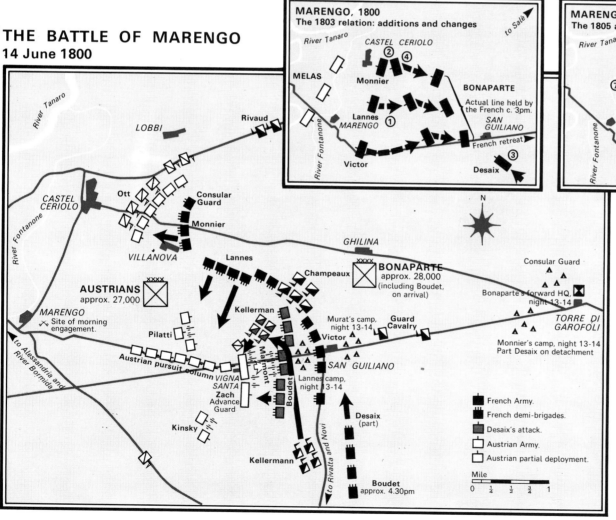

French Army.
French demi-brigades.
Desaix's attack.
Austrian Army.
Austrian partial deployment.

Louis Charles Antoine Desaix (1768-1800), who consolidated his reputation as a general during the Egyptian campaign, and in 1800 saved the day at Marengo at the cost of his life. (DAG)

ARTILLERY OF THE GRIBEAUVAL SYSTEM

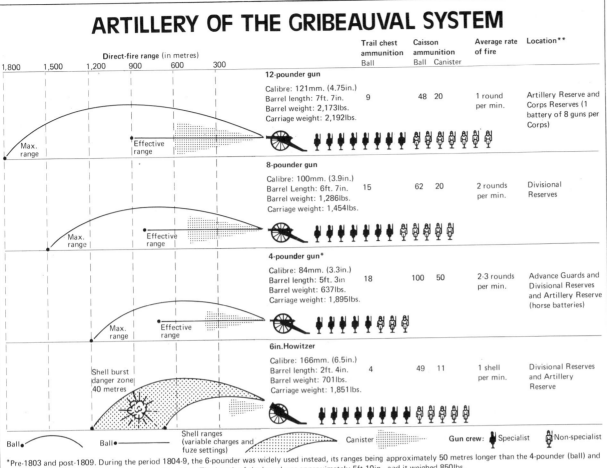

	Trail chest ammunition	Caisson ammunition		Average rate of fire	Location**
	Ball	Ball	Canister		
12-pounder gun Calibre: 121mm. (4.75in.) Barrel length: 7ft. 7in. Barrel weight: 2,173lbs. Carriage weight: 2,192lbs.	9	48	20	1 round per min.	Artillery Reserve and Corps Reserves (1 battery of 8 guns per Corps)
8-pounder gun Calibre: 100mm. (3.9in.) Barrel Length: 6ft. 7in. Barrel weight: 1,286lbs. Carriage weight: 1,454lbs.	15	62	20	2 rounds per min.	Divisional Reserves
4-pounder gun* Calibre: 84mm. (3.3in.) Barrel length: 5ft. 3in. Barrel weight: 637lbs. Carriage weight: 1,895lbs.	18	100	50	2-3 rounds per min.	Advance Guards and Divisional Reserves and Artillery Reserve (horse batteries)
6in. Howitzer Calibre: 166mm. (6.5in.) Barrel length: 2ft. 4in. Barrel weight: 701lbs. Carriage weight: 1,851lbs.	4	49	11	1 shell per min.	Divisional Reserves and Artillery Reserve

Direct-fire range (in metres): 1,800 · 1,500 · 1,200 · 900 · 600 · 300

Max. range / Effective range

Shell burst danger zone 40 metres

Ball · Ball · Shell ranges (variable charges and fuze settings) · Canister · Gun crew: Specialist / Non-specialist

*Pre-1803 and post-1809. During the period 1804-9, the 6-pounder was widely used instead, its ranges being approximately 50 metres longer than the 4-pounder (ball) and approximately 30 metres longer in the case of canister. The length of the barrel was approximately 5ft.10in., and it weighed 850lbs.

**Most guns were organized into 'companies' of eight cannon; in the case of many Divisional and Artillery Reserve formations (and some Corps Reserves) a proportion of 6 cannon to 2 howitzers was observed.

In the case of the 12-, 8-, and 6-pounders, the enployment of ricochet fire could increase the maximum range by 50—70 per cent; each bound of the cannonball theoretically decreased in distance by 50 per cent each time it hit the ground, (i.e., 600—300—150—75—37 yards).

SMALL ARMS PERFORMANCE

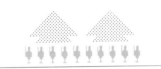

225 yards

150 yards

75 yards

Hits

From Prussian field firing trials carried out late in the eighteenth century.

Range	Hits	
225 yards	1 shot in 4	= 25 per cent
150 yards	4 shots in 10	= 40 per cent
75 yards	6 shots in 10	= 60 per cent

The target in these trials represented the equivalent of a battalion frontage. French, British and Austrian muskets were not very much better than the Prussian muskets used for these trials.

TACTICAL SEQUENCE OF FRENCH EARLY NINETEENTH CENTURY ENGAGEMENTS

In carrying through the Napoleonic grand tactical plan (see p. 98), French field formations developed a tactical sequence based on four principles: those of offensive action, relentless pressure, mobility and flexible combined attack. The following sequence was adjusted according to circumstances, but may be taken as reasonably representative. It was highly effective against an enemy drawn up in massed formation. However, Wellington's army developed countermeasures.

Phase I: Bombardment (by corps or divisional artillery)

French object: To harass the enemy, cause physical damage, and lower morale.

British counter: Place troops on reverse slopes in cover, thus minimizing casualties.

Phase II. Skirmishing fire (by voltigeur companies or tirailleurs regiments)

French object: To snipe at enemy officers, artillerymen and generally to annoy the rank and file without presenting a target worthy of a counterattack; also to gain accurate information on enemy strengths and positions which would be passed back to the local commander.

British counter: Send forward own light infantry to challenge the French and prevent their officers from reconnoitering the (concealed) British main positions.

Phase III. Cavalry attack plus horse artillery, by cuirassiers, chasseurs and lancers passing through the light infantry screen, which redeploys to harass enemy flanks

French object: To incite the enemy cavalry to counterattack and, after routing it, compel the enemy infantry to form a square, and the gunners to take shelter, temporarily abandoning their pieces. Keep the enemy infantry in squares (rarely broken by cavalry alone). Deploy horse artillery batteries at close range to decimate enemy formations, and capture or spike a number of their guns, thus protecting the approach of the infantry columns—the main attack.

British counter: Canister-shot from cannon at close ranges; last minute counter-charges by fresh squadrons against blown horsemen.

Phase IV: Infantry attack by divisional cloumns on a two-company frontage (i.e. 50 men wide by 12 ranks deep), using the bayonet and shock action rather than firepower

French object: To catch the enemy infantry still in squares, (therefore reducing firepower against the new attack) or in the process of reforming line—which would be achieved if the attacks were properly timed to coincide precisely with the cavalry withdrawal. To break the cohesion of the enemy line and create a breach through his battle-line.

British counter: Canister-shot at close range into dense columns; envelopment of columns by forward-deployed battalions in line.

Phase V: Light cavalry exploitations by hussars and dragoons

French object: To prevent the enemy from reforming a cohesive line, and to scatter his retreating formations into a horde of fugitives.

British counter: Not applicable as French break-ins were rarely achieved.

The French in the Levant, 1798-99

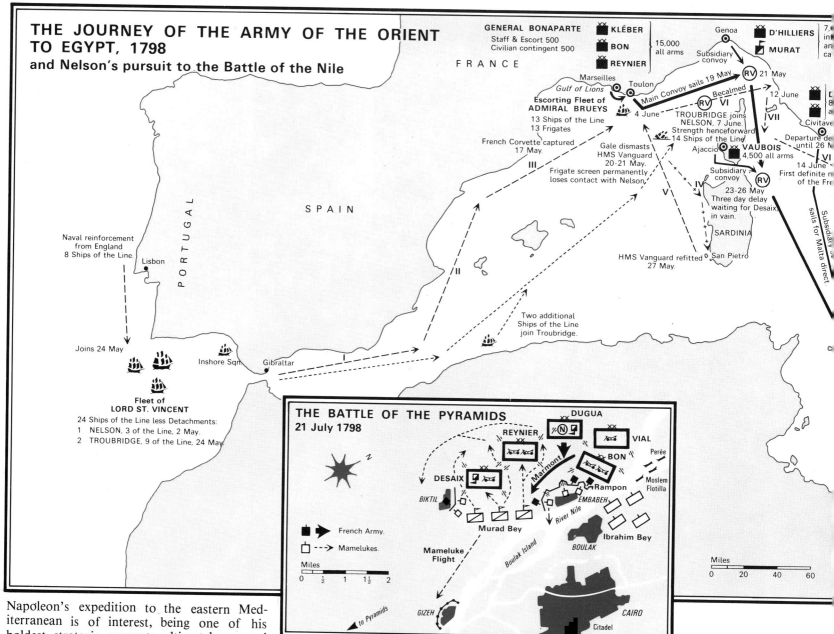

THE JOURNEY OF THE ARMY OF THE ORIENT TO EGYPT, 1798 and Nelson's pursuit to the Battle of the Nile

GENERAL BONAPARTE
Staff & Escort 500
Civilian contingent 500

KLÉBER
BON
REYNIER
D'HILLIERS
MURAT

FRANCE

Genoa
15,000 all arms
Subsidiary convoy

Marseilles
Gulf of Lions
Toulon
Main Convoy sails 19 May
RV 21 May
Becalmed
12 June
Escorting Fleet of ADMIRAL BRUEYS
4 June
RV VI
TROUBRIDGE joins NELSON, 7 June.
Strength henceforward 14 Ships of the Line
VII
Civitave

13 Ships of the Line
13 Frigates

French Corvette captured 17 May.
Gale dismasts HMS Vanguard 20-21 May. Frigate screen permanently loses contact with Nelson.
III

Ajaccio
VAUBOIS
4,500 all arms
Departure de until 26 M

14 June First definite n of the Fre

Naval reinforcement from England 8 Ships of the Line.

PORTUGAL
Lisbon

SPAIN

IV
V
23-26 May Three day delay waiting for Desaix in vain.
SARDINIA

Subsidiary convoy
RV

HMS Vanguard refitted 27 May.
San Pietro

Subsidiary sails for Malta direct

Joins 24 May
Inshore Sqn
Gibraltar
I

Two additional Ships of the Line join Troubridge.

Fleet of LORD ST. VINCENT
24 Ships of the Line less Detachments:
1 NELSON, 3 of the Line, 2 May.
2 TROUBRIDGE, 9 of the Line, 24 May.

THE BATTLE OF THE PYRAMIDS
21 July 1798

DUGUA
REYNIER
VIAL
Perée
DESAIX
Marmont
BON
Rampon
Moslem Flotilla
BIKTIL
EMBABEH
River Nile
Murad Bey
Boulak Island
Ibrahim Bey
BOULAK

French Army.
Mamelukes.

Mameluke Flight

Miles
0 ½ 1 1½ 2

to Pyramids
GIZEH
CAIRO
Citadel

Miles
0 20 40 60

Napoleon's expedition to the eastern Mediterranean is of interest, being one of his boldest strategic concepts, ultimately marred by the vital consideration of naval supremacy.

Deciding that a direct invasion of the British Isles was impracticable, in 1798 Napoleon persuaded the Directory to prepare a secret attack upon Egypt. He disguised his plan from the British by making apparent preparations for an Irish landing. Profiting from a gale, the French fleet slipped out of Toulon on 19 May and, after collecting sundry minor convoys and taking Malta on 6 June, the vast fleet headed for Egypt. The Royal Navy's pursuit—led by Nelson—was not far behind, and almost found the French on the night of 22 June; but, outstripping the slower-sailing French, reached

Alexandria ahead of Napoleon to find no sign of the enemy. Nelson sailed off to investigate other possibilities, and just one day later, on 29 June, the French hove to off Alexandria.

By 3 July the French Army, about 30,000 strong, was safely ashore. The following weeks proved them far superior tactically to the Mamelukes and felaheen of Egypt, and the conquest of Lower Egypt was clinched, after a distressing march through the desert to the Nile, by the Battle of the Pyramids on 21 July. Napoleon routed the huge but unwieldy enemy army for a loss of 390 casualties—the Mamelukes lost over 5,000—and captured Cairo.

However, on 2 August Nelson returned to destroy Admiral Brueys' fleet at the Battle of Aboukir (or the Nile), and this fatally compromised the French campaign. Not only did it isolate the Army of Egypt, but it also induced the Sultan to undertake a serious war against the invader. To forestall a planned double-offensive by land through Syria and Palestine and by sea from Rhodes, Napoleon seized the initiative on 6 February 1799, and invaded Palestine by way of Gaza and Jaffa.

Despite the ravages of plague in their ranks, the French reached Acre on 18 March and proceeded to besiege the town. The Royal Navy

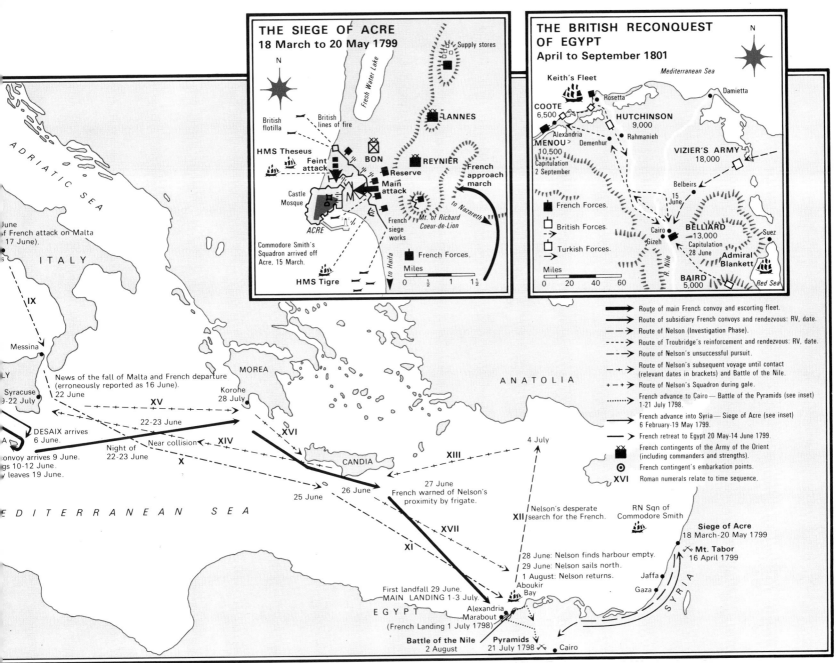

THE SIEGE OF ACRE
18 March to 20 May 1799

Fresh Water Lake

Supply stores

British flotilla

British lines of fire

LANNES

HMS Theseus

Feint attack

BON

REYNIER

Castle Mosque

Reserve

Main attack

French approach march

to Nazareth

ACRE

French siege works

Mt. of Richard Coeur-de-Lion

Commodore Smith's Squadron arrived off Acre, 15 March.

to Haifa

HMS Tigre

French Forces.

Miles
0   ½   1   1½

THE BRITISH RECONQUEST OF EGYPT
April to September 1801

Mediterranean Sea

Keith's Fleet

Rosetta

Damietta

COOTE 6,500

HUTCHINSON 9,000

Alexandria

MENOU 10,500

Demenhur

Rahmanieh

VIZIER'S ARMY 18,000

Capitulation 2 September

Belbeirs

15 June

■ French Forces.

□ British Forces.

Cairo

BELLIARD 13,000

Suez

□ Turkish Forces.

Gizeh

Capitulation 28 June

Admiral Blankett

Miles
0   20   40   60

BAIRD 5,000

Red Sea

Route of main French convoy and escorting fleet.

Route of subsidiary French convoys and rendezvous: RV, date.

Route of Nelson (Investigation Phase).

Route of Troubridge's reinforcement and rendezvous: RV, date.

Route of Nelson's unsuccessful pursuit.

Route of Nelson's subsequent voyage until contact (relevant dates in brackets) and Battle of the Nile.

Route of Nelson's Squadron during gale.

French advance to Cairo — Battle of the Pyramids (see inset) 1-21 July 1798.

French advance into Syria — Siege of Acre (see inset) 6 February-19 May 1799.

French retreat to Egypt 20 May-14 June 1799.

XX French contingents of the Army of the Orient (including commanders and strengths).

☉ French contingent's embarkation points.

XVI Roman numerals relate to time sequence.

ADRIATIC SEA

ITALY

June of French attack on Malta 17 June).

IX

Messina

Syracuse 9-22 July

convoy arrives 9 June.
ngs 10-12 June.
y leaves 19 June.

DESAIX arrives 6 June.

News of the fall of Malta and French departure (erroneously reported as 16 June). 22 June

Night of 22-23 June

Near collision

XIV

X

22-23 June

MOREA

Korohe 28 July

XV

XVI

MEDITERRANEAN SEA

25 June

26 June

CANDIA

ANATOLIA

XIII

4 July

27 June French warned of Nelson's proximity by frigate.

XVII

XI

Nelson's desperate search for the French.

XII

RN Sqn of Commodore Smith

28 June: Nelson finds harbour empty.
29 June: Nelson sails north.
1 August: Nelson returns.

First landfall 29 June. MAIN LANDING 1-3 July.

Aboukir Bay

EGYPT

Alexandria Marabout

(French Landing 1 July 1798)

Battle of the Nile 2 August

Pyramids 21 July 1798

Cairo

Siege of Acre 18 March-20 May 1799

Mt. Tabor 16 April 1799

Jaffa

Gaza

SYRIA

again intervened, not only sinking the flotilla carrying the French siege guns but also reinforcing the garrison. As the siege dragged on, the Army of the Pasha of Damascus drew near, and on 15 April 35,000 Turks were encountered by General Kléber's force of 1,500 near Mount Tabor. Kléber won time for the arrival of an alerted Bonaparte from Acre, and in the resultant battle on the 16th the 4,500 French troops again showed their superiority by thrashing the foe, inflicting 7,000 casualties for a loss of 100. The Army of Damascus fled.

Acre, however, remained uncowed, and news that Egypt was on the point of revolt and that the Army of Rhodes was putting to sea induced Napoleon to accept second-best and retreat (20 May). Returning to Egypt, he suppressed local

revolts with customary ferocity, and on 25 July met and routed the Army of Rhodes near Aboukir shortly after its landing for a loss of 970 men. The Turks sailed away. Acutely aware of his isolation, Bonaparte was not willing to mark time in far-away Egypt for long, and on 22 August he abandoned his army and secretly set sail for France. On 9 October he landed at Fréjus. A month later he was Consul.

The British Reconquest of Egypt, 1801

Back in Egypt, Napoleon's abandoned army steadily deteriorated. At last, in March 1801, a British army, convoyed by the Royal Navy, carried out a model landing against a defended shore and won the night battle of Aboukir against Menou (Kléber's successor), but this

success cost the life of the commander, Sir Ralph Abercromby. His successor, Hutchinson, supported by naval flotillas on the Nile and the arrival on the Red Sea coast of 5,000 men from India under Baird, steadily advanced on Cairo, leaving Coote to besiege Alexandria. The 9,000 British troops, after linking with 18,000 Turks on 15 June, induced Belliard to capitulate at Cairo with 13,000 men on 28 June. Subsequently, Menou capitulated with 10,500 more at Alexandria on 2 September. The French were granted repatriation, and Napoleon, victor of Marengo and, through Moreau, of Hohenlinden, induced the Allies to sign the Preliminaries of Peace at Amiens on 1 October—before news of the end of the Egyptian campaign reached London.

ORGANIZATION

STAFF

Light Cavalry Division · Sapper Train · Corps Artillery · Infantry Divisions · Optional Extra Infantry Division/Divisions

Cavalry Brigades · Divisional Horse Artillery

Infantry Brigades · Divisional Artillery Reserve

Cavalry Regiments

Infantry Regiments · 6 guns (optional)

Cavalry Squadrons

Infantry Battalion · Band and Transport · Medical Detachment

Voltigeur (Light Infantry Company) · Fusilier Companies · Grenadier Company

STRENGTH (Paper)

xxx COMMANDER: Marshal or Full General.
TOTAL STRENGTH: approx. 37,000 (depending on number of infantry divisions).

STAFF: Chief of Staff, Arm representatives, Staff officers, ADCs, escorts, clerks: approx. 150.

Corps Artillery: 8×12pdr.; Engineers (1 company) and Pontoon train.

xx CAVALRY DIVISION: approx. 5,000 men.   xx INFANTRY DIVISION: approx. 16,000 men
Divisional Horse Artillery: 6×6pdr. cannon.   Divisional Foot Artillery: 6×8pdr. 2×6inch howitzers.

x CAVALRY BRIGADE: approx. 2,400 men (occasionally more if 3 regiments). Usually 'mixed'—containing elements of 'heavies', 'medium' and 'light'—but 'light' predominated.   x INFANTRY BRIGADE: approx. 8,000 men

⋯ CAVALRY REGIMENT: approx. 1,200 men. Either Light Cavalry: 1,200-1,800, Dragoons: 1,200, or Cuirassiers: 1,040 'heavies'.   ⋯ INFANTRY REGIMENT: approx. 4,000 men. Regimental Artillery—discontinued 1802. Regimental Artillery—reintroduced 1813. Type of gun: 4pdr. or 6pdr.

CAVALRY SQUADRONS: approx. 250 men.   ⋯ INFANTRY BATTALION: approx. 1,200 men One would usually be a 'Tirailleur' or light infantry battalion.

Note
After 1812, many corps d'armée only included a single brigade of cavalry.

INFANTRY COMPANIES:
Grenadier: approx. 90 men.
Voltigeur: approx. 140 men.
Fusilier: approx. 140 men.

Advantages of the Corps System
Each corps d'armée was a self-contained combat team, or miniature army, with its own staff. It could fight alone against superior numbers for up to 36 hours, and was able to march semi-independently thereby increasing the overall speed of movement. The corps d'armée's flexible organization could be expanded or contracted and therefore confused enemy intelligence. A series of semi-independent corps in a strategic web could cover a vast area at the start of a campaign, but could rapidly concentrate on a single point when battle loomed.

Note
Total strength of corps varied immensely according to a number of variable factors—including the number of infantry divisions attached, the size of cavalry attachment, and the actual operational strength of units. The range was 20,000-70,000 men.

NAPOLEON AS A GENERAL: HIS CONTRIBUTION TO WARFARE

Napoleon's contribution to the art of warfare has often been misunderstood. He was not, essentially, an innovator of method, but a great applier of the ideas of others. Drawing many of his ideas from Frederick the Great and earlier exponents of the military art, he added to them a pace and decisiveness that in effect transformed the conduct of warfare into a form of blitzkrieg. He utilized the reforms of the late eighteenth century and the innovations of the French Revolution—mass conscription, promotion through merit, a simplified logistical concept—to out-manoeuvre and out-general his eighteenth century contemporaries. As Camon so aptly remarked: "He added little to the armies of France—except victory."

His main contributions lay in three fields: strategy, grand tactics, and man-management. In the strategic field, he strongly believed in the crushing blow aimed at an opponent's main army as the means of achieving his political objectives. Although he disliked categorizing his methods, he did in fact operate three types of strategic manoeuvre, separately or in combination. These were the 'manoeuvre of envelopment' (see diagram), which was designed to isolate a weaker foe from his depots and line of retreat, and force a

'reversed front' battle situation upon him; the 'strategy of the central position' (see diagram), in which Napoleon tackled superior foes by dividing them into sectors and then defeating each in turn with local superiority of force; and, thirdly, 'strategic penetration', a blow over a river line or a frontier which often served as the preliminary to one of the other two forms of strategic attack. Details varied every time, but the principles of surprise, out-march and hit hard remained constant.

At the grand tactical level he tried—often unsuccessfully—to achieve a 'strategic battle' situation (see diagram). This method had much in common with the 'strategy of envelopment', and was in fact a continuation of it. Using frontal pressure to attract enemy reserves into action, Napoleon would infiltrate a force round one flank and make it reveal its presence—threatening the enemy's rear—when all the foe's forces were engaged. To face the new threat, the enemy would be tempted to weaken the front nearest to the danger, and form a new line. Waiting for this event, Napoleon would have a special reserve massed behind his front at the sector concerned—ready to smash a hole and let the cavalry have full rein behind the enemy front. This type of battle system—as used in

part at Castiglione, 1796, and Bautzen, 1813—was only rarely achieved, but Napoleon saw that it could bring more decisive results than either the frontal battle of attrition fought at Borodino, or the two-field battle, primary and secondary, as at Austerlitz (see p. 100) or Jena-Auerstadt (see p. 104).

Although he had little interest in minor tactics, Napoleon did advocate the use of l'ordre mixte: troops in column and others in line. He also advocated the use of massed artillery, and heavy cavalry charges. In the field of man-management he had few rivals. He used medals, promotions and the award of titles to get his officers and men to serve him diligently, but above all it was the force of his personality and his general accessibility to his troops that achieved results. He could reprimand blisteringly as well as commend, and worked his staff into nervous wrecks.

Unremitting toil and attention to detail, together with a phenomenal memory and an acute analytical mind of a mathematical bent were also large factors in his success. He was ruthless and dynamic.

However, from 1807 onwards there was a deterioration in some of his standards and abilities. His ideas became stereotyped and predictable, enabling enemies to devise

counter-strategies to meet his methods. He payed too little attention to logistics, which was one reason for the failures in Spain and Russia. He had to rely on less trustworthy allies as his armies swelled to gargantuan proportions, and his own sense of reality began to suffer. But his fall was that of a giant amongst pygmies.

L'ordre mixte dates from the Battle of Valmy in 1792, and was the most effective French infantry attack formation. Originally designed to help inexperienced troops (two battalions of raw recruits being steadied by one of regulars) it combined shock action by troops moving forward in columns of division (forty files broad by sixteen ranks deep) with troops in line (four ranks deep) to provide covering fire. Boudet's Division at Marengo comprised three brigades totalling eight battalions. The two flanking formations were drawn up in l'ordre mixte, the central brigade wholly in line. The whole formation was deployed in oblique order (see p. 12), right flank refused, and was preceded by a battalion of skirmishers. The combination of infantry attack, supported by guns and cavalry as related, was the real secret of success in this famous attack in which some 6,000 troops effectively routed three times their own number.

THE NAPOLEONIC STRATEGIC BATTLE (schematic)

PHASE ONE
Contact and Pinning Attack

Enemy depot

1 Cavalry screen reports contact.
2 Advance guard immediately engages.
3,4 Nearest corps move up to support advance guard and extend front — thus attracting more enemy troops.

Advance Guard

III, IV, V and reserve in rear

PHASE TWO
Battle of Attrition Covers Main Moves

Enemy depot

Main line of communications

1 Frontal attack develops, drawing more enemy troops into battle.
2 Cavalry screen concealing:
3 Enveloping force moving up to attack enemy flank adjacent to his line of communications.
4 Reinforcement of front attracts last enemy forces.
5 'Masse de rupture' massing behind right flank 'en potence'.

Advance Guard

Guard in rear

PHASE THREE
Envelopment, Breakthrough and Pursuit

Enemy depot

Main line of communications

1 Renewed frontal attack pins enemy.
2 Revealed enveloping attack induces foe to weaken his left to form new line (a--a).
3 'Masse de rupture', preceded by massed artillery bombardment smashes through weakened enemy sector.
4 Light cavalry passes through gap and commences pursuit.

Advance Guard

Reserve and Guard

THE STRATEGY OF THE CENTRAL POSITION
Napoleon's 'inferiority' stratagem, as used for the first time on the Ligurian coast, April 1796

PHASE ONE
French take the initiative.

L. of C.   L. of C.

A 60,000   B 60,000

Left 30,000   Right 30,000

30,000

L. of C.

1 Cavalry and advance guard occupy central position between enemy armies A and B.
2 Main French army advances in two wings and a reserve.

PHASE TWO
French select army B as first target.

L. of C.   L. of C.

A   B

20,000   70,000

L. of C.

1,2 French wings engage respective enemy armies.
3 'Secondary' wing contains army A and detaches a division to envelop right wing of army B.
4 Part of reserve extends front of right wing.
5 Remainder of reserve forms 'masse de décision'.

PHASE THREE
French switch superior strength against army A.

L. of C.   A   B   L. of C.

L. of C.

70,000   20,000

Forced march

L. of C.

1 Right wing and cavalry assume pursuit rôle of defeated army.
2 Left wing detachment returns to parent body.
3 Part of reserve force marches to envelop army A's exposed flank.
4 Remainder of reserve countermarches to repeat 'masse de décision' rôle.

Note:
The aim is to defeat a superior enemy in detail by securing local superiority of force on each battlefield in turn, using the advantages of 'interior lines'.
French corps d'armée were individually capable of taking on superior enemy forces and holding them off for twenty-four hours.

THE STRATEGY OF ENVELOPMENT
Napoleon's 'superiority' stratagem

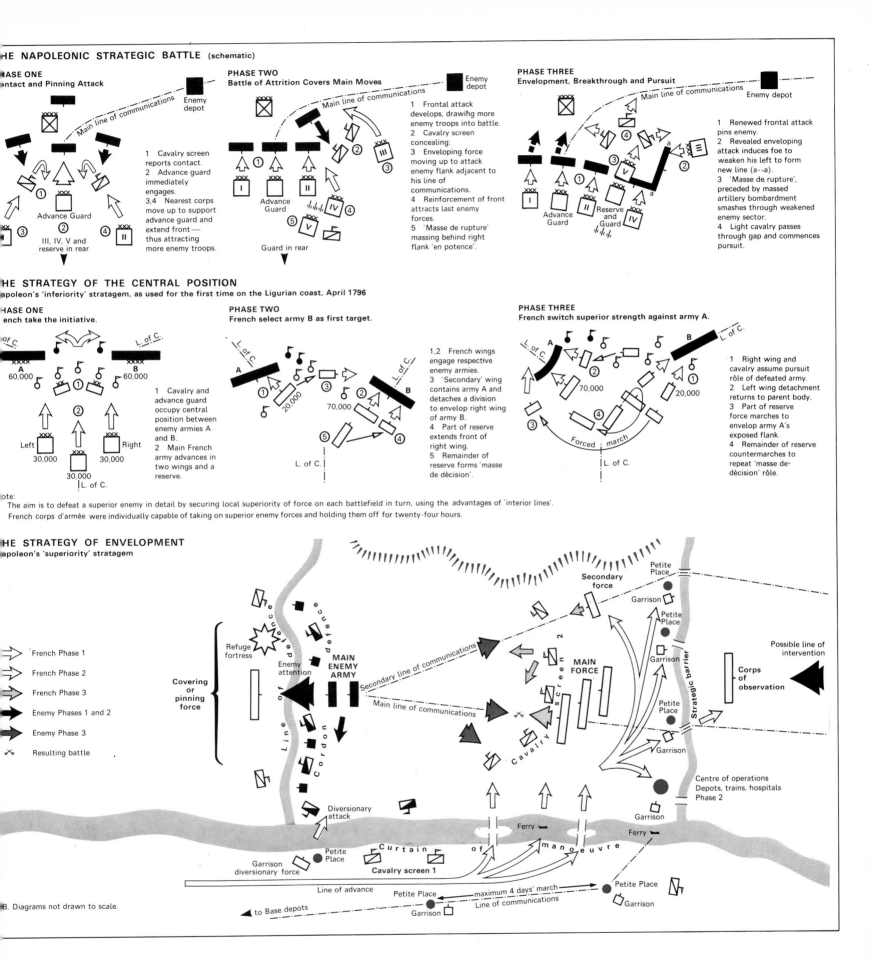

→ French Phase 1
→ French Phase 2
→ French Phase 3
→ Enemy Phases 1 and 2
→ Enemy Phase 3
⤫ Resulting battle

Refuge fortress

Enemy attention

MAIN ENEMY ARMY

Covering or pinning force

Line of defence

Cordon

Diversionary attack

Garrison diversionary force

Petite Place

Cavalry screen 1

Secondary force

Petite Place

Petite Place

Garrison

Secondary line of communications

Main line of communications

MAIN FORCE

Cavalry screen 2

Garrison

Petite Place

Garrison

Curtain   of   manoeuvre

Ferry   Ferry

Line of advance

Petite Place

Garrison

maximum 4 days' march

Line of communications

Petite Place

Garrison

to Base depots

Strategic barrier

Possible line of intervention

Corps of observation

Centre of operations Depots, trains, hospitals Phase 2

Garrison

NB. Diagrams not drawn to scale.

The Austerlitz Campaign, 1805

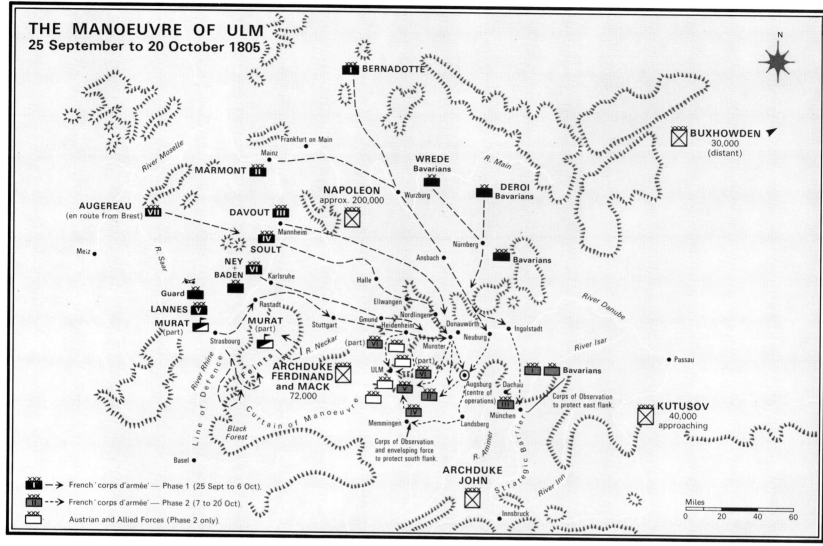

THE MANOEUVRE OF ULM
25 September to 20 October 1805

N

BERNADOTTE

BUXHOWDEN
30,000
(distant)

River Moselle

Frankfurt on Main

Mainz

WREDE
Bavarians

R. Main

MARMONT

NAPOLEON
approx. 200,000

Wurzburg

DEROI
Bavarians

AUGEREAU
(en route from Brest)

DAVOUT

Mannheim

Nürnberg

Metz

R. Saar

SOULT

Ansbach

Bavarians

NEY
+
BADEN

Karlsruhe

Halle

Guard

Rastadt

Ellwangen

River Danube

LANNES

MURAT
(part)

MURAT
(part)

Stuttgart

Nordlingen

Heidenheim

Donauwörth

Ingolstadt

River Isar

Strasbourg

R. Neckar

(part)

Munster

Neuburg

Passau

River Rhine

ULM

ARCHDUKE
FERDINAND
and MACK
72,000

(part)

Augsburg
(centre of
operations)

Dachau

Bavarians

Corps of Observation
to protect east flank.

KUTUSOV
40,000
approaching

Feints

Black
Forest

Memmingen

Landsberg

München

R. Ammer

Line of Defence

Curtain of Manoeuvre

Basel

Corps of Observation
and enveloping force
to protect south flank.

R. Lech

Strategic Barrier

River Inn

ARCHDUKE
JOHN

Innsbruck

Miles
0   20   40   60

French corps d'armée — Phase 1 (25 Sept to 6 Oct).
French corps d'armée — Phase 2 (7 to 20 Oct).
Austrian and Allied Forces (Phase 2 only).

The events between late September and early December 1805 finally established Napoleon's claim to European paramountcy as a commander. Abandoning all hope of invading England long before Trafalgar was fought, Napoleon closed the Camp of Boulogne and transferred his formations to the Rhine in order to deal with the massing forces of the new Third Coalition of Austria and Russia.

Aiming to eliminate Austria before Russian armies could materialize, on 25 September Napoleon launched 200,000 men of la Grande Armée into Central Germany in a vast sweeping manoeuvre that totally confounded his opponents, who had been lured west of Ulm in response to Prince Murat's demonstrations in the Black Forest. The French Corps—with their Bavarian allies—made good progress, their different lines of advance steadily converging as they approached the Danube which was crossed on a seventy-mile front by the end of the first

week in October. It is interesting to compare the speed of this manoeuvre, which took twelve days, with Marlborough's celebrated march to the Danube in 1704 (see p. 43).

After crossing the Danube, Napoleon used Davout, Bernadotte and the Bavarians to form a defensive flank to the east along the Isar to watch for the approach of the Russian, Kutusov, and swung the rest northwards to ensnare the bemused forces of General Mack and the Archduke Ferdinand. The days that followed held some confusions for the French as well as a number of hard-fought actions against Austrian detachments at Haslach, 11 October, and Elchingen, 13 October, but the jaws of the trap snapped shut around Ulm. General Mack, totally outclassed, decided to negotiate, and on 20 October surrendered at the head of 27,000 men. In the days that followed 30,000 more laid down their arms.

With the main Austrian army disposed of

without recourse to a major battle, the Emperor turned east and attempted to mete out similar treatment to Kutusov. That wily officer, however, managed to elude the French advance by means of precipitate retreats and, thanks to an error by Murat who insisted on heading for Vienna, by 9 November the Russians had regained the north bank of the Danube, thereby drawing close to a second Russian army under Buxhowden and further corps of Austrians. Napoleon was furious and led a hard pursuit, but his army was now rapidly dwindling through exhaustion and over-extension. Large formations had to be detached to watch for Archdukes John and Charles, and garrisons had to be provided for the growing lines of communications. By late November he had barely 53,000 men under direct command in Moravia, his foes had linked up near Olmutz to a total of almost 90,000 men and 300 guns.

By the time he reached the Austerlitz area on

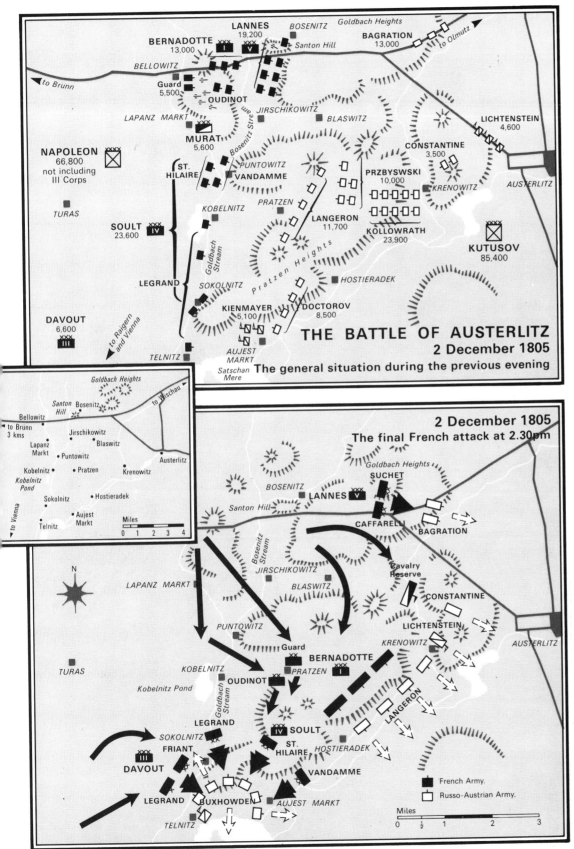

THE BATTLE OF AUSTERLITZ
2 December 1805
The general situation during the previous evening

2 December 1805
The final French attack at 2.30pm

French Army.
Russo-Austrian Army.

---

23 November, Napoleon had to call a halt. With great duplicity he feigned weakness and lured his opponents—accompanied by the Tsar and the Austrian Emperor—to advance and attack him. By 1 December, however, Napoleon had secretly summoned two more corps to bring his strength to some 72,000 (although 6,600 were still on the road from Vienna). The Austro-Russian forces occupied the dominating Pratzen Heights and jubilantly predicted a great victory. Meanwhile, the French took up their positions in the mist-filled valleys of the Goldbach and Bosenitz streams, with most strength on their left.

Long before dawn a massive Russo-Austrian force (some 40,000 men led by Buxhowden in four great columns) advanced from the Pratzen area and advanced to encircle Napoleon's weak right wing. A second force, under Bagration, headed west along the Brünn-Olmutz high road to attack Lannes, who was holding the Santon mound guarding Napoleon's left. As the mists rose a desperate battle began, and, despite some initial confusion the Allies seemed to be making progress. However, Napoleon was able to use Davout's newly-arriving III Corps and his reserve under Oudinot to check the advance of the Allied left, whilst V Corps and Murat's cavalry held their ground on the northern sector. Timing his blow with great care, at 9am Napoleon unleashed part of Soult's IV Corps (previously hidden in the fog-filled valley) to storm the Pratzen Heights which were now almost empty of Allied troops. Too late, Kutusov tried to recall men to hold this key ground. Meanwhile, Prince Lichtenstein's Allied cavalry were decisively defeated near Blaswitz to the north. By midday Soult, reinforced by Bernadotte, was master of the Pratzen, and all attempts to regain this height— including that of the Russian Guard Corps at about 1pm—were repelled by Soult's weary men backed by Bernadotte's I Corps and the cavalry of the Imperial Guard.

Their centre shattered, the Allies were in a critical position. Napoleon converted success into triumph by wheeling his Pratzen forces south to envelope Buxhowden's unwieldy wing around the frozen meres. Many Allies drowned trying to escape over the ice as French cannon balls ruptured the surface. To the north Bagration beat a hasty retreat for Olmutz with the two Emperors. By 4.30pm the battle was over. The Allies lost 26,000 men and 180 cannon; the French losses were 9,000. This decisive battle effectively destroyed the Third Coalition. The Tsar, followed by Austria, lost no time in making peace. Such was the great campaign of 1805 which brought Napoleon's powers as a soldier to their zenith.

The Campaign of 1806

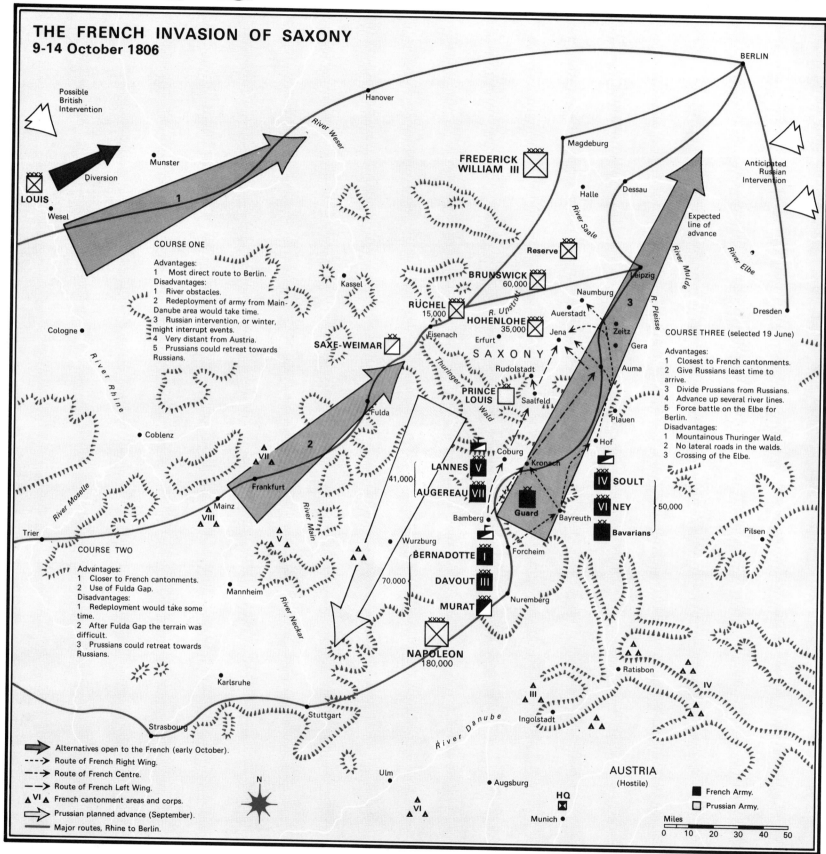

THE FRENCH INVASION OF SAXONY
9-14 October 1806

Possible British Intervention

Anticipated Russian Intervention

Diversion

LOUIS

1 — COURSE ONE

FREDERICK WILLIAM III ✗✗✗✗

Reserve ✗✗✗

Expected line of advance

COURSE ONE

Advantages:
1 Most direct route to Berlin.
Disadvantages:
1 River obstacles.
2 Redeployment of army from Main-Danube area would take time.
3 Russian intervention, or winter, might interrupt events.
4 Very distant from Austria.
5 Prussians could retreat towards Russians.

BRUNSWICK ✗✗✗
60,000

RÜCHEL ✗✗✗
15,000

HOHENLOHE ✗✗✗
35,000

SAXE-WEIMAR ✗✗✗

SAXONY

PRINCE LOUIS

COURSE THREE (selected 19 June)

Advantages:
1 Closest to French cantonments.
2 Give Russians least time to arrive.
3 Divide Prussians from Russians.
4 Advance up several river lines.
5 Force battle on the Elbe for Berlin.
Disadvantages:
1 Mountainous Thuringer Wald.
2 No lateral roads in the walds.
3 Crossing of the Elbe.

3

LANNES V
41,000
AUGEREAU VII

SOULT IV
NEY VI   50,000
Bavarians

Guard

BERNADOTTE I

2 — COURSE TWO

COURSE TWO

Advantages:
1 Closer to French cantonments.
2 Use of Fulda Gap.
Disadvantages:
1 Redeployment would take some time.
2 After Fulda Gap the terrain was difficult.
3 Prussians could retreat towards Russians.

DAVOUT III
70,000

MURAT ✗✗✗

NAPOLEON ✗✗✗✗
180,000

HQ ✗

AUSTRIA (Hostile)

Towns/places: BERLIN, Hanover, Magdeburg, Halle, Dessau, Munster, Wesel, Kassel, Eisenach, Erfurt, Naumburg, Auerstadt, Jena, Zeitz, Gera, Auma, Leipzig, Dresden, Cologne, Fulda, Coblenz, Rudolstadt, Saalfeld, Plauen, Hof, Kronach, Coblenz, Frankfurt, Mainz, Wurzburg, Bamberg, Forcheim, Bayreuth, Nuremberg, Pilsen, Mannheim, Trier, Karlsruhe, Strasbourg, Stuttgart, Ulm, Augsburg, Munich, Ingolstadt, Ratisbon

Rivers: River Weser, River Saale, River Mulde, River Elbe, River Pleisse, R. Unstrut, River Rhine, River Moselle, River Main, River Neckar, River Danube, Thuringer Wald

Key:
- ➤ Alternatives open to the French (early October).
- ---▸ Route of French Right Wing.
- -·-▸ Route of French Centre.
- --▸ Route of French Left Wing.
- ▲ᵛᴵ▲ French cantonment areas and corps.
- ⇨ Prussian planned advance (September).
- — Major routes, Rhine to Berlin.

■ French Army.
□ Prussian Army.

N

Miles
0 10 20 30 40 50

Tension between France and Prussia had been steadily rising since 1805, but after Austerlitz the Prussian government had refrained from entering the war. This politic reticence did not appease Napoleon, who throughout the first half of 1806 mercilessly bullied Frederick William III, forcing him to cede territory and to enter a one-sided treaty with France. In return Prussia received Hanover; but then, in June, Napoleon unilaterally offered to restore it to George III. This humiliating disregard for Prussian feelings led to the secret decision on war. Napoleon's agents soon divined the intention, but he refused to take it seriously until Prussia occupied Saxony and Russia declined a treaty offer.

Sensing a new hostile coalition, Napoleon took precautions and began to consider three possible courses of action. Reassured that Austria would not mobilize, the French aim was to eliminate Prussia before Russian forces could actively intervene. By 19 September Napoleon had chosen an advance on Berlin through Saxony, despite the difficulties posed by the Thuringer Wald, which was unavoidable by this route and offered only three roads. Confident in the stamina of his army, Napoleon divided it into three columns, relying on speed and surprise, backed by demonstrations by Louis Bonaparte on the Rhine, to enable him to cross the obstacle.

Led by their King and the aged Duke of Brunswick, the Prussian generals advocated conflicting plans, but eventually chose a drive on Stuttgart to sever a supposedly unsuspecting French Army from the Rhine. Events totally vitiated this scheme, for Napoleon's answer to the Prussian ultimatum was the surprise invasion of Saxony on 8 October.

Within forty-eight hours all three French columns were through the Thuringer Wald, defeating two small Prussian forces on the way, and had reached the Saale. In consternation the Prussians began to mass near Weimar, west of the river. But for several critical days Napoleon remained unaware of the exact Prussian

Napoleon's greatest military achievements date from the foundation of the First Empire (1804). In the years that followed, his mastery of strategy and grand tactics brought his armies a string of victories. (DAG)

location, and continued to head for Leipzig and the Elbe by marching north-east. When his patrols still reported no contacts north of Gera, Napoleon swung two-thirds of his army westwards towards the Saale, basing his forces on Auma, but leaving Soult and the cavalry near the Elster and Mulde in case his guess proved wrong. During the 13th, however, it seemed certain that the Prussians were in force west of the Saale, and the Emperor accordingly ordered a crash concentration of the whole Army around Jena and Naumburg. The flexibility of the French strategic formation enabled these radical changes of direction to be immediately implemented.

Joining Lannes on the Landgrafenberg west of Jena late on the 13th, Napoleon wrongly assessed that he was facing most of the Prussian Army. No longer expecting a battle near Weimar on about the 16th, Napoleon summoned the bulk of the Army to Jena while Davout's III Corps and Bernadotte's I Corps were ordered through Naumburg (or, if not in

---

Chronology of the 1806 Campaign

7 August: Prussia secretly decides on war against Napoleon.
5 Sept: French government calls up 50,000 conscripts and recalls as many reservists.
18 Sept: Napoleon learns of the Prussian occupation of Saxony.
19 Sept: Napoleon issues 102 orders, mobilizing the Army.
2 Oct: Napoleon takes over command at Würzburg; Prussian ultimatum received in Paris.
8 Oct: Prussian ultimatum expires; the French enter Saxony in three columns.
9 Oct: Action at Schleiz: Bernadotte and Murat (centre) defeat Tauenzien.
10 Oct: Action at Rudolstädt: Lannes (left) defeats Louis Ferdinand; Hohenlohe retires on Kahla and Jena; Brunswick masses at Weimar.
11 Oct: French continue north-east towards Leipzig, out of contact with the enemy.
12 Oct: Discovering the Prussian position, the French Army wheels west to the Saale.
13 Oct: All French forces converge on Jena and Naumburg; Napoleon joins Lannes.
14 Oct: Double battle of Jena-Auerstadt from which the French emerge victorious.
15 Oct: Napoleon launches full-scale pursuit of retreating Prussians.
16 Oct: Murat captures Erfurt, taking some 14,000 prisoners of war.
17 Oct: Action of Halle: Bernadotte defeats Württemberg (5,000 casualties).
20 Oct: French reach the Elbe river on a wide front. Hohenlohe heads for Stettin.
22 Oct: French cross the Elbe at two places; Ney besieges Magdeburg.
25 Oct: Davout occupies Berlin as two corps march for Oder river as covering force.
28 Oct: Hohenlohe and 14,000 surrender to Murat at Prenzlau.
29 Oct: Lasalle takes 5,000 captives at Stettin without opposition.
5/6 Nov: Blücher and 22,000 surrender to Bernadotte at Lübeck.
10 Nov: Magdeburg surrenders (22,000 prisoners); Mortier occupies Hamburg.
But Prussia refuses to sue for peace despite the annihilation of its army.

LE BATAILLON CARRÉ
and the reinforcement plan

Presumed enemy position

Cavalry Screen

24 hrs — 24 hrs
LEFT WING — ADVANCE GUARD — RIGHT WING
24 hrs — HQ
24 hrs — 24 hrs
RESERVE

Discovered enemy position

NEW RIGHT WING
orders
reports — HQ — orders
orders
confirmation
NEW ADVANCE GUARD — NEW RESERVE
NEW LEFT WING

Note the flexibility of the system — permitting a radical adjustment to the line of advance with minimum delay or confusion.

PLAN FOR FRENCH REINFORCEMENT CAPACITY AT JENA, 1806

Noon, 13 October 1806: Lannes' V Corps only: 21,000 men.
11.59pm, 13 October 1806: Lannes' V Corps plus part of the Imperial Guard: 25,000 men.
10am, 14 October 1806: V, Guard, plus VII Corps (16,500 men) and part of IV Corps (9,000 men): 50,500 men.
Noon, 14 October 1806: V, Guard, VII, part of IV plus rest of IV Corps (18,000 men), VI Corps (15,000 men) and heavy cavalry (7,000 men): 90,500 men.
4pm, 14 October 1806: As at midday, plus I Corps (20,000 men) and III Corps (27,000 men) and light cavalry (8,000 men): 145,500 men.

Thus Napoleon could mass 145,500 men at Jena (revealed position of the enemy) in 28 hours.
N.B. This build-up was not wholly achieved on 14 October, 1806, e.g. I and III Corps did not appear owing to unforeseen circumstances (see text).

The Campaign of 1806

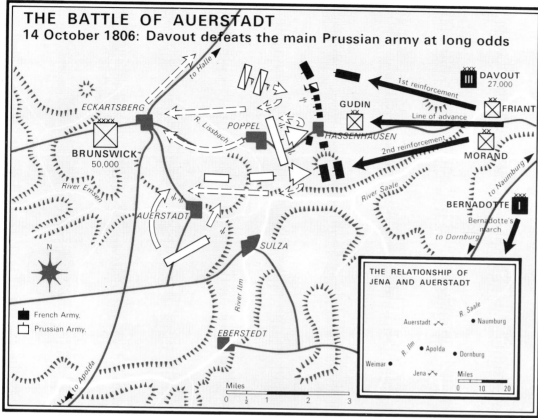

THE BATTLE OF AUERSTADT
14 October 1806: Davout defeats the main Prussian army at long odds

to Halle

DAVOUT
27,000

1st reinforcement

ECKARTSBERG

GUDIN

Line of advance

FRIANT

R. Lissbach

POPPEL

HASSENHAUSEN

BRUNSWICK
50,000

2nd reinforcement

River Saale

MORAND

River Emsen

AUERSTADT

to Naumburg

BERNADOTTE

Bernadotte's march

to Dornburg

N

SULZA

River Ilm

THE RELATIONSHIP OF JENA AND AUERSTADT

R. Saale

Auerstadt • Naumburg

R. Ilm • Apolda • Dornburg

Weimar • Jena

Miles
0   10   20

EBERSTEDT

to Apolda

■ French Army.
□ Prussian Army.

Miles
0 ½ 1   2   3

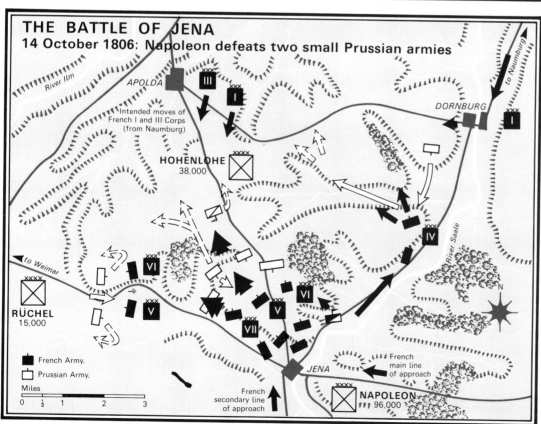

THE BATTLE OF JENA
14 October 1806: Napoleon defeats two small Prussian armies

River Ilm

to Naumburg

APOLDA

III

I

Intended moves of French I and III Corps (from Naumburg)

DORNBURG

I

HOHENLOHE
38,000

River Saale

to Weimar

IV

VI

RÜCHEL
15,000

V

VI

VII

V

N

JENA

■ French Army.
□ Prussian Army.

Miles
0 ½ 1   2   3

French secondary line of approach

French main line of approach

NAPOLEON
96,000

contact with Davout, the I Corps was to advance via Dornburg) so as to sever the foe's communications with Halle. In fact, Napoleon was only facing Hohenlohe's 38,000 near Jena; 50,000 more were moving north towards Auerstadt, while Rüchel's 20,000 were still near Weimar. Further, Bernadotte deliberately misinterpreted his orders. The sum of these errors was to place Davout and his force of 27,000 in an impossible position next day.

Napoleon's plan for Jena began with the seizure of the plateau to gain space for new formations arriving over the Saale, while Soult and Augereau accomplished outflanking manoeuvres to the north and south. Once assembled in sufficient force, the French would smash the Prussian centre, timing the breakthrough to coincide with Davout's and Bernadotte's arrival in their rear from the north.

In the event, matters worked out differently. Napoleon's plan began well enough, but Ney's premature attack in the centre caused a crisis that took some time to remedy. Nevertheless, by 12.30pm Napoleon had 96,000 men on the field, and decided to launch the coup de grâce without waiting for I and III Corps which had inexplicably not arrived. Hohenlohe's gallant but hopelessly outnumbered force cracked under the strain, and soon their flight was being shared by Rüchel's divisions arriving from the west. By 4pm Murat's cavalry were entering Weimar. This success had cost the French 5,000 casualties; they had inflicted some 25,000.

Then news arrived that Davout, alone, had in fact been fighting the larger part of the Prussian Army and miraculously had routed it despite unfavourable odds of two to one. Thick morning mist had concealed the true French weakness on the secondary battlefield, and French valour allied to Prussian confusions had resulted in an outstanding success for Davout. For the loss of 7,000 men (almost a third), III Corps had inflicted 13,000 casualties, including Brunswick himself.

Bernadotte took no part in the day's fighting, but exploited an ambiguous order to avoid serving under his rival, Davout. The Emperor almost had him court-martialled, but changed his mind as Bernadotte led the pursuit of the Prussian survivors that now followed.

The pursuit after Jena-Auerstadt forms a classical operation of war. Napoleon allowed his routed opponents no rest, and by 10 November had accounted for 125,000 out of the original 160,000 operating in Saxony. Militarily, Prussia had been brought to its knees, but its fugitive king rallied his surviving troops in east Prussia; against all probabilities, he determined to continue the struggle.

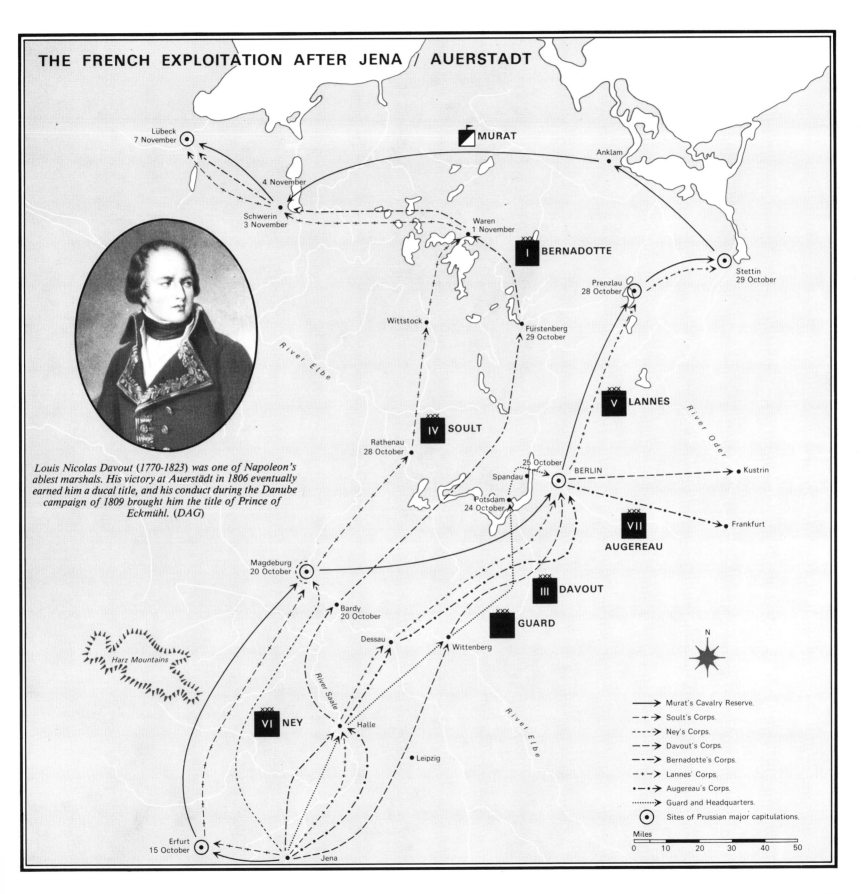

THE FRENCH EXPLOITATION AFTER JENA / AUERSTADT

MURAT

Lübeck
7 November

Anklam

4 November

Schwerin
3 November

Waren
1 November

I BERNADOTTE

Stettin
29 October

Prenzlau
28 October

Wittstock

Fürstenberg
29 October

River Elbe

V LANNES

River Oder

IV SOULT

Rathenau
28 October

25 October

Spandau

BERLIN

Kustrin

Louis Nicolas Davout (1770-1823) was one of Napoleon's ablest marshals. His victory at Auerstädt in 1806 eventually earned him a ducal title, and his conduct during the Danube campaign of 1809 brought him the title of Prince of Eckmühl. (DAG)

Potsdam
24 October

VII
AUGEREAU

Frankfurt

Magdeburg
20 October

III DAVOUT

Bardy
20 October

GUARD

Harz Mountains

Dessau

Wittenberg

River Elbe

N

River Saale

VI NEY

Halle

Leipzig

——— Murat's Cavalry Reserve.

–+–> Soult's Corps.

– – –> Ney's Corps.

– – –> Davout's Corps.

–·–·> Bernadotte's Corps.

–+–+> Lannes' Corps.

–•–•> Augereau's Corps.

········> Guard and Headquarters.

⊙ Sites of Prussian major capitulations.

Erfurt
15 October

Jena

Miles
0 10 20 30 40 50

Napoleon's First Setbacks in Battle

The Battle of Eylau, 1807

After his crushing defeat of Prussia in 1806, Napoleon still had to deal with Russia and the relics of the Prussian Army as a general pacification did not ensue. After occupying Warsaw the French went into winter quarters in the New Year of 1807. Later that month, however, the Russian Bennigsen launched a surprise offensive, and the Emperor summoned the corps for a bitter campaign in the stark continental winter.

Although Bennigsen managed to evade a trap near Inkovo (having fortuitously captured a set of French orders), and escaped from Allenstein, 3 February, Napoleon caught up with him near the township of Eylau late on 7 February. Fighting went on until well after dark, but the town fell to the French. Both armies then spent a freezing night. Orders were sent summoning Davout, Ney and Bernadotte to the field.

Next morning the battle opened amidst a snow-covered landscape. The French drew up their 45,000 men and 210 guns along a ridge at Eylau; Bennigsen deployed his 67,000 and 470 guns around Anklappen, but could not entrench his front. Seriously outnumbered, but expecting reinforcements, Napoleon let Bennigsen take the initiative. He soon regretted this, for by 9am Soult's Corps was under heavy pressure. Worse was to follow. To create a distraction, Napoleon launched Augereau's VII Corps in an attack at 10am. Losing his way in a blizzard, the Marshal blundered into the Russian main battery and his force was decimated. The Russian counterattack almost reached Napoleon's own command-post, but the day was saved by a magnificent charge by Murat with 10,700 cavalry which penetrated Bennigsen's centre, scattered the Russian gunners, and won a respite.

At 1pm Davout at last arrived from the south, and his 15,000 men pressed back the Russian left until it almost broke; but Bennigsen then received timely reinforcements in the form of Lestocq's Prussian Corps of 9,000 which had outmarched Ney to the field. By 4pm the situation was stalemate. Ney's arrival at 7pm with 14,000 was too late to be of use. Another terrible night was then experienced, during which Bennigsen—to the French relief—decided to retire. There was no

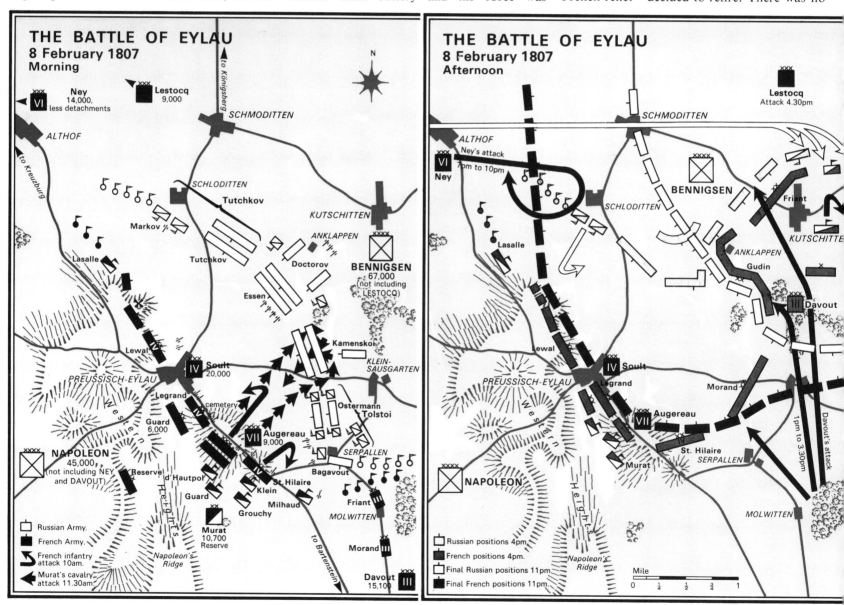

pursuit. The French had lost 20,000 men to the Russian 15,000. Although disguising the outcome, Napoleon had received his first check in battle, although his propaganda machine tried hard—but in vain—to conceal this.

The Battle of Aspern-Essling, 1809

Twenty-seven months later, Napoleon suffered an even more severe check on the north bank of the Danube. After occupying Vienna on 13 May Napoleon set out to cross the river and hunt down the Archduke Charles whose whereabouts he did not know. In fact, the Austrians (115,000 strong) were close to the Danube, and when Napoleon somewhat incautiously began to cross to the Isle of Lobau by a single bridge he soon found himself in trouble. On 20-21 May the Austrians repeatedly sent floating and blazing mills downstream to break the vital bridge and managed to isolate Massena's IV Corps and some cavalry on the north bank. At 10am Charles attacked with five columns, 90,000 men, and by 1pm a desperate battle was raging for Aspern, Essling and Grossenzersdorf. By the day's end the French had sent 31,500 into action and had retained control of the villages, but the situation was desperate.

It did not improve greatly overnight, although by dawn Napoleon had some 62,000 men and 144 guns over the river. At 7am Lannes and Oudinot attacked Charles's centre, but after making some ground were repulsed. Running short of ammunition, the French were in a sorry state, but Bessières' charges with the cavalry managed to ease the pressure. However, the vital bridge soon parted again and kept Davout out of the fray, although 73,000 French were now in action. After another breakdown at the bridge, Napoleon ordered a retreat; by 3.30am on the 23rd the French had withdrawn with great difficulty into Lobau Island. They had lost 22,000 men and inflicted 23,000 casualties, but they had been repulsed. However, Napoleon could learn from his errors. When he recrossed the river again on 4 July, careful preparations secured the vital three bridges, and surprise was achieved. In the ensuing Battle of Wagram, 5-6 July, Napoleon brought off a hard-won but decisive victory. For the loss of 32,000 casualties the French inflicted 40,000—and peace soon followed.

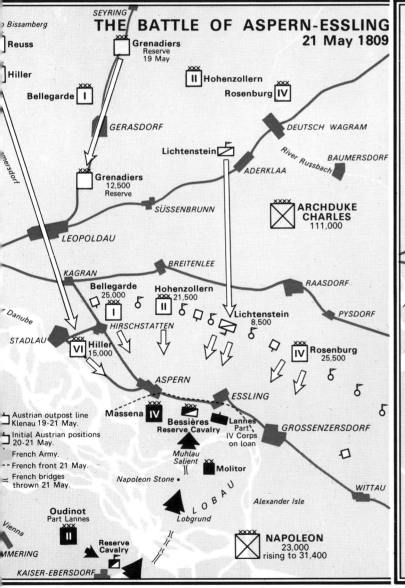

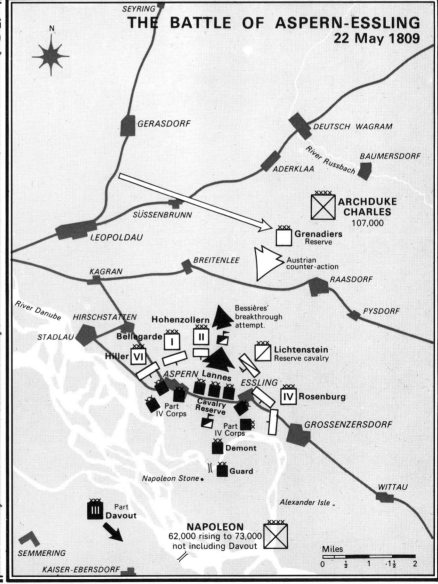

Europe under Napoleon

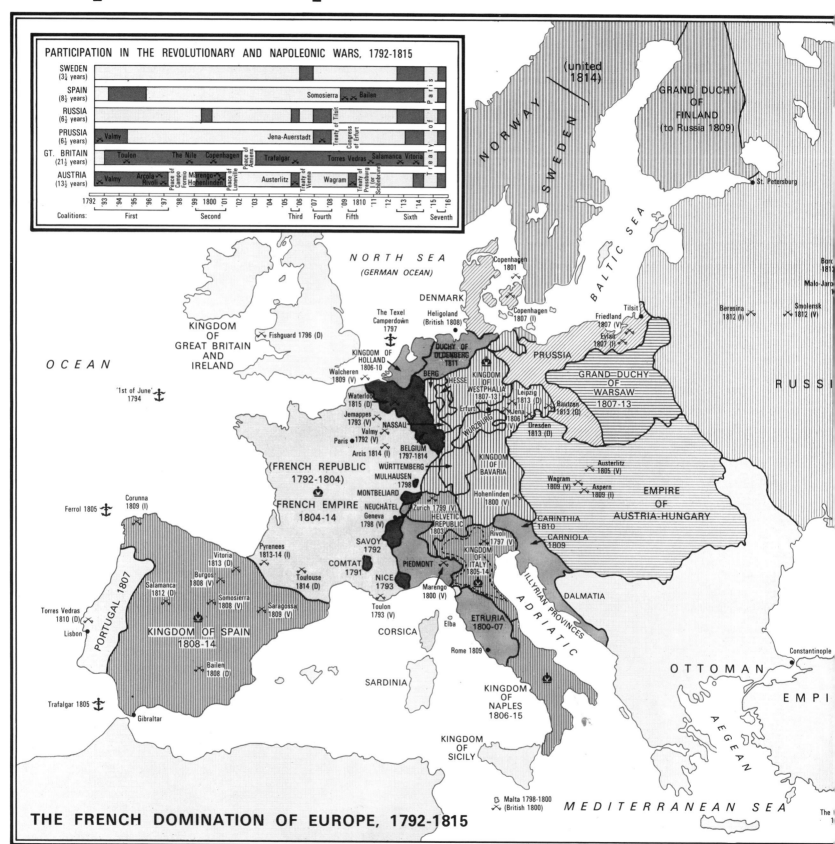

PARTICIPATION IN THE REVOLUTIONARY AND NAPOLEONIC WARS, 1792-1815

SWEDEN (3¼ years)		
SPAIN (8½ years)	Somosierra Bailen	
RUSSIA (6½ years)	Treaty of Tilsit	
PRUSSIA (6½ years)	Valmy Jena-Auerstadt Congress of Erfurt	
GT. BRITAIN (21½ years)	Toulon The Nile Copenhagen Peace of Amiens Trafalgar Torres Vedras Salamanca Vitoria	Treaty of Paris
AUSTRIA (13½ years)	Valmy Arcola Rivoli Peace of Campo Formio Marengo Hohenlinden Peace of Lunéville Austerlitz Treaty of Vienna Wagram Treaty of Pressburg (or Schönbrunn)	

| 1792 | '93 | '94 | '95 | '96 | '97 | '98 | '99 | 1800 | '01 | '02 | '03 | '04 | '05 | '06 | '07 | '08 | '09 | 1810 | '11 | '12 | '13 | '14 | '15 | '16 |

Coalitions: First Second Third Fourth Fifth Sixth Seventh

NORWAY

SWEDEN

(united 1814)

GRAND DUCHY OF FINLAND (to Russia 1809)

BALTIC SEA

St. Petersburg

NORTH SEA (GERMAN OCEAN)

DENMARK

Copenhagen 1801

Copenhagen 1807 (I)

Tilsit

Friedland 1807 (V)

Eylau 1807 (I)

PRUSSIA

Berezina 1812 (I)

Smolensk 1812 (V)

Bord 181

Malo-Jaro

RUSSIA

KINGDOM OF GREAT BRITAIN AND IRELAND

OCEAN

Fishguard 1796 (D)

The Texel Camperdown 1797

Heligoland (British 1808)

KINGDOM OF HOLLAND 1806-10

Walcheren 1809 (V)

BERG

HESSE

DUCHY OF OLDENBURG 1811

KINGDOM OF WESTPHALIA 1807-13

Erfurt

WÜRZBURG

Leipzig 1813 (D)

Jena 1806 (V)

Dresden 1813 (D)

Bautzen 1813 (D)

GRAND DUCHY OF WARSAW 1807-13

'1st of June' 1794

Waterloo 1815 (D)

Jemappes 1793 (V)

NASSAU

Valmy 1792 (V)

Paris 1792 (V)

BELGIUM 1797-1814

Arcis 1814 (I)

(FRENCH REPUBLIC 1792-1804)

WÜRTTEMBERG

MULHAUSEN 1798

MONTBELIARD

FRENCH EMPIRE 1804-14

NEUCHÂTEL

Geneva 1798 (V)

Zürich 1799 (V)

HELVETIC REPUBLIC 1803

KINGDOM OF BAVARIA

Hohenlinden 1800 (V)

Wagram 1809 (V)

Aspern 1809 (I)

Austerlitz 1805 (V)

EMPIRE OF AUSTRIA-HUNGARY

CARINTHIA 1810

CARNIOLA 1809

Corunna 1809 (I)

Ferrol 1805

Vitoria 1813 (D)

Pyrenees 1813-14 (I)

SAVOY 1792

COMTAT 1791

NICE 1793

Toulon 1793 (V)

Marengo 1800 (V)

Rivoli 1797 (V)

KINGDOM OF ITALY 1805-14

PIEDMONT

ILLYRIAN PROVINCES

DALMATIA

ADRIATIC

Burgos 1808 (V)

Salamanca 1812 (D)

Somosierra 1808 (V)

Saragossa 1809 (V)

Toulouse 1814 (D)

Torres Vedras 1810 (D)

PORTUGAL 1807

Lisbon

KINGDOM OF SPAIN 1808-14

Bailen 1808 (D)

Trafalgar 1805

Gibraltar

CORSICA

SARDINIA

Elba

ETRURIA 1800-07

Rome 1809

KINGDOM OF NAPLES 1806-15

Constantinople

OTTOMAN EMPI

AEGEAN

KINGDOM OF SICILY

Malta 1798-1800 (British 1800)

MEDITERRANEAN SEA

The

THE FRENCH DOMINATION OF EUROPE, 1792-1815

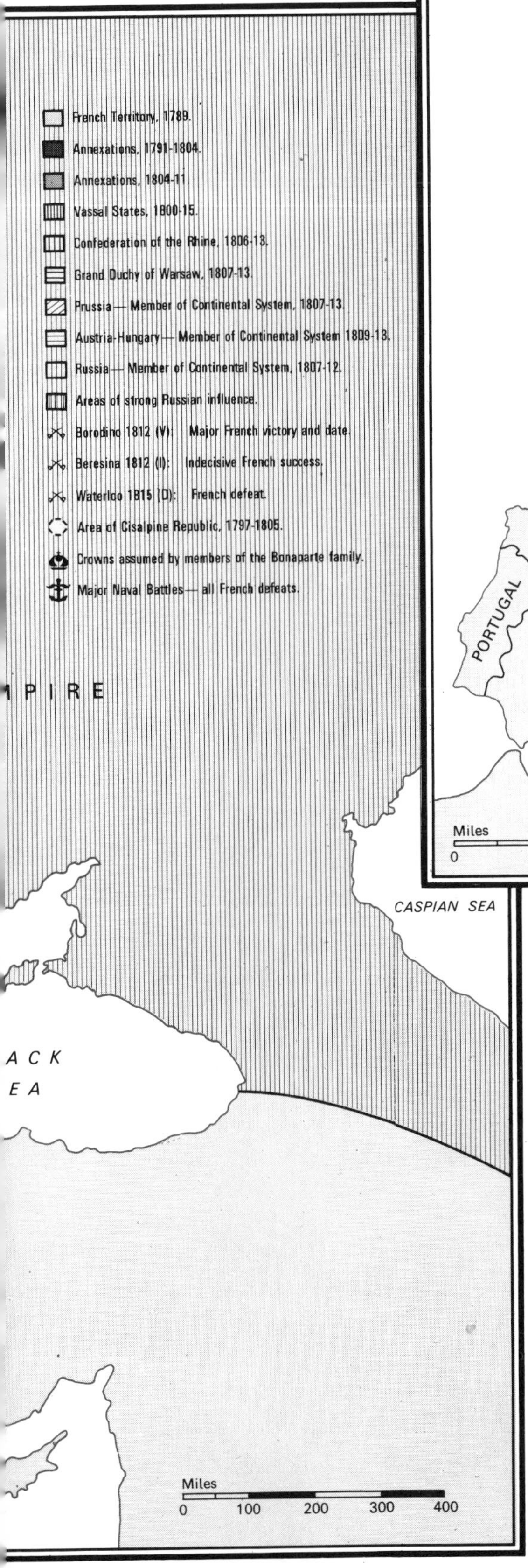

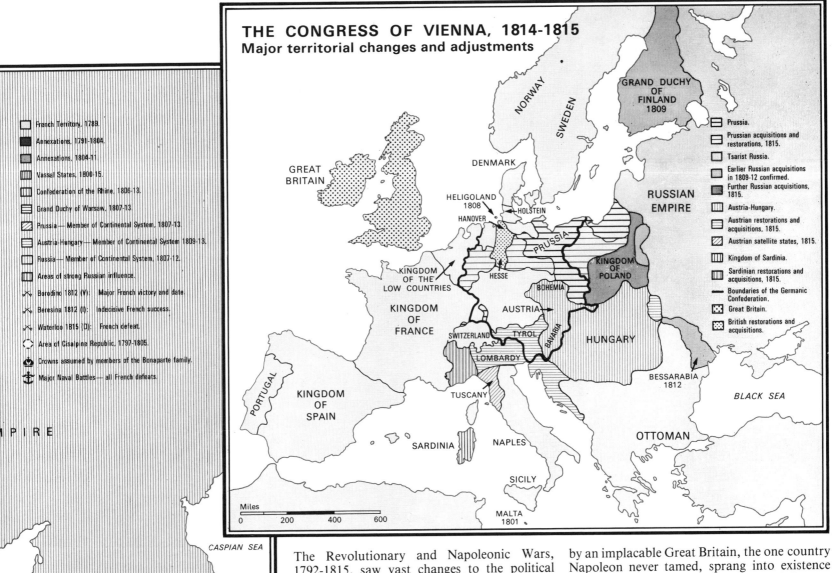

THE CONGRESS OF VIENNA, 1814-1815
Major territorial changes and adjustments

Legend (left panel):

- French Territory, 1789.
- Annexations, 1791-1804.
- Annexations, 1804-11.
- Vassal States, 1800-15.
- Confederation of the Rhine, 1806-13.
- Grand Duchy of Warsaw, 1807-13.
- Prussia — Member of Continental System, 1807-13.
- Austria-Hungary — Member of Continental System 1809-13.
- Russia — Member of Continental System, 1807-12.
- Areas of strong Russian influence.
- Borodino 1812 (V): Major French victory and date.
- Beresina 1812 (I): Indecisive French success.
- Waterloo 1815 (D): French defeat.
- Area of Cisalpine Republic, 1797-1805.
- Crowns assumed by members of the Bonaparte family.
- Major Naval Battles — all French defeats.

Legend (right panel):

- Prussia.
- Prussian acquisitions and restorations, 1815.
- Tsarist Russia.
- Earlier Russian acquisitions in 1809-12 confirmed.
- Further Russian acquisitions, 1815.
- Austria-Hungary.
- Austrian restorations and acquisitions, 1815.
- Austrian satellite states, 1815.
- Kingdom of Sardinia.
- Sardinian restorations and acquisitions, 1815.
- Boundaries of the Germanic Confederation.
- Great Britain.
- British restorations and acquisitions.

The Revolutionary and Napoleonic Wars, 1792-1815, saw vast changes to the political map of Europe. Until approximately 1806 France was basically after her traditional goals of the 'natural frontiers' (Alps, Rhine and Pyrenees), together with the reorganization of Italy, but thereafter the 'Napoleonic Empire' began to expand by leaps and bounds. New states were carved out of Central Europe to replace the moribund Holy Roman Empire: first the Confederation of the Rhine, then the Kingdom of Westphalia, as Austria and Prussia were brought to heel. Farther east, the Grand Duchy of Warsaw existed from 1807-13 under varying degrees of French patronage. At different dates large areas were annexed by France; Napoleon's brothers Joseph, Louis and Jérome, and his brother-in-law Joachim Murat received crowns. Many other powers were reduced to vassal states and were expected to provide men, money and supplies to feed the insatiable French war machine.

These French and Napoleonic expansionist ambitions bred much resistance, and no less than seven major coalitions, all master-minded by an implacable Great Britain, the one country Napoleon never tamed, sprang into existence one after another. Only for one brief year, 1802-3, was Europe even nominally at peace, and even then tensions remained and mounted, making a return to war inevitable.

Blaming all his problems on British intrigue Napoleon tried, from December 1806, to bring ruin to British prosperity by instituting an all-out economic blockade against British goods: the Continental System. This proved more of an irritant than an advantage, for the British countermeasures bit deeper and British trade found other outlets. French vassal states found the restrictions irksome and much illicit trading developed, which in turn forced Napoleon to extend his direct power to stretch from the Niemen to Dalmatia. Above all, it led to the wars in Portugal and Spain, 1807-14, and the invasion of Russia, 1812, which proved fatal to Napoleon's ambitions.

After 1814 Europe was redrawn to a large extent, but some Napoleonic changes were incorporated in the settlements agreed at Vienna by the victorious Allies.

The War in Spain and Portugal, 1807-14

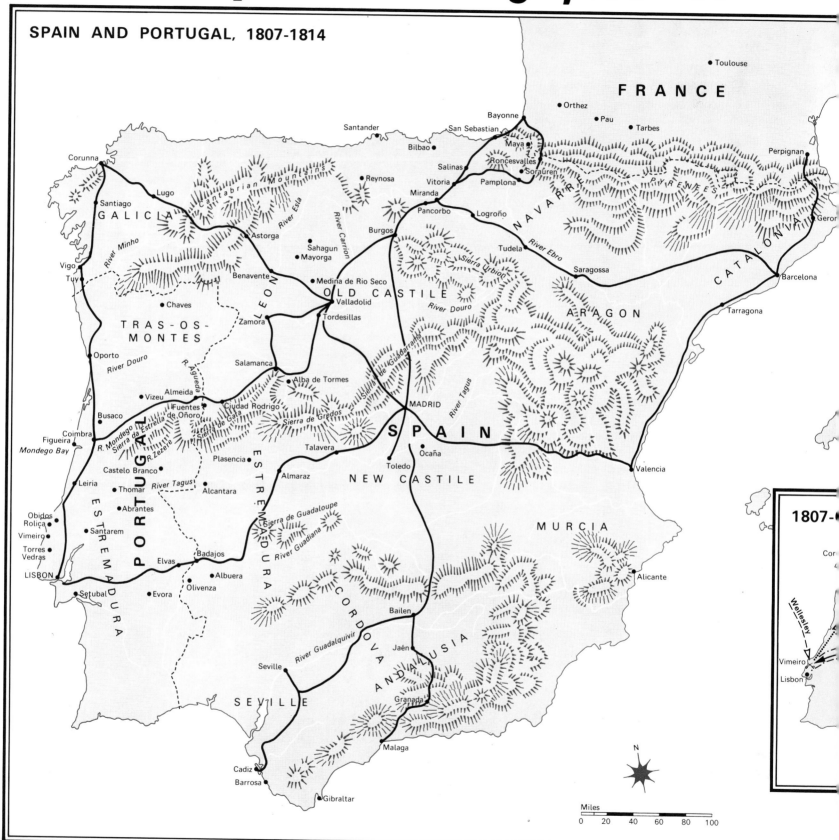

SPAIN AND PORTUGAL, 1807-1814

FRANCE

Toulouse

Orthez
Pau
Tarbes

Bayonne
San Sebastian
Maya
Roncesvalles
Salinas
Sorauren
Vitoria
Pamplona

Perpignan

Santander
Bilbao
Reynosa

Miranda
Pancorbo
Logroño

Geror

Corunna
Lugo
Santiago

GALICIA

Cantabrian Mountains

River Esla

River Carrion

Burgos
Tudela

NAVARR

River Ebro

PYRENEES

Vigo
Tuy

River Minho

Astorga
Sahagun
Mayorga

Sierra Urbion

Saragossa

CATALONIA

Barcelona

Chaves

TRAS-OS-MONTES

Benavente

LEON

Medina de Rio Seco
Valladolid
Tordesillas

OLD CASTILE

River Douro

ARAGON

Tarragona

Oporto
River Douro

Zamora

R. Agueda

Salamanca

Sierra de Guadarrama

Vizeu
Almeida
Fuentes de Oñoro
Ciudad Rodrigo
Alba de Tormes

River Tagus

MADRID

River Tagus

SPAIN

Busaco

PORTUGAL

Coimbra
Figueira
Mondego Bay

R. Mondego
Sierra da Estrella
R. Zezere

Sierra de Gata
Sierra de Gredos

Talavera

Ocaña

Valencia

Castelo Branco
Leiria
Thomar
Abrantes

River Tagus
Alcantara

Plasencia

Almaraz

Toledo

NEW CASTILE

ESTREMADURA

Obidos
Roliça
Vimeiro
Torres Vedras

Santarem

Sierra de Guadaloupe

River Guadiana

MURCIA

Alicante

LISBON

Elvas
Badajos

ESTREMADURA

Setubal

Evora

Albuera
Olivenza

CORDOVA

Bailen

ANDALUSIA

River Guadalquivir

Jaén

Seville

SEVILLE

Granada

Malaga

Cadiz
Barrosa

Gibraltar

N

Miles
0   20   40   60   80   100

1807-
Cor
Wellesley
Vimeiro
Lisbon

110

Wellington as Chef de Guerre

Napoleon's wish to see the Continental System effectively spread to include the Iberian Peninsula caused him to launch an invasion of Portugal with Spanish aid in November 1807. Junot duly occupied Lisbon. Soon the Emperor turned his attentions on his erstwhile ally, and, profiting from a division in the Spanish royal family, he sent an army under Murat to 'restore order' in the spring of 1808. At first welcomed by the population, this all changed on 2 May when Madrid rose. The flame of revolt spread to every part of Spain as it became clear that Napoleon's brother Joseph was to become king. This afforded Britain the chance to send an army back to the continent. On 19 July a Spanish force compelled General Dupont to surrender at Bailen, and on 21 August Sir Arthur Wellesley's British army inflicted a defeat on Junot at Vimeiro in Portugal. These setbacks threatened the whole French position in the Peninsula.

Napoleon decided to intervene in person. On 7 November the new campaign opened and in a few short weeks he scattered the Spanish armies and reoccupied Madrid. To aid the Spaniards, Sir John Moore, Wellesley's successor, marched towards the capital, but was fortunate to escape to Corunna when Napoleon set out to pursue him. Moore was killed there on 16 January 1809, but the army checked Soult and was evacuated by sea. Napoleon returned to France soon afterwards to counter plots and to prepare for a new struggle against Austria.

The team of Marshals left in Spain proved highly discordant, and Joseph's authority was little more than nominal. Wellesley returned to the Peninsula and, allying himself to the Spanish guerrilla movement, began to harass the French forces. Soult proved incapable of taking Cadiz, and Wellesley's incursion into Spain to Talavera, 27-28 July 1809, where he won a substantial victory over Jourdan, threw the French into disarray. The British subsequently withdrew again into Portugal.

Next year, Marshal Massena at the head of the Army of Portugal set out to expel Wellesley, now Viscount Wellington. Fighting a long retreat, Wellington won a battle at Busaco, before slipping into the secretly prepared Lines of Torres Vedras. Massena did not dare assault so strong a position, and after several months, October 1810 to March 1811, the growing starvation of his army forced him to retreat out of Portugal, harassed by Portuguese militia and Wellington's regular forces. Far to the south, Cadiz continued to defy the French.

The year 1811 saw much activity centring around the two corridors linking Spain and Portugal. Four fortresses—Almeida and Ciudad Rodrigo in the north, and Badajoz and Elvas in the south—guarded the two practicable invasion routes, and several sieges and associated battles took place; including Albuera 16 May 1811, near Badajoz, which was a narrow victory for Beresford over Soult, and Fuentes de Oñoro 3-5 May 1811, where Wellington repulsed Massena. The same pattern of events filled the last months of 1811, but on 19 January 1812 Ciudad Rodrigo fell to Wellington and on 6 April Badajoz was stormed (see p. 112). Fearful excesses followed each of these victories, but the strategic result was to open Spain to Allied liberation, particularly as the French forces were being run-down in preparation for the invasion of Russia which was to begin in June (see p. 118).

After considering the alternatives, Wellington decided to strike north-west of Madrid at Marmont. A month of complex manoeuvring ended in the great Battle of Salamanca, 22 July (see p. 112), which broke the French hold on northern Spain. Wellington occupied Madrid, but was repulsed from Burgos and had to make a winter retreat back to Portugal. Guerrilla activities continued to harass the French.

The year 1813, however, found a stronger Allied Army than ever re-invading Spain. This led to the famous Vitoria campaign and battle (see p. 114), the outcome of which broke the French hold on Spain. A great deal of hard fighting remained to be done, however, as the Allies swept towards the Pyrenees (July 1813), much of it associated with the siege of San Sebastian, which was finally taken on 9

French offensive, 1807.
British counter, 1807.
French offensive, 1808.
British counter, 1808.
French offensive, 1809.
British counter, 1809.

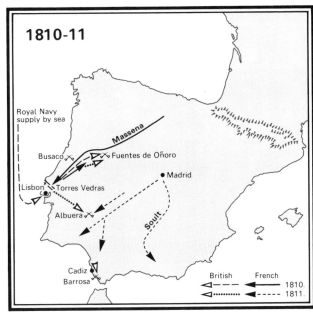

1810-11

Royal Navy supply by sea

Massena

Busaco • Fuentes de Oñoro

Lisbon • Torres Vedras

Madrid

Albuera

Soult

Cadiz
Barrosa

British   French
                    1810.
                    1811.

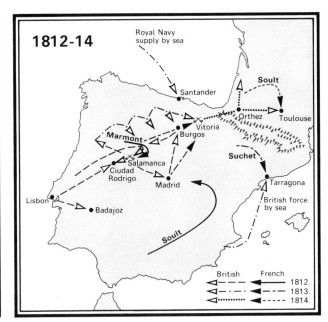

1812-14

Royal Navy supply by sea

Soult

Santander

Vitoria Orthez Toulouse

Marmont Burgos

Suchet

Salamanca

Ciudad Rodrigo Madrid Tarragona

Lisbon   British force by sea

• Badajoz

Soult

British   French
                    1812.
                    1813.
                    1814.

THE MAJOR LINES OF ATTACK AND COUNTERATTACK

The War in Spain and Portugal, 1807-14

September. Then Wellington passed over the mountains into southern France where the war ended after the Battle of Toulouse, 10 April 1814, fought unbeknowingly after Napoleon's first abdication.

Wellington on the Defensive: Torres Vedras, 1810

After his long withdrawal before Massena, Wellington positioned his army within a vast defended area outside Lisbon known as the Lines of Torres Vedras. These had been prepared since October 1809 by Colonel Fletcher, RE and seventeen fellow engineering officers with the help of some 7,000 Portuguese peasantry and militia. The fact that the French had no idea of their existence until 11 October 1810 gives some idea of the excellent security the Allies could impose, and also the hatred of the French amongst the Portuguese population at large.

As the map shows, the lines did not comprise continuous lines of trenches and redoubts but whole series of detached forts, strengthened by scarped hillsides or inundations, relying on interlocking fire-patterns and natural strength to defy an enemy. The forts were of two main types: small works designed to hold 50 men and two cannon apiece, and large ones with a capacity of 500 men and mounting six guns. Most were polygonal in shape and designed to fit the configuration of the ground. In the first line (A), running from the River Sizandro to Alhandra, were 32 redoubts mounting a total of 158 guns and requiring a garrison of 10,000 men. In the second line (B), from the River Lorenzo to the River Calandria, were 65 redoubts, 206 guns and some 14,500 men. These two successive lines of defences covered 29 and 22 miles respectively. Finally, as an ultimate reserve position guarding a possible embarkation point for the army in case of dire necessity (C), was a two-mile long third defensive area holding 4,000 men and 83 guns placed in 11 redoubts. Eventually, these lines were strengthened by the positioning of a further 42 redoubts at important points.

To garrison this deep zone of defences Wellington relied upon Portuguese troops and militia units. His main army was held in reserve, ready to march up specially constructed lateral roads to meet any serious threat. Flotillas of naval craft guarded the Tagus and Atlantic flanks, and further parties of seamen manned signal stations along the lines; in good weather a message could be passed from end to end in seven minutes. All road bridges were prepared for demolition. Finally, to the fore of these imposing defences, a wide belt of territory had been laid waste as a deliberate example of scorched earth to deprive the French of food.

Massena was astounded by these positions, and apart from a few probes near Sobral, 12-14 October, made no attempt to attack them, deeming them impregnable. However, the fact that the French were able to linger before them and in the general area (Massena retired on Santarem in November) for a total of almost six months shows that the devastation of the countryside had not been very effective. Nevertheless, the French were ultimately reduced to very low circumstances, and on 5 March Massena began a general retreat towards the northern corridor. Thus the lines represented a great strategic triumph of a defensive type for the Allies. Lisbon was completely safe, and the Royal Navy could bring in supplies, munitions and reinforcements to sustain Wellington's army. But some 40,000 Portuguese peasantry, who had taken shelter within the lines, are estimated to have died during the winter. Nevertheless, Wellington's waiting game paid a great dividend in the end. As for Massena, on 7 May 1811 he was removed from command of his army by Napoleon and replaced by Marshal Marmont.

Wellington on the Offensive: Salamanca, 1812

Although 1811 and the first months of 1812 saw mainly sieges and associated battles, by June 1812 Wellington was poised for a great offensive into French-occupied Spain. With both corridors secured after the capture of Badajoz (Mar.-Apr.) and its notorious aftermath, he decided to attack Marmont. Diverting the other French forces with secondary operations, he advanced from Ciudad Rodrigo on 13 June at the head of 48,000 men. Marmont, taken by surprise, collected his army of about 50,000 behind the River Tormes. The great city of Salamanca fell to the Allies, but three forts held out until 27 June and Marmont steadfastly refused to be drawn into an unfavourable battle situation. A month of sterile manoeuvring followed as each army sought an advantage, and Wellington began to consider a retreat to Ciudad Rodrigo. On 21 July both armies recrossed the River Tormes not far from Salamanca, and Wellington ordered his convoys to evacuate the city next morning.

Early on the 22nd, the French drove Allied outposts from several positions (A), but Wellington secured Point 901 (B) and reordered his army so that it faced south as well as west, recalling Pakenham's 3rd Division and d'Urban's cavalry to Aldea Tejada (C). He was still considering a general retreat when, early in the afternoon, he learnt that Marmont was trying to interpose part of his army between the Allies and their line of possible withdrawal, and in the process had allowed his leading divisions to draw ahead of both each other and the main body. Wellington determined to exploit this mistake to the utmost.

By 3pm he had briefed his generals. Pakenham was to attack the leading French division frontally, while Leith's 5th Division, supported by more infantry and cavalry, would attack the second French force with greatly superior strength before Marmont's main body could appear.

At 4.45pm, 3rd Division duly attacked Thomières east of Miranda de Azan (D), while Leith fell upon Maucune's brigades (E). Both attacks took the French by surprise. In the onslaught, Marmont and his second-in-command were gravely wounded. Within forty minutes a quarter of the French Army had been shattered: a superb cavalry charge by the British General Le Marchant clinching the destruction of Maucune. But Clausel—who now commanded the French Army—was made of stern stuff, and shortly after 5pm he extemporized a heavy counterattack that almost broke through Wellington's line to Las Torres (F). Wellington's 6th Division managed to stem the tide, Boyer's French Dragoons were decimated, and the day swung back in favour of the Allies. As night fell, the French divisions of Foy and Ferey successfully covered the flight of their comrades, who poured away towards Alba de Tormes (G).

The French lost 14,000 men, the Allies 5,200. Wellington had won a great offensive battle, and on 12 August he made his entry into Madrid.

The Siege of Badajoz, 1812

This famous operation was undertaken to provide the Allies with the full use of the southern 'corridor' linking Spain and Portugal (see p. 111). Wellington, fresh from capturing Ciudad Rodrigo, massed 32,000 men against General Phillipon's garrison of 5,000 French troops. Twice before the Allies had failed to take Badajoz, but this time Wellington was determined to succeed.

On 24 March the outlying Fort Picurina (A) was captured, and the Allied siege lines began to creep in towards their main objective. The weak south-east corner of Badajoz's defences were subjected to heavy fire and began to crumble. By 6 April three breaches had been battered, and Wellington ordered a general assault for that night. He was anxious lest neighbouring French forces—particularly those of Marshal Soult—intervened, so he ordered what was bound to be a costly attack. Two

THE LINES OF TORRES VEDRAS, 1810

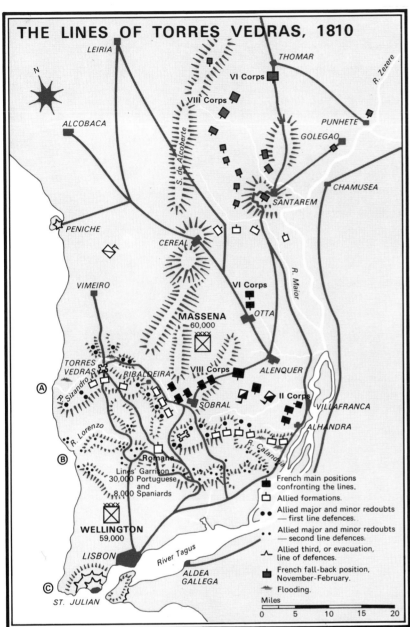

LEIRIA
THOMAR
R. Zezere
VI Corps
ALCOBACA
VIII Corps
S. de Alcoberte
PUNHETE
GOLEGAO
CHAMUSEA
SANTAREM
PENICHE
CEREAL
R. Maior
VIMEIRO
VI Corps
OTTA
MASSENA
60,000
TORRES VEDRAS
RIBALDEIRA
VIII Corps
ALENQUER
R. Sizandro
SOBRAL
II Corps
VILLAFRANCA
R. Lorenzo
ALHANDRA
R. Calandria
Romans
Lines' Garrison
30,000 Portuguese
and
8,000 Spaniards
WELLINGTON
59,000
River Tagus
LISBON
ALDEA
GALLEGA
ST. JULIAN

French main positions confronting the lines.
Allied formations.
Allied major and minor redoubts — first line defences.
Allied major and minor redoubts — second line defences.
Allied third, or evacuation, line of defences.
French fall-back position, November–February.
Flooding.

Miles
0   5   10   15   20

Sir Arthur Wellesley's string of victories in Spain and Portugal demonstrated his mastery over Napoleon's lieutenants. Created Duke of Wellington after Vitoria (1813), he ended the Peninsular War by invading Southern France. In 1815 came his crowning achievement at Waterloo. (Ken Trotman Arms Books)

THE BATTLE OF SALAMANCA
22 July 1812

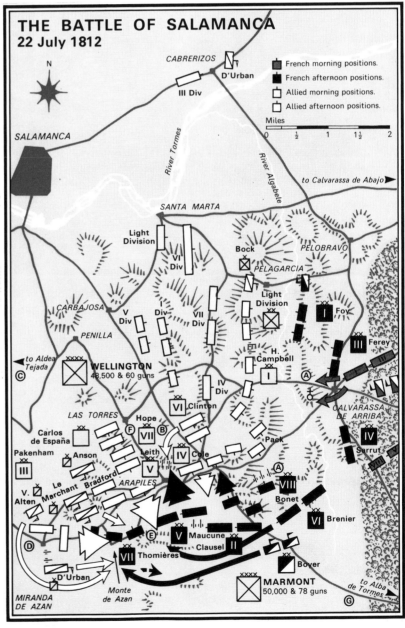

French morning positions.
French afternoon positions.
Allied morning positions.
Allied afternoon positions.

Miles
0   ½   1   1½   2

CABRERIZOS
D'Urban
III Div
SALAMANCA
River Tormes
River Algabele
to Calvarassa de Abajo
SANTA MARTA
Light Division
Bock
PELOBRAVO
VI Div
PELAGARCIA
CARBAJOSA
Light Division
V Div
Div
VII Div
Foy
PENILLA
I
H. Campbell
Ferey
to Aldea Tejada
WELLINGTON
48,500 & 60 guns
IV Div
CALVARASSA DE ARRIBA
VI Clinton
IV
LAS TORRES
Hope
Pack
Sarrut
Carlos de España
VII
Leith
IV Cole
Pakenham
Anson
V
VIII
III
Le Marchant Bradford
ARAPILES
Bonet
V. Alten
Maucune
VI Brenier
D'Urban
VII Thomières
Clausel
Boyer
MIRANDA DE AZAN
Monte de Azan
MARMONT
50,000 & 78 guns
to Alba de Tormes

1. André Massena, Duke of Rivoli and Prince of Essling (1756–1817), a soldier of vast repute, was defeated by Wellington three times in twelve months.
2. Nicolas-Jean de Dieu Soult (1769–1851) fought the British at Oporto (1808) besieged Cadiz. Later he fought a notable campaign amidst the Pyrenees but was finally defeated at Toulouse.

1.   2.

divisions—the 4th and the Light—were to storm the breaches (B), while Picton's 3rd Division attacked the castle (C) and 5th Division mounted a diversion near the bastion of San Vincente (D). The attack began after dark, and the troops storming the breaches were soon suffering horrific casualties, and making scant progress, for Phillipon's troops were ready for them. Fortunately for Wellington, however, Picton managed to enter the castle—at heavy cost—and capture the French reserve ammunition. The 5th Division also scaled the walls and, with the 3rd, took the French defending the breaches in the rear. Phillipon withdrew his survivors into Fort San Christoval, where they eventually surrendered.

The Allies, meanwhile, subjected Badajoz and its hapless Spanish population to a sacking, which Wellington deplored but could not halt for some time. Order was eventually restored after three days. To take Badajoz cost the Allies some 5,000 casualties in a single night, but its capture gave Wellington the strategic initiative, as Soult and Marmont both fell back on their respective sectors.

The Campaign and Battle of Vitoria, 1813
Although the campaign of Salamanca ended with Wellington back at Ciudad Rodrigo, the setback to his plans proved only temporary. Reinforcements reached him during the winter, and in late May a total of 70,000 Allied troops advanced into Spain; Wellington and Hill leading a force of 30,000, and General Graham the remaining 40,000 by a more northerly route. This latter move outflanked the French defences, and by mid-June Wellington was deep into Castile. A further series of outflanking 'hooks', which in conception owed much to Napoleon's 'manoeuvre of envelopment' (see p. 98), forced King Joseph and Marshal Jourdan to abandon position after position. On reaching the vicinity of Vitoria, however, the French decided to make a stand behind the River Zadorra, although they overlooked the need to destroy the bridges. They could field 43,000 infantry, 7,000 cavalry and 153 guns but were encumbered with baggage trains and refugees. Wellington enjoyed a considerable numerical advantage; he had at his disposal 27,000 British, 27,500 Portuguese and almost 7,000 Spanish troops (or a total of 70,000 men including 8,000 cavalry), but only 90 guns.

On the wet but hot morning of 21 June the Allied attack opened along a battle-front that would eventually stretch for almost 15 miles. At about 8am Lord Hill, who was on Wellington's extreme right, sent Morillo's Spaniards to seize the steep Heights of Puebla, supporting them with a British brigade (A). After a tough fight

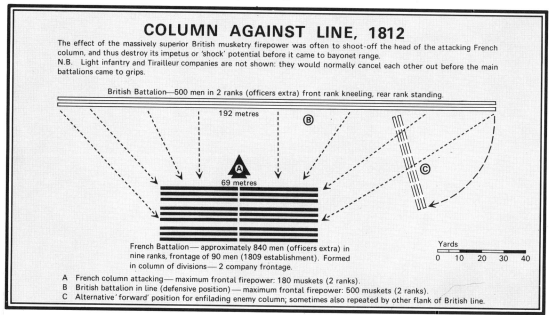

COLUMN AGAINST LINE, 1812

The effect of the massively superior British musketry firepower was often to shoot-off the head of the attacking French column, and thus destroy its impetus or 'shock' potential before it came to bayonet range.
N.B. Light infantry and Tirailleur companies are not shown: they would normally cancel each other out before the main battalions came to grips.

British Battalion—500 men in 2 ranks (officers extra) front rank kneeling, rear rank standing.

192 metres

69 metres

French Battalion— approximately 840 men (officers extra) in nine ranks, frontage of 90 men (1809 establishment). Formed in column of divisions— 2 company frontage.

Yards
0 10 20 30 40

A   French column attacking— maximum frontal firepower: 180 muskets (2 ranks).
B   British battalion in line (defensive position) — maximum frontal firepower: 500 muskets (2 ranks).
C   Alternative 'forward' position for enfilading enemy column; sometimes also repeated by other flank of British line.

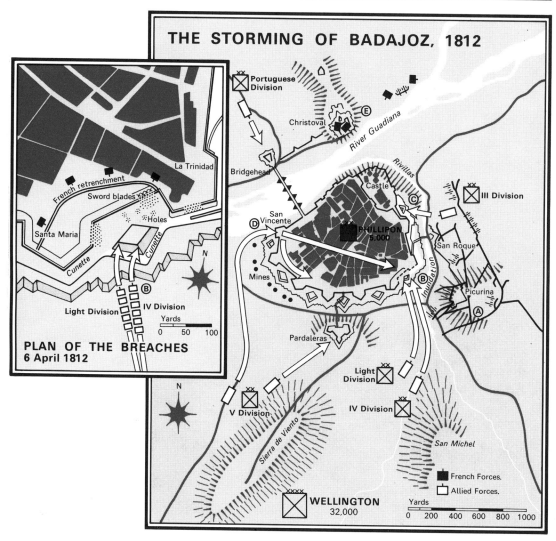

THE STORMING OF BADAJOZ, 1812

Portuguese Division
Christoval
River Guadiana
Bridgehead
Rivillas
Castle
III Division
San Vincente
PHILLIPON 5,000
San Roque
Mines
Picurina
Pardaleras
Inundation
Light Division
Light Division
IV Division
V Division
Sierra de Viento
San Michel

French Forces.
Allied Forces.

WELLINGTON 32,000

Yards
0 200 400 600 800 1000

PLAN OF THE BREACHES
6 April 1812

La Trinidad
French retrenchment
Sword blades
Holes
Santa Maria
Cunette
Cunette
N
Light Division
IV Division

Yards
0 50 100

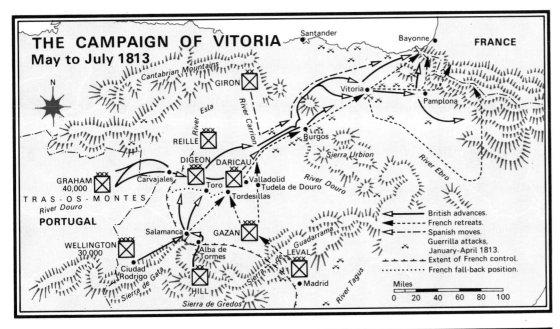

THE CAMPAIGN OF VITORIA
May to July 1813

Santander
Bayonne
FRANCE
Cantabrian Mountains
GIRON
Vitoria
Pamplona
River Esla
River Carrion
REILLE
Burgos
Sierra Urbion
River Ebro
DIGEON
DARICAU
Carvajales
Toro
Valladolid
Tudela de Douro
River Douro
GRAHAM
40,000
TRAS-OS-MONTES
River Douro
PORTUGAL
Tordesillas
Salamanca
GAZAN
Guadarrama
WELLINGTON
30,000
Alba de
Tormes
LEVAL
Ciudad
Rodrigo
Sierra de Gata
Sierra de
Madrid
River Tagus
HILL
Sierra de Gredos

British advances.
French retreats.
Spanish moves.
Guerrilla attacks,
January–April 1813.
Extent of French control.
French fall-back position.

Miles
0 20 40 60 80 100

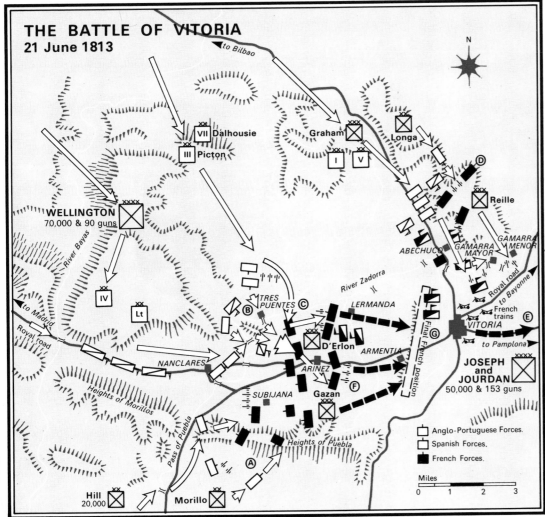

THE BATTLE OF VITORIA
21 June 1813

to Bilbao
N
VII Dalhousie
Graham
Longa
III Picton
I
V
Reille
WELLINGTON
70,000 & 90 guns
River Bayas
D
GAMARRA
MENOR
ABECHUCO
GAMARRA
MAYOR
IV
River Zadorra
LERMANDA
French
trains
VITORIA
TRES
PUENTES
B
C
Final French position
to Bayonne
Lt
E
to Madrid
Royal road
NANCLARES
D'Erlon
ARMENTIA
to Pamplona
ARINEZ
JOSEPH
and
JOURDAN
50,000 & 153 guns
SUBIJANA
F
Gazan
Heights of Morillos
Pass of Puebla
Heights of Puebla
A
Hill
20,000
Morillo

Anglo-Portuguese Forces.
Spanish Forces.
French Forces.

Miles
0 1 2 3

along the crest, the French General Gazan was forced to relinquish the ridge, whereupon Hill sent another brigade to extend his right along the ridge, thus threatening the French left flank. This danger took up much of the French attention all morning, but they never regained the ground and also lost control of Subijana.

In the centre, meantime, Wellington was pressing forward to the Zadorra. Kempt's brigade crossed this obstacle by an unguarded bridge near Tres Puentes (B) sometime after midday; farther north, on his own initiative Picton led his 3rd Division over the river (C) as Dalhousie's 7th Division was late in coming up into position. Kempt's move was hidden from the French by intervening ground, but Picton was seen.

Some time before these developments, however, the French right wing, commanded by General Reille, had come under serious attack from General Graham's two divisions and Longa's Spaniards (D). Reille gave up several villages but kept the Allies from crossing the river on this sector despite hours of bitter fighting. However, Longa's guns, near Gamarra Menor, made the 'Royal Road' towards Bayonne untenable for the French, who had to rely instead upon the road to Pamplona (E).

By early afternoon, Wellington's central forces were heavily engaged with two of d'Erlon's divisions and part of Gazan's command. Kempt's Brigade won a bridgehead over the Zadorra, and Allied troops poured over, including the 4th and Light Divisions. The French tried to hold a knoll and ridge near Arinez (F), but under pressure decided to fall back at about 3.30pm to a supporting line near Lermanda (G). They did not manage to hold this position for long. Growing anxiety for his right wing, which was still under heavy pressure from Graham, persuaded King Joseph to order a general retreat towards Pamplona. This began in orderly fashion, but the encumberments of baggage trains and non-combatants meant rich booty for the Allied soldiers. Indeed, the plundering earned the French a welcome respite, for the Allied pursuit was both badly conducted and hampered by heavy rain. It was while recounting this episode that Wellington described his army as 'the scum of the earth'.

For some 5,000 casualties the Allies had inflicted almost 8,000 on the French, decisively defeating the remnants of three French armies. It was the end of the French presence in Spain: by early July, their troops (apart from some garrisons) were behind their own frontiers. However, some stiff fighting still awaited the Allies in the Pyrenean region.

The American War of 1812

Far from the great European war fronts, a small but bitter struggle was waged across the Atlantic. Ostensibly caused by the British naval blockade and claim to the 'right of search' of United States' vessels on the high seas (in search of contraband), the war was used by the Americans for an attempt to conquer Canada. The Americans planned four invasions, but lacked the men or the experience to co-ordinate them. On the other hand, until 1814 the British were also short of men, owing to their preoccupations in Europe, but they made good use of Indian allies.

The initial American attacks ended in failure. Governor Hull's attack on Fort Malden (A) was called off, and Brigadier-General Isaac Brock's swift countermove induced Hull to surrender at Detroit on 16 August 1812. Brock then rapidly returned to the Niagara sector (B) in time to meet and rout, at the cost of his own life, van Rensselaer's invasion attempt at Queenstown Heights on 13 October. The other two proposed thrusts were called off.

On the western sector skirmishing continued, and in 1813 Commodore Perry gained naval superiority over the British Fleet on Lake Erie. Perry then ferried General Harrison's army over the water, and ultimately defeated Colonel Proctor's troops at Moravian Town (C). On the Niagara River, a combined operation enabled the American General Dearborn to make some progress, but he was replaced by Major-General Wilkinson who advanced on Quebec at the head of over 8,000 men. However, on 11 November he was confronted by Colonel Morrison with 800 men of the 89th and 49th Regiments beside the St. Lawrence river at Crysler's Farm. In a superb and salutary action (D) Morrison routed the American force.

This setback effectively ended serious American hopes of conquering Canada, and inaugurated a bitter period of atrocity and counter-atrocity. After the British regained Fort George (E) in December, the Americans burnt Queenstown, and in reprisal the British ravaged the American frontier, capturing Fort Niagara in the process. Operations in the area resumed in July 1814, when the Americans of General Brown captured Fort Erie (F) on 3 July. An attempt to exploit this success led to a sharp fight at Lundy's Lane on 25 July from which the Americans withdrew. The British General Drummond failed to retake Erie, but later in the year the Americans abandoned it once their navy lost its influence on Lake Ontario.

Retribution was now on its way for the Americans, as the end of the European struggle released large numbers of British Peninsular War veterans for service over the Atlantic. However, when in August 1814 the Governor

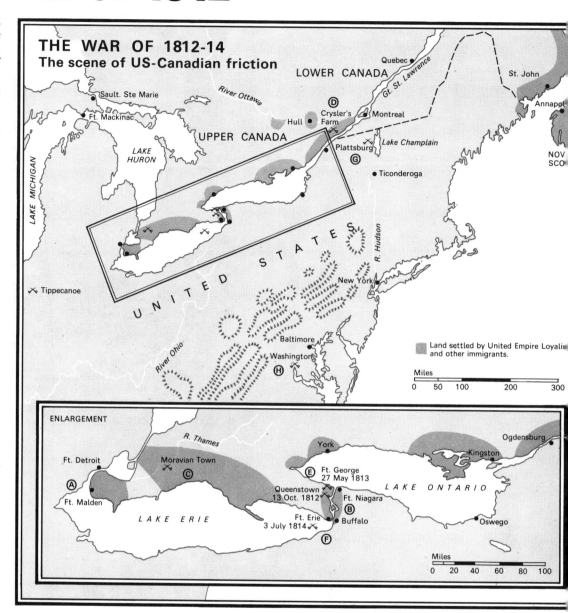

THE WAR OF 1812-14
The scene of US-Canadian friction

Land settled by United Empire Loyalists and other immigrants.

ENLARGEMENT

of Canada, Sir George Prevost, advanced to Lake Ticonderoga intent on capturing Plattsburg (G), the American naval squadron of Commodore Macdonough put paid to his plans by defeating a British squadron in the Bay.

Meanwhile, the British were carrying out further punitive operations against the United States. As early as April 1813 Admiral Cockburn had burnt Hampton on the River Chesapeake. Now, in August 1814, Admiral Cochrane's fleet brought General Ross's army to the approaches to Washington (H). Scattering the local defence forces at Bladensburg on 24 August, Ross occupied and burnt the American capital before withdrawing. The next British target was Baltimore, but they considered its defences too strong. A third target

was the city of New Orleans, far to the south. Retaking a base at Pensacola from the British, Major-General Andrew Jackson prepared to defend the area against Pakenham's 8,000 troops. A night attack failed to check the British advance, and on 1 January operations near New Orleans opened in earnest. On the 8th, Pakenham advanced to assault Jackson's position behind the Rodriguez Canal. Despite a shortage of bridging equipment, Pakenham pressed the attack bravely but was decisively beaten off with heavy losses. On 18 January the British moved back towards Mobile, but news then arrived that the war had been ended by the Treaty of Ghent on 24 December 1814. Thus, the battle of New Orleans, like that of Toulouse, was fought to no purpose.

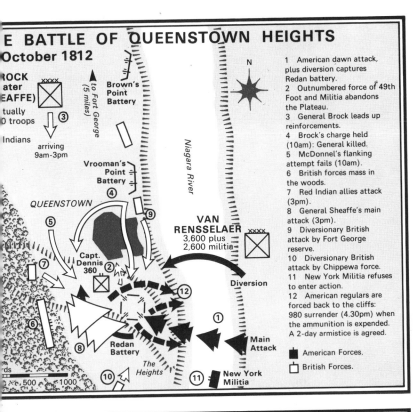

E BATTLE OF QUEENSTOWN HEIGHTS
October 1812

BROCK
ater
HEAFFE)
0 troops

to Fort George
(5 miles)

Brown's
Point
Battery

Indians

arriving
9am-3pm

Niagara River

Vrooman's
Point
Battery

QUEENSTOWN

Capt.
Dennis
360

VAN
RENSSELAER
3,600 plus
2,600 militia

Diversion

Redan
Battery

The Heights

Main
Attack

New York
Militia

500   1000

1 American dawn attack, plus diversion captures Redan battery.
2 Outnumbered force of 49th Foot and Militia abandons the Plateau.
3 General Brock leads up reinforcements.
4 Brock's charge held (10am): General killed.
5 McDonnel's flanking attempt fails (10am).
6 British forces mass in the woods.
7 Red Indian allies attack (3pm).
8 General Sheaffe's main attack (3pm).
9 Diversionary British attack by Fort George reserve.
10 Diversionary British attack by Chippewa force.
11 New York Militia refuses to enter action.
12 American regulars are forced back to the cliffs: 980 surrender (4.30pm) when the ammunition is expended. A 2-day armistice is agreed.

■ American Forces.
□ British Forces.

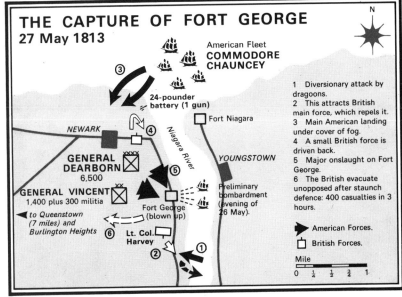

THE CAPTURE OF FORT GEORGE
27 May 1813

American Fleet
COMMODORE
CHAUNCEY

24-pounder
battery (1 gun)

Fort Niagara

NEWARK

Niagara River

GENERAL
DEARBORN
6,500

YOUNGSTOWN

GENERAL VINCENT
1,400 plus 300 militia

Fort George
(blown up)

Preliminary
bombardment
(evening of
26 May).

to Queenstown
(7 miles) and
Burlington Heights

Lt. Col.
Harvey

1 Diversionary attack by dragoons.
2 This attracts British main force, which repels it.
3 Main American landing under cover of fog.
4 A small British force is driven back.
5 Major onslaught on Fort George.
6 The British evacuate unopposed after staunch defence: 400 casualties in 3 hours.

▶ American Forces.
□ British Forces.

Mile
0 ¼ ½ ¾ 1

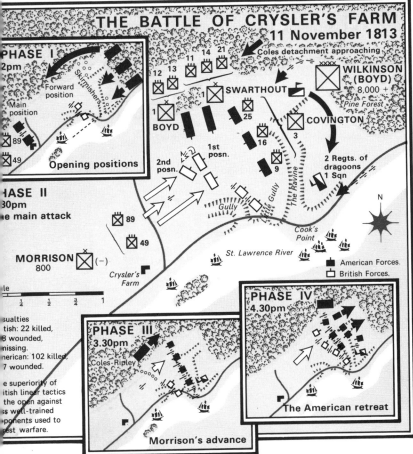

THE BATTLE OF CRYSLER'S FARM
11 November 1813

Coles detachment approaching

PHASE I
2pm

Skirmishers

Forward
position

Main
position

89

49

Opening positions

11 14 21

12 13

SWARTHOUT

1

BOYD

25

WILKINSON
(BOYD)
8,000 +

Pine Forest

COVINGTON

16

3

2 Regts. of
dragoons
1 Sqn

PHASE II
30pm
e main attack

89

49

MORRISON
800

Crysler's Farm

2nd
posn.

1st posn.

The Ravine

The Gully

Gully

Cook's
Point

St. Lawrence River

■ American Forces.
□ British Forces.

PHASE III
3.30pm
Coles-Ripley

Morrison's advance

PHASE IV
4.30pm

The American retreat

asualties
tish: 22 killed,
8 wounded,
missing.
merican: 102 killed,
7 wounded.

e superiority of
itish linear tactics
the open against
s well-trained
ponents used to
est warfare.

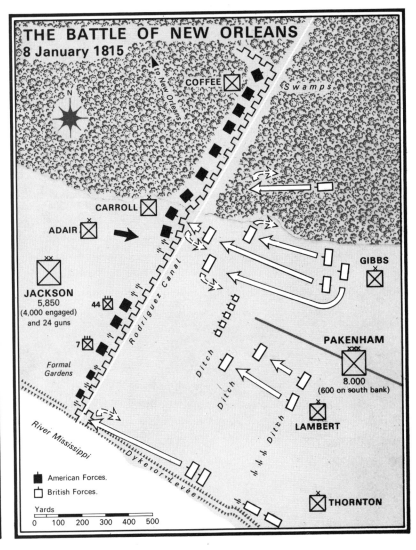

THE BATTLE OF NEW ORLEANS
8 January 1815

COFFEE

Swamps

to New Orleans

CARROLL

ADAIR

JACKSON
5,850
(4,000 engaged)
and 24 guns

Rodriguez Canal

GIBBS

44

7

Ditch

Ditch

PAKENHAM
8,000
(600 on south bank)

Formal
Gardens

LAMBERT

River Mississippi

Dyke or Levee

THORNTON

■ American Forces.
□ British Forces.

Yards
0 100 200 300 400 500

The French Invasion of Russia, 1812

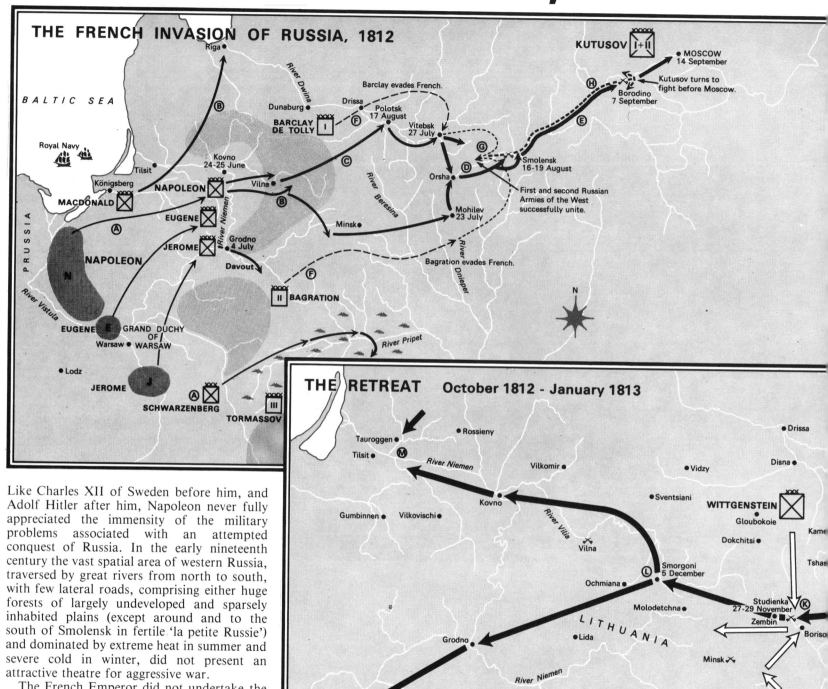

THE FRENCH INVASION OF RUSSIA, 1812

THE RETREAT   October 1812 - January 1813

Like Charles XII of Sweden before him, and Adolf Hitler after him, Napoleon never fully appreciated the immensity of the military problems associated with an attempted conquest of Russia. In the early nineteenth century the vast spatial area of western Russia, traversed by great rivers from north to south, with few lateral roads, comprising either huge forests of largely undeveloped and sparsely inhabited plains (except around and to the south of Smolensk in fertile 'la petite Russie') and dominated by extreme heat in summer and severe cold in winter, did not present an attractive theatre for aggressive war.

The French Emperor did not undertake the campaign lightly; the breakdown of 'the spirit of Tilsit' took several years, as Franco-Russian friction over the trade sanctions, the question of Poland, and the Tsar's ambitions involving the Balkans and Constantinople slowly poisoned relations. Napoleon could not forget Alexander's failure to honour the Erfurt agreements, 1808, to restrain Austria in 1809; conversely, the Tsar's ministers resented the brusque rejection of marriage to a Russian princess in 1810. Month by month war grew

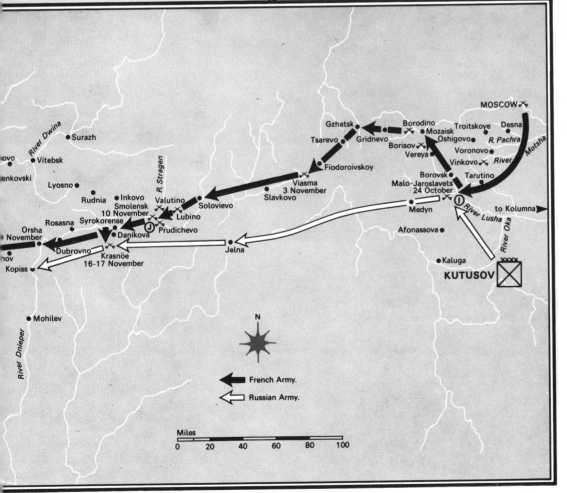

A French advance through Grand Duchy - East Prussia.

B Invasion, 24 June - 8 July.

C Advance on Vitebsk, 10-26 July.

D Manoeuvre of Smolensk, 8-19 August.

E Renewed advance on Moscow, 24 August - 7 September.

F Initial Russian retreat, June - July.

G The link-up, August.

H Renewed retreat, August - September.

French pre-invasion concentration areas.

Initial Russian positions.

Miles

50   100   150   200   250

closer, and by late 1810 both sides were making preparations. Barclay de Tolly completed his reorganization of the Russian Army along French lines, while engineers fortified the Dvina-Dnieper river lines. Napoleon began to build-up a huge army in Prussia and the Grand Duchy of Warsaw, bullying a dozen countries into providing contingents. He also created twenty-six transport battalions with which to supply them, for he was at least partly aware of the logistical problems to be faced.

By May 1812 the French had massed 650,000 troops in east Europe. It might be thought that the threat these posed would have been sufficient to cause the Tsar to seek a rapprochement if he were not determined to resist to the utmost. Still, Napoleon prophesied a quick victorious campaign of some five weeks duration. In the diplomatic offensive preceding the outbreak of war, however, the Tsar came off best. Britain promised money and a fleet in the Baltic; Sweden professed friendly neutrality; accommodations were patched up on the Finnish and Moldavian fronts, thus releasing

Russian armies. But only some 210,000 men were immediately available to face the French.

By June 1812 Napoleon was ready to strike; a central group of three armies (N, E and J) were echeloned back to Warsaw, with semi-autonomous corps on each distant flank (A). Two lines of reserve formations stretched back into central Germany. However, owing to the continuing struggle in Spain, tying down 250,000 French troops, less than half Napoleon's men were French nationals. The reliability of the larger part of this multi-national colossus was uncertain, both militarily and politically. Problems of command, supply and control were to prove daunting. Despite this, Napoleon's plan envisaged a quick victory. After selecting the region north of the Pripet Marshes for his attack (it was closer to his bases at Danzig and Königsberg, posed a double-threat towards St. Petersburg or Moscow, and would permit close supervision of Prussia), Napoleon planned to smash his way over the River Niemen near Kovno with the main army, supported by Eugène, to penetrate Tolly's extended forces and envelop them piecemeal. Meantime, Jérome's army would lure Bagration's Russians towards Warsaw to pin them down on the River Narew or Bug while part of the northern forces swept into their rear. The flanking corps would keep abreast with the major formations, pressing up the Baltic coastline or deep into Volhynia. On 24 June, the French offensive was launched.

From the first it was hampered by the unaccustomed reliance on slow-moving transport and food columns. One after another, three French traps snapped shut, only to find that their prey had eluded them each time.

The Manoeuvre of Vilna, 24 June-8 July. Murat, followed by Napoleon, successfully seized Tolly's former headquarters area (B), but were unable to exploit owing to the tardiness of supply-hampered Eugène de Beauharnais to move into position to support the flank. Farther south, Jérome proved incapable of either luring or pinning Bagration (F), who set out to retire north-east hoping to join Barclay, so Davout's lunge southwards from Vilna achieved nothing.

The Manoeuvre of Drissa, 16-26 July. A second attempt to trap Tolly on the Dvina near the fortified camp of Dunaburg and Drissa also failed (C), but the French compelled the Russians to abandon their defensive river line. To this point the Russian retreat may have been deliberate; thenceforward, it was almost certainly dictated by considerations of survival. However, Bagration's 48,000 men were steadily converging with Tolly's 110,000 despite repeated French efforts to keep them apart.

119

The French Invasion of Russia, 1812

The Manoeuvre of Smolensk, 11-26 August. After a pause to re-order his chaotic rear areas and to rest his exhausted troops, during which time the two Russian armies successfully linked up west of Smolensk (G), Napoleon hurled his forces into the River Gap (D). Using Murat to draw the Russians westwards, the bulk of the army crossed the Dnieper in a surprise operation and attempted to sweep on to Smolensk, there to sever the Russian links to the east. Further delays and a small Russian observation corps, however, gave Tolly and Bagration time to realize their danger and to retire pell-mell for Smolensk. The resultant lengthy battle, 16-19 August, was an indecisive French victory; Marshal Victor failed to cut the vital road near Lubino with the result that the Russians escaped.

So far the combination of lessened mobility, inadequate overall control, and Russian evasiveness, had thwarted Napoleon. Now he had to decide whether to press ahead after his discomfited opponent, or recast his strategy and accept a nine-month pause and a second campaign in 1813. His instinct as a gambler forced him on, although his front now stretched more than 700 miles in extent from Riga to Smolensk to the River Pripet, and his original half-million men of the central army group had dwindled through strategic consumption to 156,000 troops capable of further advance, the deadly heat of the summer causing many casualties amongst both men and horses. But Moscow lay only 280 miles ahead, and there surely the Russians would fight and be forced to accept Napoleon's terms. So the advance (E) was resumed on 28 August. Meantime, the Tsar had appointed seventy-four year old Marshal Kutusov to command the Russian armies, with strict instructions to fight for Moscow. The wily veteran selected a strong defensive position near Borodino (H), some 70 miles west of the holy city, and there made his stand.

The French in Moscow. The prize of the incomplete French victory at Borodino was Moscow, but its occupation on 14 September proved a delusion. The Russians abandoned, and probably set fire to, their religious capital, and the Tsar rejected all French approaches. New Russian armies were converging on the area (Steinheil and Wittgenstein from the Finnish frontiers, Tshitsagov from Moldavia) and the strategic initiative was fast turning against the French, who had but 95,000 men around Moscow. On 19 October—accepting that Alexander would not negotiate—Napoleon began his retreat, prompted by renewed Russian military activity to the south of Moscow where Kutusov controlled all of

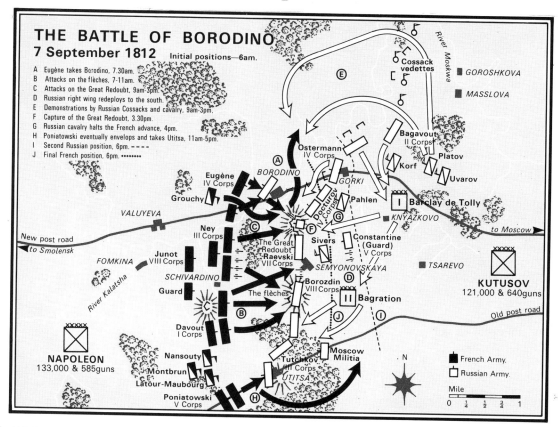

THE BATTLE OF BORODINO
7 September 1812 Initial positions—6am.

A Eugène takes Borodino, 7.30am.
B Attacks on the flèches, 7-11am.
C Attacks on the Great Redoubt, 9am-3pm.
D Russian right wing redeploys to the south.
E Demonstrations by Russian Cossacks and cavalry, 9am-3pm.
F Capture of the Great Redoubt, 3.30pm.
G Russian cavalry halts the French advance, 4pm.
H Poniatowski eventually envelops and takes Utitsa, 11am-5pm.
I Second Russian position, 6pm. ----
J Final French position, 6pm. ·······

NAPOLEON 133,000 & 585guns

KUTUSOV 121,000 & 640guns

■ French Army.
□ Russian Army.

110,000 men.

The Retreat, first phase. The first French plan was to retire on Smolensk through the prosperous countryside around Kaluga. A sharp brush with part of Kutusov's army at Malo-Jaroslavets (I), 23-25 October, deflected the army back to the ravaged route past Borodino. The 50-mile French column was continually harassed by Cossack and partisan attacks although Kutusov held off, and this relentless pressure together with the breakdown of supply arrangements soon began to convert the Grande Armée into an indisciplined and desperate rabble. Russian strategy was to confine all the French forces to a narrow central corridor, and to seize certain crucial river-crossings and supply depots ahead of Napoleon before closing in from three sides for the final kill.

The general deterioration of his position induced the Emperor to press ahead with his men through Smolensk (J). He headed for Orsha and the River Beresina crossings without waiting for Marshal Ney and the rearguard who were cut off by Russian forces. Brushing off Kutusov's clumsy attempt to intercept the French at Krasnöe, 17 November, Napoleon reached Orsha on the 19th, and was there joined by Ney who had made a miraculous escape. But evil tidings also arrived: Minsk and

its depots had fallen to Tshitsagov, and on the 22nd it was learnt that the Russians had blocked the main Beresina crossings. The earlier cold weather had given place to an unseasonable thaw, which meant that the river, normally hard-frozen by mid-November, would constitute an impassable obstacle. With Kutusov moving up slowly from the south-east and Wittgenstein closing in from the north, it seemed that the French were irretrievably doomed to total disaster. However, the discovery of an unmarked ford near Studienka (K) enabled Napoleon to win a costly four-day battle at the Beresina and to evacuate about 40,000 survivors over the river.

The Retreat, second phase. The road to Vilna and Poland now lay open before the emaciated French forces. Napoleon left the army on 5 December (L) to return to Paris, there to quash rumours of his death and to start building new armies. Commanded by a disgruntled Murat, the survivors staggered on towards Poland, encountering terribly severe winter conditions which took a heavy toll of life. The King of Naples soon departed for his Mediterranean kingdom, and the last stages of the retreat were supervized by Eugène. The Russians were content to leave the destruction of the remaining French to the climate and starvation, and made no further attempts to come to grips. In the New Year of 1813, some 94,000 French and Allied survivors—to include the flank forces which suffered slightly less drastically—came out of Russia; the half million men of the original central army group had been reduced to a mere 25,000. The legend of French invincibility was now shattered. But the Russians had also lost over 125,000 troops and many civilians.

The Battle of Borodino, 1812

At Borodino, seventy-five miles west of Moscow, the Russian Army of the recently-appointed Marshal Kutusov, 121,000 strong, at last turned at bay and gave Napoleon the opportunity of fighting a major battle. Napoleon, with some 133,000 troops available for action, opted for a straightforward battle of attrition, but his plan lacked subtlety.

At 6am the French batteries opened fire, but had to be moved forward as the range proved too great. Soon after, Eugène attacked Borodino and drove the Russians out (A) over the Kalatsha, while Davout began a whole series of heavy attacks on the flèches in the centre of the Russian line (B). This was followed by a mighty attack on the Great Redoubt (C) by Eugène and Ney, but these failed to make ground against the doughty Russian defence; at 11am the Russian line was still holding. Indeed, Kutusov had now become convinced that Napoleon's main effort was being made south of the Kalatsha. At 9am a probe towards Borodino (E) with cossacks and cavalry had delayed the first French attack on the Great Redoubt, and the Russian commander thereafter began to redeploy his right wing southwards (D) to strengthen his centre and left.

Meanwhile, on the extreme French right, Prince Poniatowski was making slow progress in his attempts to capture Utitsa and then envelop the Russian flank (H), the woodland being stoutly defended by Russian light troops. Casualties were immense in all sectors on both sides.

Shortly after 2pm, with the central flèches at last in French hands, Napoleon ordered an all-out attack on the Great Redoubt, which was the key to the Russian position. This was once again forestalled and delayed by a large Russian cavalry demonstration on the northern flank, but at about 2.30pm the great onslaught began. Murat launched two cavalry corps against the weakened Russian centre, and Eugène assaulted the redoubt. The 5th Cuirassiers broke into the fortification from the rear and then Eugène's infantry swarmed in through the embrasures; by 3.30pm the Great Redoubt (F) had fallen to the French.

Both sides were nearly exhausted, but by 4pm the Russian cavalry proved able to check the French attempt to exploit this belated success (G). Poniatowski, however, managed to encircle the Russian left (H) an hour later, and the Russians fell back to a second line (I), where the arrival of the Moscow Militia put fresh heart into the Tsar's soldiers and depressed the weary French. So the battle petered out indecisively. The French had lost 30,000 (including 48 generals) and the Russians may have suffered as many as 40,000 killed and wounded. Next day, Kutusov resumed the Russian retreat, and a week later the French occupied Moscow—but this would prove a barren triumph.

The Battle of the Beresina, 1812

On 19 October the French retreat from Moscow began. Fine weather soon turned into frost and snow, and supply breakdowns and Russian harassing tactics shook the integrity of the French Army. Early in November, the Russians sprang their trap at the River Beresina. Three armies converged on the struggling French columns, and with a watery torrent uncrossed by any bridge blocking their passage the outlook seemed bleak. Napoleon had barely 36,000 men capable of fighting while the Russians had 64,000 on the spot and a further 65,000 approaching from the east.

Fortunately for Napoleon, cavalry found a ford near Studienka on 23 November; the Emperor made a feint (A) towards Ucholodi to divert Admiral Tshitsagov's attention, while the French pontoon troops struggled in icy water to build two ramshackle bridges near Studienka. Both the trick and the bridging

1. Mikhail Illarionovich Kutusov, Prince of Smolensk (1745-1813), took command of the Russian Army shortly before Borodino in 1812. After abandoning Moscow to the French, he later proceeded to harass the Grand Army as it retreated.
2. Prince Peter Ivanovich Bagration (1765-1812) commanded the Russian Second Army in 1812. Mortally wounded at Borodino, he died seventeen days later. Popular amongst the men, he was much mourned.

attempt proved successful, and on 27 November part of the French Army crossed the Beresina (B). On the 28th the Russians attacked on both sides of the river. Despite breakdowns of the bridges, magnificent fighting by the skeleton corps of Oudinot and Ney fought off Tshitsagov (C), while Victor kept Wittgenstein at bay (D). At dusk the French rearguard began to cross the bridges, but the horde of non-combatants refused to follow. Early on the 29th the bridges were burnt and all on the east bank were lost. To save a remnant of his army Napoleon had used 55,000 lives, including those of 25,000 combatants, but now the road to Poland lay open to him.

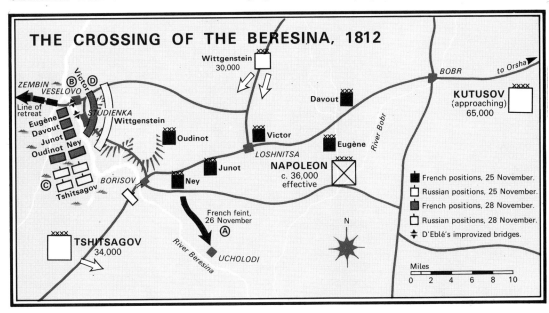

THE CROSSING OF THE BERESINA, 1812

The Campaigns of 1813-14

The defection of General Yorck and the Prussian contingent at Tauroggen (M) on 30 December 1812 was followed by that of Prussia itself on 13 March 1813. Soon Russian, Prussian and Swedish forces were closing on the Elbe, whence the remnants of the French Army had withdrawn by early March. By herculean efforts Napoleon managed to produce another 500,000 men for the defence of the Empire, although quality was far from good and the cavalry arm was particularly weak.

First Phase, April to June. Making full use of the central position around Leipzig, the Emperor won the Battles of Lützen, 2 May—against Wittgenstein—and of Bautzen, 20-21 May, against the Russians and Prussians. On both occasions his crippling shortage of cavalry and the inexperience of his conscripts robbed him of the chance of converting victory into complete triumph. Both sides were glad to accept an armistice, 1 June-16 August, although neither was prepared to negotiate a compromise settlement. The enemies of France gained most from the lull, for on 12 August Austria threw in her lot with the Coalition.

Second Phase, August to November. Once again beset on all sides by converging foes, Napoleon was committed to the strategic defensive. Leaving generally unreliable subordinates to hold the ring, the Emperor rushed 120,000 men to Dresden, 26-27 August. Although the Allies (170,000 strong in this sector) had agreed not to attempt a direct confrontation with their still-respected opponent, they were drawn into battle in spite of themselves and received a severe drubbing, losing some 38,000 casualties to the French 10,000. Once again, however, Napoleon could not pursue, but had to hurry to sustain Oudinot and Macdonald; in his absence, the Allies snatched a third small success at Kulm against Vandamme, 29 August.

Step by step the ring closed tighter around the French, despite French attempts to create a diversion towards Berlin. Further local setbacks drove the French back on Leipzig, and by this juncture all of Germany, save only loyal Saxony, deserted the French cause. New Russian and Swedish forces were arriving on the scene. The ultimate result was the great 'Battle of the Nations' around Leipzig, 16-19 October. By the last day the French were compelled to admit defeat. They had lost almost 40,000 men and inflicted 54,000 casualties. Napoleon was now forced to fall back to the Rhine, but inflicted a sharp check on his pursuers at Hanau on the 30th. The year had cost France another 400,000 lives.

The Campaign of 1814. Still Napoleon refused to accept terms, but with desperate zeal produced yet another army of schoolboys and pensioners to defend the soil of France. Allied doubts and dissensions won the French a little time, but in the south-west Wellington was pressing into the Pyrenees. When the Allied advance on Paris began in late 1813, it at first seemed that all was over, but there was still much fight left in Napoleon. Absorbing two setbacks at Brienne, 29 January, and La Rothière, 1 February, he launched a cunning offensive against Blücher's army and inflicted three severe reverses on Allied detachments between 10 and 14 February at Champaubert, Montmirail and Vauchamps. Swinging south against Schwarzenberg's Allied Army, Napoleon reoccupied Troyes and forced his opponents to adopt a precipitate retreat to the River Aube. Blücher, meantime, had been reinforced and was able to join Schwarzenberg at Méry-sur-Seine. There the Allies conferred and refurbished their resolution to conquer. Blücher returned north to resume pressure on the Marne, whilst the Austrians and Russians pressed forward along the southern axis. The ultimate target was again Paris.

Once again, by carefully balancing the situation on the two sectors, Napoleon used interior lines to force-march his small field army of barely 50,000 men to face Blücher. The resultant battles at Craonne and Laon in early March ended in a limited success for the Prussians, but Napoleon induced them to fall back against Reims by a deft stroke in their rear. However, he ordered the defences of Paris to be manned, and again moved south to face the hesitant Army of Bohemia on the Seine.

The Battle of Arcis-sur-Aube, 20-21 March, was fought between 80,000 Allies and 21,000 French. Faced by the prospect of defeat, Napoleon again—as at Laon earlier—had to break off the action and resort to his well-tried manoeuvre of a blow at the foe's communications to win him further time. Accordingly, he occupied St. Dizier, but for once the Allies were determined to call his bluff. Instead of falling back, they marched on towards Paris, virtually ignoring Napoleon's position in their rear, and on the last day of the month fighting took place on the heights of Montmartre as Mortier and Marmont tried to hold the defences of Paris.

Napoleon headed for Fontainebleau, only to learn that the capital's defenders had signed an armistice. He still spoke of continuing the struggle, but on 2 April under the leadership of Ney the Marshals rebelled. By the 6th, Napoleon had been induced to abdicate unconditionally, and two weeks later he was on his way towards the Mediterranean isle of Elba. Louis XVIII entered Paris on the 30th.

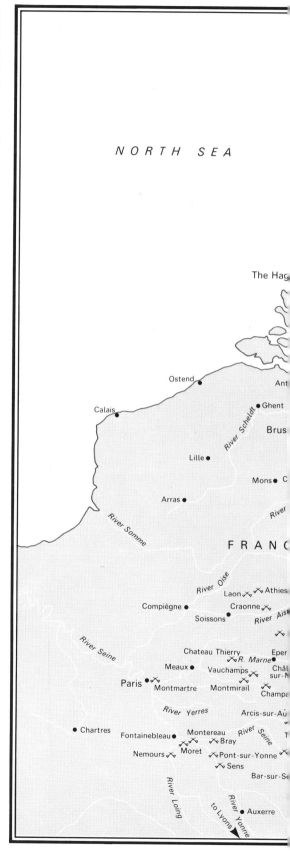

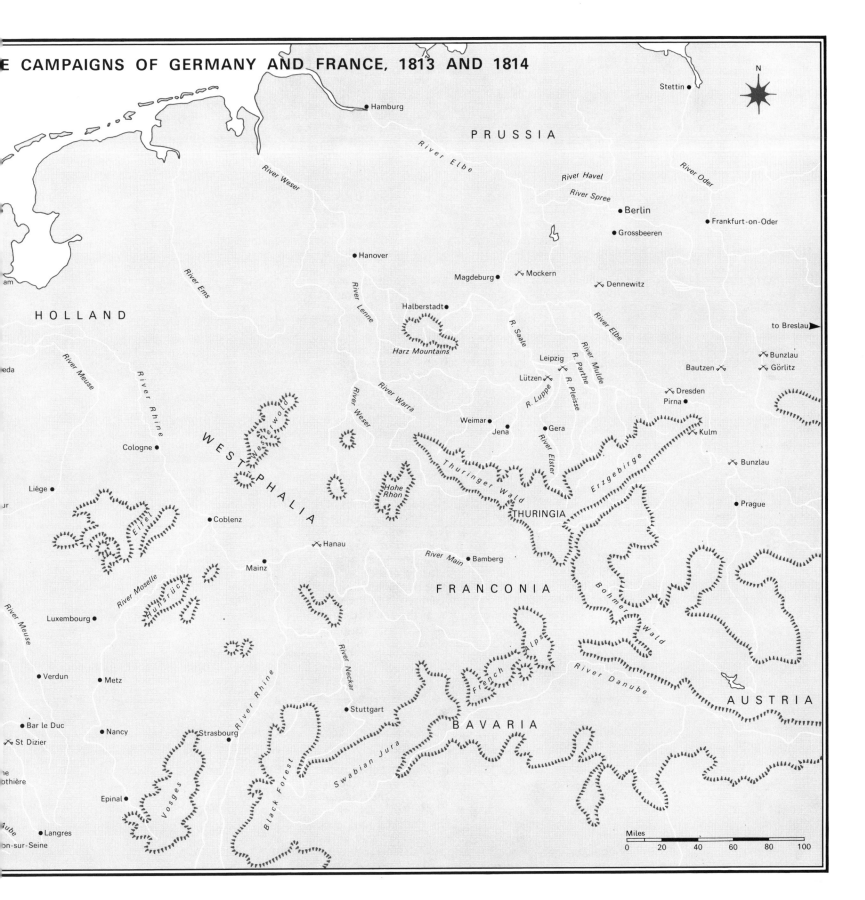

N

Stettin

PRUSSIA

River Elbe

Hamburg

River Weser

River Havel

River Oder

River Spree

Berlin

Frankfurt-on-Oder

Grossbeeren

Hanover

River Ems

Magdeburg · ✕ Mockern

Dennewitz

HOLLAND

River Lenne

Halberstadt ·

River Elbe

to Breslau ▶

R. Saale

Harz Mountains

Leipzig

River Mulde

Bunzlau

River Meuse

River Rhine

River Weser

River Warra

Lützen

R. Parthe

R. Pleisse

Bautzen ✕

Görlitz

R. Luppe

✕ Dresden

Pirna ·

Weimar ·

Jena ·

Gera

River Elster

Kulm

Westerwold

Cologne ·

WEST

PHALIA

Hohe Rhon

Thuringer Wald

Erzgebirge

Bunzlau

Liége ·

Eifel

THURINGIA

Prague

Coblenz

Hanau ✕

River Main · Bamberg

Böhmer Wald

Mainz

River Moselle

Hunsrück

FRANCONIA

Luxembourg ·

French Alps

River Danube

River Meuse

Verdun ·

Metz ·

River Rhine

River Neckar

AUSTRIA

Bar le Duc ·

Nancy ·

Strasbourg ·

Stuttgart ·

BAVARIA

✕ St Dizier

Vosges

Black Forest

Swabian Jura

he
othière

Epinal ·

Aube

Langres ·

on-sur-Seine

Miles

0   20   40   60   80   100

The Waterloo Campaign

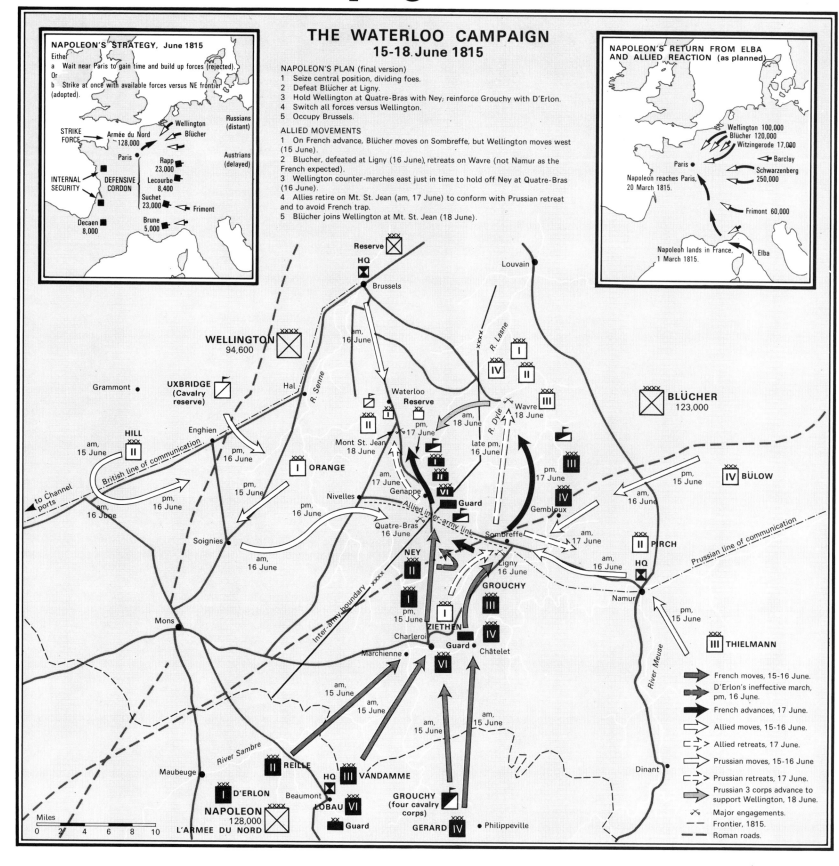

NAPOLEON'S STRATEGY, June 1815

Either
a Wait near Paris to gain time and build up forces (rejected).
Or
b Strike at once with available forces versus NE frontier (adopted).

STRIKE FORCE
Armée du Nord 128,000
Wellington
Blücher
Russians (distant)
Paris
Rapp 23,000
Lecourbe 8,400
Austrians (delayed)
INTERNAL SECURITY
DEFENSIVE CORDON
Suchet 23,000
Frimont
Decaen 8,000
Brune 5,000

THE WATERLOO CAMPAIGN
15-18 June 1815

NAPOLEON'S PLAN (final version)
1 Seize central position, dividing foes.
2 Defeat Blücher at Ligny.
3 Hold Wellington at Quatre-Bras with Ney; reinforce Grouchy with D'Erlon.
4 Switch all forces versus Wellington.
5 Occupy Brussels.

ALLIED MOVEMENTS
1 On French advance, Blücher moves on Sombreffe, but Wellington moves west (15 June).
2 Blucher, defeated at Ligny (16 June), retreats on Wavre (not Namur as the French expected).
3 Wellington counter-marches east just in time to hold off Ney at Quatre-Bras (16 June).
4 Allies retire on Mt. St. Jean (am, 17 June) to conform with Prussian retreat and to avoid French trap.
5 Blücher joins Wellington at Mt. St. Jean (18 June).

NAPOLEON'S RETURN FROM ELBA AND ALLIED REACTION (as planned)

Wellington 100,000
Blücher 120,000
Witzingerode 17,000
Barclay
Paris
Schwarzenberg 250,000
Napoleon reaches Paris, 20 March 1815.
Frimont 60,000
Napoleon lands in France, 1 March 1815.
Elba

Reserve
HQ
Brussels
Louvain
WELLINGTON 94,600
am, 16 June
R. Lasne
I
IV
II
UXBRIDGE (Cavalry reserve)
Grammont
Hal
R. Senne
Waterloo Reserve
am, 18 June
Wavre 18 June
III
BLÜCHER 123,000
HILL
am, 15 June
Enghien
pm, 16 June
II
Mont St. Jean 18 June
pm, 17 June
R. Dyle
late pm, 16 June
III
British line of communication
pm, 15 June
I ORANGE
am, 17 June
Genappe
pm, 17 June
IV BÜLOW
to Channel ports
am, 16 June
pm, 16 June
Nivelles
VI
Guard
Gembloux
IV
pm, 15 June
am, 16 June
Allied inter-army link
Quatre-Bras 16 June
Sombreffe
am, 17 June
II PIRCH
Soignies
am, 16 June
pm, 16 June
NEY
II
Ligny 16 June
am, 16 June
HQ
Prussian line of communication
Inter-army boundary
I
GROUCHY
Namur
pm, 15 June
Mons
pm, 15 June
I
III
Charleroi
ZIETHEN
IV
Guard
Châtelet
III THIELMANN
Marchienne
VI
River Meuse
am, 15 June
am, 15 June
am, 15 June
am, 15 June
Dinant
River Sambre
II REILLE
HQ
III VANDAMME
Maubeuge
I D'ERLON
Beaumont
LOBAU VI
GROUCHY (four cavalry corps)
Philippeville
NAPOLEON 128,000
Guard
GERARD IV
L'ARMÉE DU NORD

Miles
0 2 4 6 8 10

Legend
French moves, 15-16 June.
D'Erlon's ineffective march, pm, 16 June.
French advances, 17 June.
Allied moves, 15-16 June.
Allied retreats, 17 June.
Prussian moves, 15-16 June.
Prussian retreats, 17 June.
Prussian 3 corps advance to support Wellington, 18 June.
Major engagements.
Frontier, 1815.
Roman roads.

After the Treaty of Paris (1814), which restricted France to the frontiers of 1792, Europe longed for peace. But the Emperor had one last surprise to spring—the so-called 'Campaign of the Hundred Days'.

Napoleon's final gamble began early in March 1815, when he abruptly returned to France after eleven months' exile on Elba. His welcome was at first cautious, but rapidly grew, and the restored Bourbons abandoned Paris in panic. Meeting in Vienna, the European powers declared Napoleon an outlaw and planned a vast offensive against France, with almost a million men advancing from four directions.

The Emperor was soon aware that he would have to fight to survive. Rather than await the onset of his combined foes, he decided to mass most of his available troops in l'Armée du Nord, leaving only thin covering forces elsewhere, and make an all-out effort to destroy Wellington's Allied Army and Blücher's Prussians in Belgium.

By 14 June the French had massed 125,000 men near Beaumont without their opponents divining their purpose. Before dawn on the 15th the advance began over the Sambre into Belgium in two main columns backed by a reserve, intending to drive a wedge between Wellington and Blücher and then defeat each in turn with local superiority of force. (See p. 98 'the strategy of the central position'.) Despite some initial confusion things went well, but at dusk, Ney—who was commanding the left-hand column—had stopped short of the vital Quatre Bras lateral road which was the sole convenient link between the Allied and Prussian armies. The Prussians, meanwhile, were concentrating on Sombreffe—right in Napoleon's path—while Wellington's orders caused his main corps to draw slightly away from Blücher, as he was anxious for his communications to the channel coast if Napoleon attempted a typical envelopment.

Early on the 16th, while at a ball in Brussels, the Duke became aware of his error, and ordered his troops to head for Quatre Bras at once. There, a small detachment of Nassauers had bluffed Ney the previous evening and with some reinforcements continued to do so all morning.

Napoleon, meanwhile, had decided to concentrate against Blücher and destroy him, if possible, near Ligny. He expected Ney to seize Quatre Bras and hold it with one corps whilst his second, under Count d'Erlon, manoeuvred around the Prussian flank to create the opportunity for the French reserves, in conjunction with the French right wing under Grouchy, to gain a crushing victory. Unfortunately for the Emperor, Ney took no steps to secure Quatre Bras until early afternoon, by which time the battle was rising in intensity as more Allied troops reached the field. Nor was he informed of Napoleon's intentions for d'Erlon. Consequently, when at last he attacked Quatre Bras (recently reinforced by Wellington's reserve from Brussels) at about 2pm, he was thunderstruck to see d'Erlon marching away towards Sombreffe. D'Erlon had been ordered to do this by an imperial aide, but Ney furiously countermanded the order.

Meanwhile, at Ligny, Grouchy attacked the three Prussian corps present at about 2.30pm under the Emperor's eye. The battle was bitter around the Ligny brook and villages, but Napoleon was about to launch the Guard in the coup de grâce when he became aware of an unidentified force near his left flank. This was in fact d'Erlon, who was marching on the wrong objective, but no sooner had this been ascertained than Ney's order of recall caused the newcomers to countermarch back towards Quatre Bras. This confusion had two bad effects for the French: it meant that d'Erlon's men took no part on either battlefield that day, although its presence at either would have been decisive; and secondly, Napoleon lost much vital time at Ligny.

Eventually, Napoleon launched his grand attack at Ligny and broke through the exhausted Prussians, who incurred many casualties. But the onset of night and the absence of d'Erlon made an effective follow-up impracticable. At Quatre Bras, meanwhile, Wellington had been able to fend off each of Ney's attacks, as more Allied troops arrived, and the crossroads remained in Allied possession in a drawn but costly fight.

During the night, the Prussians—temporarily bereft of their commander—drew away north towards Wavre. Von Gneisenau wished to retreat to Namur, but Blücher, restored to his headquarters, refused to agree and sent assurances to Wellington that he would march to his aid with at least two corps were he attacked. Napoleon spent a bad night, and expended most of the morning of the 17th viewing the battlefield before, at about 11am, heeding Grouchy's pleas to be allowed to pursue the Prussians properly. By that time the Prussians were well away; worse for the French, Napoleon assumed his foes would be heading for Namur. So the pursuit began, not only delayed but in the wrong direction.

Wellington, aware that Blücher had received a drubbing at Ligny, thinned out his men at Quatre Bras and headed north towards Waterloo. Ney, once again wholly quiescent, made no attempt to renew his attacks on the crossroads, until a furious Napoleon spurred up at about 2pm and hounded him into action. By then it was too late to catch Wellington, whose rearguard fought a running battle with the pursuing French amidst heavy thunder and torrential rain. By dusk, Wellington had 68,000 men drawn up on his pre-selected position atop the ridge of Mont St. Jean; and Napoleon, with 72,000 men, halted between La Belle Alliance and Le Caillou after contenting himself that Wellington was no longer retreating. Grouchy, meanwhile, had discovered the general Prussian whereabouts and changed his march accordingly, but news of this only reached Napoleon early on the 18th. Both armies spent a wretched night.

On the morning of the 18th the Emperor did not hasten to attack: partly because he scorned Wellington, and partly because the ground was too wet for his guns. Wellington, aware that Blücher would eventually come to his aid, had taken up a strong position, with strongpoints in front of his main array and the main weight of his army behind his right, or western, wing. However, as a precaution the Duke left a substantial force near Hal to the west in case he had to retire.

By 10am Napoleon had at last sited his 266 cannon and issued his orders. The plan was far from subtle: a preliminary attack was to be made against the Hougoumont position (in the hope of attracting Wellington's reserves); a massed battery was then to pound the Allied centre and left, prior to a mass attack by d'Erlon's Corps; and, to clinch it all, the reserves of cavalry and the Imperial Guard, with Count Lobau's Corps, would hold themselves ready to smash through the centre and head for Brussels. It was to be a battle of attrition rather than one of manoeuvre.

The action began at about 11.30am. (The main stages of the celebrated battle can be traced on the accompanying map.) Napoleon made several vital errors. First, he failed to recall Grouchy from near Wavre until it was too late for him to be able to come. Secondly, by delaying the opening of the battle Napoleon gave Blücher's three corps time to draw near to the battlefield whilst his fourth kept Grouchy safely in play. Thirdly, he decided to place Ney in charge of the actual fighting. The Marshal proceeded to pile error on error. The Hougoumont attack got out of hand, but drew no Allied reserves. The attack by d'Erlon was made in the wrong formation without adequate cavalry support, and failed. Next, Ney launched huge unsupported cavalry attacks against Wellington's right centre in the erroneous belief that the Duke was falling back. These, in turn, were beaten off. Meanwhile, from 2pm, Napoleon had become aware of the

The Waterloo Campaign

approach of the Prussians, and by 4.30pm had been forced to extemporize a new front on his right to keep the Prussians from taking Plancenoit. This absorbed all his reserves, including, for a time, the Imperial Guard. So it was that when Ney eventually managed to take La Haie Sainte at about 6pm and Wellington's centre was wavering, all appeals for reinforcements had to be ignored, and the last chance of French victory passed. By the time the Guard was back into central reserve and the situation around Plancenoit under some form of control, Wellington had been given time to reinforce his centre and the crisis was past.

However, it was still touch and go for the Allies; Ziethen's Prussians were only prevented from marching to the wrong point by the intervention of an intelligent officer; and at 7.30pm, heartened by a deliberate falsehood spread by the Emperor that these troops were in fact

Grouchy's wing, the French began a last all-out attack to support the dreaded advance by eight battalions of the Middle Guard. Wellington, however, was ready by this time and the Guards' attack was repulsed. French morale sank to zero when they realized that the famous veterans were retreating and that Grouchy's supposed troops were opening fire on their comrades. The word 'sauve qui peut' spread and the French fled. The Old Guard carried out a superb fighting withdrawal, but all was over for Napoleon at Waterloo. The Prussian cavalry swarmed forward in pursuit and Napoleon was almost taken. By dusk the French had lost 33,000 men and 220 guns (or 60,000 losses in all since the 15th), whereas the Allies and Prussians had lost 22,000 men at Waterloo (or 55,000 since the opening of the campaign). Grouchy made a good retreat from Wavre, but the die was cast.

Napoleon spoke of rallying a new army and reversing the outcome, and indeed as the Allies advanced into France they became dangerously strung out, inviting defeat in detail. But Napoleon's political as well as his military credibility had run out, and the French politicians enforced his second abdication. On 15 July, Napoleon surrendered to Captain Maitland on HMS *Bellerophon*, and was soon on his way to exile on the remote South Atlantic island of St. Helena. So ended his fantastic career.

Waterloo marked the end of an era of warfare as well as of the great Napoleonic adventure. The forces of part of the Seventh Coalition had at last mastered the greatest military genius of modern history, and the four closely-linked battles of Quatre Bras and Ligny, Waterloo and Wavre, form a dramatic page in the annals of military history.

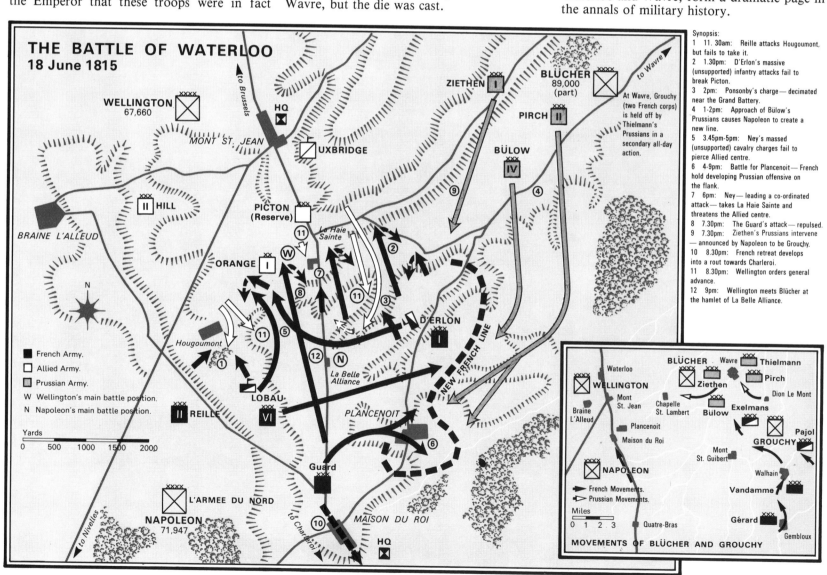

THE BATTLE OF WATERLOO
18 June 1815

WELLINGTON 67,660

ZIETHEN

BLÜCHER 89,000 (part)

PIRCH

BÜLOW

At Wavre, Grouchy (two French corps) is held off by Thielmann's Prussians in a secondary all-day action.

to Brussels

HQ

MONT ST. JEAN

UXBRIDGE

HILL

PICTON (Reserve)

La Haie Sainte

ORANGE

Hougoumont

La Belle Alliance

D'ERLON

NEW FRENCH LINE

BRAINE L'ALLEUD

N

LOBAU

REILLE

PLANCENOIT

Yards
0 500 1000 1500 2000

■ French Army.
□ Allied Army.
▨ Prussian Army.
W Wellington's main battle position.
N Napoleon's main battle position.

Guard

L'ARMÉE DU NORD
NAPOLEON 71,947

to Nivelles

to Charleroi

MAISON DU ROI

HQ

Synopsis:
1 11.30am: Reille attacks Hougoumont, but fails to take it.
2 1.30pm: D'Erlon's massive (unsupported) infantry attacks fail to break Picton.
3 2pm: Ponsonby's charge — decimated near the Grand Battery.
4 1-2pm: Approach of Bülow's Prussians causes Napoleon to create a new line.
5 3.45pm-5pm: Ney's massed (unsupported) cavalry charges fail to pierce Allied centre.
6 4-9pm: Battle for Plancenoit — French hold developing Prussian offensive on the flank.
7 6pm: Ney — leading a co-ordinated attack — takes La Haie Sainte and threatens the Allied centre.
8 7.30pm: The Guard's attack — repulsed.
9 7.30pm: Ziethen's Prussians intervene — announced by Napoleon to be Grouchy.
10 8.30pm: French retreat develops into a rout towards Charleroi.
11 8.30pm: Wellington orders general advance.
12 9pm: Wellington meets Blücher at the hamlet of La Belle Alliance.

MOVEMENTS OF BLÜCHER AND GROUCHY

BLÜCHER Wavre Thielmann
Waterloo Ziethen Pirch
WELLINGTON Dion Le Mont
Mont St. Jean Chapelle St. Lambert Bülow Exelmans
Braine L'Alleud
Plancenoit GROUCHY
Maison du Roi Pajol
NAPOLEON Mont St. Guibert
Walhain
Vandamme

→ French Movements.
➤ Prussian Movements.

Miles
0 1 2 3 ■ Quatre-Bras

Gérard
Gembloux

THE EVOLUTION OF GENERAL STAFF ORGANIZATIONS

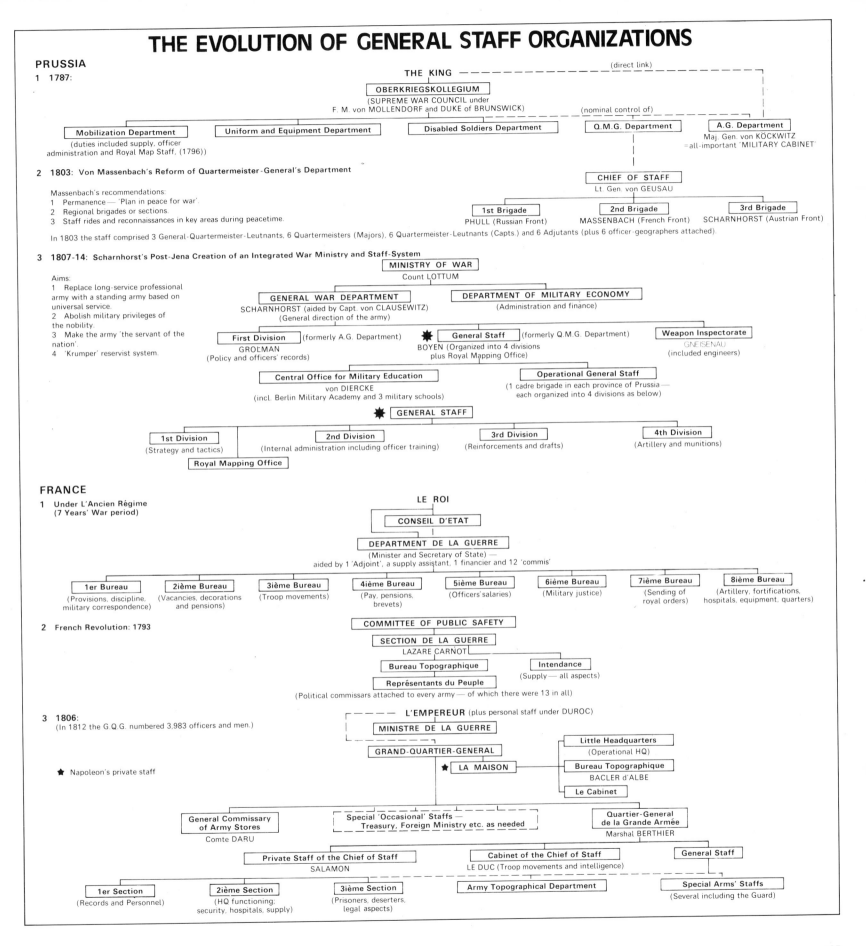

PRUSSIA

1 1787:

THE KING ———————————————————— (direct link)

OBERKRIEGSKOLLEGIUM
(SUPREME WAR COUNCIL under
F. M. von MOLLENDORF and DUKE of BRUNSWICK)

(nominal control of)

| **Mobilization Department** | **Uniform and Equipment Department** | **Disabled Soldiers Department** | **Q.M.G. Department** | **A.G. Department** |

(duties included supply, officer
administration and Royal Map Staff, (1796))

Maj. Gen. von KÖCKWITZ
=all-important 'MILITARY CABINET'

2 1803: Von Massenbach's Reform of Quartermeister-General's Department

Massenbach's recommendations:
1 Permanence — 'Plan in peace for war'.
2 Regional brigades or sections.
3 Staff rides and reconnaissances in key areas during peacetime.

CHIEF OF STAFF
Lt. Gen. von GEUSAU

| **1st Brigade** | **2nd Brigade** | **3rd Brigade** |
PHULL (Russian Front) | MASSENBACH (French Front) | SCHARNHORST (Austrian Front)

In 1803 the staff comprised 3 General-Quartermeister-Leutnants, 6 Quartermeisters (Majors), 6 Quartermeister-Leutnants (Capts.) and 6 Adjutants (plus 6 officer-geographers attached).

3 1807-14: Scharnhorst's Post-Jena Creation of an Integrated War Ministry and Staff-System

Aims:
1 Replace long-service professional army with a standing army based on universal service.
2 Abolish military privileges of the nobility.
3 Make the army 'the servant of the nation'.
4 'Krumper' reservist system.

MINISTRY OF WAR
Count LOTTUM

| **GENERAL WAR DEPARTMENT** | **DEPARTMENT OF MILITARY ECONOMY** |
SCHARNHORST (aided by Capt. von CLAUSEWITZ) | (Administration and finance)
(General direction of the army) |

| **First Division** (formerly A.G. Department) | ★ **General Staff** (formerly Q.M.G. Department) | **Weapon Inspectorate** |
GROLMAN | BOYEN (Organized into 4 divisions | GNEISENAU
(Policy and officers' records) | plus Royal Mapping Office) | (included engineers)

| **Central Office for Military Education** | **Operational General Staff** |
von DIERCKE | (1 cadre brigade in each province of Prussia —
(incl. Berlin Military Academy and 3 military schools) | each organized into 4 divisions as below)

★ **GENERAL STAFF**

| **1st Division** | **2nd Division** | **3rd Division** | **4th Division** |
(Strategy and tactics) | (Internal administration including officer training) | (Reinforcements and drafts) | (Artillery and munitions)

Royal Mapping Office

FRANCE

1 Under L'Ancien Régime
(7 Years' War period)

LE ROI

CONSEIL D'ETAT

DEPARTMENT DE LA GUERRE
(Minister and Secretary of State) —
aided by 1 'Adjoint', a supply assistant, 1 financier and 12 'commis'

| **1er Bureau** | **2ième Bureau** | **3ième Bureau** | **4ième Bureau** | **5ième Bureau** | **6ième Bureau** | **7ième Bureau** | **8ième Bureau** |
(Provisions, discipline, military correspondence) | (Vacancies, decorations and pensions) | (Troop movements) | (Pay, pensions, brevets) | (Officers' salaries) | (Military justice) | (Sending of royal orders) | (Artillery, fortifications, hospitals, equipment, quarters)

2 French Revolution: 1793

COMMITTEE OF PUBLIC SAFETY

SECTION DE LA GUERRE
LAZARE CARNOT

| **Bureau Topographique** | **Intendance** |
| | (Supply — all aspects)

Représentants du Peuple

(Political commissars attached to every army — of which there were 13 in all)

3 1806:
(In 1812 the G.Q.G. numbered 3,983 officers and men.)

L'EMPEREUR (plus personal staff under DUROC)

MINISTRE DE LA GUERRE

GRAND-QUARTIER-GENERAL

| **Little Headquarters** |
(Operational HQ)

★ Napoleon's private staff

★ **LA MAISON**

| **Bureau Topographique** |
BACLER d'ALBE

Le Cabinet

| **General Commissary of Army Stores** | **Special 'Occasional' Staffs —** Treasury, Foreign Ministry etc. as needed | **Quartier-General de la Grande Armée** |
Comte DARU | | Marshal BERTHIER

| **Private Staff of the Chief of Staff** | **Cabinet of the Chief of Staff** | **General Staff** |
SALAMON | LE DUC (Troop movements and intelligence) |

| **1er Section** | **2ième Section** | **3ième Section** | **Army Topographical Department** | **Special Arms' Staffs** |
(Records and Personnel) | (HQ functioning; security, hospitals, supply) | (Prisoners, deserters, legal aspects) | | (Several including the Guard)

Part III, 1816-78: Towards Total War

For almost forty years after the great holocaust of the Napoleonic Wars there were no major struggles between the great powers (see diagram Major Wars). The reason for this lull was a combination of sheet exhaustion, a renewed repugnance to the idea of settling problems by recourse to war, and preoccupation with internal troubles and reforms. Yet the times remained very unsettled: the forces of nationalism and liberalism—released by the French Revolution and Empire—increasingly came into conflict with reactionary governments (especially in Europe and South America) and, in due course, much of the pent-up national spirit of Great Britain and France was diverted to a competitive hunt for new colonies—especially in Africa (see pp. 134-136).

This period produced several notable attempts to outlaw war. After Waterloo the great powers attempted to achieve a measure of international cooperation through what was known as 'the Congress System' (a series of summit conferences to discuss points at issue), but the participants doggedly clung to their selfish national interests and the attempt failed. On 2 December 1823, the American president

(James Monroe) proclaimed the famous 'Monroe Doctrine', which decreed that the American continents should be closed to European interference. This instrument of policy greatly facilitated the liberation of the South American Spanish colonies (see pp. 132-133) and remains influential to the present day. These were genuine attempts to restrict the frequency, scale and effects of wars, and were backed by the greater number of nineteenth century philosophers who—Karl Marx apart—fervently believed in the virtues of pacifism. On a more practical level, the powers quickly reduced their armed forces to mere shadows of their former sizes. The British Army (685,000 in 1815) was down to 100,000 by 1821; the French Bourbon governments retained conscription, but reduced their forces to 180,000 (although this figure was doubled after 1830); Prussia called up only a mere third of each year's available conscripts. Only Russia maintained her Army at full strength, keeping 750,000 men under arms and devoting a third of her annual budget to the armed forces (see diagram, Approximate Army Strengths on Specified Dates, p. 130). Full employment for these forces was found with Russia's rapid expansion

to the Pacific and the borders of Afghanistan after the débâcle of the Crimean War.

An atmosphere of bourgeois respectability dominated British politics and society. Money spent on improving weapons was considered a complete waste, whilst the Duke of Wellington (and after him the Duke of Cambridge) stoutly opposed most attempts to modernize the organization or training of the British Army. Consequently, the old standards lapsed, slackness and boredom set in and, once again, soldiers were recruited from the riff-raff of society and all too often officered by incompetent and idle aristocrats. And the grave problems of the Crimean War (see pp. 142-147) and the Indian Mutiny (see pp. 156-157) were the results. In fact, the British Army was out of step with industrial development; the Royal Navy—the pride and joy of the Victorian age—was greatly reduced in size and power, but remained just strong enough to maintain the 'Pax Britannica' on the high seas and, eventually, with great misgivings, started to change from sail to steam propulsion.

By the 1850s, however, after the widespread upheavals of the 'Year of Revolutions' (1848), most of the liberal movements in Europe had

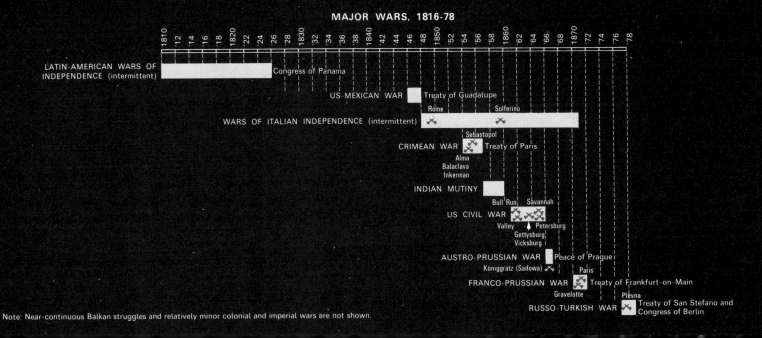

MAJOR WARS, 1816-78

LATIN-AMERICAN WARS OF INDEPENDENCE (intermittent) — Congress of Panama

US-MEXICAN WAR — Treaty of Guadalupe

WARS OF ITALIAN INDEPENDENCE (intermittent) — Rome / Solferino

CRIMEAN WAR — Sebastopol / Treaty of Paris
Alma
Balaclava
Inkerman

INDIAN MUTINY

US CIVIL WAR — Bull Run / Savannah
Valley / Petersburg
Gettysburg
Vicksburg

AUSTRO-PRUSSIAN WAR — Peace of Prague
Königgratz (Sadowa)

FRANCO-PRUSSIAN WAR — Paris / Treaty of Frankfurt-on-Main
Gravelotte

RUSSO-TURKISH WAR — Plevna / Treaty of San Stefano and Congress of Berlin

Note: Near-continuous Balkan struggles and relatively minor colonial and imperial wars are not shown.

been successfully contained or assimilated, at least for the present, leaving the powers free to turn their attentions to wider issues, and international rivalries once again appeared on an increasing scale to endanger the peace of Europe. The wars that resulted were, however, decidedly limited in extent. The Crimean War (1854-56) served only to expose the chronic inefficiency of the regular armies of the day, and is mainly of interest because it saw the introduction of effective military nursing (Florence Nightingale) and modern-type Press reporting (Howard Russell of *The Times*). The Italian War of Liberation (see pp. 158-161), the major phase of which proceeded by fits and starts from 1859 to 1870, was a more serious affair. Napoleon III (nephew of the great Emperor) provided an effective army for this 'liberal', inspired struggle, and a technological development of great importance in the shape of rifled artillery (see pp. 186-187). With its aid, the French and Italian forces got the better of the Austrian Army. Similarly, the progressive decline of Turkey was to some extent deliberately cushioned by Britain and France and, in 1878, by Germany.

Nevertheless, a very articulate section of public opinion remained devoted to the works of peace. In 1864, the Red Cross Society was founded at Geneva; twenty-six nations eventually signed the Convention, which attempted to lay down regulations for the conduct of war, especially regarding the treatment of non-combatants and prisoners-of-war. The influence of international law was considerably strengthened, and a host of general regulations were promulgated, restricting types and specifications of weapons and explosives and abolishing the 'Right of Search' of neutral shipping on the high seas.

At the same time, however, other philosophers were hard at work trying to analyze the success of Napoleon I, and studying the whole subject of warfare as a philosophical concept, thus laying the foundations for future, more total, wars. Two men, in particular, stand out: the first was the Swiss Baron Jomini, who did valuable work in categorizing the various levels of warfare (strategy, tactics, etc.) and tried to set down basic principles for the guidance of generals. On the other hand, he was basically a

reactionary, and tried to explain away Napoleonic warfare in terms of the eighteenth century, failing to realize that a new age had dawned. The second man, whose influence was far more pervasive, was the German Karl von Clausewitz. He was the first to realize that political revolutions had inculcated a new spirit into warfare, and in his famous book *On War*, he laid down the foundations on which strategy must be built. His most important tenet— variously translated and interpreted—was that "War is the continuation of policy" (i.e., diplomatic pressure) "by other means". That is to say, he placed warfare fairly within its political, social and economic context. He also taught that 'total' war—the nation-at-arms— was the military ideal, but admitted that, in practice, wars could never approach real totality—at least, not in the context of the nineteenth century. However, he was no dogmatist, and perhaps his most important contribution to the theory of the art of war was his conviction that there can be no unalterable set of strategic or tactical principles. Clausewitz was destined to have the profoundest influence on Prussian military thought and, after von Moltke's successes in 1868 and 1870, his ideas spread rapidly to much of the world.

If Clausewitz was one of the prophets of total war, the conditions of his time were also suitable for the expansion of the scale of warfare. The industrial and agrarian revolutions were fast proceeding all over Europe and North America, providing production means for new weapons and food surpluses for the feeding of large conscript armies. Changes in the iron industry were perhaps the most important: not only did the processes invented by Bessemer and Gilchrist revolutionize the materials available for cannon-forging, but the advent of railways (see pp. 166-167) changed the whole face of the problem of transportation, strengthened the cohesion of nation states, and added a new feature to strategy. Mechanical skills and precision engineering led to the manufacture of breech-loading rifles with interchangeable parts, and the introduction of metal cartridges and cylindrical bullets (see p. 186). Moreover, there was a great boom in the size of populations (see diagram Approximate Army Strengths on Specified

Dates), a factor that eventually produced the manpower for the new armies. Britain's population trebled during the century, and that of the USA increased fifteen-fold (largely through immigration, see pp. 162-165). A combination of industry and larger populations produced new urban centres, and with them the sinews of war. The gradual spread of public education brought literacy and the preparation of youth for 'duty to the Fatherland'. So, war and its preparation at last passed out of the hands of the aristocracy onto a broader and more professional social basis.

The first major conflict of the truly industrial age was the American Civil War (1861-65) (see pp. 168-185). In some ways it was a reversion to the Napoleonic concept of the nation-in-arms, but it was also an economic struggle fought out between two very different societies: the northern industrialists, preaching economic development, federal control, and tight commercial policies; and the southern cotton-growers, clinging onto an aristocratic way of life based on slavery, states' rights and free trade. In the end it was the 'big battalions' of northern manpower and industrial resources that won. From the point of view of the art of war, this conflict contributed several new concepts. It was the first war in which railways played an important role, in which the electric telegraph was used tactically (see p. 167), and in which rifled cannon and rifled small-arms were commonly employed, between them inflicting horrific casualty figures. This led to the abandonment of rigid line and column tactics in favour of individual skirmishing and loose brigade rushes (see p. 187). Troops began to dig-in as a regular feature of campaigning, and the art of field fortification (see pp. 140-141), took on greater importance. On the naval side, the first iron-clad equipped with a revolving gun-turret appeared in the form of the Federal *Monitor*, and the first practicable (though short-lived) submarine, the USN *Housatanic*, was put through its paces. It was also a war of conscript soldiers. Apart from some 500 regular officers (who divided pretty equally between the two sides), and perhaps 20,000 enlisted men (who mainly spent the war fighting Indians in the far west) there was no regular army to speak of, but by the end of the war the North

Towards Total War

alone had put some two and a half million men into uniform. Lack of experience in both officers and men led to huge, costly, but largely indecisive, battles on the eastern front. But in the west, along the Mississippi, General Grant eventually waged a brilliant war of manoeuvre to cut the Confederacy in half; an aim achieved when Vicksburg (see p. 182) surrendered on 4 July 1863. In the following year, General Sherman again bisected the eastern half of the Confederacy when he conducted his famous march 'From Atlanta to the Sea' (see pp. 170-171). However, this phase of the war was also notorious for the Federal Army's deliberate policy of terrorizing the population as it passed through the secessionist states.

Each side produced a number of very able generals, amongst them the famous Con-federate commanders Robert E. Lee and 'Stonewall' Jackson: the victors of the Valley Campaign (see pp. 175-178) Second Bull Run (1862) and Chancellorsville (1863) (see p. 179). However, the severe mauling undergone by the Army of North Virginia at Gettysburg (1-3 July 1863) at the hands of the Federal General Meade (see pp. 180-181) made it impossible for the South to win the war in the east, and, from that time on, the superior resources in men and material of the North played a decisive part in determining the outcome of the conflict.

Certain problems faced by democracies at war became highlighted during this period. The North set its face against Press censorship and, consequently, military security was often adversely affected. Time and again in the first years of the war, the North's opponents received advance warning from the newspapers of what was afoot. Similarly, the use of the electric telegraph caused considerable friction between the President and the Union generals. As never before, the political authorities were able to interfere in the conduct of operations, expecting daily reports from the front and sending up endless suggestions and instruc-tions. This interference was strongly resented by many senior soldiers, and the fact that Lincoln had to replace a whole series of army commanders before he hit upon the utterly dependable Grant as commander-in-chief in late 1863 proves the extent of the irritant. It was becoming clear, however, that the direction of the overall war effort was a civil, and not a purely military responsibility—hence Lincoln's ill-conceived dabbling in matters he could not

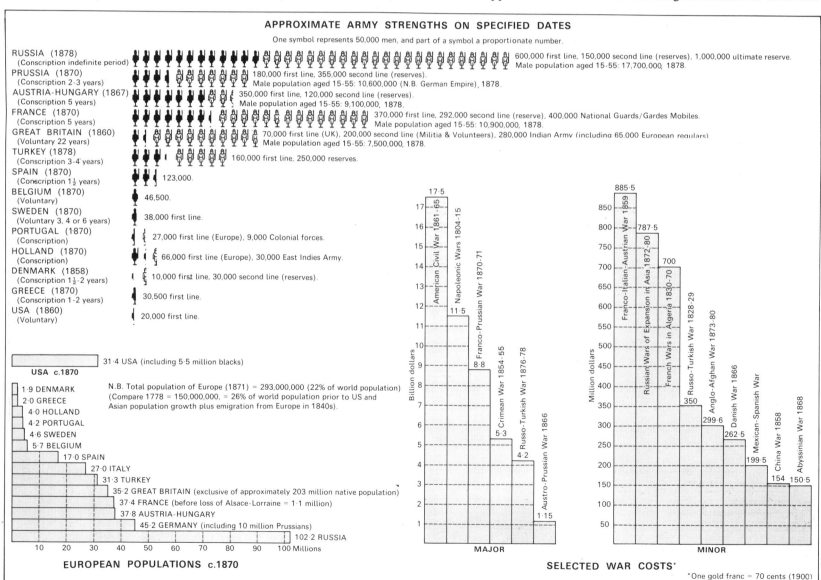

understand—and Congress went so far as to appoint a Joint-Committee of Senators and Representatives to supervize the soldiers. This problem of democratic versus military control descended as low as regimental level in the Union forces; at the outset of the war, indeed, the greater number of junior officers—and even regimental commanders—were elected by the rank and file, and throughout the war any number of 'political' appointments were made, sometimes with disastrous military results. This trouble did not affect the Confederacy to such a degree. By appointing an ex-military man (Jefferson Davis) to their presidency, the worst clashes of interest were avoided by the Southern States. Nevertheless, both sides were forced to resort to conscription to fill the ranks. In this way the American Civil War was very much a struggle between citizen armies, and thus a pointer to the total wars of the future.

The events of the American Civil War were studied with great interest by the military men in Europe, but it is significant that they drew more lessons from the events in the eastern states, and virtually ignored what had taken place on the Mississippi front. This would lead to the conviction that "the defensive is the stronger form of war" (as Clausewitz had claimed earlier) and, ultimately, to the stalemate of the First World War on the Western Front, which would largely stem from this mis-appreciation of what was significant. Nevertheless, one expanding European power was evolving a formidable military system, Prussia. Under the brilliant leadership of the elder Moltke (Chief of Staff from 1857), the first truly modern Staff System was perfected. Realizing that in an age of large armies, railways and conflicting national ambitions, direct personal control of all troops in the field was a thing of the past, he set out to produce a method of centralized command exerted from a distance through the medium of the telegraph. Above all, he was determined to avert the dangers of inadequate military preparation in time of peace. Consequently, he used his Staff System to provide all Prussian generals with a common background of principles and training, and further insisted on the formulation of detailed contingency plans ready to meet any future emergency. Although in time this led to a dangerous rigidity of military policy and planning, the Prussian General Staff (which significantly opened a Railway Section in 1857) had no equal (see p. 193).

The efficiency of Moltke's work was adequately demonstrated in two wars that took place soon after the end of the American struggle. Eager to unify the North German kingdoms into a strong national state, the Prussian Chancellor, Otto von Bismarck, committed his country to a sharp but very short war with his rival for German hegemony—Austria—in 1866, crushed her armed forces at the Battle of Königgrätz, and brought the struggle to a close in the space of six weeks (see pp. 190-191). He then imposed a very mild peace treaty, thus demonstrating his great statesmanship, for the Prussian victory left surprisingly little rancour and no desire for revenge at Vienna. Four years later there occurred the celebrated Franco-Prussian War (1870-71), which further demonstrated German military superiority (see pp. 192-195). Napoleon III's France was the only power strong enough to challenge the creation of a united Germany, and Bismarck deliberately goaded him into declaring war. Then the Prussian armies swept into action. Using the railways to speed his mobilization, Moltke had 400,000 men in the field in the space of three weeks, whilst his opponents only managed to produce 100,000 in the same period. Napoleon III's reliance on the supposed methods of his uncle (complete centralization of command and all-out attack) led him into grave errors, and the Prussian Army was able to carry out a vast concentric turning movement based on external lines (again making full use of railways to increase their mobility), which resulted in the victories of Gravelotte-St. Privat and Spicheren (see pp. 194 and 193) and the encirclement and ultimate surrender of Napoleon III and his last army at Sedan. Although the French people fought on for a further period under the aegis of the newly-created Third Republic, and refused to give in until Paris had been forced to surrender after a long siege (see p. 195), the Prussians had demonstrated their military superiority for all the world to see. Less sensibly, however, the eventual terms accorded to France by a triumphant Prussia, newly become the First German Empire, were severe, and left a scar that poisoned Franco-German relations for the future and was one cause of the First World War.

The Prussian military achievement was irrefutable. A combination of a highly trained Staff corps, speedy mobilization and the fullest use of railways had reinforced the concept of the nation-at-arms and made it into an effective military system. Prussian weapons were also superior to the French. Although potentially the French Chassepot was the better weapon, the Prussian breech-loading needle-gun proved far more effective in action because the troops were more familiar with its use (see pp. 186-187). Similarly, the French possessed a rudimentary machine-gun in the secret weapon, the Mitrailleuse, but in the interests of security failed to use it effectively. In the realm of artillery, however, the breech-loading field pieces and long-range heavy howitzers proved far superior to anything the French could produce. Prussian tactics also outclassed the French. Although they learnt the hard way, losing terrible casualties in the face of French rifle-fire, the Prussians soon realized the need to abandon the massed attack and, instead, substituted fighting in skirmishing order, using the combination of fire and movement to overcome their less imaginative opponents. The ultimate lesson of the Franco-Prussian war was that firepower was undoubtedly the vital factor in war, and this again stressed the importance of industrial potential.

The effect of the Prussian victories was immediate. All other armies immediately put in train the processes of reform. Although Britain still clung to the idea of a small standing army, Edward Cardwell (Secretary of State for War, 1868-74) centralized administration, abolished the sale of commissions and allotted two battalions to each regiment for alternative service at home and abroad. (His work was later supplemented by that of Haldane, who in the early 1900s would organize the British Army into divisions, set up the Imperial General Staff (1908), and establish the combination of regular and territorial forces, which would enable Britain to move fifteen divisions onto the Continent at the outbreak of hostilities in 1914.) Similarly, the United States created a General Staff under the guidance of Secretary Elihu Root, but the ancient American distrust of powerful armed forces persisted, and the President remained commander-in-chief and Congress continued to control the purse-strings. The raising of the National Guard (an approximate equivalent to the British Territorial Army) was another symptom of military reorganization.

The recurrent Balkan problem of 1877-78 gave the German Empire (acting as mediator between Turkey and Russia) the opportunity to emerge as the dominant power on the diplomatic scene at the Congress of Berlin (see pp. 196 and 197). Next year, the close alliance with Austria created a formidable central bargaining power bloc, practically generating future war and a diversity amongst races.

In this way the great powers of the nineteenth century laid the foundations for total war. In 1914 a complete social order and several empires would disappear into the maelstrom of Armageddon and, before it ended, many million men—an entire generation, no less—would be destroyed. But most of the ingredients of that terrible tragedy were present on the international scene from 1879.

The Latin-American Wars of Independence

Inevitably, the events of the American War of Independence and the French Revolution had important repercussions in South America. For a time, the conservatism of the Creoles, reinforced by the influence of the Church and the fear of Indian risings, held secessionist impulses in check. First tidings of the French invasion of Spain in 1808 caused a loyalist outburst in favour of King Ferdinand VII, but as the dissent-ridden Supreme Junta at Cadiz gave place to a Council of Regency in early 1810, the first great revolts against the tenuous links with Spain began. On Good Friday of that year, the Creole oligarchy of Caracas deposed the Spanish governor, and soon the new state of Venezuela was proclaimed. The example was soon copied. (Much of the needed leadership and inspiration was provided by Simon Bolivar (1783-1830), the son of a rich Caracas family, who had been sent to Spain to complete his education.) Inspired by Simon Bolivar and General Francisco de Miranda, both returned from living in Europe, revolts against Spanish rule quickly spread to Chile, Buenos Aires, Bogota and Cartagena; and in July 1811 the Supreme Junta of Caracas declared full independence from Spain. Similar moves in Mexico (begun as early as 1808) were complicated by first a Creole reaction and then an Indian revolt.

The independence movement was quickly challenged. In Mexico, the patriot Hidalgo y Costilla was hunted down and executed in July 1811, but revolt flared up again that October. In Venezuela, Miranda's 5,000 troops were soon on the defensive against the Spanish royal army. Bolivar had to recommence the liberation of the country from Cartagena and, in August 1813, he returned to Caracas to found the Second Republic. But again, loyalists in the interior under José Boves ruined his achievements, and forced 'the Liberator' to flee to New Granada and then to Jamaica in 1815. Miranda surrendered to the Spaniards, and died in prison in Cadiz the next year. Revolt was similarly stamped out, for the time being, in Mexico. Repression seemed to be triumphing in most quarters, but Bernardo O'Higgins kept the flame of revolt alight in Chile, whilst Juan José de San Martin, retreating from Buenos Aires to Mendoza, at the foot of the Andes, joined forces with the Chilean patriots. The arrival of 10,000 Spanish royal troops from Europe under General Morillo led to a bloody 'White Terror' which threatened to exterminate the independence movements. However, hard economic conditions induced Britain to espouse the cause of the rebels in the hope of securing valuable trade outlets and, in 1822, the Royal Navy effectively prevented the possibility of a collective European intervention in South American affairs. President Monroe of the United States riposted with his celebrated Declaration in 1823, but America was not strong enough to be able to impose its terms.

The new liberal revolt of 1820 in Spain against the increasingly authoritarian Ferdinand VII effectively signalled a major turning-point in the second struggle of independence in South America, by reducing the number of royal troops that could be sent there, and by frightening the conservative Creoles with the prospect of a liberal and republican régime in Spain. In consequence, the idea of independence again gained ground in the colonies. Already, in late 1816, San Martin had led 4,200 men over the high Andes, accompanied by O'Higgins, to fall upon Chile. After winning a decisive victory at Chacabuco (1817), the insurgents effectively liberated the country by April 1818. Two years later, determined to strike at the heart of Spanish colonial power, and aided by the squadron of Lord Cochrane, San Martin and O'Higgins invaded Peru and occupied Lima. In July 1821, independence was duly proclaimed, and this proved a death-blow to the Spanish empire in the Americas. A few months later, San Martin's forces made contact with Bolivar's men as the latter advanced towards Quito (capital of Ecuador) from the north.

Bolivar had rebuilt an army in Jamaica, and he landed in Venezuela once more in late 1816. Two years later he was strong enough to liberate the valley of the Orinoco, but he faced bitter fighting against the able Morillo, who at one point almost crushed the patriots. Scorning inactivity, in 1819 Bolivar transferred his attentions to New Granada and, after crossing the Andes with 2,500 men, he entered Columbia in time to rally the hard-pressed patriots there. This led to the proclamation of the United States of Columbia and, in December 1819, Bolivar was proclaimed its president and military dictator. Morillo opened negotiations in 1820, but the war soon recommenced. However, after Morillo's recall to Spain, Bolivar won the second Battle of Maracaibo on 24 June 1821; a success that clinched the liberation of Venezuela. In August, a Third Republic was proclaimed in Caracas, and Bolivar became its President.

Meanwhile, Bolivar's lieutenant, Antonio Sucre, liberated Quito in Ecuador, which merged with Venezuela and Columbia to form the new Republic of Greater Columbia. At this time, in July 1822, San Martin and Bolivar met at Guayaquil. To avoid a clash, San Martin selflessly stood down and left for Europe, where he died in obscurity in 1850. His departure emboldened the Spanish troops to reoccupy Lima, but Sucre regained control after the Battle of Ayacucho on 9 December 1824. Next year he liberated the region of Charcas, which assumed the name of Bolivia. The last Spanish garrison surrendered at Callao in January 1826. Apart from the islands of Cuba and Puerto Rico, Spanish America was now almost a thing of the past.

New Spain, or Mexico, was also severing its

Simon Bolivar (1783-1830), known as 'the Liberator'. He led the fight against Spain in South America, and created six new countries. His mastery of both guerrilla and conventional warfare was superb.

José de San Martin (1778-1850) was, after Bolivar, the greatest commander to engage in the wars of independence. His military talents as the liberator of the southern half of South America were rewarded by death in obscurity.

SOUTH AMERICA, 1800-30

Cartagena
Caracas
Maracaibo
1821
R. Orinoco
PANAMA
VENEZUELA
1811 1830
GUIANA
Bogota
NEW GRANADA
1811 1831
R. Japura
Quito
EQUADOR
1822 1830
Guayaquil
R. Negro
River Amazon
R. Xingu
RIO NEGRO
R. Madeira
R. Tapajos
R. Parnaiba
R. Purus
Pernambuco
PERU
1821 1821
EMPIRE OF BRAZIL
1822
BAHIA
Callao
Lima
Salvador
(Bahia)
The last Spanish
fortress, 1826.
Ayacucho
1824
La Paz
R. Araguaia
R. Tocantins
BOLIVIA
1825
R. Paraguay
Rio de Janeiro
PARAGUAY
1811
R. Parana
ARGENTINE
CONFEDERATION
1810
R. Uruguay
Ituzaingu
1827
CHILE
1810 1818
Sarandi
1825
Chacabuco 1817
Mendoza
Maipu 1818
Santiago
URUGUAY
Rancagua 1814
1814 1828
Buenos Aires
Montevideo
R. Negro
PATAGONIA
Magellan's Strait

Republic of Greater Columbia, 1819-30,
separated into Independent states.

1810 Date of independence from Spain.

1810 Date of separate statehood.

Battles in the wars of independence,
1810-27.

Miles
0   200   400   600   800   1000

last links with Spain. The events of 1820 in Spain had provoked a strong reaction amongst the Creoles, fearful of a new popular rising, and, in 1821, it was proposed to create an independent kingdom. The Emperor Augustin I was proclaimed and crowned in 1822, but his incompetence led to an army revolt, under General Antonio de Santa Anna, which deposed him. Guatemala, which Mexico had originally hoped to join to its own lands, was in 1823 recognized as independent. Next year, Mexico became a republic with a federal form of constitution, but ahead lay many years of turbulence.

In 1826, Bolivar summoned a congress of all American states at Panama. His hopes of creating a Holy Alliance of Liberty, linking the thirteen former Spanish colonies together, were quickly dashed. Before his death in December 1830 he would see anarchy spread over most of South America, which he had done so much to free. Nevertheless, he had seen the achievements of independence by virtually all Spain's Latin-American colonies.

The history of Portugal's great Viceroyalty of Brazil was on a different plane. The arrival of Dom John and the court of the House of Braganza in Bahia (1808), having escaped from Lisbon (with British assistance) in the face of the invading French, effectively made Rio de Janeiro the capital of Portugal. Prosperity came with British trade, and Brazil's status did not change with the end of the Napoleonic struggle. However, the new King, John VI (the former regent Dom John), succeeding to the Portuguese throne in 1815, seemed little disposed to leave the prosperity of Brazil for the uncertainties of Europe, and grave problems soon arose. An economic recession in the north-east of the country led to a serious revolt around Pernambuco, which was repressed by troops from Rio, in 1817. Then, in March 1821, John VI at last left for Lisbon, leaving his son Pedro as Regent. Social problems dominated politics and, in 1820, a party favouring independence from Portugal collected around José e Silva, a proponent of constitutional monarchy. Dom Pedro (under growing pressure from Lisbon which he resented) appointed Silva his minister in 1822, and in June accepted the title of 'perpetual defender' of Brazil. Three months later, Pedro decided that full independence from Portugal was inescapable, and on 7 September this was proclaimed. Next year the last Portuguese troops were evicted from Brazil with the help of Lord Cochrane's vessels. Portugal accepted the situation in August 1825. So ended the long history of the Portuguese empire in South America.

The century after Waterloo would see an immense explosion of European energy into Africa, Asia and Australasia, and corresponding developments in Canada, the USA and Tsarist Russia. Although the greater part of the "rush for Africa" took place after 1878 and before 1914, and thus strictly speaking lies outside the coverage of the present Atlas, it seems logical to illustrate the scene of the greatest nineteenth century colonial rivalry, at least in outline form, to round off the subject of colonial expansion (see also pp. 148-155). By 1914 the British Empire and Commonwealth ('on which the sun never sets') was reaching almost its greatest extent and, although the post-First World War settlement would add a substantial number of Germany's former colonies to its boundaries, in fact the processes of imperial decline were already setting in.

The impetus for overseas expansion sprang from several interrelated sources. In the case of Great Britain, the economic growth associated with the developing industrial revolution caused a growing demand for sources of cheap raw materials and also for 'captive' markets capable of absorbing manufactured goods in return. The current splendid isolation as regards Europe reinforced this attitude. Instead of "trade following the flag" it was the reverse situation until at least 1870, for the early and mid-Victorian governments only unwillingly accepted responsibility for further slices of 'the white man's burden'. The need to acquire suitable coaling stations throughout the oceans was appreciated, but political and public protests followed the acquisition of each new land area, save in the case of the Indian Sub-Continent which was regarded as a totally different case. The demands of merchants, traders, manufacturers and missionaries could not go wholly unheeded, however, and step by step the Empire grew, and military as well as commercial expeditions multiplied (see map Growth of World Empires, and for the situation in Africa, p. 155). The eagerness with which Napier withdrew from Abyssinia in 1869 (see p. 154) was a final reflection of this widely prevalent negative view of Empire, but from 1870 a wave of jingoistical imperialism caused a radical change of British mood—linked to a growing sense of rivalry with French, Russian and later German colonial aspirations—and the Empire grew by leaps and bounds.

French colonial growth was at first directed towards north-west Africa and what would become known as French Indo-China, but it would later develop into Central and West Africa (leading to some clashes with British interests) as well as envelop Madagascar and numbers of Pacific islands required as coaling

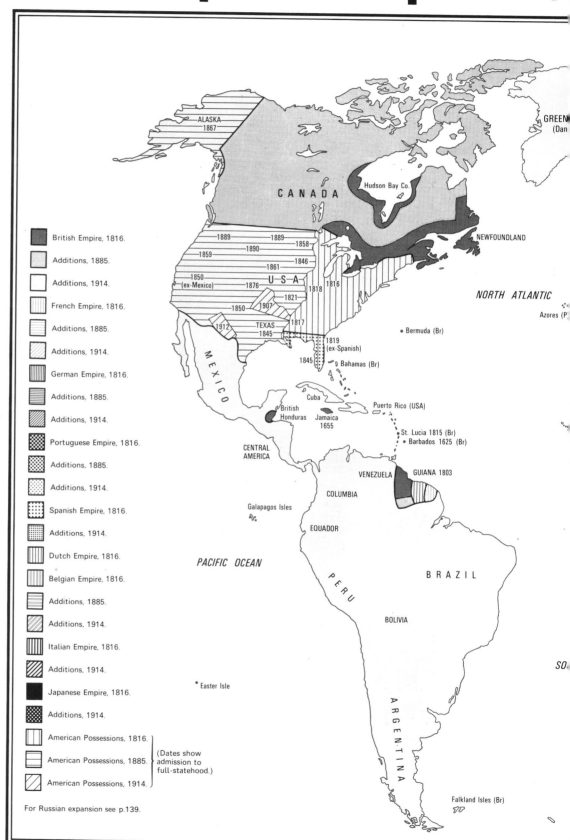

British Empire, 1816.	
Additions, 1885.	
Additions, 1914.	
French Empire, 1816.	
Additions, 1885.	
Additions, 1914.	
German Empire, 1816.	
Additions, 1885.	
Additions, 1914.	
Portuguese Empire, 1816.	
Additions, 1885.	
Additions, 1914.	
Spanish Empire, 1816.	
Additions, 1914.	
Dutch Empire, 1816.	
Belgian Empire, 1816.	
Additions, 1885.	
Additions, 1914.	
Italian Empire, 1816.	
Additions, 1914.	
Japanese Empire, 1816.	
Additions, 1914.	
American Possessions, 1816.	
American Possessions, 1885.	(Dates show admission to full-statehood.)
American Possessions, 1914.	

For Russian expansion see p.139.

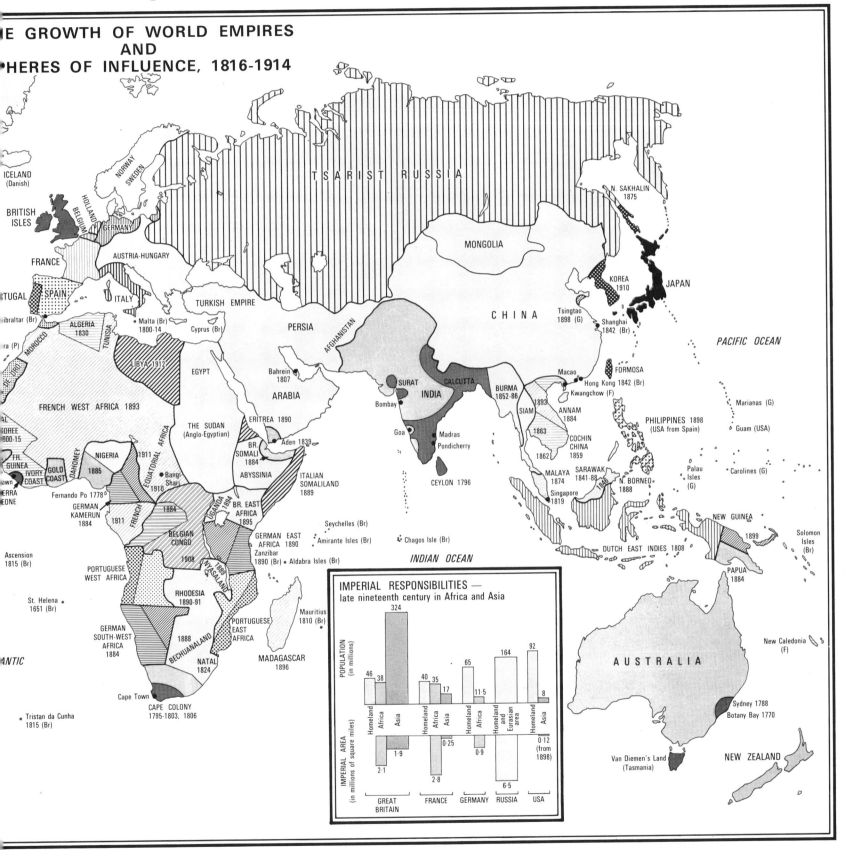

ICELAND
(Danish)

BRITISH
ISLES

NORWAY
SWEDEN

HOLLAND
BELGIUM
GERMANY

FRANCE

AUSTRIA-HUNGARY

TSARIST RUSSIA

MONGOLIA

N. SAKHALIN
1875

KOREA
1910

JAPAN

RTUGAL SPAIN
ITALY

PORTUGAL

ibraltar (Br)

ra (P)

ALGERIA
1830

TUNISIA

Malta (Br)
1800-14

Cyprus (Br)

TURKISH EMPIRE

PERSIA

AFGHANISTAN

CHINA

Tsingtao
1898 (G)

Shanghai
1842 (Br)

PACIFIC OCEAN

MOROCCO

LIBYA 1912

EGYPT

ARABIA

Bahrein
1807

SURAT

CALCUTTA

INDIA

Macao

Hong Kong 1842 (Br)

Kwangchow (F)

FORMOSA

Marianas (G)

FRENCH WEST AFRICA 1893

Bombay

THE SUDAN
(Anglo-Egyptian)

ERITREA 1890

BR.
SOMALI
1884

Aden 1839

Goa

Madras
Pondicherry

BURMA
1852-86

SIAM

1893

ANNAM
1884

1863

COCHIN
CHINA
1859

PHILIPPINES 1898
(USA from Spain)

Guam (USA)

DREE
800-15

FR.
GUINEA

IVORY
COAST

GOLD
COAST

DAHOMEY

NIGERIA

1885

FRENCH
EQUATORIAL AFRICA

1911

1910

Bangi
Shari

1884

ABYSSINIA

ITALIAN
SOMALILAND
1889

CEYLON 1796

1862

MALAYA
1874

SARAWAK
1841-88

N. BORNEO
1888

Singapore
1819

1846

Palau
Isles
(G)

Carolines (G)

wn
ERRA
EONE

Fernando Po 1778

GERMAN
KAMERUN
1884

1911

FRENCH

UGANDA
1894

BR. EAST
AFRICA
1895

Seychelles (Br)

GERMAN EAST
AFRICA 1890

Zanzibar
1890 (Br)

Amirante Isles (Br)

Aldabra Isles (Br)

Chagos Isle (Br)

INDIAN OCEAN

DUTCH EAST INDIES 1808

NEW GUINEA

1899

PAPUA
1884

Solomon
Isles
(Br)

Ascension
1815 (Br)

BELGIAN
CONGO

1908

PORTUGUESE
WEST AFRICA

St. Helena
1651 (Br)

GERMAN
SOUTH-WEST
AFRICA
1884

NYASALAND

RHODESIA
1890-91

1888

BECHUANALAND

PORTUGUESE
EAST
AFRICA

NATAL
1824

Mauritius
1810 (Br)

MADAGASCAR
1896

New Caledonia
(F)

AUSTRALIA

ANTIC

Cape Town

CAPE COLONY
1795-1803, 1806

Tristan da Cunha
1815 (Br)

Sydney 1788
Botany Bay 1770

Van Diemen's Land
(Tasmania)

NEW ZEALAND

IMPERIAL RESPONSIBILITIES —
late nineteenth century in Africa and Asia

POPULATION
(in millions)

	Homeland	Africa	Asia
GREAT BRITAIN	46	38	324
FRANCE	40	35	17
GERMANY	65	11·5	
RUSSIA	164 (Homeland and Eurasian area)		
USA	92		8

IMPERIAL AREA
(in millions of square miles)

	Homeland	Africa	Asia
GREAT BRITAIN	2·1		1·9
FRANCE	2·8		0·25
GERMANY	0·9		
RUSSIA	6·5 (Homeland and Eurasian area)		
USA			0·12 (from 1898)

The Growth of World Empires, 1816-1914

Left: The French attack on the bridge at Pa-Li-Chian, eight miles from Peking in September 1860.

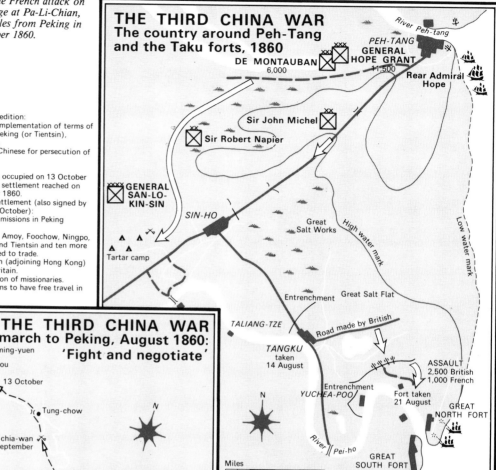

THE THIRD CHINA WAR
The country around Peh-Tang and the Taku forts, 1860

DE MONTAUBAN 6,000

PEH-TANG GENERAL HOPE GRANT 11,500

Rear Admiral Hope

Sir John Michel

Sir Robert Napier

GENERAL SAN-LO-KIN-SIN

SIN-HO

Great Salt Works

High water mark

Low water mark

Tartar camp

Entrenchment

Great Salt Flat

TALIANG-TZE

Road made by British

TANGKU taken 14 August

ASSAULT 2,500 British 1,000 French

Entrenchment

YUCHEA-POO

Fort taken 21 August

GREAT NORTH FORT

River Pei-ho

GREAT SOUTH FORT

Miles 0 ½ 1 1½ 2

Aim of expedition:
To secure implementation of terms of Treaty of Peking (or Tientsin), June 1858.
To punish Chinese for persecution of Europeans.

Peking was occupied on 13 October 1860 and a settlement reached on 24 October 1860.
Terms of settlement (also signed by French, 26 October):
A Foreign missions in Peking accepted.
B Canton, Amoy, Foochow, Ningpo, Shanghai and Tientsin and ten more ports opened to trade.
C Kowloon (adjoining Hong Kong) ceded to Britain.
D Protection of missionaries.
E Europeans to have free travel in China.

THE THIRD CHINA WAR
The march to Peking, August 1860: 'Fight and negotiate'

● Yuen-ming-yuen
● Ping-cheou

13 October
PEKING

Tung-chow

Chan-chia-wan 15 September

N

Gnan-ping

San-lo-kin-sin

River Pei-ho

Yang-tsun

HOPE GRANT
(accompanied by LORD ELGIN)

TIEN-TSIN

Siang-chwang-chi

Rear Admiral Hope
March starts 24 August

Koku

Taku

Miles 0 5 10 15 20

stations. Belgium latterly developed substantial African interests, Spain clung on to the last pathetic relics of her once vast Empire (losing the Philippines and Florida to the USA), whilst Portugal—the original European colonial power—retained her African possessions and the last reflection of her empire in the East Indies in the form of Goa and Macao. Dutch aspirations remained centred upon the riches of the former Portuguese possessions in the East Indies, which she had absorbed by conquest as early as the seventeenth century.

In the case of Tsarist Russia (see pp. 138-139), the thwarting of her ambitions to take over Constantinople led to the diversion of her expansionist energies eastwards through Central Asia towards the Pacific and southwards in the direction of Afghanistan; the latter development leading to considerable friction with Britain, anxious for her Indian hegemony.

In North America, the expansion westwards by both Canada and the United States (see p. 162) was closely linked with the development of railroads and the arrival in the New World of large numbers of European immigrants: fugitives from the Irish famines and the central European turbulence of the 1840s and 1850s. The taking over of the vast prairies and open plainlands of the central areas of the Continent was only achieved, in the case of the United States at least, at the expense of the unfortunate Red Indians, but the expansionist zeal of the waggon train trail-blazers—which even the bitter Civil War did not divert—followed by the railroad company bosses, was not to be denied, and further gains to the south and west were made at the expense of Mexico (see p. 164) and, to a lesser degree (in the north-west), Canada. Friction between the USA and her neighbour to the north was low-key but endemic throughout

the nineteenth century, involving Britain to a degree. This factor, together with the problems of assimilating the French communities of Quebec and the north-east provinces, led to the stationing of a substantial British imperial garrison in eastern Canada.

British penal settlements in Australia gradually grew into permanent settlements and then into states, and ultimately into a single mighty federation. New Zealand also became firmly colonized, once the Maori challenge had been overcome by a combination of force and negotiation. In the case of China, several European powers led by Britain and France tried to break through the barriers, fighting three China Wars in the process, of which the Anglo-French expedition of 1860, centring around the capture of the Taku Forts and the subsequent march on Peking certainly was the most effective. By early in the 1900s, after the Boxer Revolt, large concessionary areas would be wrung from the Manchu Emperors, but until then China retained its integrity—as did Japan.

The Sick Man of Europe*

The decline of the Turkish Empire, signalled by the Battle of Belgrade (see p. 56) in 1718, had continued slowly throughout the eighteenth century in a series of fluctuating struggles against the Russians and the Austrians. Nevertheless, the power of the Ottoman Turks had survived after a fashion. In 1789, their Empire included, at least on the map, the Balkans area south of the Danube (comprising the Greeks, Bulgars, Serbs, Rumanians and Albanians), Asia Minor, the Levant, and the Mediterranean islands of Crete and Cyprus; Syria, Palestine, Egypt and the greater part of the North African littoral as far west as Morocco, although in these areas control was little more than nominal.

However, as the internal problems of the Ottoman Empire grew, so did the aspirations of the subjected populaces in the hope of achieving independence. The result was continuous intrigue and turmoil, which was often encouraged by both Russia and Austria, who had hopes of securing control of the Balkan peninsula and the vital straits leading from the Black Sea. These ambitions alarmed Britain and France, who attempted to maintain at least the semblance of a balance of power in the region once the Napoleonic wars were finally a thing of the past. To catalogue all the wars would be tedious, but the major developments can be summarized.

In 1799, the small state of Montenegro, with Russian aid, secured its independence from Turkey. The British reconquest of Egypt from the French in 1801 (see p. 97) slightly restored Turkish control over Syria and the Nile delta, but not for long in the latter case, for the rise of Mohammed Ali in Egypt from 1805 led to virtual autonomy. Meanwhile, in the Balkans, from 1804 to 1813, a bitter struggle raged in Serbia, but the revolt collapsed in 1812. Simultaneously (1806-12), a serious war between Russia and Turkey (encouraged by France) over the future of Wallachia, Moldavia and Bessarabia, was brought to an end through British influence on 28 May 1812, and, by the Treaty of Bucharest, Turkey ceded Bessarabia to the Tsar. Three years later, a new revolt flared up in Serbia, but was repressed by 1817. However, a half century of almost continuous strife would follow.

The Greek War of Independence, 1821-32. The massacre of 10,000 Turks in the Morea by Greeks led to harsh reprisals, which in turn inspired a full-scale Greek revolt. On 22 January 1822 the patriots declared their independence, and the struggle rapidly grew amidst an atmosphere of atrocity and counter-atrocity. The Greeks forced the Turks to abandon the first siege of Missalonghi (1822-

23), but then internal squabbles ruined their achievements, and the Turks returned to capture the fortress after a second siege (1825-26) by forces under Reshid Pasha. The Turks were aided by an Egyptian army under Ibrahim, son of Mohammed Ali, and the capture of the Acropolis in Athens signalled the near-triumph of the Ottomans, but then Britain, France and Russia intervened. On 20 October 1827 the naval battle of Navarino resulted in the sinking of three-quarters of the Turkish fleet, and growing pressures led to the Treaty of London in 1832, by which Turkey was induced to recognize Greek independence.

War with Persia, 1821-23. Turkish assistance to certain rebel tribes led to the Persian invasion to Lake Van. The Turkish counter-attack was repulsed; they lost the Battle of Erzerum and, by the treaty of the same name, both sides accepted the status quo.

War with Russia, 1828-29. Linked with the Greek struggle, Russian forces occupied coveted Moldavia and Wallachia and, by the Treaty of Adrianople (16 September 1829), were ceded control of the mouth of the Danube and the east coast of the Black Sea. Russia occupied Moldavia and Wallachia until 1834.

The First Egyptian War, 1832-33. Mohammed Ali, dissatisfied with the grant of Crete in return for Egyptian support against Greece, demanded Syria as well; Egyptian forces occupied Syria and later invaded Anatolia. In desperation, Constantinople appealed to Russia for aid and a defensive alliance was concluded. This development

alarmed Britain and France, and, largely at their instigation, Turkey was persuaded to sign the Convention of Katahia in May 1833, which ceded Syria and Adana to Egypt, but rancour remained.

The Second Egyptian War, 1839-41. This conflict grew out of The First Egyptian War. A Turkish army tried to regain Syria, but was defeated at Nezib. The Turkish fleet surrendered at Alexandria and the Ottoman Empire seemed in extremis, but her total collapse was prevented by the intervention of the powers, and a British and Austrian fleet bombarded Beirut and Acre. In 1840, by the Convention of Alexandria, Egypt was persuaded to abandon Syria and restore the Turkish fleet. In July 1841, by the Straits Convention, the powers agreed that, except in time of war, the Bosphorous and Dardanelles should be closed to all foreign warships.

Wallachian Revolt, 1848. Rapidly suppressed by Russian troops with Turkish approval.

The Crimean War, 1854-56. Britain, France and Turkey ally against Russia (see pp. 142-147). Once again, the effect was to bolster up the crumbling Turkish authority for a little while longer. Nevertheless, despite her political and economic weakness, Turkey still produced formidable armies and good generals—although the equipment tended to be seriously out-dated and the logistic concepts rudimentary in the extreme.

Cretan Insurrection, 1866-68. Instigated by Greece, it was savagely repressed.

*Also see pp. 196-197 for The Balkan Crisis of 1877-78.

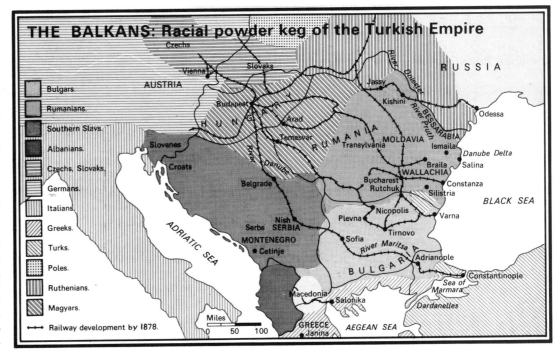

THE BALKANS: Racial powder keg of the Turkish Empire

Bulgars.
Rumanians.
Southern Slavs.
Albanians.
Czechs, Slovaks.
Germans.
Italians.
Greeks.
Turks.
Poles.
Ruthenians.
Magyars.
Railway development by 1878.

Miles
0   50   100

The Expansion of Tsarist Russia

The military achievements of Peter the Great (see pp. 59-61) led to considerable territorial gains in the eastern Baltic region at the expense of Sweden (see map, Strategic Effects of the Treaties of 1713 to 1721, p. 64), and some expansion also took place in the Black Sea and Caspian regions; although the débâcle of the Pruth (see p. 61) in 1711 at the hands of the Turks led to a temporary setback in the area of Azov. Peter's work was pursued by later rulers of the House of Romanov, whose guiding principles were to secure full control over the peoples within their sphere of territorial influence and, above all, to secure access to seas open to navigation for the entire year. The building of the new capital at St. Petersburg, adjoining the Baltic, was symptomatic of this ambition.

Under the Empress Anna (1730-40), the region around Azov was regained from the Turks in 1739, together with substantial gains to the west of the River Dnieper adjoining the southern boundaries of Poland. Her successor, the Empress Elizabeth (1741-62), made minor gains along the northern shore of the Gulf of Finland by the Treaty of Abo with Sweden, and also linked Azov to the Dnieper at the expense of the Turk. The notable reign of the great Empress Catherine (1762-96), however, saw the most dramatic advances on the western and southern frontiers of Russia. In the west, the successive dismemberments of Poland in 1772, 1793 and 1795 (see map Europe, 1740-95, and inset, p. 68), in which Prussia and Austria shared, saw Russia gain Lithuania and vast areas of east Poland, extending her control to the River Niemen and the upper reaches of the River Bug. In the south, a series of wars and treaties with the Sultan led to the assimilation of the Crimea (1783) and expansion to the River Dniester consolidated by the Treaty of Jassy (1792). To the east of Azov, Russia took over most of the Kuban between 1774 and 1783 (see inset opposite). These substantial gains (particularly those in the Baltic region) led to considerable friction with Sweden and the European powers, including Great Britain, but Catherine generally had her way. Confirmed as a power on the Black Sea, Russian aspirations for a warm-water port increasingly came to concentrate on Constantinople.

The Napoleonic Wars saw further advances. Although Russian hopes to secure a Mediterranean foothold in Malta were thwarted, and Constantinople remained beyond Tsar Alexander I's reach, Finland was taken over in 1809, making Russia the greatest power in the Baltic (Sweden being compensated with Norway from 1815); the Congress of Vienna also (after much wrangling with Austria and Prussia) awarded Tsar Alexander much of the Grand Duchy of Warsaw, which became the new kingdom of Poland—separate from Russia but with Alexander as its ruler. Twice during the nineteenth century—in 1830 and 1863—Russia repressed large-scale Polish revolts as patriots vainly strove to regain their independence. After the first revolt, Tsar Nicholas I withdrew the remnants of the constitution granted by Alexander I, and Russian rule over the Poles became ever more tyrannical. After the second, Poland was merged with Russia (1867).

Catherine II, the Great, Empress of Russia (1729-96), a strong ruler with a boundless appetite for both lovers and territorial expansion. Much of Poland was absorbed during her long reign.

Throughout the first half of the nineteenth century, Russia's major ambition remained the securing of entry to the Mediterranean via the Dardanelles. British and French suspicion of such a development led to their support of Turkey against Russian claims. This resulted in the Crimean War (see pp. 142-147), which once again thwarted Russia's aspirations concerning Constantinople. From 1855, Tsar Alexander II (1855-81) gave up Mediterranean ambitions and, instead, turned to three alternative lines of development: through the Caucasus towards Persia; towards Afghanistan; and towards Manchuria and the Far East. The first and second of these lines of advance soon brought British suspicions and protests, as the route to India seemed involved; the third would in due course lead to a major confrontation with Japan early in the twentieth century.

The Caucasus region was incorporated within a decade of the Crimean War; the east in 1859, the west by 1864, strengthening the links with Georgia and Daghestan, which had already been colonized early in the century (see inset opposite). The advance into Central Asia continued with the conquest of most of Turkestan in 1865; three years later, the Khan of Bokhara submitted after the Battle of Irgai, and in 1868 came the fall of Samarkand. In 1873, the ancient khanate of Khiva on the River Oxus was absorbed, followed by that of Khokand in 1876. Earlier, in 1860, the Province of Amur on the distant Asian shores had been occupied, and the great port of Vladivostok established.

Not all of Russia's expansionist energies were directed away from Europe, however. Despite the de-militarization of the Black Sea by the Peace of Paris (see p. 142), the restrictions of the Black Sea Clauses were virtually evaded by agreement with Bismarck from 1863 (and finally abrogated in 1871), and Russia protested four years later over some of the conditions of the Peace of Prague (1866). Meanwhile, Russia developed her interests in the Balkans, desiring to protect the Christian Slavs and improve her own paramountcy in the Black Sea region. In 1861 the Tsar supported the union of Moldavia and Wallachia into Rumania and, in 1863, recognized George as King of the Hellenes. In 1867, Turkey was persuaded to evacuate Belgrade, and Russian intrigue involved Serbia the following year (see map, The Dismemberment of the Ottoman Empire in Europe, 1683-1878, p. 196). Nevertheless, Russia's ability to achieve a degree of amity and co-operation with the other major powers was demonstrated by her inclusion in the Drei-kaiserbund (Three Emperors' League) signed with Austria and the new German Empire in 1872, whilst the same year saw an agreement with Britain delimiting the frontiers of Afghanistan, and the Russians pulled back from a substantial area south of the Caspian Sea. However, the Balkan Crisis of 1877-78 led to considerable territorial adjustments in the Treaty of San Stefano, as adjusted by the Treaty of Berlin in 1878, which, nevertheless, confirmed Russia's gains in the Caucasus and Bessarabia (see pp. 196-197).

By this time, the Tsarist Empire had almost reached its final delimitation, and attempts were made to create a reasonably cohesive country. The Trans-Siberian railway, completed in the 1890s, was an expression of this desire, but ahead lay only international humiliations and internal unrest which would lead to the revolution of 1917.

THE EXPANSION OF TSARIST RUSSIA UP TO 1914

NOVA ZEMLYA

NORWAY
SWEDEN
FINLAND 1809
BALTIC SEA
St. Petersburg

WHITE
RUSSIA

S I B E R I A

River Lena

Yakutsk

River Ob

River Yenisei

Okhotsk

KAMCHATKA

Tomsk
Krasnoyarsk

SAKHALIN
1875

River Volga

Moscow
Samara
Omsk
Irkutsk

PROVINCE
OF
AMUR
1860

1815
POLAND

River Bug

River Irtysh

1763

River Amur

MANCHURIA
(occupied 1900-05)

1860

River Danube

Odessa
1878

BLACK SEA

CASPIAN SEA

1864

1825
River Oxus
Khiva
Irgai
1867
1868
Bokhara
Samarkand
1856
1872

1865
T U R K E S T A N

Tashkent
Khokand
1895

Gained 1871 PROVINCE
Restored 1881

SINKIANG

MONGOLIA

River Hwang-ho

Vladivostok

Constantinople

AFGHANISTAN

River Indus

Miles
0   250   500   750   1000

ENLARGEMENT
1783
1783

Azov
1739

ASTRAKHAN

CASPIAN SEA

SEA OF AZOV

K U B A N
Stavropol

1784
Derbent

CRIMEA
1783

C I R C A S S I A
1864
1829
1854

1774

1859
1806

D A G H E S T A N

CAUCASUS MOUNTAINS
MINGRELIA

1805
Baku

BLACK SEA

Batum
1878

1804
Tiflis
Kars
Erivan
1828

GEORGIA
1813
1801

TURKEY

N

PERSIA

Miles
0   50   100   150   200

Russia, 1725.

Acquisitions, 1726-1815.

Acquisitions, 1816-1855.

Acquisitions, 1856-1914.

Areas relinquished.

Trans-Siberian Railway.

Trans-Caspian Railway.

Chinese Eastern Railway.

Alexander II, Tsar of Russia (1818-81). His reign, which opened in the last year of the Crimean War, saw much Russian expansion in Asia, and also a war in the Balkans (1877). He fell victim to an assassin four years later. (National Army Museum)

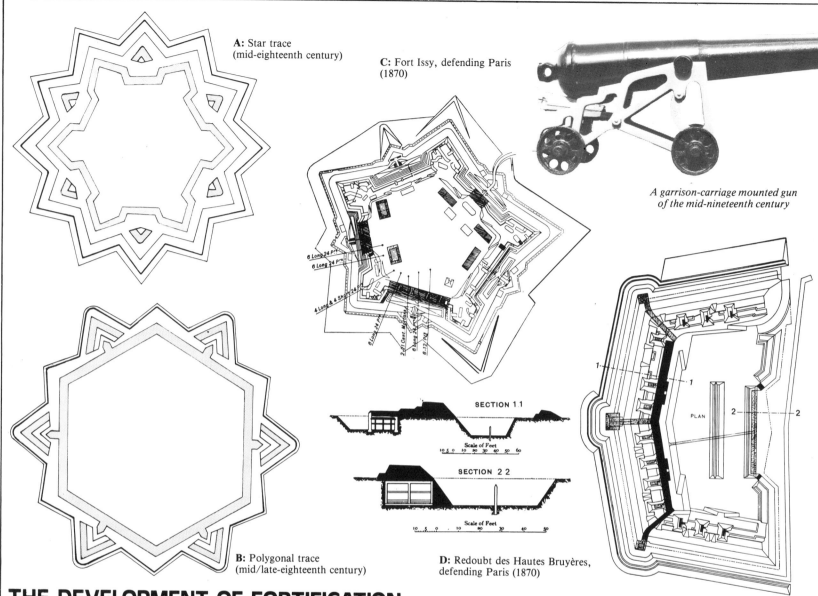

A: Star trace
(mid-eighteenth century)

C: Fort Issy, defending Paris
(1870)

*A garrison-carriage mounted gun
of the mid-nineteenth century*

SECTION 1.1

Scale of Feet

SECTION 2.2

Scale of Feet

PLAN

B: Polygonal trace
(mid/late-eighteenth century)

D: Redoubt des Hautes Bruyères,
defending Paris (1870)

THE DEVELOPMENT OF FORTIFICATION

For nearly two centuries after his death in 1707, the name Vauban was practically synonymous with the development of fortification (see pp. 40 and 41). His principles for the conduct of sieges and for the defence of fortresses were spread by his immediate disciples, and rapidly became standard practice (with local variations) throughout the European world. Bernard Forest de Belidor continued his work in France until the mid-eighteenth century, whilst a pupil of Vauban designed the fortress of St. Peter and St. Paul at St. Petersburg, establishing a tradition that was fostered by Münnich and other engineers imported into Russia. Fort George, in the Scottish Highlands, indicates Vauban's influence on British military engineering, whilst the fortress of Louisbourg in French Canada and that at Arcot in distant India illustrate the extent of his influence worldwide.

So deeply did Vauban's ideas permeate the world of military engineering that many of his successors contented themselves with copying his methods wholesale. Indeed, new ideas tended to be regarded as heretical: "To attempt novelties in fortification is to demonstrate one's ignorance", wrote Chasseloup-Laubat in the 1750s. Nevertheless, certain improvements in both profile and trace did emerge over the next century and a half; most of them made necessary by the considerable advances in the range and power of artillery over this period.

One school of engineers perverted Vauban's ideas and preached the need to develop a purely geometrical approach, which resulted in the standardized 'star trace' (see diagram A). Such systems reduced the volume of outwards fire in the search for strong cross-fire patterns, and exposed the parapets to an increased danger of enfilading by largely ignoring Vauban's cardinal principle that all fortifications should be designed to exploit the actual configuration of the ground.

Other concepts, however, proved more successful. Chasseloup-Laubat, despite his caution, experimented to good effect with the placing of ravelins beyond the glacis to increase the area of fire-swept ground, at the same time designing longer parapet fronts to increase the volume of outward fire. Most significant of all, however, was the gradual development of the 'polygonal trace' (see diagram B), which dispensed with the angle-bastions, replacing them with bomb-proof caponiers at the foot of the ditch. The Prussian engineer, Landsberg the Younger, was one of the first to press this idea, but its strongest champion was Marc-René Montalembert. He was also the greatest protagonist in his day of 'active' rather than 'passive' defence associated with carefully-devised patterns of fire, incorporating long stretches of low parapet with tiered casements holding the guns employed on distant firing.

These ideas were simplified by Lazare Carnot in his book *De la défence des places fortes*, published in 1810, but the polygonal school reached its fullest development under the German school. The use of protruding caponiers at the foot of the ramparts, protected by ravelins, supplemented the fire of the counterscarp-galleries (see diagram) and did much to carry out those tactical functions of the old angle-bastions that were associated with the fire-domination of the ditch. However, although there was a growing realization that simplicity and strength were the most important principles underlying the art and science of fortification, a number of French engineers remained intensely conservative in outlook, and continued to employ the ancient bastion-trace until as late as the Franco-Prussian War.

The Emergence of the Detached Fort System. Long before 1870 it was apparent that the bastion-trace was rapidly becoming obsolescent in the face of rifled artillery (see p. 186) with improved ranges of 1,500 yards or more, and a new form of

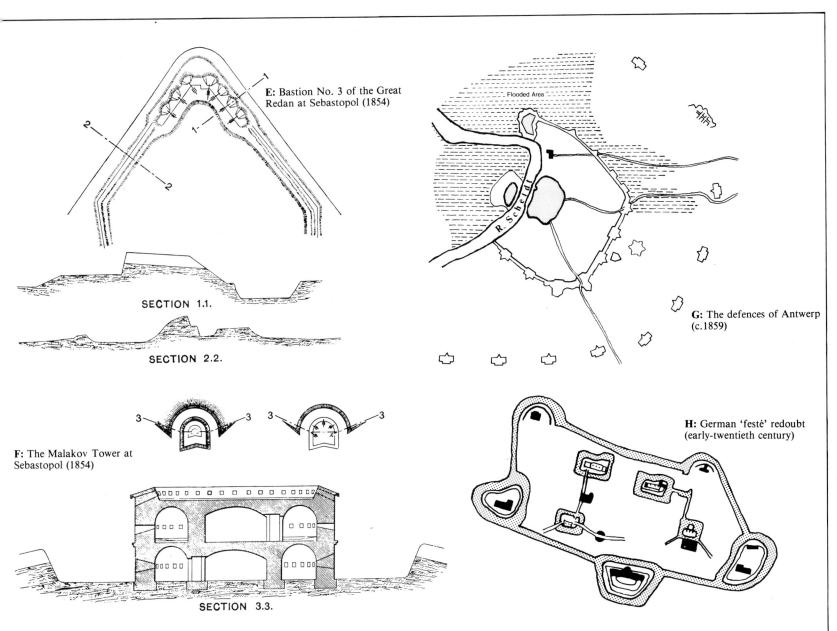

E: Bastion No. 3 of the Great Redan at Sebastopol (1854)

SECTION 1.1.

SECTION 2.2.

F: The Malakov Tower at Sebastopol (1854)

SECTION 3.3.

G: The defences of Antwerp (c.1859)

H: German 'festé' redoubt (early-twentieth century)

permanent fortification was in the process of evolution. This system was based upon detached forts. To negate the power of the improved guns, the outlying defences had to be placed at ever-greater distances from the heart of the defence system. This compelled the enemy to commence his siege operations at a greater distance from the ultimate objective and inevitably on a larger scale. Conversely, the vast increase in the area of the defensive perimeter to be covered made the construction of a continuous regular enceinte prohibitively costly. So, the use of chains of detached forts came into existence, each representing a self-contained island of resistance, relying on firepower and, in some cases, linking entrenchments to command the intervals between them.

Such an idea had been experimented with by the great Vauban at Toulon in the 1690s, but it was Wellington's famed Lines of Torres Vedras (see p. 112), built outside Lisbon between 1809 and 1810, that formed

the most striking early example of the use of strings of such works, instead of a continuous line of ramparts and ditches. This was extremely effective in sealing off the thirty-mile peninsula from the Atlantic to the Tagus against Massena's army, and it cost only a little over £100,000 to build.

Torres Vedras was, in fact, a little ahead of its time, for the detached fort did not really come into its own until 1830, when Austria constructed the Linz towers. Between 1840 and 1844, Paris was refortified with a girdle of fourteen forts (see diagram C), three miles apart and between two and three-miles distant from the bastioned enceinte. These defences provided the basis for those employed to defend the capital during the Prussian siege of 1870-71 (see p. 195 and map Siege of Paris). The notable engineer Todleben extemporized a similar system at very short notice for the Russians holding Sebastopol during the Crimean War; it was observed that hardly any masonry was used

in his construction of such positions as the Great Redan and the Malakov (see diagrams E , F). Earthworks, revetted with gabbions and fascines and supported by fire trenches, were becoming the order of the day, and the length of time it took the Allies to overcome these defences proved their effectiveness.

The detached fort system reached its fullest development with the construction of the defences of Antwerp (begun in 1859), which comprised a nine-mile polygonal enceinte divided into eleven fronts with a citadel to the north of the city, supported by fourteen detached forts at approximately one-mile intervals placed two miles out from the city, each of which was designed to hold 120 pieces of artillery, five mortars and 1,000 men (see diagram G). The mighty River Scheldt, supplemented by flooded areas, completed these redoubtable defences, within which an army of 100,000 men could find shelter. The line of forts defending Portsmouth from the North are

of the same generation.

By the time of the American Civil War in the early 1860s, the defences of such important positions as Vicksburg, on the Mississippi (see p. 182), and Petersburg (see p. 185), in the eastern theatre of war, were almost entirely based on a series of mutually-supporting strongpoints, sunk into the earth, linked by a series of fire-trenches. The same was true of the defences of Plevna in the later 1870s (see p. 196). In its turn, therefore, the day of the detached fort was drawing to its close, with the advent of heavy long-range artillery firing powerful high explosive shells. In 1904, the six forts of Port Arthur would crumble under the bombardment of Japanese eleven-inch howitzers, and the writing would be on the wall. The future lay with complex barbed-wire protected all-round defensive positions of the type known as the 'festé' redoubt (see diagram H), developed in Germany at Metz and elsewhere in the last years of the century.

The Crimean War, 1854-56

The roots of this struggle were buried in the traditional Russo-Turkish antipathy, and the overt Russian desire for a warm-water port. Friction between the two had been rising since the 1840s. Tsar Nicholas I, as 'protector' of Greek Orthodox Christians in the Islamic Turkish Empire, and confident of Russia's ability to impose her terms on the continental European powers (Russia alone had escaped the turmoil of the 'Year of Revolutions' in 1848), was determined to benefit from what he believed to be the impending break-up of Turkish cohesion. Britain had been suspicious of Russian intentions since 1844—and there had been a naval confrontation in the Dardanelles in 1849—and by 1853 the British government had declared its intention of maintaining Turkish territorial integrity. The immediate crisis that led to war came in 1852, when the Tsar protested against the Sultan's agreement to recognize France as protector of the Holy Places in Jerusalem. Napoleon III was determined not to be brow-beaten over the issue. Then, in April 1853, Russia delivered an ultimatum to Constantinople, which was rejected by the Sultan the following month after British support had been promised.

On 22 June 1853, a Russian army marched into the Principalities (Moldavia and Wallachia), but war ensued only after diplomatic negotiations had broken down over the issue of the Greek Christians. Russia refused to evacuate the invaded territory and on 5 October, Turkey declared war. Omar Pasha led a Turkish army over the Danube and won a victory at Oltenitza (4 November). In the previous month both Britain and France had sent fleets to the Dardanelles in support of Turkey, and renewed their promises of support if the Tsar refused reasonable offers of mediation. The sinking of a Turkish squadron by Russian vessels at Sinope (20 November) outraged British public opinion and, on 3 January 1854, the Anglo-French fleet sailed into the Black Sea. Both powers signed a formal alliance with Turkey on 12 March, and fifteen days later Britain and France declared war on Russia.

In June, a Russian attempt to take Silistria was thwarted by Butler and Naismith acting in conjunction with the Turks, and in August Russia evacuated the Principalities. That same month, Admiral Sir Charles Napier led an Allied naval expedition to capture Bomarsund (16 August) in the Baltic. Next, in order to deny Russia the free use of the great naval base of Sebastopol, and to create a military diversion to drain Russian resources, it was determined to land an Allied Army in the Crimea.

In September 1854, the Allies prepared to invade the Crimea under command of Lord Raglan and Marshal St. Arnaud, whose troops were eager for action. After massing at Malta, the 27,000 British had spent some time first at Gallipoli (where they had been joined by 50,000 French) and Scutari, and later at Varna in Bulgaria, which had been occupied in May. A terrible epidemic of cholera had decimated the ranks in the camps around Varna, and when news of the Russian withdrawal came in August there was great frustration. Consequently, the decision to attack Sebastopol was welcomed as an opportunity for action.

Some 600 ships sailed from Varna on 7 September carrying 60,000 men, but leaving behind most of the heavy guns, the tentage and the pack-horses of the supply train. Arriving off Cape Tarkan, Raglan selected Calamita Bay for the landing, and on 14 September the troops disembarked unopposed. Leaving the Turks near Eupatoria (see map opposite), the Allied Army then marched south towards Sebastopol, hugging the coast and escorted by the protecting fleet. The Russians, under Menshikov, attempted to intercept the advance, but were defeated on 20 September at the Battle of the Alma (see map on p. 144). However, it was all of three weeks before the siege lines could be drawn about Sebastopol, and this delay gave the famous engineer Todleben time to improve the fortifications, adding to the Malakov Fort and the Redan (see p. 145). The death of the aged St. Arnaud in late September led to General Canrobert succeeding to the command of the French. Next, on 25 October, Menshikov again challenged battle (this time from the east), resulting in the indecisive but famous action fought at Balaclava (see p. 145). Nothing if not a trier, Menshikov attacked again on 5 November, but was prevented from breaking through to relieve Sebastopol at Inkerman (see p. 145). These battles cost both sides many casualties, and the Allies were not strong enough to assault Sebastopol, so a winter campaign became inevitable.

The loss of vital stores and equipment in the great storm that wrecked many British transports was one reason for the horrors suffered by the troops over the ensuing months (as reported to the British public by the correspondent William Howard Russell of *The Times*). The French and the Turks fared somewhat better; the latter winning a small battle against another Russian attack at Eupatoria on 17 February 1855.

During the winter, negotiations were held at Vienna, but they foundered over the issue of the neutrality of the Black Sea. In January, Sardinia and Sweden declared for the Allies, and Cavour sent 15,000 troops to the Crimea

from Italy. On 16 May, Canrobert resigned his command, being replaced by Pelissier.

With the return of reasonable weather, the progress of the siege became the focal point of attention (see map on p. 147). A storming attempt on 18 June was repulsed, and ten days later Raglan died. The British command then passed to General Simpson.

The siege dragged on through the summer. A Franco-Sardinian force was attacked at T'Chernaya on 16 August, and what proved to be the last Russian attempt to raise the blockade failed. The climax came in early September. On the 8th, the French stormed the Malakov, whilst the British first took, then lost, the Redan. Next day, however, the Russians decided to abandon Sebastopol, and the Allies at last took possession of the city and its naval arsenal. The war was now practically over, but on 26 November the Russians captured Kars on the Caucasus front after a brave defence sustained by General Fenwick Williams.

The preliminaries of peace were settled at Vienna on 1 February 1856, and the final terms were signed in Paris twenty-four days later. By the terms of the Treaty of Paris, the Black Sea was neutralized, and no warships whatsoever were to be allowed there. The navigation of the Danube was to be supervized by international commissioners. As for Turkey, its territorial integrity was formally guaranteed by Great Britain, France and Austria. The Sultan's right to settle internal affairs was restated. Part of the Danubian delta was restored to Turkey, and a part of Bessarabia was added to Moldavia.

To a degree, Russian power was checked, but Napoleon III came out best, with greatly enhanced prestige. The imperfections of the British Army system had been demonstrated, and hastened the work of reform. But the Treaty of Paris would prove a dead letter by 1878, when none of the powers intervened on Turkey's behalf against Russia (see p. 196). Anglo-Russian tensions were also far from being things of the past, for the Tsar's government turned his expansionist interests to the East, and rivalry would soon follow in Afghanistan and along the North-West Frontier of India. Thus, the Crimean War achieved very little and ended in an unstable peace. To achieve such a result had cost 300,000 lives on each side.

The Battle of the Alma, 1854

As the cholera-stricken Allied Army advanced southwards from Calamita Bay towards Sebastopol, news arrived that Prince Menshikov and 36,400 men were advancing to meet them. On 19 September a brisk artillery

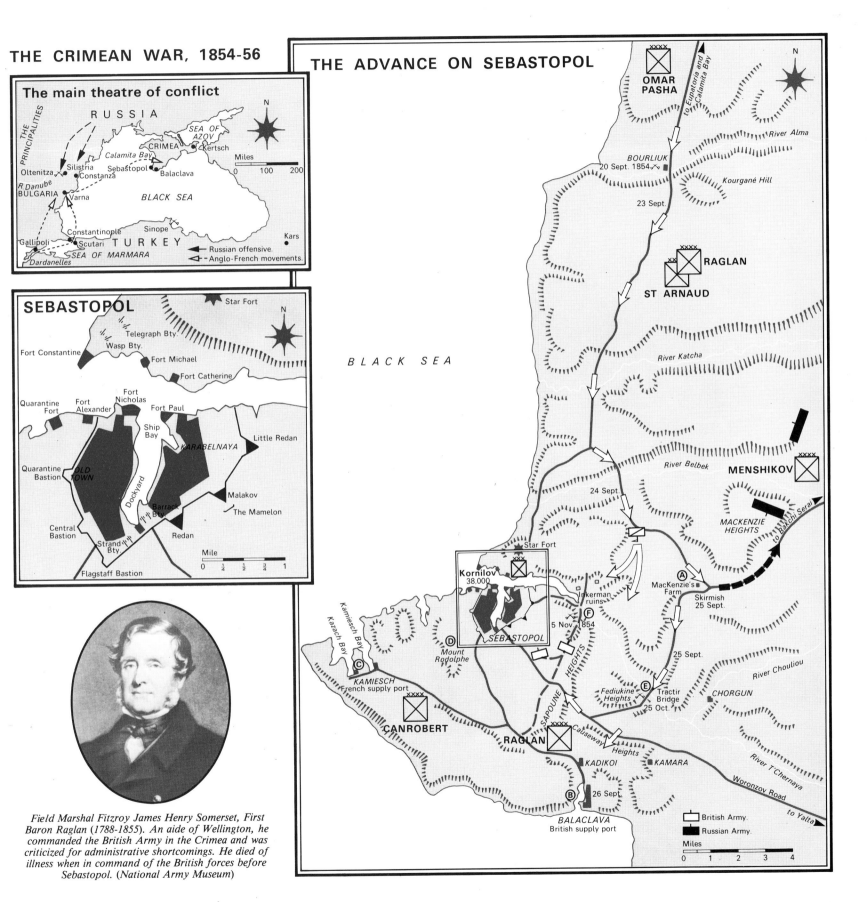

THE CRIMEAN WAR, 1854-56

The main theatre of conflict

RUSSIA

THE PRINCIPALITIES

SEA OF AZOV

CRIMEA
Kertsch

Calamita Bay
Sebastopol
Balaclava

Oltenitza
Silistria
Constanza

R.Danube
BULGARIA
Varna

BLACK SEA

Constantinople
Scutari
Sinope

Gallipoli
Dardanelles
SEA OF MARMARA

TURKEY

Kars

Miles
0   100   200

N

→ Russian offensive.
⇨ Anglo-French movements.

SEBASTOPOL

Star Fort

Telegraph Bty.
Wasp Bty.
Fort Constantine
Fort Michael
Fort Catherine

Quarantine Fort
Fort Alexander
Fort Nicholas
Fort Paul

Ship Bay

KARABELNAYA

Little Redan

Quarantine Bastion
OLD TOWN
Dockyard

Malakov

Barrack Bty.
The Mamelon

Central Bastion
Strand Bty.
Redan

Flagstaff Bastion

N

Mile
0   ¼   ½   ¾   1

Field Marshal Fitzroy James Henry Somerset, First Baron Raglan (1788-1855). An aide of Wellington, he commanded the British Army in the Crimea and was criticized for administrative shortcomings. He died of illness when in command of the British forces before Sebastopol. (National Army Museum)

THE ADVANCE ON SEBASTOPOL

OMAR PASHA

to Eupatoria and Calamita Bay

River Alma

BOURLIUK
20 Sept. 1854

Kourgané Hill

23 Sept.

RAGLAN
ST ARNAUD

River Katcha

BLACK SEA

River Belbek

MENSHIKOV

24 Sept.

MACKENZIE HEIGHTS

to Bakchi Serai

Star Fort

Kornilov
38,000

MacKenzie's Farm

Inkerman ruins
5 Nov. 1854

Skirmish 25 Sept.

Kamiesch Bay

Kazach Bay

SEBASTOPOL

HEIGHTS

25 Sept.

River Chouliou

Mount Rodolphe

D

C

KAMIESCH
French supply port

CANROBERT

SAPOUNE HEIGHTS

Fediukine Heights

E Tractir Bridge 25 Oct.

CHORGUN

RAGLAN

Causeway Heights

KAMARA

KADIKOI

River T'Chernaya

B

26 Sept.

Woronzov Road

to Yalta

BALACLAVA
British supply port

British Army.
Russian Army.

Miles
0   1   2   3   4

N

143

The Crimean War, 1854-56

duel induced the Russians to retire beyond the River Alma, where they deployed on each side of the causeway, fortifying with redoubts (A) the approaches to Kourgané Hill (B) the key to their position. At midday, the Allied fleet (C) opened fire, but the Russians had very few troops deployed on their Black Sea flank.

Raglan and St. Arnaud had only the vaguest plan of attack. The approach of the British army on the left had been leisurely, and they were only drawn up in line of battle from 11.30am, and ready to advance at 1pm. The French, on the right, had been waiting for some time, and St. Arnaud was keen to unleash his men over the river and up the heights in order to turn the left wing of Menshikov's position. As the British advanced towards the Alma they extended into line as they came under artillery fire, whilst the French columns, General Bosquet's division in the lead, successfully

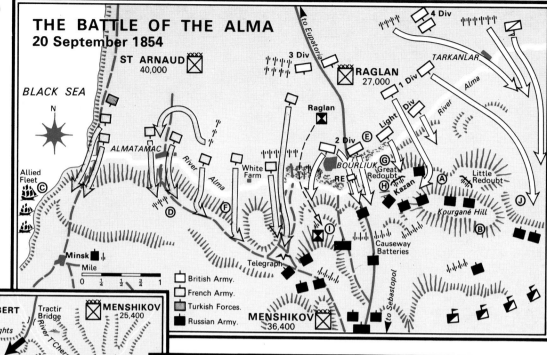

THE BATTLE OF THE ALMA
20 September 1854

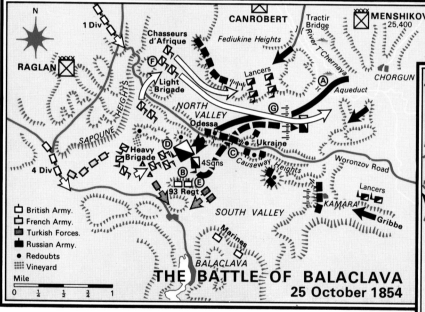

THE BATTLE OF BALACLAVA
25 October 1854

negotiated the river and the heights beyond (D), taking twelve guns with them. Menshikov was horrified by this development, and hurried over with eight reserve battalions.

The British advance was slow and confused, the Light and 2nd Divisions becoming partially entangled (E). The French also became overcrowded as they tried to reinforce their forces on the southern heights, and Prince Napoleon's division came under heavy fire (F). Fortunately, Menshikov did not counterattack at this juncture.

At about 2pm the British leading formations were at last over the Alma (G) and heading uphill for the Great Redoubt, which the Light Division succeeded in capturing (H). Unfortu-

Prince Alexander Sergeievich Menshikov (1789-1869), senior Russian commander in the Crimea. (National Army Museum)

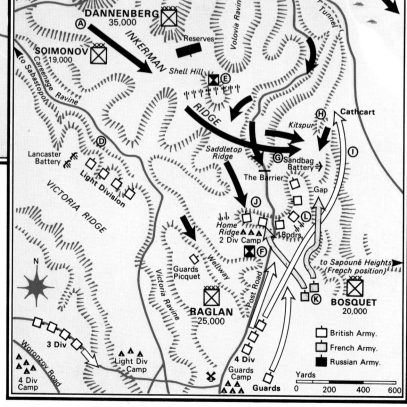

THE BATTLE OF INKERMAN
5 November 1854

nately, the 1st Division was not in close support and, in some confusion, the British abandoned the position, which the surprised Russians gratefully reoccupied. Lord Raglan, meanwhile, had cooly ridden forward with his staff to a hill-top some 800 yards inside the Russian lines. Two 9-pounders joined him on this position (I), and this pair of guns caused much havoc to the enemy, and effectively cleared the way for the delayed advance by the 2nd and 1st Divisions, inducing the Russians to pull back their causeway batteries. The Guards Brigade of the Duke of Cambridge's 1st Division was repulsed from its frontal attack on the Great Redoubt; General Gorchakov's counterattack was decimated, and then the Highland Brigade managed to get round behind the main Russian position (J), and Menshikov ordered a general retreat. The Allies lost 3,300 men to the Russians 5,000, but there was no attempt at an immediate pursuit.

The Battle of Balaclava, 1854
From the outset of the siege of Sebastopol on 28 September, the British lines of communication stretched from the siege lines to the port of Balaclava. In late October, Prince Menshikov advanced at the head of 25,400 men and 78 guns to threaten them, in the hope of interfering with the siege. This move took Lord Raglan by surprise, and the Russians were able to pass 3,000 cavalry over the River T'Chernaya (A) unchallenged. To check the advance, Raglan ordered the 93rd Highlanders under General Sir Colin Campbell to take post on a low ridge in the South Valley (B), and alerted Lord Lucan's cavalry division, comprising the Light and Heavy Brigades (1,600 cavalrymen plus an attached squadron, 150 strong, of the Chasseurs d'Afrique), whilst urgent messages were sent to summon the 1st and 4th Divisions and further French and Turkish troops to the scene of impending conflict.

The Russians routed a Turkish force holding part of the Fediukine Heights, and then captured a number of redoubts on Causeway Heights (C), whilst their cavalry moved forward into South Valley. This force was attacked by the Heavy Brigade under Sir James Scarlett (about 900 strong) and repulsed with heavy loss (D). Four Russian squadrons that wheeled south to head for Balaclava were met by the thin red streak of the 93rd and, although some neighbouring Turkish formations fled, the cool Scotsmen's volleys (E) repelled the Russians, who abandoned the captured redoubts and fled north.

Meanwhile, Lord Raglan noted that the Russian infantry were in the process of evacuating some Turkish guns they had captured from the redoubts, and sent Captain Nolan with an order to Lord Lucan for the Light Brigade (F) to be sent forward to prevent this. Owing to poor visibility, Lord Lucan misconstrued the intention, and ordered his hated rival and brother-in-law, Lord Cardigan, to charge down the North Valley against the only guns he could see, namely the main Russian battery (G). The result was the famous charge of the Light Brigade into 'the Valley of Death'. Nolan tried to inform Cardigan of his true objective but was shot down, and the 673 light cavalrymen charged down the valley, engaged by Russian artillery on three sides. "C'est magnifique, mais ce n'est pas la guerre" was General Bosquet's comment. The horsemen reached the guns and scattered the gunners before retiring back up the valley. Of the 673 men in the charge, 247 had become casualties and all of 475 horses were lost.

This gallant but futile charge ended the battle. The Russians (who had lost 627 men to the Allied total of 615 casualties) retained the Causeway Heights and the important Woronzov Road, needed by the Allies to move supplies from Balaclava to the siege lines five miles to the north.

The Battle of Inkerman, 1854
The arrival of reinforcements within Sebastopol and an awareness that the British links with Balaclava were still exposed induced the Russian high command to attempt another attack on Raglan's position in early November. Three Russian formations were to be involved: General Soimonov, with 19,000 men, was to advance on the Inkerman Heights from Sebastopol (A), and General Paulov (16,000) was to attack over the River T'Chernaya (B), and once on the ridge both these forces would be commanded by General Dannenberg. In the meantime, General Gorchakov (22,000) was to make a diversionary attack (C) against the French forces holding the Sapouné Heights, and also operate against the British right flank.

The Russians began to move forward, unnoticed by the British outposts, at 2am on the 5 November. At dawn, a thick mist replaced the early-morning rain, and whilst this hampered the Russian deployment it also helped the surprise of their overall operation, which was almost complete. The first fire was exchanged on the left of the Light Division's position (D), but soon a number of scrappy disconnected actions were taking place amidst the fog-bound ridges and re-entrants of Inkerman ridge. The Russians managed to position a large battery (E) on Shell Hill, which bombarded the 2nd Division's camp beyond Home Ridge (F) with considerable effect. Offers of French assistance were declined by the British, but despite the very scrappy nature of the battle, the Russians were still making ground (G) by sheer weight of numbers, and the 2nd Division was driven off the Kitspur feature (H). The Guards Brigade came to the 2nd Division's aid, and General Cathcart of the 4th Division sent in a brigade in a counter-attack (I), which, however, soon ran into trouble.

By 9am it was becoming clear that the major Russian attack (J) was developing towards Home Ridge, as Dannenberg led forward 6,000 men against only 3,000 defenders. The 2nd and 4th Divisions were tiring, and there were no British reserves immediately available. Raglan decided, at last, to accept Bosquet's earlier offer of assistance, and soon a division of Zouaves and Algerians under General Bosquet were sweeping into the fray (K). This proved the turning point of the day, and lost ground on the Kitspur was regained. Two British 18-pounders (L) found the range of the Russian battery and caused considerable havoc. At about 12.30pm General Dannenberg began to make an orderly withdrawal. The Allies did not pursue. The Russians had lost 10,300 casualties to the British 2,400 and the French 880. It had been a "soldier's battle", but Raglan was promoted Field Marshal.

The Siege of Sebastopol, 1854-55
The great Russian naval arsenal at Sebastopol was the focal point of the main operations in the Crimean War. Its rather inadequate defences were greatly reinforced by the celebrated engineer Count Franz Todleben, but comprised strengthened earthworks rather than formal fortifications, there being only two stone-built towers: one in the old town, and one embracing part of the Malakov Redoubt. (The main positions, comprising the Star Fort to the north of the harbour estuary with associated positions, and the chain of six redoubts defending the southern and eastern approaches to the city, are shown on the inset to the map The Advance on Sebastopol, p. 143.) The garrison at first comprised only 10,000 men, but by early October had risen to 38,000. Admiral Kornilov led the defence until his death, when Prince Menshikov and, ultimately, Prince Gorchakov (operating mainly from outside the city at the head of the Russian relieving army positioned to the north-east upon the Mackenzie Heights), assumed the main responsibility. The city was never wholly cut off by the besiegers.

Missed Opportunities: 28 September to 19 October 1854. After the Battle of the Alma, Sebastopol lay open for the Allied forces to occupy, but their advance was slow and

The Crimean War, 1854-56

indecisive. Raglan's desire to storm the northern defences without delay met with obstruction from the ailing Marshal St. Arnaud and his own senior engineer, Burgoyne, who argued that it was essential to first occupy the port of Balaclava. Accordingly, the Allies moved round Sebastopol in a wide sweep, fighting a cavalry engagement with Russians at MacKenzie's Farm (A) on 25 September to cross the River T'Chernaya and occupy Balaclava (B) the following day (see map The Advance on Sebastopol, p. 143). Raglan and Canrobert (St. Arnaud's successor) decided that Balaclava was not large enough to support both the British and the French armies, and it was agreed that the French should base themselves on Kamiesch and Kazach bays (C), farther west. Balaclava was far from an ideal harbour, being cramped and landlocked and with poor roads leading up to the camps on the heights. Its occupation also implied British responsibility for the open eastern flank. Once again, Raglan's advocacy of an immediate assault was ignored, and preparations for a formal siege were set in hand, even though the city had open communication to the north.

The First Siege Period: 17 to 24 October. On 17 October, 126 Allied guns opened fire against the landward defences. The 49 French guns on Mount Rodolphe (D) ceased fire when a Russian shell exploded a powder magazine; the naval bombardment began very late and had scant effect; only the British batteries caused notable damage to the Redan. However, by the 20th the Russians had lost 2,000 killed, including Admiral Kornilov, but as no Allied

assault was attempted the damage was soon made good.

Russian Attempts to Raise the Siege: 25 October to 6 November. Realizing the susceptibility of the British supply routes to Balaclava, Menshikov made two determined attacks in rapid succession, leading to the Battle of Balaclava (25 October) (E, and see map on p. 144) and, after a further raid on the 26th against the British 2nd Division on Home Ridge, the Battle of Inkerman (F, and see map on p. 144); both battles ended in partial Russian withdrawals. Meanwhile, the bombardment of Sebastopol continued, but without much effect.

The Winter Phase: November 1854 to February 1855. A fierce gale on 14 November sank twenty-one British ships in Balaclava harbour and badly damaged eight more. This setback led to administrative chaos that persisted throughout the bitter winter. Items in shortest supply were fuel and fodder for the horses and mules. Balaclava held much food and many stores, but forward distribution proved hopeless. Morale sank, and the descriptions of the administrative shortcomings published in *The Times* (as revealed by the pen of W. H. Russell), caused a furore in England and, in February, Lord Aberdeen's government was replaced by Palmerston's. Meantime, Florence Nightingale was working to improve

Below: The year-long siege of Sebastopol (1854-55), saw much suffering in the trenches during the hard winter months. The French fared better than the British, whose pitiable condition was reported in The Times *by the correspondent Russell. (DAG)*

the appalling medical conditions prevailing in the rear-hospitals at Scutari; to such good effect that a death rate of 44 per cent eventually dropped to 2.2 per cent.

As spring approached, the situation around Balaclava slowly improved. The completion of the first stage of the railway from the port (see G on map, The Siege of Sebastopol: The siege lines, May 1855) and the arrival of large numbers of Spanish mules aided the distribution of supplies, whilst a Turkish labour corps improved the roads. Nevertheless, in February 1855 the British Army had only 11,000 fit men, and all of 23,000 sick and wounded, on the muster rolls. By the same date, French strength had risen to 90,000 men, and the Turks had some 50,000 more (under Omar Pasha) around Eupatoria near Calamita Bay. However, Russian strength was now in excess of 100,000 troops, and that they were still full of fight was demonstrated on 22 February when they took possession of the Mamelon (H) to the fore of the Malakov Redoubt and fortified it, and again on 22 March when a strong sortie (I) succeeded in damaging French trenches on Victoria Ridge.

The Second Siege Period: 23 March to 18 June. Allied reinforcements arrived in a steady stream. In April, 20,000 Turks were brought over from Eupatoria, whilst German, Swiss and Polish Legions produced a further 13,500 men. In May, 15,000 Sardinians under General La Marmora appeared, and a further 20,000 French troops. However, in the belief that Napoleon III was about to arrive in person, General Canrobert vetoed plans to retake the Mamelon. A heavy Allied bombardment from

9 to 19 April caused 6,000 Russian casualties, but did not lead to any assault. Todleben rapidly made good most of the damage, except to Flagstaff Bastion, which was abandoned (J).

The electric telegraph, linking Varna to headquarters, permitted Napoleon III (although he did not come to the Crimea) to meddle in events; an attack on Kertsch in the Sea of Azov was called off on 2 May at his insistence. Canrobert asked to be replaced, and took over the French 4th Division whilst Pelissier assumed command of the French army on 16 May. The new French commander proved of sterner stuff than his predecessor, and at once began a more active policy. On 22 May, a night-attack captured some Russian trenches, and the Kertsch expedition was resumed successfully. Two more periods of heavy bombardment of Sebastopol followed and, on 8 June, the French at last retook the Mamelon and the British closed nearer to the Redan.

On 17 June the fourth bombardment preceded a major assault against the Malakov and the Redan. Confused timings, disagreement between Pelissier and General Bosquet and the fact that the Allied intention had leaked to the Russian high command led to failure. By midday, the French and British had lost 1,500 casualties apiece, and both attacks were recalled. This failure contributed to hasten the death of Lord Raglan, who was succeeded by General Sir James Simpson.

The Third Siege Period: 19 June to 11 September. The routine of the siege was inevitably resumed. In late July, the Tsar, Alexander II (who had succeeded Nicholas I in March), replaced Prince Menshikov with Prince Gorchakov. On 16 August, Gorchakov, to show his mettle and in an attempt to regain the Fediukine and Sapouné Heights (K), launched a heavy attack over the River T'Chernaya. This thrust, which proved to be the last attempt to break the Allied siege by major field operations, was defeated by the French and Sardinians, and driven off after losing 6,000 casualties.

On 8 September the long-delayed climax at last arrived. After a heavy bombardment, French and British troops advanced once again against the Malakov (M) and the Redan (L). The former attack was crowned with success and the Malakov was occupied, but the British were eventually driven back from the Redan after losing 2,500 casualties. However, Gorchakov had now decided that the defence of Sebastopol was hopeless and, after blowing up many military installations and the naval docks, he abandoned the city on the 9th, eleven months after the opening of the siege. This event virtually ended the war.

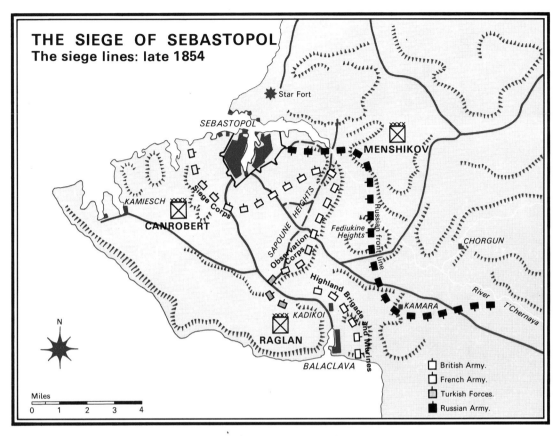

THE SIEGE OF SEBASTOPOL
The siege lines: late 1854

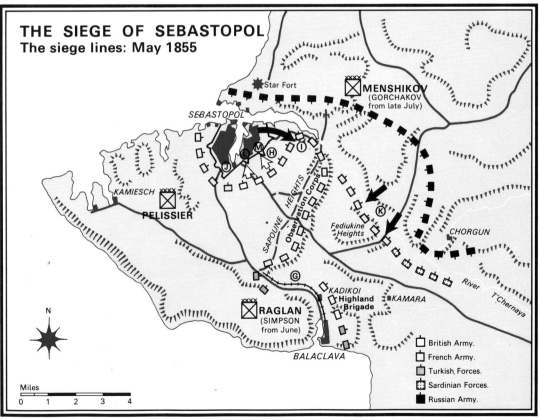

THE SIEGE OF SEBASTOPOL
The siege lines: May 1855

The Growth of the 'Second' British Empire

In 1815 it appeared that Britain had at last triumphed in the century-old struggle with France for colonial and commercial supremacy, more than compensating for the loss of the thirteen colonies in North America a generation earlier. Paradoxically, however, the whole concept of Empire was in the process of changing, and the triumph of the mercantilist view (whereby colonies existed to supply raw materials and local produce in exchange for the industrial products of the mother-country) came at the very moment when it was about to be replaced by the doctrine of free trade. So vast had been the expansion of British industry that her industrialists and merchants now demanded access to all foreign markets, as the old closed-market system could no longer either supply its requirements or absorb its products. Thus, by the 1840s, the possession of an Empire was widely regarded as unprofitable and irksome, and the processes of creating the 'Second' British Empire were hesitant in the extreme. The main markets for British goods were in Europe, America, the Far East and South America, and only British interests in India were comparable in scale or importance.

The Indian sub-continent, in fact, became the main raison d'être for the British Empire, and for much of British foreign policy. The main roles of the British Army in the post-Waterloo period became colonial garrisoning, protection and imperial expansion. Little wars, usually against native peoples, became the rule. The paramount requirement was to keep open and secure the trade routes leading to India, whether round the Cape of Good Hope or through the Mediterranean and Red Seas. The former route led to an increasing involvement in Africa; the latter, long before the opening of the Suez Canal in 1869, to deeper involvement in Mediterranean, Balkan and Turkish affairs. Suspicion of Russian intentions vis-à-vis Afghanistan and India was sufficient to ensure British interest in the maintenance of the decrepit Ottoman Empire, and led to the only struggle fought against a European power during the nineteenth century: the Crimean War.

The need to guard the routes to India led, long before the overt expansion of Empire that began in the 1880s, to steady (albeit piecemeal) growth involving Africa, South-East Asia, Australasia and the Far East—as well as India itself, and the immediate neighbouring countries. Allied to this motivation was a growing sense of religious purpose: the duty to shoulder 'the white man's burden' and to bring the benefits of British civilization to ". . . the lesser breeds beyond the Pale''. However outdated and even repugnant this view may seem today, it was very strongly held and championed by the nineteenth-century British citizen.

Control of affairs within India itself was one pre-requisite for imperial stability. Time and again, the need for military action against corrupt or warlike princes and rajahs led to the further extension of political control, and these in turn created challenges as new potentates were encountered beyond the redrawn frontiers. The requirements of Indian security led inexorably to involvements beyond its boundaries: north into Tibet and China, north-west via the frontier into Afghanistan and Persia, east towards Burma, Malaya and Indo-China, south-east into Ceylon and west into Arabia and the Persian Gulf. Much of this expansion was unwillingly undertaken—with much carping and criticism from London—but the process continued inexorably. If much of the military force needed to underpin India was the responsibility of the armies of the three Presidencies—Madras, Bombay and Bengal—

there was also a need for British troops to a considerable degree: the British-led Sepoy forces raised by the East India Company were not sufficient in themselves.

The spread of British influence over India down to 1817 has already been illustrated (see map, Expansion of British Rule over India, p. 79). Subsequent events can only be summarized

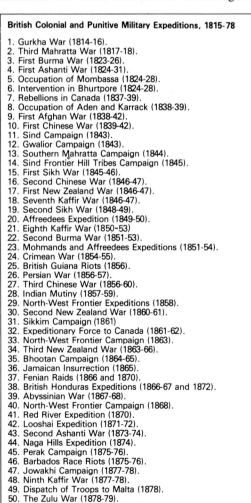

British Colonial and Punitive Military Expeditions, 1815-78

1. Gurkha War (1814-16).
2. Third Mahratta War (1817-18).
3. First Burma War (1823-26).
4. First Ashanti War (1824-31).
5. Occupation of Mombassa (1824-28).
6. Intervention in Bhurtpore (1824-28).
7. Rebellions in Canada (1837-39).
8. Occupation of Aden and Karrack (1838-39).
9. First Afghan War (1838-42).
10. First Chinese War (1839-42).
11. Sind Campaign (1843).
12. Gwalior Campaign (1843).
13. Southern Mahratta Campaign (1844).
14. Sind Frontier Hill Tribes Campaign (1845).
15. First Sikh War (1845-46).
16. Second Chinese War (1846-47).
17. First New Zealand War (1846-47).
18. Seventh Kaffir War (1846-47).
19. Second Sikh War (1848-49).
20. Affreedees Expedition (1849-50).
21. Eighth Kaffir War (1850-53)
22. Second Burma War (1851-53).
23. Mohmands and Affreedees Expeditions (1851-54).
24. Crimean War (1854-55).
25. British Guiana Riots (1856).
26. Persian War (1856-57).
27. Third Chinese War (1856-60).
28. Indian Mutiny (1857-59).
29. North-West Frontier Expeditions (1858).
30. Second New Zealand War (1860-61).
31. Sikkim Campaign (1861)
32. Expeditionary Force to Canada (1861-62).
33. North-West Frontier Campaign (1863).
34. Third New Zealand War (1863-66).
35. Bhootan Campaign (1864-65).
36. Jamaican Insurrection (1865).
37. Fenian Raids (1866 and 1870).
38. British Honduras Expeditions (1866-67 and 1872).
39. Abyssinian War (1867-68).
40. North-West Frontier Campaign (1868).
41. Red River Expedition (1870).
42. Looshai Expedition (1871-72).
43. Second Ashanti War (1873-74).
44. Naga Hills Expedition (1874).
45. Perak Campaign (1875-76).
46. Barbados Race Riots (1875-76).
47. Jowakhi Campaign (1877-78).
48. Ninth Kaffir War (1877-78).
49. Dispatch of Troops to Malta (1878).
50. The Zulu War (1878-79).
51. Second Afghan War (1878-80).

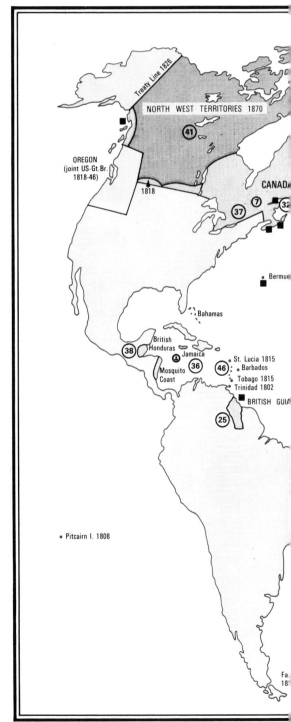

here. Lord Hastings' Gurkha War (1815-16) had ended in amicable agreement and alliance rather than conquest, but this was to prove the exception rather than the rule. Hastings' Pindari and Mahratta War (1817-19) led to the cession of vast areas of territory in central India, and his successor, Lord Amherst, took over Assam and the Arakan after the First Burma War (1823-26) on the eastern frontiers. A period of peace followed, during which Lord William Bentinck introduced many administrative reforms—including the land-settlement of the North-West Provinces and the suppression of Suttee (wife-burning) and Thuggee (ritual murder) practices—but from 1837 a more active period followed. The Persian invasion of Afghanistan in 1837 led to British intervention and Lord Auckland's unwise replacement of Dost Mohammed by Shah Shuja. In 1841 came a massive revolt and the disasters of the First Afghan War, which were only retrieved by the occupation of Kabul and the restoration of Dost Mohammed by Lord Ellenborough in 1842. British prestige had suffered a great blow, but it

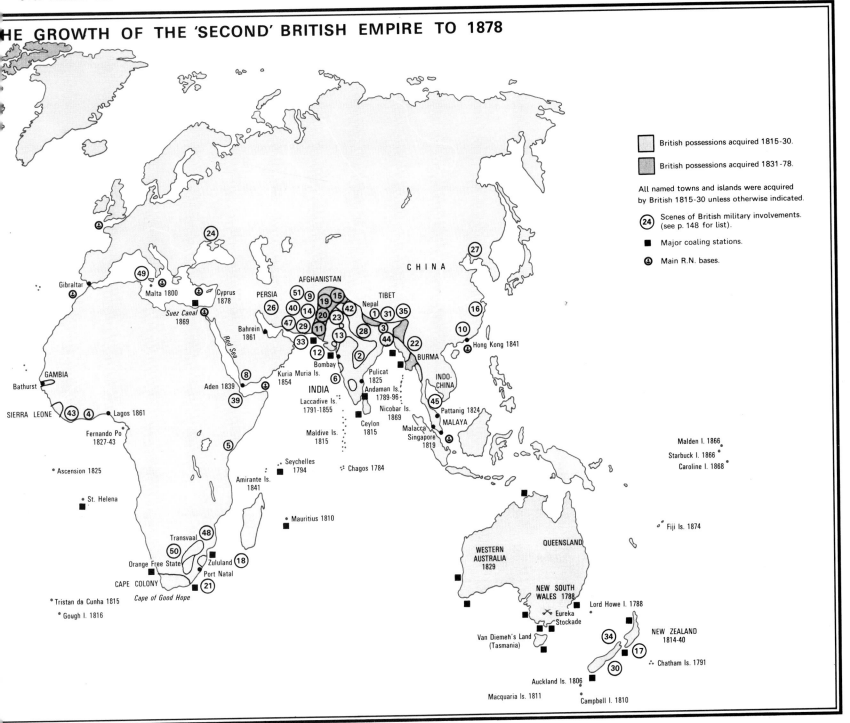

THE GROWTH OF THE 'SECOND' BRITISH EMPIRE TO 1878

British possessions acquired 1815-30.

British possessions acquired 1831-78.

All named towns and islands were acquired by British 1815-30 unless otherwise indicated.

⓴ Scenes of British military involvements. (see p. 148 for list).

■ Major coaling stations.

⚓ Main R.N. bases.

CHINA

Gibraltar

Malta 1800

Cyprus 1878

AFGHANISTAN

TIBET

PERSIA

Nepal

Suez Canal 1869

Bahrein 1861

Red Sea

Hong Kong 1841

BURMA

INDO-CHINA

GAMBIA

Bathurst

Aden 1839

Kuria Muria Is. 1854

Bombay

INDIA

Pulicat 1825

Andaman Is. 1789-96

Laccadive Is. 1791-1855

Nicobar Is. 1869

Ceylon 1815

Pattanig 1824

MALAYA

Malacca

Singapore 1819

SIERRA LEONE

Lagos 1861

Fernando Po 1827-43

Maldive Is. 1815

Malden I. 1866

Starbuck I. 1866

Caroline I. 1868

Ascension 1825

St. Helena

Seychelles 1794

Chagos 1784

Amirante Is. 1841

Fiji Is. 1874

Mauritius 1810

Transvaal

Orange Free State

Zululand

Port Natal

CAPE COLONY

Cape of Good Hope

Tristan da Cunha 1815

Gough I. 1816

WESTERN AUSTRALIA 1829

QUEENSLAND

NEW SOUTH WALES 1788

Lord Howe I. 1788

Eureka Stockade

Van Diemeh's Land (Tasmania)

NEW ZEALAND 1814-40

Chatham Is. 1791

Auckland Is. 1806

Macquaria Is. 1811

Campbell I. 1810

149

The Growth of the 'Second' British Empire

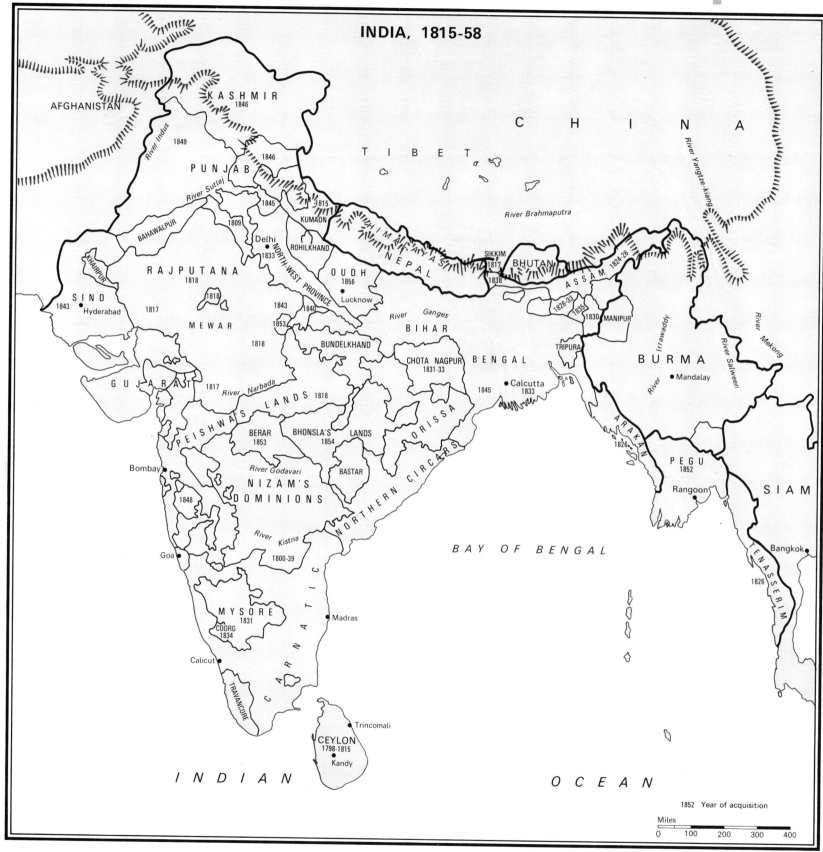

INDIA, 1815-58

AFGHANISTAN

CHINA

TIBET

KASHMIR 1846

River Indus 1849

PUNJAB 1846

River Sutlej 1845

1815

KUMAON

BAHAWALPUR 1809

Delhi 1833

ROHILKHAND

NORTH-WEST PROVINCE

River Brahmaputra

NEPAL

SIKKIM 1817

BHUTAN

1838

ASSAM 1824-26

River Yangtze-kiang

KHAIRPUR

RAJPUTANA 1818

OUDH 1856

Lucknow

1826-33

1835

1830 MANIPUR

SIND 1843

Hyderabad

1817

1818

1843

1840

1853

River Ganges

BIHAR

TRIPURA

River Irrawaddy

Mandalay

MEWAR

BUNDELKHAND

CHOTA NAGPUR 1831-33

BENGAL

BURMA

River Mekong

1818

River Salween

GUJARAT 1817

River Narbada

1818

PEISHWA'S LANDS

ORISSA

1845

Calcutta 1833

ARAKAN 1826

BERAR 1853

BHONSLA'S 1854

LANDS

NORTHERN CIRCARS

PEGU 1852

Bombay

River Godavari

BASTAR

NIZAM'S DOMINIONS

SIAM

Goa

1848

River Kistna

1800-39

Rangoon

Bangkok

BAY OF BENGAL

TENASSERIM 1826

MYSORE 1831

Madras

CARNATIC

COORG 1834

Calicut

TRAVANCORE

CEYLON 1798-1815

Kandy

Trincomali

INDIAN

OCEAN

1852 Year of acquisition

Miles
0   100   200   300   400

150

was partially restored by Sir Charles Napier's annexation of Sind in 1843; an unprovoked act of expansionism causing grave suspicions amongst the remaining independent Indian princes. This gave rise to the revolt of Gwalior, late in 1843, and the brief Maharajpur campaign which returned the area to order. The next challenge came from the Punjab, where the 'Khalsa' or Sikh Army—the brilliant creation of the great Ranjit Singh—launched a campaign over the River Sutlej in an expansionist bid. The result was the First Sikh War (see p. 152), with its important Battles of Ferozeshah and Sobraon, which defeated the Sikhs and led to treaty limitations being imposed upon their military power. Three years later, however, came the Second Sikh War (1848-49) with its climax at the Battle of Gujerat (see p. 152), which led to the formal annexation of the Punjab.

Lord Dalhousie's period as Governor-General saw a great expansion of British India. No sooner had the Punjab been organized into a province under Sir Henry and Sir John Lawrence, than the Mahratta state of Satara was annexed. In 1852, after the Second Burma War, the province of Pegu was taken over. Various reasons were put forward to justify the assimilation of Jhansi adjoining the North-West Provinces and of the great central Indian state of Nagpur, besides several more minor princedoms. Finally, in 1856, the Moslem kingdom of Oudh was annexed by Dalhousie on the grounds of the misrule of its kings, and a fresh tremor of alarm ran through the remaining independent Indian princes.

British East India Company rule never seemed more powerful than at this moment of, in fact, its greatest peril. The dire events of the Indian Mutiny (1856-58)—see p. 156—led to the powers of the Company being rescinded and India transferred to the British Crown, represented by a viceroy. In 1877 Queen Victoria was proclaimed Empress of India and, although future years would hold much more campaigning on the North-West Frontier, the evolution of British control over the subcontinent was virtually complete.

The opening of the Suez Canal in 1869 (and Disraeli's purchase of a controlling number of shares in the company) was a very significant strategic move which shortened the journey between Britain and India. This crucial international waterway would figure large in twentieth-century military strategy, especially as reliance on Persian oil to power the new generation of shipping and vehicles (both land and air) rapidly grew.

The British involvement in Southern Africa sprang from the retention of the Cape of Good Hope, which was occupied in 1795 and again in 1806 when its governors became allies of France. The arrival of British settlers, laws and customs followed, and the Dutch colonists became increasingly restive under British rule. The final straw for these strict Calvinists came when Britain abolished slavery in 1833, and the Boers set out with their families and possessions to make the 'Great Trek' inland in the hope of escaping from British influence. In 1836 they formed three new states: the Orange Free State, the Transvaal and Natal, but eight years later Britain took over Natal. The Seventh Kaffir War (1846-47) was followed by the Eighth (1850-53), the independence of the Transvaal in 1852 and, in 1854, that of the Orange Free State. However, in 1877, owing to the threat posed by the Zulus, Britain annexed the Transvaal, and the Zulu War (1878-79) followed. This annexation was greatly resented by the Boers and, in 1880, would result in the First Boer War. Further major trouble would ensue at the close of the nineteenth century.

The British connection with Australasia dates from 1770. Eighteen years later a penal colony was planted at Botany Bay, and as prisoners were freed and new settlers arrived from England, the state of New South Wales began to develop around Sydney, initially under a military governor. In 1812, Tasmania was made a separate colony. Australia was established in 1829, followed in 1834 by Victoria and South Australia. Queensland (initially, a penal settlement) dates from 1826. Real expansion started in the 1850s, when the punishment of transportation to the colonies was discontinued, and the process received a great boost with the discovery of gold (1851). This led to Australia's only nineteenth century military engagement: the storming of Eureka Stockade (December 1854). In 1854, a large measure of responsible self-government was granted to the colonies, and five years later Queensland was separated from New South Wales and assumed its own status as a colony. Federation would come in the first year of the next century.

New Zealand was annexed to New South Wales from 1839 to 1850, with the assent of the Maori peoples embodied in a treaty. The settlers were increasingly at risk as racial tension grew, but the Governor, Sir George Grey, kept the situation under control during the First Maori War of 1846-47. In 1852, the colony was granted responsible government, but the Maori war was renewed (1860-61) and quite severe fighting took place. Unrest continued for a number of years, and the Maoris were only brought to full submission to the government in 1871 by the Peace of Waitingi.

Left: Sir James Andrew Brown, Tenth Earl and First Marquis of Dalhousie (1812-80). As Governor-General of India from 1847 to 1856 he ruthlessly occupied the Punjab, fought the Second Burma War, and annexed Oudh. Right: General Hugh Gough, First Viscount (1779-1869) was an experienced Victorian commander who saw much service in India, and emerged the victor from two Sikh Wars. He was known to his troops as 'Old White Coat'.

One part of the Old Empire also made rapid, and mainly peaceful, progress over the nineteenth century. The influx of loyalists from the United States in the late eighteenth century to mix with the predominantly French colonists led to the Canada Act of 1791, which divided the colony into Upper Canada (mostly British) and Lower Canada (mainly French), each being allowed its accustomed legal system. American attacks were thwarted during the War of 1812 (see pp. 116-117). Over the succeeding decades, however, government remained in the hands of the Governor and of Crown nominees, and this, together with increasing Anglo-French friction, led to Papineau's insurrection (1837-39), which had to be quelled by troops. Lord Durham's celebrated Report, led to the Reunion Act of 1840, which rejoined Upper and Lower Canada and made the executive responsible to the colonial legislature. Troops were sent to Canada in 1861, and Fenian raiding from the USA had to be repelled in 1866 and 1870. In 1867 the British North America Act established the Dominion of Canada as a type of federation, practically self-governing except in matters of foreign policy, with power shared between central and provincial governments. All other North American colonies, save only Newfoundland, in due course joined the Dominion. In 1870, troops took part in the Red River Expedition to quell a revolt by independent fur-trappers led by Louis Riell.

Thus, the growth of the 'Second' British Empire was not devoid of military activity between 1815 and the third quarter of the nineteenth century.

The Growth of the 'Second' British Empire

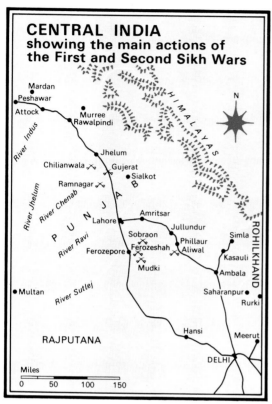

CENTRAL INDIA showing the main actions of the First and Second Sikh Wars

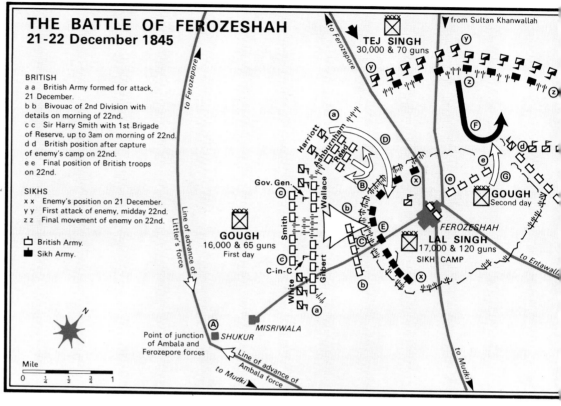

THE BATTLE OF FEROZESHAH 21-22 December 1845

BRITISH
a a British Army formed for attack, 21 December.
b b Bivouac of 2nd Division with details on morning of 22nd.
c c Sir Harry Smith with 1st Brigade of Reserve, up to 3am on morning of 22nd.
d d British position after capture of enemy's camp on 22nd.
e e Final position of British troops on 22nd.

SIKHS
x x Enemy's position on 21 December.
y y First attack of enemy, midday 22nd.
z z Final movement of enemy on 22nd.

British Army.
Sikh Army.

The Battle of Ferozeshah, 1845

The events leading up to the First Sikh War have been summarized briefly on p. 148. Summoning Sir John Littler's force from the post of Ferozepore, General Sir Hugh Gough, fresh from the action at Mudki on 18 December, managed to concentrate 16,000 men (including 6 British battalions) and 65 guns at the village of Shukur (A) by 1.30pm on 21 December. Facing his mainly Sepoy army were the 17,000 Sikhs and 120 guns of Sardar Lal Singh, manning breast-high entrenchments running along 3,000 yards of the defensive perimeter surrounding the village of Ferozeshah. The Governor-General, Lord Hardinge, experienced difficulty in restraining the fiery Gough until his Army of Bengal was completed by the arrival of Littler's two brigades to form his left wing, but at 3.30pm the attack could at last begin. Littler advanced Reed's brigade, supported by Ashburnham's three battalions, to attack the breastworks (B); but although they were accompanied by two horse artillery batteries (10 guns) and part of a field battery— all of which moved forward by bounds to cover the infantry—they met such a storm of fire from perhaps 50 guns of the finely-trained and organized Sikh artillery that the attack failed, the 9th Foot suffering 260 casualties. Gough then advanced his centre and right (C) and, aided by a superb but costly charge by the 3rd Light

Dragoons (D) which defeated a Sikh counter-attack and penetrated into the Sikh rear, he managed to clear part of the entrenchments before the early onset of darkness ended operations at about 5pm. Gough fell back 400 yards for the night (to bb) and artillery fire punctuated the darkness.

The attack was resumed early on the 22nd, Sir Harry Smith's reserve brigades taking the lead, and Gough soon retook the entrenchments which were only lightly defended, penetrating deep into the Sikh camp (D) where he took 72 guns.

At this juncture, a second Sikh army— 30,000 men and 70 guns strong under Tej Singh —came upon the scene from the direction of Ferozepore, and the weary Army of Bengal, jubilant at its earlier success but short of ammunition, had to redeploy (at ee) to face the new threat. The Sikh guns blasted Gough's lines, and their irregular horse threatened (F) to attack the army's open right wing; but these horsemen refused to charge the bayonet-bristling British squares, and a new charge by the intrepid 3rd Light Dragoons (G) put them to flight at about 3pm. This repulse was enough for Tej Singh, and he began to withdraw his troops, to the fury of his ferocious and well-disciplined rank and file. By 4pm the Battle of Ferozeshah was over; for a loss of 2,415 casualties (including 700 killed), Gough had

inflicted perhaps 3,000 losses on the Sikhs. However, over half the Army of Bengal's losses were from the British forces present; the Sepoy units having been noted to hang back on both days, and Hardinge remarked that "another such victory will cost us the Empire". Nevertheless, the Sikh forces fell back from East India Company territory over the River Sutlej. Thus ended the first phase of the First Sikh War, but further bitter fighting was still to be faced against a determined and skilled foe.

The Battle of Sobraon, 1846

After their defeat at Ferozeshah, the Sikhs adopted a policy of smaller raiding to gain supplies whilst their main army fortified Sobraon on the south bank of the River Sutlej and built a pontoon bridge over the river. The British were waiting for supplies and siege guns, but Sir Harry Smith was sent to check the Sikh marauding. His 4,000 men fought a sharp action against Ranjodl Singh at Baddowal (21 January), followed by a larger engagement at Aliwal (28 January) which proved a substantial victory. Reinforced to a strength of 10,000, he economically defeated 14,000.

Gough determined to exploit the psychological advantage won at Aliwal and, as his heavy guns had now come up and Smith had rejoined the main army, he ordered an all-out attack on the fortified bridgehead of Sobraon,

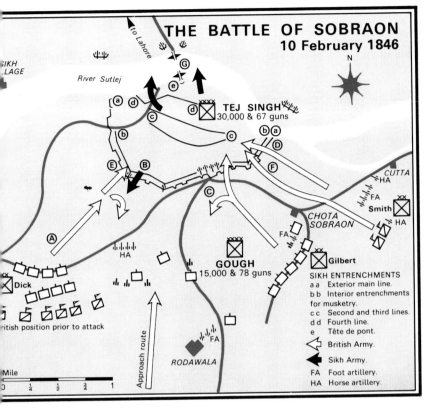

THE BATTLE OF SOBRAON
10 February 1846

to Lahore

SIKH VILLAGE

River Sutlej

TEJ SINGH
30,000 & 67 guns

CUTTA
HA

FA

Smith
HA

CHOTA
SOBRAON

FA

GOUGH
15,000 & 78 guns

Gilbert

SIKH ENTRENCHMENTS
a a Exterior main line.
b b Interior entrenchments
for musketry.
c c Second and third lines.
d d Fourth line.
e Tête de pont.

British Army.

Sikh Army.

FA Foot artillery.
HA Horse artillery.

Dick

British position prior to attack

Approach route

FA

RODAWALA

Mile

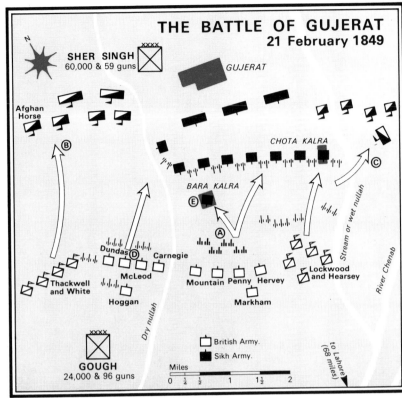

THE BATTLE OF GUJERAT
21 February 1849

N

SHER SINGH
60,000 & 59 guns

GUJERAT

Afghan
Horse

CHOTA KALRA

BARA KALRA

Dundas

Carnegie

McLeod

Mountain Penny Hervey

Lockwood
and Hearsey

Thackwell
and White

Hoggan

Markham

Stream or wet nullah

River Chenab

British Army.

Sikh Army.

GOUGH
24,000 & 96 guns

Miles

0 ¼ ½ 1 1½ 2

to Lahore
(68 miles)

---

garrisoned by 30,000 Sikhs under Tej Singh's command. The Sikh position comprised some 4,000 yards of strong trenches and breastworks in a semi-circle, both flanks resting on the River Sutlej, with several interior lines of fortification defending the bridgehead. Some 67 guns were mounted in the defences.

A heavy bombardment was opened against Sobraon at dawn on the 10th, but when the fire began to slacken as ammunition stocks ran low, Gough ordered Major-General Sir Robert Dick to launch the main attack (A) against the right of the Sikh position, supported by diversions against the centre and left by Gilbert's and Smith's divisions. At 10.30am Dick's attack began, but it was halted by heavy Sikh fire 300 yards from its objective, and then repulsed by a strong Sikh counterattack (B). The British supporting attacks were not properly synchronized—which largely accounted for Dick's setback—and when Gilbert eventually approached his sector he diverged towards the strongest part (C) of the Sikh defences, and was also forced to retire. Three times this attack was renewed before a foothold was achieved. Meanwhile, on the right flank, Smith was also checked (D) for a time. However, Dick's renewed onslaught at the head of his five British and eight Sepoy battalions at last stormed the outer lines (E), this event coinciding with Gilbert's successful attack in

the centre. The Sikhs fell back to their second line of defences (bb), and Smith's column was aided to its objective by the sending forward of the 3rd Light Dragoons, who entered the main position through some gaps in the 16ft-thick earthen ramparts blasted by the pioneers to reform within and then charge. This attack overwhelmed several Sikh batteries (F).

The three divisions now converged on the river. Both sides fought ferociously, giving no quarter, but the Sikhs faced disaster when they discovered that their commander had removed part of the pontoon bridge behind them, and that a spate had raised the water level by seven feet. A massacre followed, and many of the estimated 10,000 Sikh casualties drowned in the Sutlej (G). Gough lost 300 killed and 2,000 wounded. This battle effectively ended the First Sikh War, leading to the Treaty of Kasure signed with Ghulab Singh, later Maharajah of Kashmir, on March 1846.

The Battle of Gujerat, 1849
Almost three years of intrigue and growing unrest in the general region of the Punjab ensued. The new Governor-General, Lord Dalhousie, almost deliberately provoked a Sikh uprising to justify outright annexation. In October 1848, a number of troops mutinied in Peshawar and elsewhere, and soon the main Sikh army under Chattar Singh marched south

to join his son, Sher Singh, near Ramnagar on the River Chehab as a preliminary to an advance on Lahore. On 9 November, General Gough crossed the Sutlej moving north and, after a number of actions around Ramnagar from 21 November, he pressed ahead to the River Jhelum, where the hard-fought and sanguinary Battle of Chilianwala was fought on 13 January 1849, which ended in only a technical British victory. This outcome was to lead to Gough's supercession as commander-in-chief by Sir Charles Napier.

However, the capture of Multan in late January freed troops for transfer to Gough's army, which was built up to 24,000 men and 96 guns (11 regiments of cavalry and 23 battalions of infantry, three of the former and eight of the latter being European units). Sher Singh, running short of supplies, moved away eastwards towards the River Chenab, hoping to force a way over towards Lahore. Gough followed and, on 20 February, made contact with the Sikhs near Gujerat. Sher Singh had some 60,000 troops (many of them newly-raised owing to the restrictions imposed on the Sikh forces in March 1846) and 59 guns, drawn up in a large crescent with cavalry on the flanks, relying upon the fortified villages of Bara Kalra and Chota Kalra—set to the fore of his centre—to break up any British attack.

At 7am on the morning of 21 February,

153

Gough advanced for two miles, determined to draw the Sikh fire. When the guns blazed out at about 1,200 yards, the infantry were ordered to lie down in their ranks, and the artillery was rushed to the front. A large battery of ten 18-pounders and as many 8-inch howitzers was established (A) opposite the centre of the Sikh line, and this soon silenced, and then destroyed, numbers of the revealed Sikh guns in a three-hour bombardment that lasted from 9am until midday. Then Gough ordered a general advance. The British and Sepoy cavalry, accompanied by horse artillery batteries, dominated the hovering wings of Afghan and Sikh horsemen (B and C), whilst the infantry, accompanied by the lighter guns, marched forward into heavy Sikh fire (D). The strength of the garrison within Bara Kalra (E) had not been estimated correctly, and there the 2nd Bengal Europeans lost 148 out of 900 men. But the Sikhs were soon in full flight towards the North-West Frontier (where most surrendered on 14 March) pursued by the seven battalions that had borne the brunt of the action and the cavalry. The victory had cost Gough's army 800 casualties (including many gunners), Sikh losses are uncertain, but all 59 guns were captured. In April, Dalhousie announced the annexation of the Punjab. So ended the Second Sikh War.

The Expedition to Abyssinia, 1867-68

Not all Victorian military campaigns led to territorial accessions. Some were inspired by humanitarian as well as punitive considerations; General Sir Robert Napier's large-scale march to Magdala in Abyssinia being a major example. Since 1864 the crisis there had escalated, with the half-mad Emperor Theodore III using imagined slights by the British government to justify his retention of several consular officers and missionaries as hostages. When successive concessions and gifts failed to secure the release of the prisoners, in June 1867 the British government, at length and with great reluctance, decided on military intervention, and warned the army of the Bombay Presidency to be ready to act if a final ultimatum was rejected.

The problems were daunting. Firstly, a 400-mile approach march over rugged and roadless terrain was involved, much of it almost waterless; secondly, Theodore's army, although primitive, was known to be large; thirdly, the tyrant might well massacre his captives as the expedition approached; and lastly, the expedition would need to develop a base on the African coast before proceeding inland. Napier tackled all these difficulties in turn. After a reconnaissance he selected Zula rather than Massawa for the base, and in mid-December the bulk of his 13,000 combat troops (about one third British, the rest Indian) and five batteries of artillery arrived off Annesley Bay, conveyed in 291 vessels (including 75 steamships). Also shipped were 36,000 animals: mules, horses, camels, bullocks and 44 elephants, a railway comprising six engines and 60 trucks, a Royal Engineer Telegraph Company (besides 8 companies of sappers) and a total of 47,000 servants, bearers and workmen.

The base established at Zula proved unhealthy and short of drinking water, but although many animals perished, three naval condensers helped overcome the immediate difficulties. Napier and his staff arrived on 2 January 1868, and soon his advance guard had

Field Marshal Sir Robert Cornelis Napier (1810-90), who fought with distinction in India and China before commanding the successful expedition sent into Abyssinia to rescue the hostages of the demented Emperor Theodore. (National Army Museum)

penetrated inland as far as Adigrat. Meanwhile, Theodore had moved his 59 hostages, 8,000 warriors, 6 guns and 70-ton monster mortar 'Theodorus' into his mountain fastness at Magdala, far away in the interior.

Aware of the need to woo local Ethiopian princes en route, if his extensive lines of communication were to be safe, Napier made his preparations with care, and only on 29 January was the advance able to start. He planned to establish a forward base at Antalo, 200 miles into the interior; bulk supplies would be passed over the first miles to Koomayli and the Soroo Pass at the foot of the hills by railway but this proved slow to construct. The aim was to reach Magdala with 5,000 combat troops, rescue the hostages, and then evacuate the country.

On 14 February the advance guard established the forward base at Antalo, but move-ment was proving so difficult that much baggage and personal equipment had to be left there. The mutinous attitude of many of the Indian mule-drivers caused further trouble, but the advance went on. On 25 February, Prince Kassai of Shoa was bribed into supporting the expedition and, by 2 March, the main army had closed up to Antalo. On the 12th the selected strike force set out for the second part of the campaign. The Prince of Lasta was induced to give his aid on the 30th, and the rate of the advance quickened. The 45th Foot, for example (which with the 4th and 33rd comprised the British units in the forward division), covered 300 miles in 24 days, which was commendable given the terrain. At last, on 8 April, the River Bashillo was reached, just 12 miles from Magdala, and scouts brought news that Theodore was in position on the Mount Fahla massif at the head of some 10,000 Abyssinian warriors, and that he had ominously butchered some 600 native prisoners. Napier decided that he must attack without delay.

On Good Friday, 10 April, the force advanced to the attack. The 1st Brigade, supported by artillery and naval rocket-tubes, received the fire of Theodore's cannon (although 'Theodorus' exploded after the first discharge) and then 500 chiefs and 6,000 natives swarmed forward. Blasted by artillery fire, rockets and musketry, the warriors made little effect on the British and Indian troops. For a loss of 20 British casualties and under 60 Indian, Napier killed 700 warriors and wounded 1,200 more. Theodore withdrew into Magdala, and tried to negotiate on the 11th and 12th—freeing most of his prisoners—but Napier would accept nothing but a full surrender. In despair Theodore attempted suicide, but failed. At last, on 13 April, Napier ordered the 2nd Brigade to storm the fortress. Covered by artillery fire, 10th Company RE, followed by his three British regiments, the 33rd to the fore, stormed up the narrow approach track to the gates. The sappers discovered that they had overlooked the need to bring blasting charges, but men of the 33rd found a way over the palisades (earning two Victoria Crosses), opened the gates, and the action was almost over. Theodore shot himself rather than face capture. The remaining hostages were found to be safe, and Napier had completed his task.

After blowing up and burning Magdala (after the evacuation of the population), the expeditionary force began a long and painful withdrawal to the coast. Antalo was evacuated on 15 May, Senafe on the 24th. On 2 June, Zula was reached, and 16 days later the last troops were embarked. The campaign had cost 35 British and 512 Indian lives over nine months.

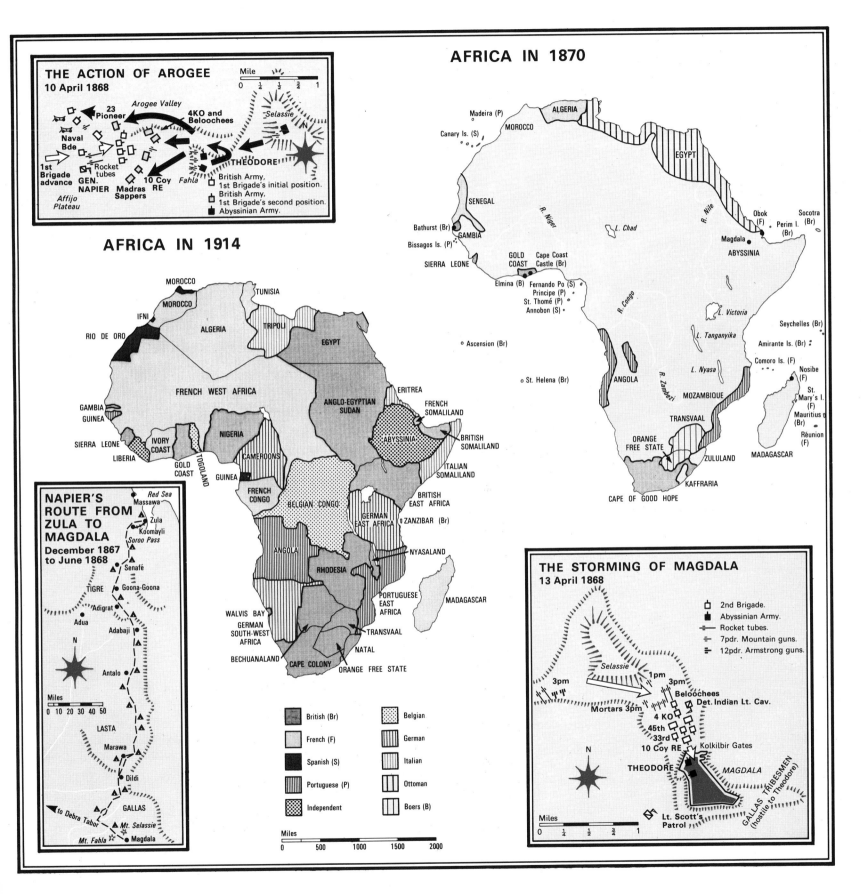

AFRICA IN 1870

Madeira (P)
Canary Is. (S)
ALGERIA
MOROCCO
EGYPT
SENEGAL
R. Niger
L. Chad
R. Nile
Obok (F)
Socotra (Br)
Perim I. (Br)
Bathurst (Br)
GAMBIA
Magdala
ABYSSINIA
Bissagos Is. (P)
SIERRA LEONE
GOLD COAST
Cape Coast Castle (Br)
Elmina (B)
Fernando Po (S)
Principe (P)
St. Thomé (P)
Annobon (S)
R. Congo
L. Victoria
Seychelles (Br)
Amirante Is. (Br)
Comoro Is. (F)
Nosibe (F)
Ascension (Br)
L. Tanganyika
L. Nyasa
St. Mary's I. (F)
Mauritius (Br)
Réunion (F)
ANGOLA
R. Zambezi
MOZAMBIQUE
MADAGASCAR
St. Helena (Br)
TRANSVAAL
ORANGE FREE STATE
ZULULAND
KAFFRARIA
CAPE OF GOOD HOPE

THE ACTION OF AROGEE
10 April 1868

Mile
0   ¼   ½   ¾   1

Arogee Valley
23 Pioneer
4KO and Beloochees
Selassie
Naval Bde
THEODORE
1st Brigade advance
GEN. NAPIER
Rocket tubes
Madras Sappers
10 Coy RE
Fahla
Affijo Plateau

British Army, 1st Brigade's initial position.
British Army, 1st Brigade's second position.
Abyssinian Army.

AFRICA IN 1914

MOROCCO
TUNISIA
MOROCCO
IFNI
RIO DE ORO
ALGERIA
TRIPOLI
EGYPT
FRENCH WEST AFRICA
ERITREA
FRENCH SOMALILAND
GAMBIA
GUINEA
ANGLO-EGYPTIAN SUDAN
ABYSSINIA
BRITISH SOMALILAND
SIERRA LEONE
IVORY COAST
NIGERIA
LIBERIA
GOLD COAST
TOGOLAND
GUINEA
CAMEROONS
ITALIAN SOMALILAND
FRENCH CONGO
BELGIAN CONGO
BRITISH EAST AFRICA
GERMAN EAST AFRICA
ZANZIBAR (Br)
ANGOLA
NYASALAND
RHODESIA
PORTUGUESE EAST AFRICA
MADAGASCAR
WALVIS BAY
GERMAN SOUTH-WEST AFRICA
TRANSVAAL
NATAL
BECHUANALAND
CAPE COLONY
ORANGE FREE STATE

NAPIER'S ROUTE FROM ZULA TO MAGDALA
December 1867 to June 1868

Red Sea
Massawa
Zula
Koomayli
Soroo Pass
Senafé
TIGRE
Goona-Goona
Adigrat
Adua
Adabaji
Antalo
LASTA
Marawa
Dildi
GALLAS
to Debra Tabor
Mt. Selassie
Mt. Fahla
Magdala

Miles
0 10 20 30 40 50

British (Br)
French (F)
Spanish (S)
Portuguese (P)
Independent
Belgian
German
Italian
Ottoman
Boers (B)

Miles
0   500   1000   1500   2000

THE STORMING OF MAGDALA
13 April 1868

2nd Brigade.
Abyssinian Army.
Rocket tubes.
7pdr. Mountain guns.
12pdr. Armstrong guns.

Selassie
3pm
1pm
3pm
Beloochees
Det. Indian Lt. Cav.
Mortars 3pm
4 KO
45th
33rd
10 Coy RE
Kolkilbir Gates
THEODORE
MAGDALA
GALLAS TRIBESMEN (hostile to Theodore)
Lt. Scott's Patrol

Miles
0   ¼   ½   ¾   1

The Indian Mutiny, 1857-59

The spread of British rule through India has already been described (p. 78 and p. 148), with special reference to the conquest of the Punjab and the other achievements of Lord Dalhousie, but the events of 1857 were to shake European equanimity to its very core, and lead to major administrative reorganization.

Causes of the Mutiny

Suspicion that the British intended to undermine local cultures and impose Christianity on Moslems and Hindus alike lay at the heart of the growing unease that eventually led to the explosion of violence in northern India. The independent princes feared for their possessions after the annexations of the Punjab and Oudh. Unwittingly, the British authorities also prepared the way for disaster. The number of Sepoy troops was multiplied to cope with the new annexations, but the total of European units actually fell owing to the demands of the Crimean War, and were not subsequently replaced. Furthermore, a high proportion of the remaining units were absorbed into the Punjab area, leaving very few to garrison Oudh and the North Central provinces.

Disaffection amongst the Sepoys rapidly rose when Lord Canning became Governor-General in 1856, and decreed that, henceforth, the Bengal Army's native units would be liable for service overseas. This would involve Hindus in losing caste. The Moslems, meantime, were being infiltrated by agitators working for the restoration of the Moghul dynasty—and the ancient Moslem ascendancy. In January 1857, rumours that new cartridges being issued were greased with pig or cow fat—the former breaking Moslem taboos of cleanliness, the latter assaulting Hindu beliefs in the sacredness of the cow—spread through the camps from Dum-Dum, near Calcutta. To cap it all, there was talk of a prophecy that British rule would collapse one hundred years after Clive's victory at Plassey (see p. 78) in 1757.

Outbreak and Spread

The authorities took scant heed of the few indications that reached them. There were reasonable numbers of European troops in Lower Bengal, but between Patna and the Punjab frontier (a distance of eight hundred miles along the great Ganges plain) there were only five white regiments and a number of batteries of artillery. At first, the mutinous incidents were isolated and contained, with native regiments that refused to handle the cartridges being disarmed and disbanded at Berhampur (February), Barrackpur (March and May) and Lucknow (early May). But then, suddenly, on 10 May there came a major

mutiny at Meerut to the north-east of Delhi, followed by a massacre of the local white population. The mutineers marched on Delhi, which rose on their approach, and proclaimed the restoration of the Moghul Empire. Europeans were again butchered, but a telegraph operator managed to pass a warning message to the Punjab and another party managed to blow up the arsenal. However, the flame of revolt gathered momentum as the 'chapati' signal was passed. Some stations—including Lucknow, Agra and Peshawar—succeeded in disarming the local Sepoy troops, but elsewhere the situation ran amok. General Anson, commander-in-chief, set out with a small force for Delhi, but died of cholera on 28 May. His successor, the sick Reid, continued the advance, and won a battle against the mutineers on 8 June at Badli-ke-serai before taking up a position on the ridge overlooking

General Sir Henry Havelock (1795-1857), experienced British soldier who was recalled from Persia at the outbreak of the Indian Mutiny, and succeeded in relieving Lucknow at the third attempt shortly before dying of illness. He became a hero of the Victorian age. (National Army Museum)

Delhi, where he awaited reinforcements. He also died of disease and was replaced by General Wilson. Elsewhere, Sir Henry Lawrence brought the Residency at Lucknow to a state of defence, and the tiny garrison of Cawnpore under General Wheeler began its defiance of the huge forces of Nana Sahib. But over fifty cantonments mutinied (see map, The Indian Mutiny, opposite) and the Sepoys in the Ganges' districts above Patna were in revolt and tales of atrocities abounded. The Rajputana, Central India and parts of Bengal were also affected.

By late June, the force outside Delhi had won a second small action on the 28th (the anniversary of Plassey), but was having difficulty holding its own. The siege of Cawnpore (4-25 June) had ended in the massacre of the men of the garrison during a negotiated withdrawal by river, and the incarceration of the women and children ensued. The siege of the Residency at Lucknow—undertaken by native forces fluctuating between 40,000 and 100,000 men—began on 30 June after the defeat of a breakout attempt at

Chinat. Henry Lawrence was killed five days later, but a British relief force, organized and commanded by Henry Havelock, was ready to advance from Allahabad in an attempt to reach Lucknow.

Outside the Ganges basin, the Sepoy revolt spread to the areas south and south-west of Agra, although the Mahratta prince remained loyal at Gwalior. This was also true of most other Hindu princes, for they had no desire to see the restoration of a Moslem Empire. The overall effect was to contain the Mutiny in North Central India, the incipient revolts in the Punjab and in Lower Bengal having been nipped in the bud.

Reinforcements were on their way to the ridge above Delhi. The controversial figure of John Nicholson arrived from Peshawar at the head of a flying column on 7 August, and the long-awaited siege train, sent by Sir John Lawrence from Lahore, reached the camp on 4 September. Thus strengthened, General Wilson ordered an assault on Delhi on 14 September. Nicholson was killed near the Kashmir Gate, but the outer defences were overrun, and after six days of fierce street fighting Delhi was cleared of the mutineers, many of whom escaped towards Lucknow. The old Moghul Emperor, Bahadur Shah II, was captured outside the city. Much damage was done inside the town, with looting and a massacre perpetrated against the inhabitants.

The Relief of Lucknow, 1857

The besieged garrison, despite their slight numbers, had proved more than a match for their less-skilled besiegers. The engineers forestalled many mining attempts with countermines; only one telling explosion damaged the defences, and even then the damage was made good before the Sepoys tried to attack. However, the loyal Sepoys were beginning to lose heart as the siege entered its third month. Fortunately, help was on the way in the form of General Havelock, who was advancing by way of Cawnpore. At his approach, Nana Sahib massacred the white women and children. The troops arrived on 17 July, too late to save them. His troops raging to avenge the innocent victims, Havelock pressed on for Lucknow, but had to fall back to wait for reinforcements before attempting to reach the Residency. Reinforced in mid-September by General Sir James Outram, Havelock resumed the advance, and on the 25th broke through the besiegers' lines to join the garrison. The siege was at once resumed, but the garrison was now strong enough to withstand all pressures and supplies were also sufficient. A terrible vengeance was wrought on the mutineers by the British soldiers

—one that has not been forgotten or wholly forgiven after 120 years.

The Reconquest of North Central India

General Sir Colin Campbell, the new commander-in-chief, had earlier reached Calcutta where he organized an army for the regular reconquest of the insurgent areas. Fresh troops had arrived from England, others had been diverted en route for the China War. In late October he marched for Cawnpore, but, ignoring the army of Tantia Topi for the time being, he pressed on for Lucknow. On 17 November he reached the Residency and evacuated the perimeter, leaving Outram in command of a powerful force in a strong neighbouring position; Campbell saw the non-combatants to safety and, thereafter, began to prepare the reconquest of Oudh. In March 1858, Campbell returned at the head of a large force, and undertook the siege of the mutineers within the city of Lucknow. By 22 March, after a prolonged assault, the town was in British hands. Meanwhile, advancing from the west, General Sir Hugh Rose had relieved a loyal garrison in Saugor (3 February), and pressed ahead against the cunning and able Tantia Topi and the embittered Rani of Jhansi, winning a number of engagements and ultimately capturing the city and fortress of Gwalior on 20 June. The remaining mutineers, still led by the elusive Tantia Topi, resorted to guerrilla warfare amongst the jungles for the rest of the year. Topi was caught, tried and hanged in April 1859. The Mutiny was at last declared over. In August of the previous year, the rule of India had been transferred from the East India Company to the British Crown, thus inaugurating a new era in the history of the British Empire in India.

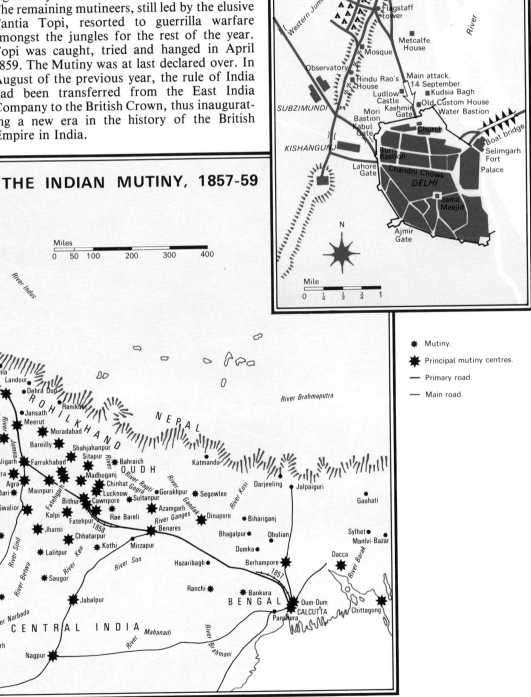

THE SIEGE OF DELHI
June to September 1857

THE INDIAN MUTINY, 1857-59

157

The Risorgimento: The Italian Struggle for Ind

Until 1796 Italy was a random collection of eleven states, comprising a haphazard group of kingdoms, duchies, republics and the Papal States, the northern part of the peninsula being under Austrian domination. Napoleon's defeat of Austria (see pp. 90-93) in 1796-97 and 1800 led to the first, though transient, steps towards Italian unity. The peninsula was at first divided into four republics, but later France ruled Piedmont, Parma, Tuscany, Rome and the Illyrian Provinces directly by various annexations carried out between 1802 and 1809. Napoleon also created the Kingdoms of Italy (1805) and of Naples (1806); the former comprising Lombardy, Venetia and (from 1808) part of the former Papal States, the latter being formed from the Kingdom of the Two Sicilies (Sicily and Naples), less Sicily itself.

This settlement did not survive Waterloo, and the Congress of Vienna recreated eight of the original states; Austrian influence was paramount or strong in Lombardy, Venetia, Parma, Modena and Tuscany. The rulers proved reactionary, and popular risings took place in 1820 in Sicily, Naples and Piedmont, and ten years later in Parma, Modena and the Papal States—but all these attempts failed. The patriot Mazzini decided to form 'Young Italy' and worked on towards liberty. The next major effort came in 1848, the 'Year of Revolutions', when risings in Sicily, Naples, the Papal States, Tuscany and Sardinia succeeded in gaining the grant of constitutions, but lack of united aims proved disastrous in the end. The Milanese expelled the Austrians in March 1848 (except from the fortresses of the Quadrilateral); Parma and Modena disposed of their rulers the same month; Venice again became a republic, followed by Rome in February 1849 under the inspiration of Mazzini. Many Italians hoped that Charles Albert, King of Sardinia, would assume the lead, and achieve a united country.

That ruler, however, was defeated by the armies of Austria at Custozza and Novara; the Habsburgs regained Lombardy, and the other revolts eventually failed. The Prince-President of France, Louis Napoleon, sent troops to Civitavecchia, inaugurating a significant period of French intervention in Italian affairs, and Rome was besieged and eventually regained for the Pope by French troops under General Oudinot.

The main work of Italian unification really commenced in 1859 in Lombardy (see map, The Unification of Italy, 1848-70). The statesman and patriot, Count Camillo di Cavour, had negotiated an agreement with France, whereby 200,000 French troops came to the assistance of Piedmont and the nationalists. The joint-army crossed the River

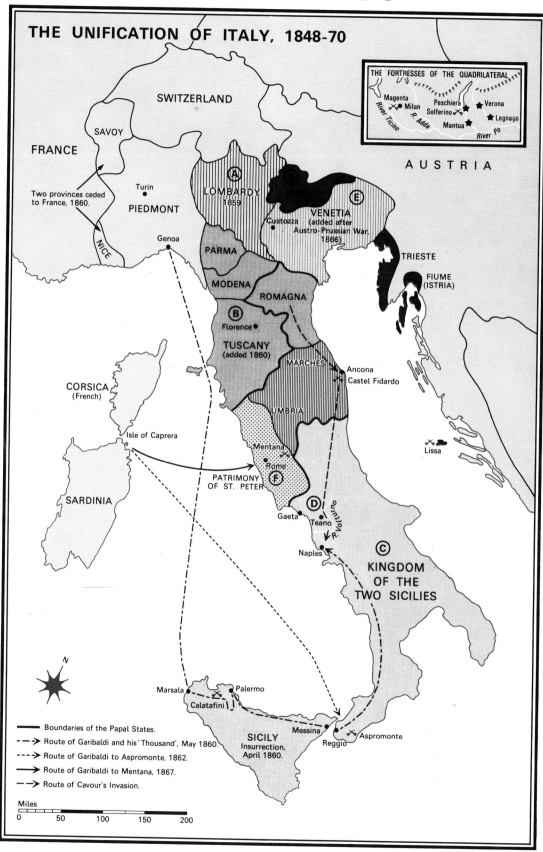

THE UNIFICATION OF ITALY, 1848-70

THE FORTRESSES OF THE QUADRILATERAL
Magenta · Milan · Peschiera · Verona
Solferino · Legnago
Mantua
River Ticino · R. Adda · River Po

SWITZERLAND

FRANCE
SAVOY
NICE
Two provinces ceded to France, 1860.
Turin
PIEDMONT
Genoa

LOMBARDY 1859
AUSTRIA
Custozza
VENETIA (added after Austro-Prussian War, 1866)
TRIESTE
FIUME (ISTRIA)

PARMA
MODENA
ROMAGNA
Florence
TUSCANY (added 1860)
MARCHES
Ancona
Castel Fidardo

CORSICA (French)

Isle of Caprera
UMBRIA
Mentana
Rome
PATRIMONY OF ST. PETER
Lissa

SARDINIA
Gaeta
Teano
Naples
KINGDOM OF THE TWO SICILIES

N

Marsala · Palermo
Calatafini
Messina
Reggio · Aspromonte
SICILY Insurrection, April 1860

——— Boundaries of the Papal States.
– – –► Route of Garibaldi and his 'Thousand', May 1860.
····► Route of Garibaldi to Aspromonte, 1862.
———► Route of Garibaldi to Mentana, 1867.
– ·► Route of Cavour's Invasion.

Miles
0   50   100   150   200

Ticino into Austrian-held Lombardy (A) in the spring of 1859 and, under command of Marshal Macmahon, defeated 50,000 Austrians under General von Clam-Gallas in a confused engagement at Magenta on 4 June. Both sides suffered heavy casualties, but the Austrians were driven back to the refuge of the Quadrilateral fortresses: Mantua, Peschiera, Verona and Legnago on the Rivers Mincio and Adige. Twenty-five days later, under command of Napoleon III, the Franco-Piedmontese won the great but gory Battle of Solferino (see p. 161 and map) and, by an agreement signed at Villafranca, the Austrians agreed to hand over all Lombardy except Mantua and Peschiera.

Next year, after further negotiations, the Duchies of Parma, Modena and Tuscany (B) severed their links with the house of Habsburg, and, together with the Papal State of Romagna, became part of the Kingdom of Sardinia (as Piedmont-Sardinia was now called). Also in 1860, at the opposite end of the Italian Peninsula, the patriot Giuseppe Garibaldi landed on 11 May at Marsala in western Sicily accompanied by 1,000 Redshirts and placed himself at the head of the local insurrection against Naples. Four days later, at Calatafini, the rebels defeated the Neapolitan forces; Palermo fell to their arms on the 27th. Garibaldi at once crossed the straits and advanced on Naples (C), declaring that his intention was to crown Victor Emmanuel II in Rome. This project alarmed Cavour, for if Garibaldi continued his advance it was likely to antagonize France. Accordingly, the Piedmontese Army was sent south through the Papal States, defeating the Pope's forces at Castel Fidardo en route, in order to intercept the victorious Redshirts. Some time was gained by the fact that Garibaldi had been held up for two weeks at the River Volturno by the Neapolitan Army and, by the time this force had been crushed, Garibaldi was induced to meet Victor Emmanuel at Teano. There, Garibaldi agreed to hand over Southern Italy to the King, and the immediate crisis was over (D). On 17 March 1861 the Kingdom of Italy was solemnly proclaimed, although Rome was still excluded.

The processes of unification were now in full flood, but Austria still remained the obstacle. Venetia, Trentino, Istria, Trieste and, above all, Rome still stood outside the new nation. Here, paradoxically, Pope Pius IX was supported by a French force and, as negotiations were making no progress, Garibaldi again took unilateral action and came out of retirement to lead a march on the Holy City. To avoid an open clash with France, Victor Emmanuel sent a force to halt the rebels and, at

Aspromonte on 29 August, Garibaldi was defeated by his erstwhile allies. Garibaldi was wounded and captured, but soon freed under an amnesty.

In 1866, on the eve of the Austro-Prussian War (see pp. 190-191), Bismarck signed an alliance with the Kingdom of Italy, which had already persuaded France to withdraw its troops from Rome. Then, to liberate Venetia (E) from Austrian rule, on 20 June Victor Emmanuel declared war. Four days later, at the second Battle of Custozza, General La Marmora was defeated by the Archduke

Napoleon III, Prince President and then Emperor of France (1808-73), whose desire to emulate the martial achievements of his famous uncle led him to successes in the Crimea and Italy but ultimately to cataclysm in the Franco-Prussian War.

Giuseppe Garibaldi (1807-82). Italian revolutionary and patriot, who led his famous 'Redshirts' to capture Rome and on many more famous exploits of mixed fortune in the interests of Italian liberation and unification.

Albert. However, disasters in Germany induced the Emperor to cede Venetia (but not Southern Tyrol) to Italy.

The question of Rome still remained a major problem. The indefatigable Garibaldi, exploiting the withdrawal of French troops in 1866, next year marched a second time on Rome (F). Napoleon III at once sent back a French force to Pope Pius IX's defence in October 1867, but Garibaldi—despite having been denounced by Victor Emmanuel—pressed ahead for Rome, only to be completely defeated at Mentana near the city on 3 November. Garibaldi was again captured and sent into exile on the Isle of Caprera, and the French troops remained in Rome. Three years later, however, Napoleon III recalled all his troops to fight in the Franco-Prussian War (see pp. 192-195), and Italian patriots promptly stormed Rome. On 3 October the Holy City became part of Italy, and Victor Emmanuel made it capital of his kingdom. Apart from Trentino (G), Istria and Trieste, which remained Austrian, Italy was now at last a united and independent nation.

The Siege of Rome, 1849

As the struggle between Austria and Piedmont continued, on 9 February 1849 Rome was declared a republic by a group of politicians and patriots, amongst them Giuseppe Mazzini. Prior to this event, Pope Pius IX had fled the Vatican and taken up residence at Gaeta. His appeals for international assistance were heeded in the French Second Republic, which saw a chance to challenge Austrian influence. Accordingly, on 25 April, a French expeditionary force of some 10,000 men, commanded by General Nicolas Oudinot (son of the Napoleonic Marshal), landed at Civitavecchia, some forty miles north-west of Rome, and prepared to march on the city after establishing a base. The morale of the republicans was greatly rallied on the 27th by the arrival in the city of the famous patriot and guerrilla leader, Garibaldi, at the head of his 1,300-strong Legion, mainly recruited over the past six months in the Romagna. This reinforcement raised the strength of Rome's defenders by late on the 29th to 9,000, most of them only partially-trained and equipped, but almost all wildly enthusiastic republicans. There were 2,500 former Papal troops, and a number of Carabinieri; Garibaldi's First Italian Legion was the second component; another 1,400 volunteers raised in the city and the provinces had experienced some active service in Lombardy; and the balance was formed of further volunteer bodies (National Guards, students and ordinary citizens) from within Rome.

THE SIEGE OF ROME
30 April and 3-29 June 1849

Yards
0   250   500   750   1000

OUDINOT
10,000
(30 April)

Vatican Hill
Vatican Palace
St. Peter's
Porta Angelica
Porta Pertusa (closed) (A)
Porta Cavalleggieri (C)
Castle of S. Angelo
Porta Angelica (B)

to Palo and Civitavecchia

Wall of Urban VIII
River Tiber
Trastevere

GARIBALDI
9,000

Aurelian Wall
Quarter

Villa Valentini
Northern Bastion
Vascello (I)
Villa Pamfili
Villa Corsini
Porta San Pancrazio (D)
Casa Giacometti
Janiculum Mount (F)
S. Pietro in Montorio
Villa Spada
Aurelian Wall
Convent of S. Pancrazio
Central Bastion
Casa Barberini
Wall of Urban VIII

OUDINOT
25,000
(3 June)

Pamfili Chapel
Monte Verde
Maison des Volets Verts (L)

French attacks.
Garibaldi's sortie.
French batteries.

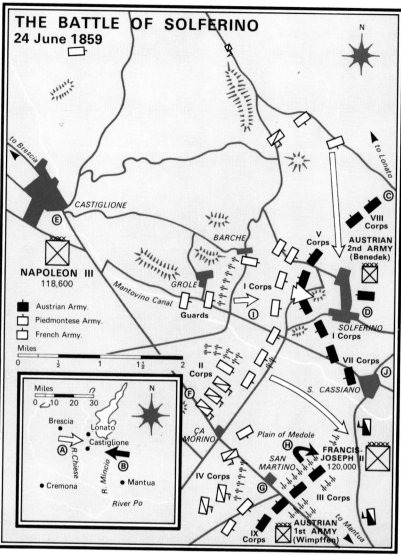

THE BATTLE OF SOLFERINO
24 June 1859

to Brescia
to Lonato

CASTIGLIONE

NAPOLEON III
118,600

Austrian Army.
Piedmontese Army.
French Army.

Miles
0   ½   1   1½   2

BARCHE
GROLE
Guards
Mantovino Canal

I Corps
II Corps
V Corps
VIII Corps
AUSTRIAN 2nd ARMY (Benedek)

SOLFERINO
I Corps
VII Corps
S. CASSIANO

ÇA MORINO
Plain of Medole
SAN MARTINO
IV Corps
III Corps
FRANCIS-JOSEPH II
120,000
to Mantua

IX Corps
AUSTRIAN 1st ARMY (Wimpffen)

Miles
0   10   20   30

Brescia
Lonato
Castiglione
Cremona
Mantua
R. Chiese
R. Mincio
River Po

The French attack would come against the Janiculum and Vatican hills, which were protected by relatively-recent fortifications (see map, The Siege of Rome). Oudinot approached on 30 April at the head of about 7,000 infantry and 12 field-guns, without any siege equipment. He anticipated little opposition, but was to be given a rude surprise. A first attempt to gain a footing within the Vatican failed at the walled-up Porta Pertusa (A), and probes against the Porta Angelica (B) and Porta Cavalleggieri (C) were also repulsed by the defenders by midday. Judging his moment with nicety, the watchful Garibaldi now sallied out from the Porta San Pancrazio (D) and, after some mixed fortunes, defeated a French Regiment amongst the villas and vineyards and recaptured the Villas Corsini and Pamfili (E). Garibaldi was wounded, but the French were repulsed, and retired towards Civitavecchia after losing 865 casualties. The

defenders were jubilant.

French operations now lapsed for several weeks, but Garibaldi was soon sent south to meet a Neapolitan incursion from the south, and to fight and win the Battles of Palestrina (9 May) and Velletri (19 May) amidst the Sabine and Alban Hills. These operations repulsed the Neapolitans, and by the end of May Garibaldi was back in Rome.

During his absence, Oudinot had been negotiating with the Triumvirs (or republican leadership), and an agreement was signed on 31 May, but this was mainly a blind on the part of the French to win time for the arrival of reinforcements and to move up close to the city. By 1 June, 20,000 French troops with six batteries of artillery and some siege guns were close at hand, with 10,000 more troops on the way from France. The same day, Oudinot denounced the armistice, and gave notice of attack for 4 June.

In fact the French assault came in on the 3rd, concentrated against the Janiculum (F). First they needed to clear the villas and outlying suburbs facing Porta San Pancrazio, and at dawn on 3 June the Villas Pamfili and Corsini were stormed successfully. At 5.30am, however, Garibaldi led out his legion to retake the hill-top Corsini position, with its four-storey villa. Despite long odds, the Legion made itself master of the position for a time, although only 3,000 republican troops were available for both the assault and the defence of the main city walls in this area. The Villa Corsini (G) changed hands several times, but the North Italian Bersaglieri (light infantry) was decimated and Garibaldi eventually fell back, calling upon his guns to demolish the Villa Corsini. Later the same day, the villa again passed, briefly, into his hands, but could not be retained. At nightfall, the republicans fell back, abandoning the Villa

Valentini (H) but retaining Vascello outside the walls (I). They had lost about 1,000 men, the French some 270.

Nevertheless, Oudinot had not brought off his intended surprise storming of the city, and he now had no option but to have recourse to a regular siege, which would last from 4-29 June. By the start of this phase, French strength was 25,000 men, equipped with a siege train and a corps of engineers commanded by General Vaillant. Whilst detachments watched the farther sides of the city, the French concentrated their efforts against the Janiculum. Some supplies and individuals could still reach Rome, but the siege lines crept remorselessly forward towards the bastions guarding the walls. The heroic defence of the Villa Vascello outside the main defences considerably delayed the full implementation of the French intention; indeed, its intrepid garrison would hold out until the 30th, even after the city walls had been taken behind their backs. However, possession of the Corsini hill enabled the French to open trenches against the Central Bastion (J) and the bastion of Casa Barberini (K), aided by a large battery on Monte Verde (L).

On the night of 21-21 June a French attempt at a storming was frustrated by the vigilance of the sentries in the Casa Giacometti (M), but the following day French columns stormed and captured both bastions. To the fury of Mazzini, Garibaldi refused to counterattack, but spent his time extemporizing a new line along the ancient Aurelian Wall (N), which was held for the next nine days. But the writing was now on the wall for the city as the French guns poured shot and shell against San Pietro in Montorio and the Trastevere Quarter beyond. Buildings collapsed, casualties mounted, and recriminations between Garibaldi and Mazzini almost resulted in the withdrawal of the Legion, but it returned to its post on the Janiculum, the rank and file wearing the famous red shirts for the first time, on the 28th, in time for the final crisis.

Oudinot selected the 30th for the final assault. Fighting was severe, but the French ultimately gained their objectives, and the Janiculum hill, overlooking much of Rome, was at last in their possession. On the 30th, Garibaldi met the Assembly of the Republic in solemn conclave. He stated that the possibility of continuing the defence was unrealistic, and advised evacuation of the Republican Army. Despite strenuous opposition from Mazzini, this course was approved.

The French were to make their formal entry into Rome on 3 July. The night before, Garibaldi led 4,000 men out of the city by the Lateran gate. Rome was in the hands of the French, and Garibaldi's famous defence of the short-lived Roman Republic was over.

The Battle of Solferino, 1859

Close by the Napoleonic battlefield of Castiglione, in late June 1859 the great Emperor's nephew, Napoleon III, commanding the 118,600 men and 320 guns of the Franco-Piedmontese Army, defeated the Emperor Francis Joseph of Austria at the head of 120,000 men and 451 guns, and by so doing conquered the greater part of Lombardy for Piedmont, and Savoy and Nice for France.

The circumstances of France's intervention in Italian affairs have already been described (see p. 158). The Austrian Army had been defeated at Magenta, and was in retreat, when the Chief of Staff, General Hess, learnt that the pursuing forces had become strung out. Hoping to drive the advance guard towards the River Chiese (A) and then to trap and wipe out the main body, he ordered the Austrian Army to recross the River Mincio (B). In fact, Napoleon III was already crossing the Chiese in strength in the belief that only a strong Austrian rear-guard remained west of the Mincio. Both sides blundered into an encounter battle of far greater gravity than either initially appreciated.

In the mid-morning of the 24th the Austrian crossing of the Mincio began, but progressed slowly. Eventually, the Austrian Second Army accompanied by the 8th Corps moved off to take up a position towards Lonato on the right (C), whilst in the centre three more corps (1st, 5th and 7th) moved past Solferino (D) towards Castiglione (E); away to the south, the Austrian First Army moved towards the Chiese, with General Jellacic's division on its extreme left. The French Emperor, meantime, had sent the four Piedmontese divisions of his left wing to advance from Lonato towards Peschiera. The four French corps were routed for Solferino, Cavriana, Medole and Castel Goffredo respectively, whilst headquarters and the Guards moved towards Castiglione slightly in rear of the remainders. In total, the Franco-Piedmontese Army comprised 146 French and 71 Piedmontese battalions, besides 72 French and 16 Piedmontese squadrons.

Finding a battle situation developing on ground not of their own choosing, the Austrian high command ordered the right wing and centre to prepare field defences in order to pin the French advance, whilst the Austrian left made an enveloping attack towards Monte Medolano (F) and the highway to Castiglione beyond. In all, they could deploy 151 battalions and 52 squadrons.

The action commenced with outpost skirmishes at about 6am. The French 4th Corps busily fortified Medole whilst Napoleon and Marshal Macmahon reconnoitred the field and adjusted their dispositions, requesting King Victor Emmanuel II to reinforce the centre. At 9am Piedmontese troops were repulsed from San Martino (G). At much the same time the Austrian General Wimpffen launched his 3rd Corps over the plain of Medole (H) in order to begin the envelopment, but his men were decimated by the 96 French rifled-cannon (which soon demonstrated their superiority over the 144 smoothbore Austrian guns on this sector) as they entered the open ground. First the 3rd, and then the 9th Corps and the Second Army cavalry failed to cross the plain towards Ça Morino.

These setbacks notwithstanding, the Emperor Francis Joseph still hoped to win the day. At 11.45am he ordered the battered First Army to swing northwards and the distant 8th Corps to move southwards in support of the troops holding Solferino. Napoleon III, aware that the crisis was about to take place in the centre, ordered his out-numbered 4th Corps to hold the Austrian left wing in check at all costs whilst the Guards and the 2nd Corps moved up (I) behind 1st Corps opposite Solferino, where it was facing the Austrian 1st, 5th and part of the 7th Corps.

An epic struggle soon raged along the battle line. The Austrian First Army failed to achieve the advance against the French right, while away to the north, Benedek, who was checking three Piedmontese attacks, proved incapable of moving southwards as ordered. The major French blow thus fell on Solferino, which fell to their onslaught. At 2pm, Wimpffen warned the Emperor Francis Joseph that he would have to give ground in view of French superiority around Solferino, to his right. From this point the battle was decided, but not until after 5pm was the Austrian First Army able to extricate itself, covered by staunch fighting by the 7th Corps on the Heights of Cavriana (J). Similarly, on the northern flank, Benedek's 8th Corps remained in action against the Piedmontese until almost 8pm, losing five guns.

The French and Piedmontese were too exhausted to launch an immediate pursuit. They had suffered 15,000 casualties and inflicted 14,000 on the Austrians, besides taking some 8,000 prisoners. Napoleon III was so sickened by the scenes of slaughter that he proposed negotiations and, by the Treaty of Prague, Austria ceded most of Lombardy to Piedmont and two Ligurian provinces to France. Thus, despite the French withdrawal from the war, the processes of Italian unification were considerably advanced.

The Expansion of the United States of America

The huge Louisiana purchase (1803), by which the USA practically doubled its land area to the west of the River Mississippi, proved but the first stage in a policy of national growth during the nineteenth century (see map inset, Territorial Expansion of the Continental United States). In 1819, Florida was purchased from Spain, and as the settlers' waggon-trains pressed ever farther westwards, so the appetites and ambitions of Washington grew. Tension with Great Britain over Canada led to several Fenian raids, but that frontier was in fact maintained.

In the west, however, the original agreement whereby Oregon was open to joint-American and British settlement was gradually eroded. After a period of hectic American colonization, the 'Oregon Claim' was accepted by Great Britain, and from 1846 London recognized that the 49th Parallel should form the international frontier.

In a similar way, the acquisition of southern Oregon brought the Americans to the northern frontiers of Mexican territory. The War of Texan Independence (1835-36)—with its siege of the Alamo and Battle of San Jacinto—had strained US-Mexican relations. The subsequent annexation of the Republic of Texas in 1845 (at the request of its inhabitants) brought the American settlers into contact with Mexico's eastern frontiers. A revolt by American colonists near San Francisco in June 1845 led to armed intervention by General Fremont. This became the excuse, but hardly the reason, for the United States' successful Mexican War of 1846-47, as America pressed ahead in pursuit of its concept of 'Manifest Destiny'. Contention over territorial control of lands north of the Rio Grande was the main cause.

1. General Antonio Lopez de Santa Anna (1794-1876), the flamboyant Mexican soldier and politician, who was defeated by the US armies of Generals Zachary Taylor and Winfield Scott in 1847.
2. General Zachary Taylor (1784-1853), commander of the US Army in Texas and victor at Buena Vista over the Mexican leader, Santa Anna. He subsequently became the short-lived 12th President of the United States.
3. General Winfield Scott (1786-1866), the senior commanding general of the US Army from 1841, who conquered Mexico and later produced, in the Anaconda Plan, the winning Northern strategy for the Civil War. His nickname—'Old Fuss and Feathers'—belied a brilliant mind.

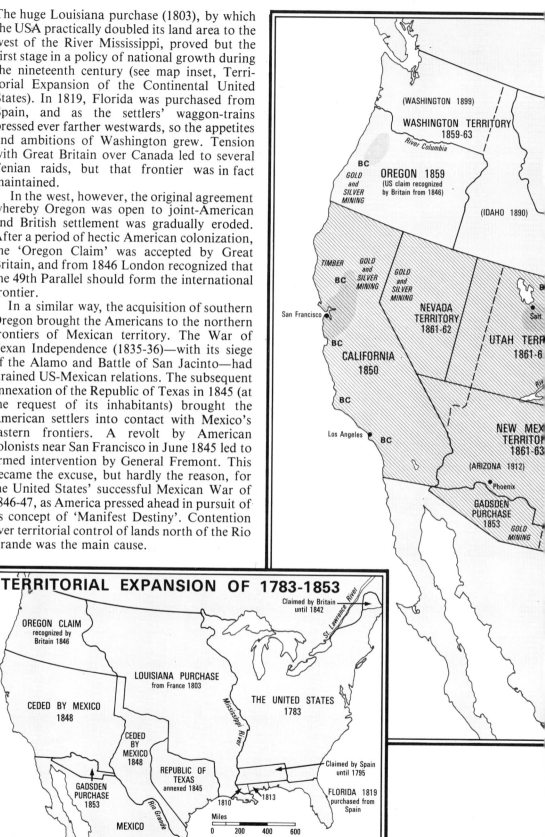

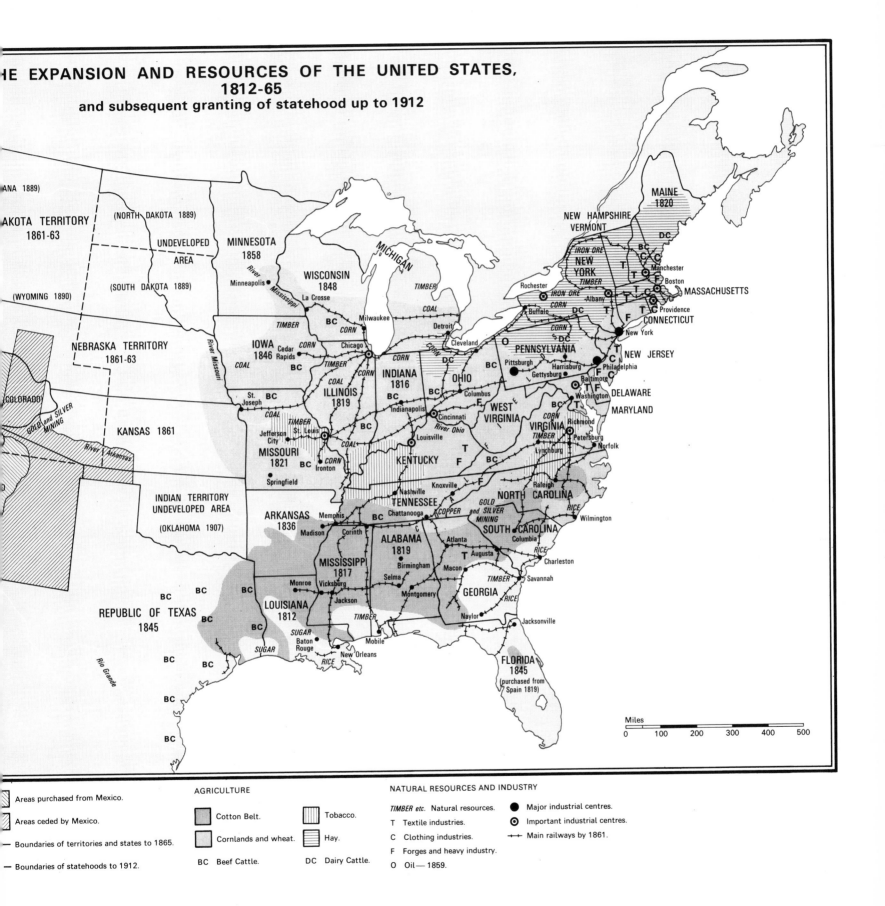

HE EXPANSION AND RESOURCES OF THE UNITED STATES, 1812-65
and subsequent granting of statehood up to 1912

(ANA 1889)

AKOTA TERRITORY 1861-63

(NORTH DAKOTA 1889)

UNDEVELOPED AREA

(SOUTH DAKOTA 1889)

(WYOMING 1890)

NEBRASKA TERRITORY 1861-63

(COLORADO)

GOLD and SILVER MINING

KANSAS 1861

River Arkansas

INDIAN TERRITORY UNDEVELOPED AREA

(OKLAHOMA 1907)

REPUBLIC OF TEXAS 1845

Rio Grande

MINNESOTA 1858

Minneapolis
La Crosse

River Mississippi

WISCONSIN 1848

TIMBER

Milwaukee

MICHIGAN

TIMBER

COAL

BC

IOWA 1846
Cedar Rapids
CORN
Chicago

COAL

BC

TIMBER

CORN

Detroit

MAINE 1820

NEW HAMPSHIRE
VERMONT

DC

IRON ORE

NEW YORK

TIMBER

Rochester

IRON ORE
Albany

Buffalo

CORN

DC

New York

Manchester
Boston
MASSACHUSETTS
Providence
CONNECTICUT

Cleveland

O

PENNSYLVANIA

NEW JERSEY

St. Joseph

BC

COAL

ILLINOIS 1819

TIMBER

CORN

INDIANA 1816

CORN

BC

Indianapolis

OHIO

Columbus

BC

CORN

DC

BC

Pittsburgh

Harrisburg
Gettysburg

Philadelphia

DC

Baltimore

Washington

DELAWARE

MARYLAND

Jefferson City
St. Louis

COAL

Cincinnati

River Ohio

Louisville

WEST VIRGINIA

F

VIRGINIA

Richmond

CORN

TIMBER

Petersburg

Norfolk

MISSOURI 1821

BC

CORN

Ironton

KENTUCKY

T

F

Lynchburg

BC

Springfield

Knoxville

F

Nashville

TENNESSEE

Chattanooga

BC

GOLD and SILVER MINING

NORTH CAROLINA

Raleigh

RICE

Wilmington

ARKANSAS 1836
Memphis

Madison

Corinth

COPPER

SOUTH CAROLINA

Columbia

RICE

Atlanta

BC

BC

BC

ALABAMA 1819

Birmingham

Selma

Augusta

T

Charleston

MISSISSIPPI 1817

Monroe
Vicksburg

Jackson

Montgomery

Macon

GEORGIA

TIMBER

RICE

Savannah

BC

BC

LOUISIANA 1812

TIMBER

Naylor

Jacksonville

BC

SUGAR

Baton Rouge

Mobile

BC

SUGAR

New Orleans

RICE

FLORIDA 1845
(purchased from Spain 1819)

BC

Miles
0   100   200   300   400   500

Areas purchased from Mexico.

Areas ceded by Mexico.

— Boundaries of territories and states to 1865.

— Boundaries of statehoods to 1912.

AGRICULTURE

Cotton Belt.

Cornlands and wheat.

BC Beef Cattle.

Tobacco.

Hay.

DC Dairy Cattle.

NATURAL RESOURCES AND INDUSTRY

TIMBER etc. Natural resources.

T Textile industries.

C Clothing industries.

F Forges and heavy industry.

O Oil — 1859.

● Major industrial centres.

◉ Important industrial centres.

+++ Main railways by 1861.

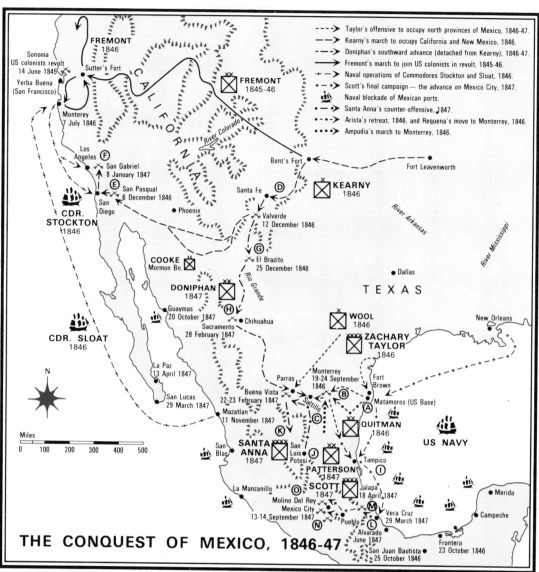

THE CONQUEST OF MEXICO, 1846-47

Map legend:
- - - ▷ Taylor's offensive to occupy north provinces of Mexico, 1846-47.
- - ▷ Kearny's march to occupy California and New Mexico, 1846.
- - - ▷ Doniphan's southward advance (detached from Kearny). 1846-47.
——▷ Fremont's march to join US colonists in revolt, 1845-46.
+ — ▷ Naval operations of Commodores Stockton and Sloat, 1846.
+ - ▷ Scott's final campaign — the advance on Mexico City. 1847.
⛴ Naval blockade of Mexican ports.
+ - - ▷ Santa Anna's counter-offensive, 1847.
• - • ▷ Arista's retreat, 1846, and Requena's move to Monterrey, 1846.
• • • ▷ Ampudia's march to Monterrey, 1846.

The American-Mexican War of 1846-47

One month after General Zachary Taylor had established a camp near Matamoros (A), Mexican troops defeated a small force of US Dragoons on 25 April 1846, and war began. After frontier skirmishing, Taylor with 6,000 men occupied Matamoros, and then moved on Monterrey, where a sharp action (B) lasting five days ended in the capitulation on terms of General Pedro de Ampudia and 10,000 men. Washington denounced the agreed armistice, and Taylor again advanced, absorbing Brigadier-General Wool's column at Saltillo (C), adding 3,000 men to his strength.

Meanwhile, far away to the north, Colonel Kearny marched from Fort Leavenworth to Santa Fe (D) at the head of 1,700 men. Pressing on, Kearny heard that California had been pacified and detached Colonel Doniphan with

850 men to ride south to Chihuahua, while he rode on for southern California. Kearny discovered that resistance was not yet over on 6 December (E) at the sharp action of San Pasqual, where 500 Mexican cavalrymen blocked his way to San Diego. Overcoming this obstacle, Kearny met Commander Stockton USN, who had landed a force from the Pacific, in early December. After crossing the Rockies, the two commanders fought and won an action at San Gabriel (F) on 9 January 1847 not far from Los Angeles, and this broke the Mexican hold on California.

In the meantime, Colonel Doniphan, who had been detached by Kearny in December, was advancing on Chihuahua. After a skirmish at El Brazito on 25 December (G), he fought a more serious action against a far larger Mexican force, which he routed (killing 600 for

the loss of seven) at Sacramento on 28 February 1847 (H). Doniphan occupied Chihuahua, then marched on to rendezvous with Taylor at Saltillo on 21 May, thus completing a truly epic march of over 2,000 miles, much of it across desert terrain.

Over the preceding winter, acrimony had grown between General Zachary Taylor and Washington over the future conduct of the war. Ultimately, Taylor refused to march over the desert to San Luis Potosi in order to reach Mexico City, and this forced President Polk to replace him in command by General Winfield Scott, despite serious political disagreements, and to accept Scott's masterplan for an advance based upon Vera Cruz. As a preliminary step, Scott sailed from New Orleans to Tampico (I), where despite administrative confusion he absorbed 5,000 of Taylor's veterans to build up his force to 10,000 men. Taylor was ordered to stay on the defensive in Central Mexico, whilst Scott prepared to advance on Mexico City from the Caribbean.

However, the Mexican President, General Santa Anna, had learnt of the overall American plan, and he determined to crush Taylor before Scott could land. Accordingly, Santa Anna advanced north from San Luis Potosi (J) with 20,000 Mexican troops, and arrived near Saltillo in mid-February after losing 4,000 men in the desert. There followed the hard-fought Battle of Buena Vista (22-23 February), which Santa Anna might well have won but for the fine performance of the US artillery, which broke up the main attack whilst Colonel Jefferson Davis defeated an attempt to turn Taylor's left flank. Santa Anna then pulled back (K) having lost 1,500 men to the American 750 casualties. This event ended the war in North Mexico, apart from troublesome guerrilla raiding by die-hard Mexicans.

All attention now switched to Central Mexico, where Scott's offensive was due to begin. On 9 March his force arrived off Vera Cruz (L), landed without opposition, and was soon besieging the city, which surrendered on 27 March when the garrison of 5,000 men laid down their arms. Marching inland, Scott encountered Santa Anna and 12,000 Mexicans at the gorge of Cerro Gordo near Jalapa (M), but outflanked them and drove them back. No less than 4,000 Mexicans were killed or taken, together with some 40 guns. On 15 May, Scott occupied Puebla, and there he paused. His force was now barely 6,000 strong, as 4,000 men had returned to the USA on the completion of their agreed period of service, but by August he had been reinforced to 11,000 fit men besides a further 3,000 sick. Leaving his sick at Puebla, Scott set out for Mexico City on

164

7 August. He was faced by Santa Anna at the head of 30,000 Mexicans, but by bold manoeuvring he won two battles (N) on 20 August (inflicting almost 10,000 losses on the Mexicans) to the south-west of the city. A two weeks' armistice broke down and, after another two actions at Molino del Rey on 8 September (O) and at neighbouring Chapultepec five days later, Mexico City was occupied on 14 September following Santa Anna's withdrawal during the preceding night. He resigned the Presidency soon after.

A Mexican force descended upon and besieged Puebla from 14 September to 12 October, but a relief column from Santa Cruz relieved the small American garrison left to guard Scott's sick. Active operations were now almost at an end, and protracted negotiations began. Scott was recalled home on trumped-up charges, but he received a hero's welcome from both Congress and people, and President Polk had to drop his vendetta. Finally, on 2 February 1848, by the Treaty of Guadalupe, the south-western boundaries of Texas were confirmed, and Mexico ceded territory (see map, Territorial Expansion of the Continental United States) comprising the later states of California, Nevada, Utah, parts of Arizona and New Mexico, together with areas of Colorado and Wyoming, receiving $15,000,000 in return. At length, American forces withdrew from Mexico City on 12 June 1848, and the treaty was ratified 33 days later. On 2 August, US forces evacuated Vera Cruz.

The war of 1846-47 demonstrated the voracious appetite for expansion on the part of the USA. And many soldiers who would rise to great prominence in the US Civil War made their first reputations during its course. Elsewhere, there were bitter struggles against the Red Indians, as the trail-blazers, followed by the Army, settled the broad expanses of the West.

In the four years between 1844 and 1849 the United States had almost doubled in size, adding an area larger than her original extent in 1776. In 1853 the Gadsden Purchase adjusted the southern border of New Mexico. In 1867, the purchase of Alaska from Russia completed the main stages of United States expansion in North America. But the great task of development could only begin once the agony of the Civil War (1861-65) had been fought and overcome.

DEVELOPMENTS IN MINOR TACTICS

The Napoleonic period dominated concepts of tactics until at least the Crimean War, if not even later. The British retained their two-deep linear formations from the days of Wellington to those of Kitchener with scant adjustment, although by 1880 the concept of fighting in open-order was at last appearing. The French clung on to the élan of the attack in column until the adoption of the Chassepot induced them to start experimenting with adaptations. The Austrians retained the Archduke Charles's two-company 'divisional groups', deployed in chess-board formation with a skirmisher screen to the fore, right through the age of Field Marshal Radetsky who remained in command until 1858, by which date he was ninety-two years old. The Prussians, too, were very conservative, despite the heroic efforts of Helmuth von Moltke to reform their thinking in tactical matters. Their sub-unit tactical system was completely patterned on Napoleonic practice, and von Wrangel determinedly resisted reform of any sort until his shortcomings became clear in the Danish War of 1864. Even then, it took the fearful casualties meted out by the French Chassepot in the first battles of 1870 to inspire a change of tactical outlook in the Prussian Army in mid-campaign, reducing the hitherto sacred belief in shock-action in favour of a more balanced concept. From 1815 until 1859, most continental armies had little direct experience of large-scale warfare. Even the Crimean struggle had degenerated into a Vaubanesque siege war. For the rest, European regular soldiers (apart from the occasional French colonial adventures in North Africa) had little experience of anything more than anti-revolutionary street fighting in their respective capitals.

The American Civil War—despite the attempts of early commanders to employ Napoleonic tactics—soon proved that such evolutions were largely beyond the wholly

The assault of Longstreet's corps on the Union position at Cemetery Ridge, during the second day's fighting at Gettysburg, provided a terrible demonstration of the firepower of modern weapons: only 150 men reached the crest of the Ridge, from which they were rapidly ejected.

inexperienced citizen-soldiery, which Moltke would describe as "two armed mobs". This factor, together with the increased killing-power of artillery and (eventually) the range and accuracy of the Northern rifles, induced the Conferacy to fight in open-order swarms rather than formal lines. The Union troops retained linear battalion and company formations longer than their opponents (and took heavier casualties in consequence), but from 1864 began to employ more flexible formations. As always in warfare, tactics lagged behind weaponry. Rifle-fire caused so much havoc amongst gun-crews that artillery soon had to be withdrawn from close-range positions. Shells could wreak fearful havoc; at Shiloh (see p. 171), for example, a joint-total of almost 24,000 casualties were sustained out of a combat-strength of 103,000 over two days of gory fighting. Prussian losses at battles such as St. Privat were equally horrific; at that battle the Prussian Guard Corps reputedly lost 8,000 casualties in the space of just twenty minutes.

Gradually, such blood-lettings taught the necessary lessons. It was realized that the breech-loading rifle would permit soldiers to fire lying down from behind cover, and that when natural cover was not available, trenches could be dug to provide it. Colourful uniforms became increasingly unpopular, and drab and muted khakis, field-greys and blues became the order of dress, at least on campaign (although even in 1914 the French Army still possessed no field uniform). Fighting in skirmisher-order gradually became the norm rather than the exception.

Cavalry tactics also changed over the years. The advent of the rifle and, even more, the prototypes of machine-guns, spelt the end of the great cavalry charge except in isolated cases. Instead, cavalry became increasingly regarded as mounted infantry, using their mounts to achieve mobility and range, but dismounting to fight with carbines, Winchesters or Spencers. However, the splendid cavalry charge was still employed in the Crimea, with success at Balaclava in the charge of the British Heavy Brigade, and with disaster at the same battle in the better-known charge of the Light Brigade. At Rézonville in 1870, the French and Prussian cavalry had a traditional encounter, but at Sedan the fate of the Chasseurs d'Afrique was the true portent. The US Civil War saw the increasing use of cavalry for reconnaissance and interdictive raiding—which would be the main trend for the future—rather than splendid charges in battle.

THE RAILWAY AGE

The coming of the railway to Europe, America and other areas was to exert an important influence on military strategy. Paradoxically, this influence was both forward-looking and, in certain respects, retrospective at one and the same time. Nevertheless, by the turn of the twentieth century Germany possessed a comprehensive railway and mobilization system (see map, 'Railway Development in the German Empire, c. 1890', p. 199) in which the emphasis on the assumed French threat— which underly von Moltke's, and his successor, Schlieffen's strategic planning— is quite clear.

In Prussia, made up of a scatter of smaller states, the railway became a major means of welding a nation out of many parts. Although later into the field of railway development than Britain, Belgium and France, Prussia's railway system was, from the first, designed for military as well as economic purposes, and the Great General Staff exerted much influence on the decisions on where to lay new track. Prussian commanders were amongst the first to appreciate that the railways could be used to speed both mobilization and concentration of force at the desired sector of the frontiers. The great von Moltke (see p. 198) inherited a developing railway system, and set out to exploit the advantages it could confer on those able to appreciate them in his wars of 1866 (see p. 190) and 1870 (see p. 192). He set up the

special railway Abteilung or office of the General Staff, and great pains were taken to work out timetables and mobilization procedures in the greatest detail. Thus in 1866, he used the three railways approaching Bohemia to position his three armies along a 400-mile front, accepting a great risk of being defeated in detail by the centrally-placed Austrian main army (see map, The Campaign in Bohemia, p. 191), as his own commanders apprehensively appreciated, but relying upon—and not without reason—the advantages of speed and surprise conferred by his daring dispositions allied to the railways, to mesmerize Prince Benedek into inaction whilst the Prussian armies converged upon him to encompass his doom. Thus was truly born the age of military strategy closely linked to the railway systems.

However, railways were not without their problems. Even the thorough professionalism of the Great General Staff could not prevent major supply problems developing after the initial blow had been launched in both 1866 and 1870. There were bottlenecks and major confusions, and the deeper armies advanced into hostile territory the more pronounced these became. Reliance on captured railway lines for logistical support could lead to serious snags. Differences of railway gauges requiring transfer of stores and equipment to new waggons was obviously one; another was the tempting interdictive targets that

railway bridges and goods yards presented to partisan forces or special raiding parties. (These problems were experienced by the Prussians in France during the siege of Paris (see p. 195) when their armies almost starved through dislocation of their supply arrangements in a generally static situation.) Moreoever, Moltke's successors paid too little attention to his repeated insistence that a readiness to extemporize and achieve fluidity were vital ingredients of victory. Lesser men became totally hide-bound in their approaches to military problems, and placed far too much reliance on railway timetabling and inflexible planning rather than trying to find novel and imaginative solutions to their military problems. (See map on pp. 198-199 for the disposition of German forces in relation to the railroads c. 1890.)

During the American Civil War, the impact of railways was less than has sometimes been represented. The North had twice the railroad mileage of the South, but with certain exceptions made slight use of its superiority. Generals tended to cling to lines of advance following a railway line, rather than seeking bold areas of manoeuvre; thus their operations often became predictable. This was tacitly encouraged by Lincoln and the Washington government, who enjoyed the ability to keep a close finger on the pulse of events at the front conferred by the railway telegraph system. However, the concept of inter

dictive raiding became something of a Southern art; General Forrest being only one of a number of dashing cavalry commanders who made a speciality of attacking railway targets far to the rear of the main Union armies. The North eventually learnt the need to mount similar operations, and during his famed 'march to the sea' Sherman's army devised a machine capable of twisting torn-up rails into fantastic shapes and precluding their re-laying and re-use. Certain commanders made imaginative use of railways; the Confederate build-up before and during First Bull Run (see p. 172) was achieved by rail, and Stonewall Jackson also used travel by train to speed part of his operations in the Valley (see pp. 174-178). Such rail centres as Chattanooga and Corinth came to play an important role in the strategies of both armies.

Back in Europe, the Italians made great use of railways to achieve Italian unification under the guidance of Cavour (see p. 158), and the Mont Cenis links into France and the Brenner Pass complex towards Austria assumed great importance as strategic railheads. Meanwhile, the Austrian government also employed railway lines to try to weld together the diverse parts of the Habsburg Empire. Links from Vienna to Prague and Budapest, and to Trieste, served military as well as political and economic needs. The Brenner Pass gave access into North Italy;

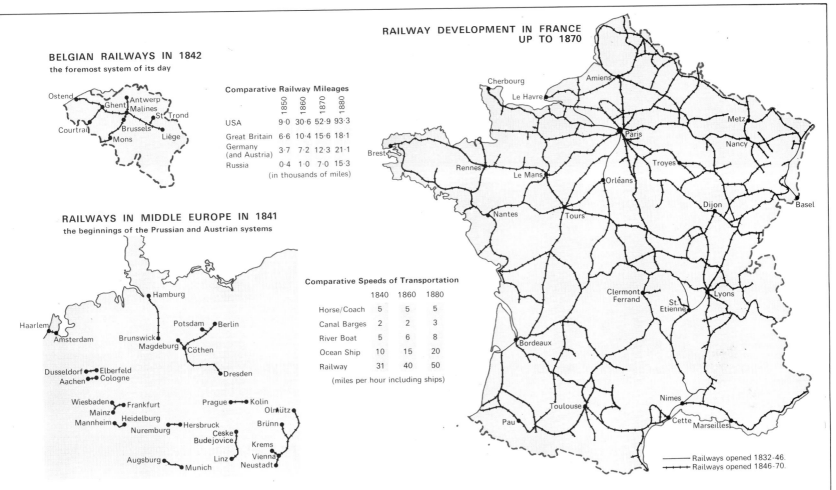

RAILWAY DEVELOPMENT IN FRANCE UP TO 1870

BELGIAN RAILWAYS IN 1842
the foremost system of its day

Ostend · Ghent · Antwerp · Malines · St Trond · Courtrai · Brussels · Liège · Mons

Comparative Railway Mileages

	1850	1860	1870	1880
USA	9·0	30·6	52·9	93·3
Great Britain	6·6	10·4	15·6	18·1
Germany (and Austria)	3·7	7·2	12·3	21·1
Russia	0·4	1·0	7·0	15·3

(in thousands of miles)

RAILWAYS IN MIDDLE EUROPE IN 1841
the beginnings of the Prussian and Austrian systems

Haarlem · Amsterdam · Hamburg · Brunswick · Potsdam · Berlin · Magdeburg · Cöthen · Dusseldorf · Elberfeld · Aachen · Cologne · Dresden · Wiesbaden · Frankfurt · Prague · Kolin · Mainz · Heidelburg · Olmütz · Mannheim · Hersbruck · Brünn · Nuremburg · Ceske Budejovice · Krems · Augsburg · Linz · Vienna · Munich · Neustadt

Comparative Speeds of Transportation

	1840	1860	1880
Horse/Coach	5	5	5
Canal Barges	2	2	3
River Boat	5	6	8
Ocean Ship	10	15	20
Railway	31	40	50

(miles per hour including ships)

Railways opened 1832-46.
Railways opened 1846-70.

railways towards South Hungary and Croatia led towards the Balkans; links with Galicia were well-suited to meet the possibility of a Russian attack. The French, on the other hand, despite their great mileage of railroads (see map, Railway Development in France up to 1870), tended to be obsessed with the need to move troops to the frontiers (particularly from Paris to Strasbourg), and their lateral links, running north-south, were weak and would remain so until 1914. The fearful confusions encountered in mounting the campaigns in North Italy in the 1850s were only equalled by those experienced in the grim days of 1870, and whole sections of the French General Staff became as purblind and inflexible as some of their Prussian opponents.

The Tsars had learnt the need for strategic railway lines as early as the Crimean War, when the problems of resupplying and reinforcing the forces around and within Sebastopol (see p. 145) led to heavy losses in both men and animals, particularly during the bitter winter of 1854-55. Gradually, a complex system was built-up (arguably the best on the Continent by 1914) which placed great reliance on links with Galicia—the route towards long-coveted Constantinople. The link with the Far East (the famous Trans-Siberian Railway) was only in the process of development towards the end of the period covered in this Atlas, and would not be

completed until the early twentieth century. The Russo-Japanese War would then demonstrate the patent shortcomings of a system dependent on a single track (with frequent sidings) for strategic support of an army operating thousands of miles away from White Russia.

Lastly, Great Britain was at one and the same time the most advanced and the most retarded military power of the period where the use of railways was concerned. The skill that underlay such feats as the construction of the railway in the Crimea (see p. 146) to supply the camps before Sebastopol was the earliest example of a custom-built railway line wholly military in intention. Similarly, the shipment of a complete railway system (engines, rolling-stock and track) to Annesley Bay in the Red Sea by Napier's expedition invading Abyssinia (see p. 154) was an astounding achievement. The RE Railway Companies that were eventually raised and based about Chatham and Woolwich were the most skilled railway troops in the world. At the same time, however, Britain had little idea about truly strategic railway design. Her insular state precluded the likelihood of foreign invasion, and it was by chance rather than design that many lines ran towards the Channel and East Coast ports (Dover, Folkestone and Harwich), natural lines of advance for forces preparing to move to the Continent. However, the railway system developed in India—

particularly after the hard lessons of the Mutiny (see p. 156)—was better designed to cope both with internal unrest, and also with the defence of the vulnerable North-West Frontier, where Afghanistan soon became a major source of Anglo-Russian rivalry, both open and overt conflict.

Thus the railway age, from slender beginnings in the 1830s, soon came to have an important effect on military strategy, and figured large in the calculations of General Staffs. Prussia (and later the German Empire), with its central position within Europe, was provided with the greatest opportunities for transferring troops from one theatre to another by virtue of its geographical position, and at the same time was most exposed to the perils (real and imagined) of encirclement by having to fight on two fronts at once.

The Electric Telegraph
Closely linked to the matter of railways was the development of another instrument destined to revolutionize the conduct of war, the electric telegraph. Without it, the coordination of large-scale military movements would have been impossible, given the increased mobility brought about by railways. By 1850 many major countries were linked by submarine cables, although remoter areas of Europe were rarely so provided. At the tactical level, messages were still carried by mounted messengers; although late 1862 saw the first laying of a

telegraph line in the field—between Fredericksburg and the approaches to Chancellorsville, in the United States. Similarly, the British expedition to Abyssinia in 1868 also included a telegraphic section which laid line deep into Abyssinia stretching some 200 miles inland. However, the most commonly available telegraph system was that associated with existing railways for signalling purposes. Abraham Lincoln made the fullest use of this facility to keep in close contact with his field commanders, expecting daily reports and issuing "helpful advice" each night, which many commanders regarded with downright suspicion if not distaste. Never before had distant governments been able to exert control over their generals so promptly: an important and not uncritical stage in civil-military relations had been reached. As early as the Crimean War, the British and French governments had been able to keep in touch with Lord Raglan and St. Arnaud in the Crimea by cable. In 1870, General Bazaine tried (vainly as it proved) to direct the Battle of Spicheren from the telegraph office of St. Avold. Later that year, when Paris was blockaded and all lines cut, recourse had to be made by the besieged to the use of balloons and pigeon-post to pass messages over the besieger's lines; not until the advent of wireless sets would armies overcome reliance on line communications.

The American Civil War, 1861-65

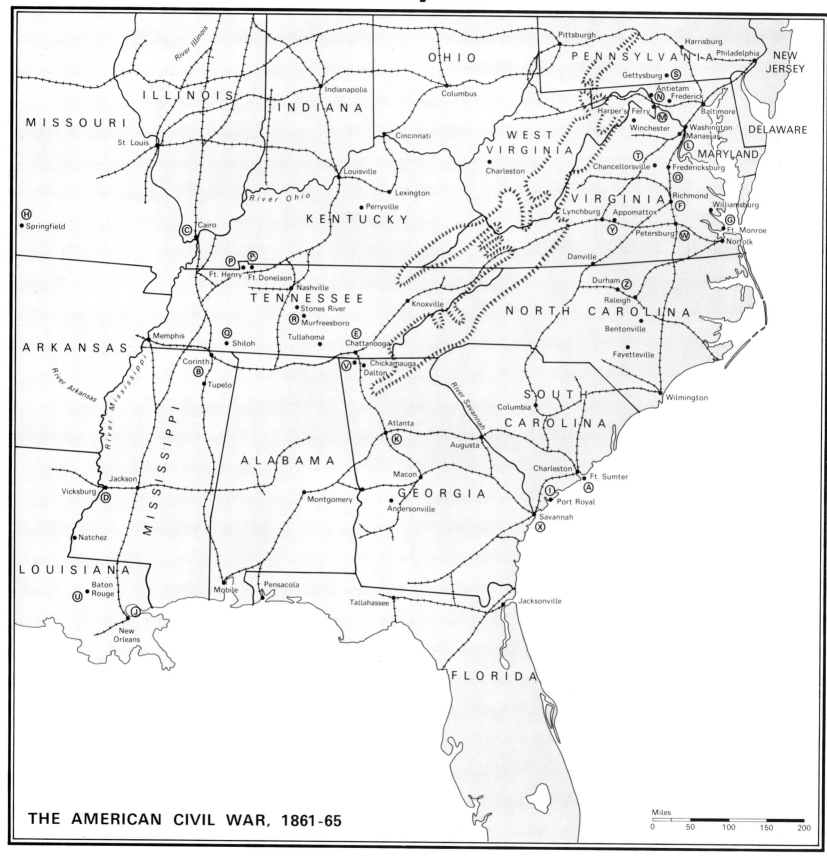

The agony of civil strife was the outcome of a long period of rising tensions born out of certain fields of clashing interests that divided the Northern from the Southern states. The basic issues were threefold. Politically, the argument concerned the ultimate site of power: was it with the central Federal government at Washington, or did it lie (at least in many respects) in the various state capitals? Economically, the contention centred around the desirability, or otherwise, of strict nationally-controlled protectionist policies (as demanded by the industrializing Northern seaboard states); the South, whose economy centred around 'King Cotton' and tobacco, required free trade in order to be in a position to gain the best possible prices for these staple crops, in which it enjoyed almost a world-monopoly. Socially, the most explosive issue was that of slavery, its retention or abolition. Regarded as an economic necessity by the Southern states (one third of whose populations were black), the institution was reviled in the emancipated North, which was far less dependant on cheap agricultural labour; although there were all of one million slaves in Northern states when the struggle began, and most were still there at its close. To some degree, two types of white society were moving into conflict: the descendants of the business-motivated Calvinistic settlers of the original New England colonies, reinforced by the great immigrant flow from Germany, Southern Europe and Ireland, coming into collision with the more aristocratic and hierarchical society to be found in the earlier Elizabethan and Stuart colonies. The third social conflict—between Red Indian and white man—was also raging in the new West, but did not directly impinge on the Civil War. These issues, particularly that of slavery, were exacerbated by undue fanaticism on both sides through the medium of the press, and a degree of incomprehension amongst short-sighted politicians and pious reformers, whose radicalism precipitated the struggle.

In some ways, 'the spoils of Mexico' brought matters to a head over the slavery issue, because it had to be decided whether the new areas gained for the United States (see map, The Expansion and Resources of the United States, 1812-65 and inset, Territorial Expansion of 1783-1853, pp. 162-163) should be based upon a slave economy or a free economy. The crisis centred around California. By the Missouri Compromise of 1850, Henry Clay and Daniel Webster forced agreement through Congress that California should be a free state, that other new territories should decide the issue for themselves, and that the Federal government would assume responsibility for hunting down and returning fugitive slaves. This preserved the peace for a time, although from 1852 *Uncle Tom's Cabin* by Harriet Beecher Stowe brought new fervour to the cause of the abolitionists. In 1857 the Missouri Compromise was surprisingly declared unconstitutional, and the rights of slaves of any sort denounced. The crisis steadily mounted, and in 1859 the fanatic John Brown almost brought matters to a head by seizing the government arsenal at Harper's Ferry and attempting to raise the slaves in revolt in the Shenandoah Valley. His movement was suppressed, but both sides were now drawing their battle lines. The election of the Republican President, Abraham Lincoln, in November 1860 implied the triumph of the abolitionists and the proponents of strong, centralized government. Six weeks later, South Carolina left the Union. There was now little hope of averting conflict. Although the US Constitution declared the right of secession to any state so-minded, Washington would not accept the situation without a struggle, and felt confident of its ability to quell the secessionists. But the movement gained momentum and, early in 1861, Mississippi, Florida, Alabama, Georgia, Louisiana and Texas followed South Carolina's lead, their leaders establishing a Confederate capital at Montgomery (Alabama), and electing Jefferson Davis as President. The time for definite Federal action had come, but lack of military resources caused a delay. On 15 April, Lincoln called for 75,000 volunteers for ninety days' service; two days later, the great state of Virginia (less some of its western counties) declared for the Confederacy, followed by Arkansas, Tennessee and North Carolina. The battle lines were now drawn (see map, The Civil War, 1861-65). The Confederates decided (unwisely) to move their capital to Richmond. Already, between 12 and 14 April, the Confederates had bombarded and taken Fort Sumter (A).

The Balance of Resources and Rival Strategies
At the outset, the North, albeit unprepared, enjoyed many advantages. All of twenty-three loyal states stood by the Union, as against the eleven that wished to secede. The North had a total population of twenty-two millions (including one million slaves) to the South's 9,105,000, fully 3,654,000 of whom were slaves. Eventually, almost three million men fought for the North, as against one and a quarter million who bore arms for the South. The advantage of the 'big battalions' (once mobilized) thus lay decidedly with the Union. Similarly, in industrial and financial power the North held huge advantages: producing ninety-six per cent of the nation's railroad equipment and ninety-seven per cent of its firearms. In military terms, the tiny US Army of 17,000 men was mostly deployed in western garrisons, but over a third of the officers joined the Confederacy. The US Navy comprised ninety vessels (only forty-two being in commission) but they were widely dispersed, and one sixth of the 1,400 commissioned officers resigned from the Federal Navy. This paucity of military resources accounts for the incompetent steps taken initially by both sides. Indeed, the South mobilized 100,000 volunteers before Lincoln's 75,000 had materialized.

The reasons why the war was to last for four years were various. First, neither side would be truly mobilized for over a year. Secondly, the South enjoyed a measure of European support until late 1862, and could draw supplies from across the Atlantic, as the North was too weak in naval terms to impose an effective blockade. Thirdly, the cold and mud of winter restricted most campaigning by the largely amateur armies to the spring, summer and early autumn months—much as had been the case in Europe during the seventeenth century. Furthermore, the strategic conditions did not favour a rapid end to the struggle, owing to certain decisions taken in Washington.

From the fact that the Confederacy had declared its independence arose the first strategic point: namely that it would basically fight on the defensive, to retain what it had taken. Defensive operations in defence of hearth and home were easier to mount for the Confederate armies, and more economical in terms of manpower. For the South, everything depended on maintaining the integrity of its member-states until the great European powers could be persuaded to recognize the Confederacy as an independent country. On the other hand, apart from occasional raids into northern territory, the Southern forces had little scope for initiative, and many of the Confederate state governors proved recalcitrantly independently-minded. But providing that the three key posts of Corinth (B), Cairo (C) and Vicksburg (D), and the link through Chattanooga (E), could be held inviolate, the political and military cohesion of the South would be maintainable.

The North, on the other hand, was committed to an offensive strategy: the physical conquest of the South to preserve the intregrity of the Union. The strategy of the aged General Winfield Scott, as it emerged in April—May 1861 in the Anaconda Plan, was to prove the correct one but was deemed unacceptable at the outset of the war. He called for an effective naval blockade to sever the Confederacy from

The American Civil War, 1861-65

European aid, the immediate raising of an army of 300,000 men, and a major effort down the Mississippi to the gulf of Mexico, to isolate and then strangle the Confederacy. But this seemed too time-consuming a concept to receive support in Washington in 1861. In the hope of snatching a quick victory, Lincoln would opt instead for a series of drives in the east hoping to capture Richmond (F). Only from late 1862 would a slow realization dawn that Scott's plan was indeed the only way to achieve the Union's objectives (see map, The North's Winning Strategy, right). The emancipation of the Southern slaves in 1862 would be Lincoln's master-stroke.

The Campaigns of 1861

The first nine months of the war held a series of setbacks for the Union armies. The campaign from Ohio into the loyal areas of what would become the state of West Virginia was successful, but elsewhere the Union faced failure. General Butler's sortie from Fort Monroe (G) in early June was repulsed near Yorktown, and McDowell's advance on Richmond was routed at Manassas (L) or First Bull Run (see p. 172 and maps, The Campaign of First Bull Run and The Battle of First Bull Run). This success bred over-confidence amongst the Confederates and, conversely, convinced Washington that the war would be harder and longer than originally thought. This impression was reinforced by a further defeat in the west near Springfield (H) in August, followed by a reverse at Ball's Bluff in Virginia in October and a setback at Belmont the next month. The establishment of a small bridgehead at Port Royal (I) in South Carolina on 7 November was small solace, and a Congressional Committee was appointed by Lincoln to examine the conduct of the war. However, he had much faith in his new commander-in-chief on the Potomac, General McClellan, who was training a force of 100,000 volunteers in the camps around Washington.

The Campaigns of 1862

The mighty mountain range of the Appalachians effectively divided North America into two theatres of war: east and west. Obsessed with the occupation of key towns and communication centres, the Union armies sought to seize Richmond (F) in the east and New Orleans (J), Vicksburg (D) (both on the Mississippi) in the west, and the key railway centres of Chattanooga (E) and Atlanta (K) on the central front, but these attempts met with little success.

In the east, operations centred around the double-thrust against Richmond from the

THE NORTH'S WINNING STRATEGY

1 1860-65: Advances on Richmond.
2 1860-65: Federal naval blockade (spasmodic).
3 1862: Capture of lower Mississippi.
4 1863: Capture of middle Mississippi.
5 1864-65: The 'March to the Sea' and the swing north.

Peninsula and Washington respectively. McClellan's and McDowell's campaign was brought to nought by the strategic skill of Robert E. Lee, who brilliantly used Jackson in the Shenandoah Valley to confuse and thwart the Union intention (see pp. 174-178 and maps, The Confederacy's Peninsula and Shenandoah Valley Strategy, 1862, p. 174), and then proceeded to defeat McClellan in the Peninsula itself (see maps, Jackson's Valley Campaign, pp. 177-178), forcing him to evacuate by sea in late July. Then, when General Pope led up a further Union army towards Richmond, Lee and Jackson met the threat at the Second Battle of Bull Run or Manassas (L) between 28 and 30 August, inflicting another heavy defeat. Lee decided to take the offensive, hoping that a great success might earn European recognition for the Confederacy. His Army of North Virginia crossed the Potomac on 5 September, and whilst Jackson took a force to capture Harper's Ferry (M), Lee advanced westwards

towards Sharpsburg, where Jackson joined him on 16 September. The next day the gory Battle of Antietam (N) was fought; Lee holding his ground for the loss of 9,000 men, and inflicting 12,000 casualties on McClellan. But Lee decided to fall back into Virginia. On 22 September, Lincoln issued his famous Emancipation Proclamation, declaring that from the next New Year all slaves in the Confederacy would become free men. This act of statesmanship finally dissuaded the European powers from recognizing the Confederacy; the struggle had now developed a humanitarian, ideological purpose. Despairing of McClellan, on 5 November Lincoln replaced him with General Burnside, who launched a desperate attack against strong Confederate positions overlooking Fredericksburg (O) on 13 December, but achieved nothing except the loss of 10,000 men. A few weeks later, Lincoln replaced Burnside in his turn by General Hooker.

Meanwhile, a complementary struggle had been raging in the west. In February 1862, General Grant attacked the over-extended Confederates at Fort Henry (6 February) (P) and Fort Donelson (16 February) (P₁), and won both battles, thereby gaining control of the River Cumberland. Farther west, more Federal successes were won in Arkansas and Missouri during March, and by the end of that month Grant was advancing towards the borders of Mississippi. The result was the two-day Battle of Shiloh (6-7 April) (Q); a hard-won victory that cost Grant 13,000 casualties to the Confederate 10,700. General Albert Johnston was killed in the battle, and General Beauregard fell back to Corinth (B), where General Bragg took over command of the dispirited Confederate forces in the area.

Far to the south, the Union enjoyed further success. Admiral Farragut captured the mouth of the Mississippi and took New Orleans (J) in May, but a river-borne attack on Vicksburg (D) failed. Elsewhere, each side mounted disruptive raids, but more important was the Confederate General Bragg's invasion of Kentucky, which enjoyed considerable success until Bragg became unduly cautious and fell back out of the state in early October, returning to Tennessee. Grant then began the first of what would prove several abortive attempts to take Vicksburg, advancing down the Mississippi in November. The only results of two assaults were humiliating failures and mounting casualties, and Grant decided to reconsider his strategy. Finally in the west that year, on 31 December General Bragg began a four-day battle against General Rosecrans, resulting in a drawn outcome at Murfreesboro (R) in Tennessee along the River Stone.

The Year of Decision: 1863
Two great Union successes—at Gettysburg (S) and Vicksburg—virtually decided the outcome of the Civil War at the climax of this year's fighting. In the east the year began with a successful Confederate cavalry raid to Halifax Court-House in early March, matched by General Stoneman's sweep through Virginia in late April and early May. Stoneman's absence, however, contributed to General Hooker's massive defeat at the hands of Lee and Jackson at Chancellorsville (T) (see p. 179); a success that cost the Confederacy dear in the mortal wounding of Jackson. However, sensing a possible chance to rally European opinion behind Richmond, Jefferson Davis sent Lee and his army on a second invasion of the North. The result was the Campaign and Battle of Gettysburg (S), 1-3 July (see pp. 180-181), which inflicted irreplaceable casualties on the

Army of North Virginia and proved one of the major turning points of the war. Lee retreated back into Virginia, followed rather than pursued by the Union general, Meade, and both armies faced one another over the River Rapidan for the remainder of the year, engaging in desultory skirmishing.

In the west the year had opened with a number of cavalry raids by each side: Grierson's Union troopers reaching Baton Rouge (U), and Bedford Forrest's Confederates attacking Tennessee. The great event of the year in the theatre, however, was Grant's celebrated and daring Vicksburg (D) campaign (see pp. 182-183). The eventual surrender of Pemberton's garrison on 4 July brought Union control of the Mississippi, and drove a strategic wedge through the Confederacy, isolating Missouri, Arkansas, Texas and much of Louisiana from the remaining secessionist states. This was a second critical setback for the South.

In the meanwhile, Rosecrans captured the key railway centre of Chattanooga (E) for the Union, but when he pursued the retiring General Bragg he was badly mauled at Chickamauga (V) (19-20 September). It was the turn of the Union forces to pull back, but Bragg attempted no pursuit, and the arrival of Grant to take over command soon led to a series of battles in late November which re-established the Federal initiative. So ended the most important year of the war.

The Campaigns of 1864
Lincoln appointed Grant his overall military commander on 9 March, and the North's great superiority in men and resources was at last brought to bear fully and effectively on a collapsing Confederacy. Grant moved to command the Army of the Potomac in the east, and command in the west was deputed to General Sherman. Despite certain local setbacks (including another successful raid by the hard-riding and hard-swearing Forrest), the double-pronged Union offensive opened in early May (see p. 185). Sherman struck south from Chattanooga (E) at the head of 100,000 men, and captured Atlanta (K) on 2 September. Grant, leading 118,000 men, drove south through 'the Wilderness' (name given to area of rough terrain near Chancellorsville) towards Richmond. Progress was slow, but although the Confederates could still inflict checks and even win small victories, their resources were fast running out.

Soon, in the east, however, military events were concentrated around the nine-month siege of Petersburg (W) (see p. 185 and map, The Petersburg Campaign, p. 184), although serious

fighting also took place once more in the Shenandoah Valley between May and October; by which latter date Federal control was re-established.

In distant Atlanta, meanwhile, Sherman had decided to launch his 'march to the sea'. Detaching General Thomas to watch the Confederate Hood and his Army of Tennessee, Sherman advanced with 68,000 men, wreaking havoc along his path, and on 21 December he captured Savannah (X) on the Atlantic coast of Georgia. Once again, the Confederacy had been bisected. The sands were fast running out for Jefferson Davis.

1865: The Coup de Grâce
On 1 February, Sherman marched north through the Carolinas, pushing J. E. Johnston before him. Moving via Charleston and Columbia, he won a series of small victories over the exhausted Confederates. Grant increased the pressure against Lee's over-extended defences around Petersburg (W) and, on 1 April, the Federal attack at Five Forks, sixteen miles south-west of Petersburg, proved the beginning of the end. The Confederate line buckled and broke. Lee had no option but to abandon both Petersburg and Richmond; Jefferson Davis moving his capital to Danville whilst Lee retreated westwards hoping to link up with Johnston's army. Then, he estimated, he might hope to defeat Sherman and Grant in turn. But it was not to be; harassed by Grant's pursuing forces, the end came on 8 April near Appomattox Court-House (Y), when Lee found his path blocked by a powerful force under Sheridan. Next day, Lee formally surrendered to Grant, and on the 12th his remaining 28,000 men of the Army of North Virginia laid down their arms. Two weeks later J. E. Johnston followed suit near Durham (Z), North Carolina. With the surrender of isolated Confederate forces west of the Mississippi, the long and gory war at last came to an end on 26 May 1865.

The cost of this long, fratricidal struggle had been high. Of some 1,234,000 Confederates who had borne arms, all of 895,600 had become casualties: 94,000 killed in battle, 164,000 dying of disease, a further 175,000 suffering wounds, and a final total of 462,600 being made prisoners. Of the Union 2,890,000 who wore uniforms, 110,000 had been killed in action, 250,000 had died of disease, a further 275,000 had been wounded, and 211,400 taken prisoner; in all, over 848,000. The cohesion of the United States had been preserved, but at a bitter price. The wounds of the struggle would take many years to heal, if indeed they can be deemed truly healed over a century later.

The Campaign and Battle of First Bull Run, 186

The proximity of the two rival capitals—Washington and Richmond—had a strong influence on the events of the US Civil War on the eastern front. Underestimating the military prowess of the Confederate Army, Lincoln and his advisers pressed General McDowell into launching a premature attack with poorly-trained troops against the Confederate forces holding Bull Run. The President was also influenced by knowledge that the service time of some '90-day' volunteers was reaching its expiry date.

The Union plan was for General Patterson to contain the Confederate J. E. Johnston in the Shenandoah Valley, whilst McDowell manoeuvred to turn the left flank of General Beauregard's 20,000 men by means of a frontal diversion to pin the enemy, associated with an envelopment carried out in the Napoleonic fashion (see p. 98) by the main body. Unbeknown to McDowell, however, the Union plan had been revealed to Confederate intelligence, and, even worse, 12,000 men of J. E. Johnston's army had been spirited away from the Valley and were being brought by railway to Manassas Junction (A). The Union army, therefore, would find itself outnumbered.

McDowell's final plan was good on paper, but beyond the capabilities of his inexperienced troops to carry out. A brigade under Generals Richardson and Davies was to demonstrate (B) opposite Beauregard's position covering Union Mills and Manassas Junction in the hope of pinning his forces there, whilst a secondary attack—three brigades under General Schenk (C) by way of Stone Bridge—threatened a local tactical outflanking attack. Meantime, the main attack (D), five brigades commanded by Generals Hunter and Heintzelman, were to execute a circuitous five-mile march to Sudley Springs, and then, after crossing Bull Run, this force was to roll up the Confederate left flank.

After advancing to Centreville, and wasting two valuable days through administrative confusion, McDowell launched his attack on 21 July. Nothing, however, went according to plan. The feint attack proved far too weak to fool Beauregard, and soon he was able to transfer troops westwards (E), led by General Early, to reinforce the developing battle. Similarly, Schenk failed to press his secondary attack with any vigour, although eventually he did find a way over Bull Run (F). As a result, General Evans of the Confederate Army was soon able to spot the slow progress of the main Union outflanking column, and by the time this had made its ponderous way towards Henry House Hill (G), substantial reinforcements under Ewell and Holmes (H) were well on their way to the sector. Beauregard, however,

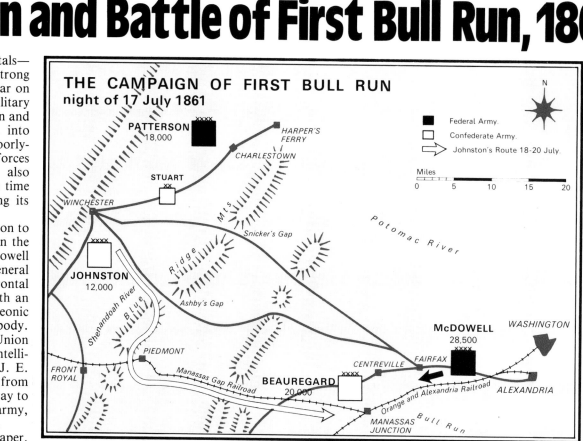

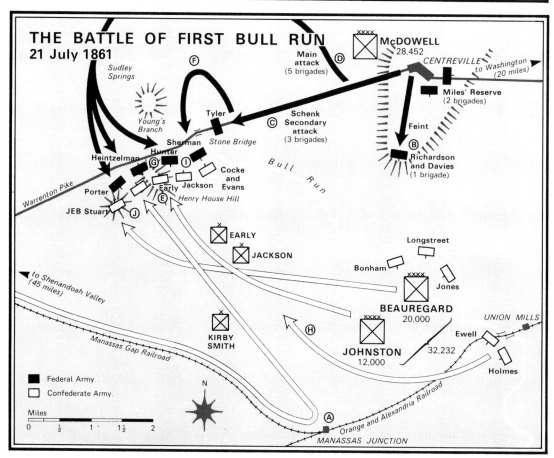

decided to abandon his original plan for an advance on Centreville, the position of McDowell's reserves.

The fight was fierce enough for several hours, and at one moment, assailed by General Sherman, the Confederates around Henry House Hill seemed on the verge of disaster. They lost their positions along Young's Branch and fled uphill. There, however, they were rallied behind General Jackson's stalwart brigade (I)—who earned his nickname 'Stonewall' for his dogged fighting on this occasion—and the Confederate line was stabilized. The arrival of a new brigade under Kirby Smith (straight off the train at Manassas) extended the Confederate line (J) until it overlapped the Union right wing. Colonel J. E.

B. Stuart then led a successful cavalry attack which overran a Union battery, and at 4pm the weary Northern troops began to give ground. McDowell tried to conduct an orderly retreat, but when bottle-necks were encountered his men panicked and the defeat became a rout. Only a single battalion under Colonel Sykes carried out an effective covering operation to aid the escape of their colleagues, but the road back to Washington was choked with troops, guns and horses all intermingled with crowds of Washington sightseers who had come out with picnic-baskets to watch the battle.

The Confederate army was itself in a state of total confusion, and no pursuit proved possible —or Washington might have been occupied that evening. This first battle of the Civil War displayed the incompetence of both sides, earning von Moltke's cutting comment about "...two armed mobs...". The Union army came off decidedly the worse from the clash, losing 2,700 casualties after inflicting about 1,980. But the Union fixation with the need to advance on Richmond to crush the centres of rebellion and compel the dissident states to rejoin the United States remained for a further considerable period. Lincoln relegated McDowell to the command of a division, and appointed General George McClellan to the chief command in the army. From November, the 34-year-old general replaced the veteran Winfield Scott, hero of the Mexican War and inventor of the Anaconda Plan (see p. 169) as commanding general over the Union armies.

1. *Abraham Lincoln, 16th President of the United States (1809-65), who was determined to return the Secessionist Southern States into the Union. He succeeded in this aim after a bitter Civil War, only to fall to an assassin's bullet.*
2. *Jefferson Davis (1808-89), President of the Confederacy throughout the American Civil War. His failure to gain European recognition proved fatal to his cause. Tried for treason, he was ultimately discharged.*
3. *General Pierre Gustave Toutant Beauregard (1818-93). As commander of Confederate forces he captured Fort Sumter and won the Battle of First Bull Run early in the American Civil War. He later fought at Shiloh and in defence of Petersburg.*
4. *Lieutenant-General James Longstreet (1821-1904), he became known as 'Lee's Old Work Horse' by the men. He had several successes, but is always remembered for his slowness in attacking on the second day of the Battle of Gettysburg.*
5. *General James Ewell Brown Stuart (1833-64) was the outstanding Confederate cavalry leader of the American Civil War. His raids caused the Union side much inconvenience.*

The Shenandoah Valley & Peninsular Campaign

Of all the many famous events of the American Civil War, none were more notable than those that took place in the Shenandoah Valley between March and June 1862 and, thereafter, in the Peninsula to the east of Richmond (the Confederate capital) during June and July. The mastermind behind both campaigns was Robert E. Lee, using General 'Stonewall' Jackson as his chief executive officer. The way in which greatly superior Union forces were misled and repulsed time and again forms a classical example of mid-nineteenth century warfare at its most effective.

The situation facing the Confederacy in early 1862 is shown on the map. The double threat to Richmond posed by the Union's 150,000 men was grave, as barely 40,000 Confederates were available, under General J. E. Johnston, to defend the capital. President Jefferson Davis's main military adviser, General Robert E. Lee, produced a masterly scheme to use Jackson's small force in the Valley to create a series of diversions. By posing an apparent threat to the security of Washington DC, Lee felt confident that Jackson would cause President Abraham Lincoln to divert McDowell's advance towards Richmond to sustain General Banks in protecting the approaches to the Union capital. This would enable Johnston to pay full attention to the main Union attack by General McClellan's 110,000 men, who were being brought by sea from their camps along the Potomac to land in the Peninsula and attack Richmond from the east. Indeed, distant events in the Valley might disrupt the entire Union plan for the eastern theatre in 1862—and so it would prove.

The geography of the Valley favoured the Confederacy's strategic plan. The lie of the

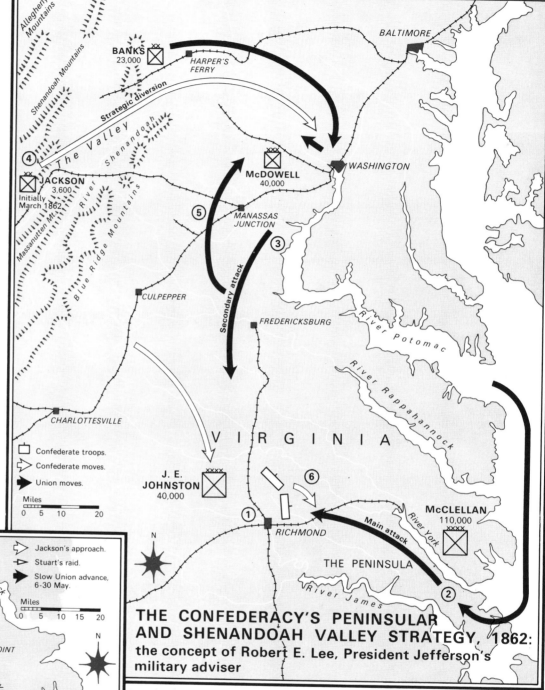

THE CONFEDERACY'S PENINSULAR AND SHENANDOAH VALLEY STRATEGY, 1862:
the concept of Robert E. Lee, President Jefferson's military adviser

THE PROBLEM

1 Richmond— target of Federal forces— defended only by J. E. Johnston's inferior forces.
2 General McClellan's 110,000 Union troops attacking up the Peninsula in association with:
3 General McDowell's 30,000 Union troops advancing from Washington.

Impossible to defy both threats.

THE SOLUTION

4 Use Jackson's small force to create threats towards Washington (revealed as Lincoln's anxiety point in 1861).
5 Thus divert McDowell to the defence of the Valley and approaches to Washington.
6 Enable Johnston (later Lee) to hold off main threat by confusing McClellan with disruption of overall Union plan.

THE PENINSULAR CAMPAIGN

For explanation of numbers see text.

1. General Robert Edward Lee (1807-70) was the famous commander of the Confederate Army of North Virginia, which he led to many victories against the Union. Defeated at Gettysburg, he fought to the bitter end, surrendering at Appomattox Court-House in 1865. He was much admired by both armies.

2. General Thomas Jonathan Jackson (1824-63) was the most famous Confederate commander (apart from Lee). His conduct of the Valley Campaign in particular singled him out and, although his record varied, the South could ill-afford his loss—the victim of a Confederate sentry's bullet during the Battle of Chancellorsville. He was nicknamed 'Stonewall' at First Bull Run, 1861.

3. General Richard S. Ewell (1817-72) was a lieutenant of 'Stonewall' Jackson, and shared with him much of the credit for the Peninsular campaign of 1862.

land ran from south-west to north-east: from Staunton to Harper's Ferry (barely fifty miles from Washington) on the River Potomac. Furthermore, the Shenandoah Valley was dominated by parallel lines of mountains: the Blue Ridge, the Shenandoah, and the Alleghennies. Half-way down its valley the Shenandoah River was divided into two waterways—the North and South Forks—by Massanutten Mountain. (The railways are indicated, and also the main gaps in the ranges and the subsidiary waterways.) The area was cultivated intensively and very fertile, forming the 'granary of the Confederacy', and the loyalties of its population were firmly behind the Southern war effort. All these features would be used to the uttermost in the execution of Lee's master-stategy by General Jackson.

The geography of the Peninsula was less favourable to the Confederates. Unless they could contain the Union landing and bridgehead between the Rivers James and York (see inset, The Peninsular Campaign), the routes to Richmond were fairly open. The Chickahominy River was not a serious obstacle in its own right. On the other hand, there were few roads in the area, and considerable stretches of woodland and dense undergrowth; features that might assist the defence.

The Peninsular Campaign, 1862

Owing to General McClellan's over-caution, his 110,000-strong army wasted April and much of May. The Federal advance up the Peninsula was hindered by General Magruder's combination of bluff and force. By the time the real campaign opened, General J. E. Johnston's original 40,000 Confederates had been built up to 56,000 by Jackson's arrival from the Valley. Prior to that welcome reinforcement, Johnston had fought a chaotic battle at Fair Oaks (see 1

on inset to map opposite) on 30 May, and despite being wounded he had repulsed the Union forces; this setback to McClellan's hopes being largely due to Longstreet's failure to reach his appointed sector in the battle-line.

General Robert E. Lee now took over command of the Army of North Virginia, and set about discomfiting his opponent. General J. E. B. Stuart's cavalry executed a ride right round the Union Army (12-15 June), and ten days later Lee began a series of small but sharp attacks which resulted in local Confederate successes at Mechanicsville on 26 June (2), Gaines Mill (3) next day, and Savage Station (4)

Chronology of the 1862 Campaign

January: Jackson uses his 11,000 men on raiding operations, including against Romney.

February: Jackson abandons Winchester and falls back to Woodstock.

23 March: Jackson force-marches north and attacks Kernstown; repulsed, he retreats to New Market.

19 April: Jackson moves to Conrad as Gen. Banks marches to Staunton, with support.

29 April: Jackson hijacks two trains and heads west to surprise Gen. Milroy.

8 May: Jackson defeats Milroy at McDowell, and saves Staunton.

23 May: Jackson, after moving north, surprises Union force at Front Royal.

25 May: Jackson defeats General Banks at Winchester; Lincoln cancels McDowell's march on Richmond.

28 May: Jackson reaches Harper's Ferry on the River Potomac. Union forces mass to trap him.

2 June: Jackson, at last aware of his dire peril, retreats south at speed, hotly pursued.

8 June: Jackson defeats Fremont's advanced formations at Cross Keys.

9 June: Jackson switches his men to defeat Shield's advance guard at Port Republic.

10 June: Both Union armies retire northwards whilst Jackson rests his men.

Mid-June: Jackson transfers his men east to join Gen. J. E. Johnston (replaced by Lee) in the Peninsula.

26 June: Lee wins action of Mechanicsville.

27 June: Lee wins action at Gaines Mill.

29 June: Lee wins action at Savage Station. Union army of McClellan heads for bridgehead.

1 July: Union rearguard repulses Lee and Jackson at Malvern Hill. Union army enters strong defensive perimeter.

Late July: McClellan's army evacuates by sea to Washington area.

on the 29th. By this time Jackson was proving an awkward colleague to control, and only slowly did he learn the need to obey his superior implicitly. However, the demoralization of the large Union army became increasingly pronounced as they fell back mile by mile towards Harrison's Landing on the James River. A spirited rearguard action at Malvern Hill (1 July) (5) allowed the main army to enter a strong, prepared perimeter (6), but there was no further thought of offensive action by McClellan, and late in July the Union Army was evacuated by the Union Navy and returned to the River Potomac.

For the time being Richmond was safe, and European opinion was clearly beginning to swing behind the Confederacy as news of Lee's and Jackson's artistry filled the newspapers. Perhaps the South was nearer to achieving its objective of gaining European recognition than ever before. Any such hopes, however, would be dashed in September when Lincoln decreed the freeing of the Southern slaves. His action created a pro-Union climate within England, France and Germany which further Confederate military successes could not shelve.

The Events of the Valley Campaign and their Effects

The first two phases of Jackson's campaign are shown on the map on p. 177. At the outset, Major-General J. F. G. Jackson had merely 3,600 men in quarters around Woodstock on the North Fork of the Shenandoah. The previous December he had commanded about 11,000 men from his base at Winchester, and used them to unsettle the Union General Rosecrans by hit-and-run raiding operations in the northern area of the Valley during January 1862, including a raid on Romney; but

The Shenandoah Valley & Peninsular Campaign

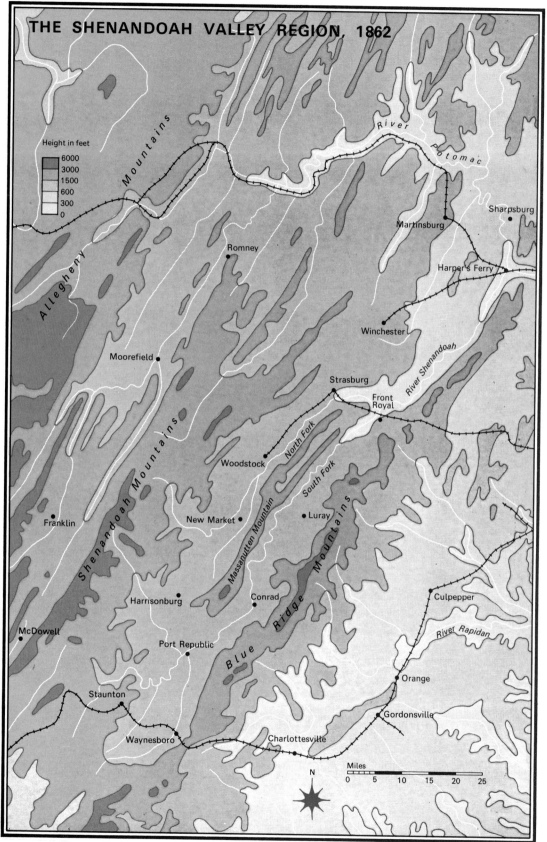

THE SHENANDOAH VALLEY REGION, 1862

Height in feet
6000
3000
1500
600
300
0

then Jackson had quarrelled with his subordinates and ultimately seen 7,000 of his men transferred to other armies in March. This reduced his strength, in numerical terms, to a bare brigade. However, his remaining troops were tough and devoted to their commander. When J. E. Johnston, learning of the preparations for the double-onslaught against Richmond, fell back to the Rappahannock so as to draw nearer to his capital, Jackson abandoned Winchester and retired to Woodstock. The Union General Nathaniel Banks with 38,000 men thereupon crossed the Potomac and occupied the Winchester area.

Jackson's orders were to prevent Banks from being able to transfer his command eastwards to participate in the onslaught against Richmond, but to avoid large-scale fighting if possible. In mid-March, learning from Colonel Ashby's watchful cavalry screen that Banks was moving east, Jackson decided on a bold pounce. Starting from Woodstock on 23 March (1 on map, Jackson's Valley Campaign, Phases One and Two), Jackson force-marched his men north to attack Kernstown, where he believed a mere four Federal regiments to be in position. In fact, General Shields had all of 7,000 men available, and Jackson's total force of 4,200 was tactically defeated at the Battle of Kernstown (2) on the 23rd, losing 1,000 men and two guns, and inflicting 590 casualties. However, as Jackson retreated to Mount Jackson and thence to New Market, the Union high command decided that Banks must return in strength to the Valley. Thus, a tactical failure had nonetheless achieved Lee's and Jackson's strategic purpose.

After two weeks rest, Jackson moved to Conrad (19 April) (3), and prepared his next move. Unaware of Jackson's position, Banks marched all the way down the Valley to the approaches to Staunton by way of the North Fork, stationing his main force, 15,000 strong, around Harrisonburg. A force of 8,000 men then moved forward towards Staunton where only 3,000 Confederates stood in their path. Meantime, the Union General Fremont was being summoned from the Alleghennies with his corps to join Banks, whilst south of Washington McDowell (4) was starting to advance on Richmond at the head of 40,000 men. J. E. Johnston was in full retreat for Richmond (5) as McClellan had now reached the Peninsula by sea.

Reinforced to 14,000 men by the welcome arrival of General Ewell, Jackson determined to prevent Fremont reinforcing Banks in the main Valley. As the latter was too strong for Jackson to attack, he determined to strike Fremont's advance guard (General Milroy) in the west,

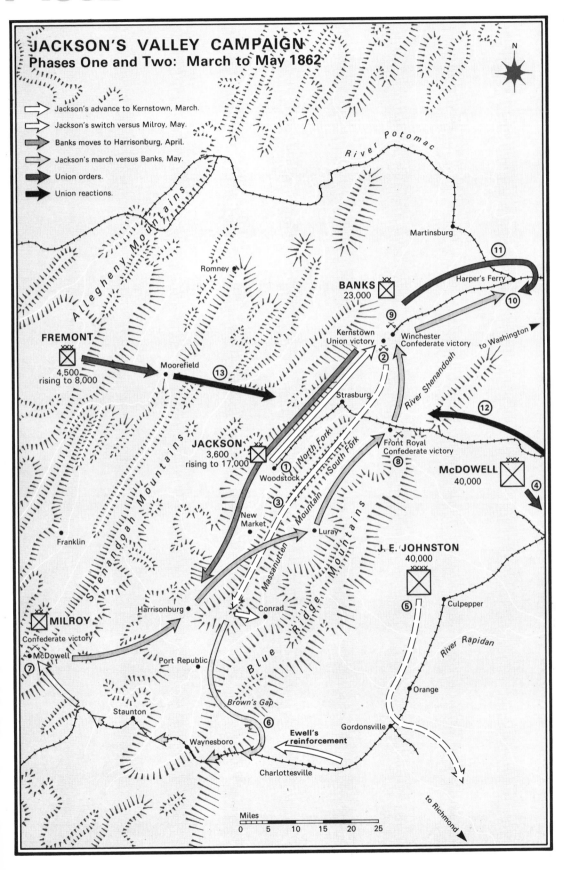

JACKSON'S VALLEY CAMPAIGN
Phases One and Two: March to May 1862

Jackson's advance to Kernstown, March.
Jackson's switch versus Milroy, May.
Banks moves to Harrisonburg, April.
Jackson's march versus Banks, May.
Union orders.
Union reactions.

River Potomac

Martinsburg

BANKS
23,000

Harper's Ferry

Romney

FREMONT
4,500
rising to 8,000

Moorefield

Kernstown
Union victory

Winchester
Confederate victory

to Washington

River Shenandoah

Strasburg

JACKSON
3,600
rising to 17,000

Woodstock

Front Royal
Confederate victory

McDOWELL
40,000

New Market

Franklin

Luray

J. E. JOHNSTON
40,000

MILROY
Confederate victory

Harrisonburg

Conrad

Culpepper

McDowell

Port Republic

River Rapidan

Staunton

Brown's Gap

Orange

Ewell's reinforcement

Gordonsville

Waynesboro

Charlottesville

to Richmond

Miles
0 5 10 15 20 25

and thus confuse the enemy by doing the unexpected. Marching through the Blue Ridge Mountains via Brown's Gap (6) on 29 April, Jackson hijacked two trains west of Charlotteville, packed his men on board, and steamed through Waynesboro to Staunton moving west, picking up the 3,000 men at Staunton, bringing his strength up to 17,000. Then marching into the mountains, he surprised Milroy's 2,500 Union troops at the town of McDowell (7) on 8 May, totally routing them with 9,000 Confederates at a cost of 498 casualties. This was expensive, but the strategic effect of this action was to persuade Fremont not to move to Banks's aid. Jackson had thus saved Staunton for the time being, and he pursued Fremont as far as Franklin before returning to the Valley proper. Meanwhile, in the Peninsula McClellan was not pressing forward, but allowed himself to be fooled by Magruder's token force into awaiting the advance of McDowell from the north. Thus the Confederacy was winning valuable time.

Intelligence now reached Jackson that Banks was about to release General Shields's division to march and join McDowell's strong corps at Fredericksburg, prior to the renewal of the projected march on Richmond that the supine McClellan was awaiting. This would leave Banks with merely 8,000 men in the Valley, and Jackson determined to exploit his new numerical superiority and play the trump-card: the threat to Washington, which would hopefully disrupt McDowell's entire move. So, moving from Harrisonburg, Jackson rapidly marched to New Market, passed over Massanutten Mountain and the South Fork to Luray, and the pounced on a Union detachment of 1,000 men at Front Royal (8) on 23 May, accounting for some 757 of them. Greatly alarmed by this sudden attack on an outpost, Banks hastened to concentrate 5,500 men and twelve guns at Winchester, where, on 25 May, he was defeated (9) in a battle that began in a dense morning mist at 7am, losing 1,500 killed and wounded and 3,000 prisoners. This shattering blow had the desired effect. Lincoln indeed cancelled McDowell's proposed march on Richmond, and ordered him to transfer 20,000 into the Valley forthwith to crush this Confederate menace operating on the flank of the main theatre of war once and for all.

Unaware of this triumph for Confederate strategy, Jackson pressed ahead to the banks of the River Potomac, near Harper's Ferry (10). At this juncture, Banks and his remaining 17,500 men were ordered by Washington to prepare to counterattack.

Lincoln and his advisers at Washington planned to spring a trap around the impudent Jackson. Banks was to attack on the Potomac

The Shenandoah Valley Campaign of 1862

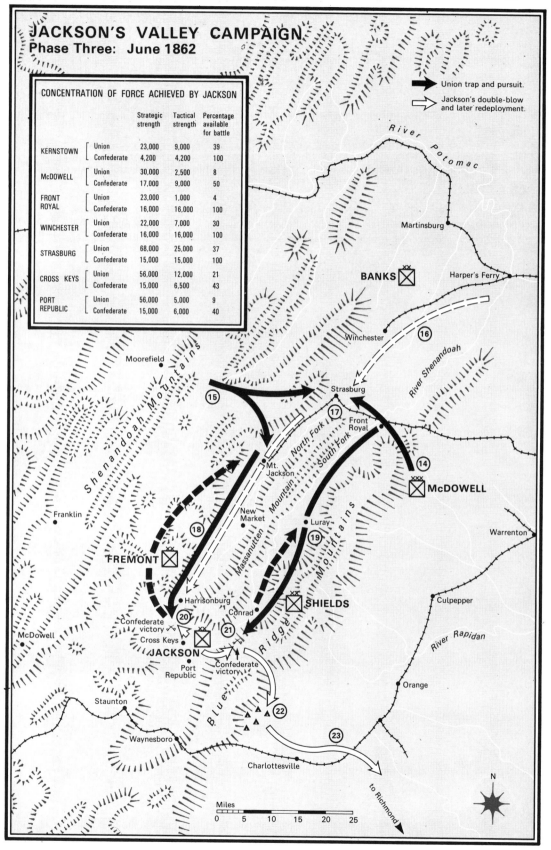

JACKSON'S VALLEY CAMPAIGN
Phase Three: June 1862

Union trap and pursuit.

Jackson's double-blow and later redeployment.

CONCENTRATION OF FORCE ACHIEVED BY JACKSON

		Strategic strength	Tactical strength	Percentage available for battle
KERNSTOWN	Union	23,000	9,000	39
	Confederate	4,200	4,200	100
McDOWELL	Union	30,000	2,500	8
	Confederate	17,000	9,000	50
FRONT ROYAL	Union	23,000	1,000	4
	Confederate	16,000	16,000	100
WINCHESTER	Union	22,000	7,000	30
	Confederate	16,000	16,000	100
STRASBURG	Union	68,000	25,000	37
	Confederate	15,000	15,000	100
CROSS KEYS	Union	56,000	12,000	21
	Confederate	15,000	6,500	43
PORT REPUBLIC	Union	56,000	5,000	9
	Confederate	15,000	6,000	40

(11), McDowell's detachment was to head for Strasburg to cut the Confederate line of retreat from the east (12), and Fremont was to advance east with 8,000 men from Moorefield (13) to complete the trap. Jackson was to be surrounded and eliminated, all of 68,000 Federal troops being involved.

In the nick of time, Jackson learnt of his peril from friendly civilians and his cavalry screen (see map, Jackson's Valley Campaign, Phase Three). As McDowell moved in from the east (14) and Fremont from the west (15), Jackson began a rapid retreat down the Valley (16), hoping to pass through Strasburg before the Federal forces could close their trap. By the narrowest of margins he won the race (17) and headed south, hotly pursued by Fremont down the North Fork (18) and Shields down the South Fork (19), leaving Banks to hold the north of the Valley. Divided by Massanutten Mountain, the Federal forces had little knowledge of each other's exact position when Jackson turned at bay south of Massanutten Mountain. On 8 June, he used 6,500 men to defeat Fremont's leading 12,000 men at Cross Keys (20), and next day rushed every man east to meet and defeat Shields's advance guard of 5,000 troops at Port Republic (21). Both Union forces recoiled up their lines of communication in complete disarray, whilst Jackson marched his tired but jubilant men over the Blue Ridge Mountains to camp and rest, and await further orders (22).

Prior to these two latest defeats, McDowell was about to be recalled from the Valley to resume his much-postponed march on Richmond, where an unholy calm still reigned in the Peninsula. However, news of the latest setbacks in the Valley caused Washington to finally abandon this movement. The Confederate strategy thus finally triumphed.

Jackson's skill at using a small force operating in friendly and well-known terrain to dominate far larger enemy forces had been a classical exposition of the art of war. High morale, dauntless purpose and great toughness lay behind his achievement. His skill at managing to mass a superior force on each battlefield in turn (save only at Kernstown), can be seen by reference to the table provided. His 'foot-cavalry'* were not to be allowed much rest, however. In mid-June he was ordered to transfer the bulk of his men eastwards (23) to share in the campaign around Richmond. Jackson carried out this move as unobtrusively as before, and the Federal forces entrenched themselves apprehensively in the north of the Valley, unaware that their scourge had gone.

*A phrase coined by Stonewall Jackson, meaning infantry who were always making forced-marches.

The Chancellorsville Campaign, 1863

Lincoln appointed General 'Fighting Joe' Hooker to replace the unsuccessful Burnside in command of the Army of the Potomac in January 1863. His army was north of the River Rappahannock opposite the city of Fredericksburg (A), scene of the heavy Union defeat of the previous December. In the months that followed his appointment, Hooker trained his men hard, reorganized them into seven infantry and one cavalry corps, each with a fair share of Union artillery, and set up an efficient intelligence organization. The Union camps were reconstructed, the hospitals cleaned, and many other administrative improvements introduced. Unfortunately, Hooker did not see eye to eye with his corps commanders. He was the army's fourth commander in a year and a half, and was aware that morale was suffering, but he was still full of confidence in his ability to master Robert E. Lee, who had his army camped along Mary's Heights above Fredericksburg (B). Lee's reputation was at its height: the year 1862 had seen his successful defence of Richmond and his victories at Second Bull Run and Fredericksburg. But the campaign about to open would prove his tactical masterpiece—and also his last great success.

As the weather improved, Hooker started to implement his plan. Four corps were to execute a wide outflanking movement of the Confederate position by way of Kelly's Ford (C) and US Ford (D), under Generals Slocum and Crouch respectively, whilst General Sedgwick mounted an attack on Fredericksburg (E) to pin the attention of the Army of North Virginia. As Lee had only 53,000 men, the Union superiority of strength was great, and there seemed to be a real chance that the Confederates would be trapped between the two pincers. On 27 April the outflanking force set off secretly, and two days later Slocum led three corps into the Chancellorsville (F) area to threaten Lee's rear.

On 29th, General Crouch also crossed the Rappahannock over US Ford to swell the main force's strength to 54,000 men, whilst below Fredericksburg General Sedgwick began to develop the diversionary attack with his 40,000 troops. The Army of North Virginia appeared to be pinned between two strong Union forces, as Stopford's cavalry swept southwards to threaten Lee's line of retreat towards Richmond.

Robert E. Lee was more than equal to the emergency, however. Leaving General Early with barely 10,000 men to defy Sedgwick, Lee marched his remaining 43,000 men west to face Hooker's main army. Hooker moved with great circumspection, and it was only on 1 May that a scrappy action developed in an area of rough

terrain known as 'the Wilderness' (G). The outcome was far from clear, but Hooker's determination wilted and he fell back to the area around Chancellorsville which he began to fortify, thus effectively handing all initiative to the Confederates.

Lee sensed his opportunity, and in a daring move sent off Stonewall Jackson and his complete II Corps (26,000 men) on a fourteen-mile march (H) to encircle the Union right flank, then to mount a surprise attack. Jackson carried out this task with some difficulty (the day was hot and the going hard) and only at 6pm on the 2nd were his men able to open their attack on Howard's XI Corps (I). The Confederates had made a little progress by nightfall, but then Jackson was tragically shot down in error by a picquet of his own men (he would die of his wound eight days later). However, J. E. B. Stuart took over command because Jackson's replacement, Ambrose Hill, had also been wounded. Early on the 3rd, Stuart assaulted the position held by Sickles and Slocum. Hooker (slightly wounded) continued to contract his position, and fell back into a shorter defensive line (J); a move that permitted the two wings of the Confederate Army to link up.

On 3 May, Sedgwick also began an advance towards Chancellorsville from his position twelve miles away to the east. It cost him four assaults against Mary's Heights before he could dislodge the tenacious Early. This delay gave Lee warning of what was afoot, and, with great skill and no little risk, he ordered McLaws to take 20,000 men from the main battlefield to

march east and confront Sedgwick, leaving Stuart with only 25,000 to face Hooker's force which had now been reinforced to almost 70,000 men. In mid-afternoon McLaws reached Salem Church (K) and after hard fighting was able to halt Sedgwick's advance. Next morning (4 May), Lee had sent back more men under Anderson, and Early was back in the battle, reoccupying Mary's Heights in Sedgwick's rear. With only 19,000 Union troops under immediate command, Sedgwick was now hard pressed in his turn, but managed to repel the Confederate attacks and, under cover of darkness, withdrew his men back over the Rappahannock to rejoin the remainder of his force.

Hooker had remained wholly inactive during the Salem Church battle on the afternoon of the 3rd and during the 4th, throwing away his best chance of a success against Stuart's wing of the army. By the 5th, Lee was back to reinforce his subordinates facing Hooker. By this time the Union commander-in-chief had only one wish: to call off the battle. During the night of the 5th-6th, he began to withdraw his men to the north bank of the Rappahannock. The long battle had cost the Army of the Potomac almost 17,300 casualties; the Army of North Virginia had not come off lightly, having lost almost 13,000 killed and wounded. Nevertheless, Lee had given a virtuoso performance of cool and inspired generalship, and won arguably his greatest battle. The loss of Jackson, however, was a grave matter, as the events of the Gettysburg campaign were soon to demonstrate.

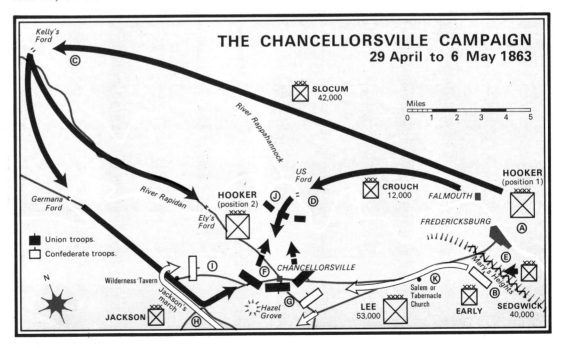

THE CHANCELLORSVILLE CAMPAIGN
29 April to 6 May 1863

The Gettysburg Campaign, 1863

By the late spring of 1863, Jefferson Davis and the Confederacy were facing a deteriorating war situation. As hopes of European intervention declined, and with Vicksburg closely besieged on the Mississippi (see p. 182), the Southern President decided to mount a last great gamble in the eastern theatre of war, which had been the scene of such dramatic but ultimately fruitless successes as Chancellorsville (see p. 179) earlier that year. Accordingly, General Lee was ordered to launch a new offensive into Union territory in the hope of shaking Lincoln's resolve, impressing Europe, and possibly earning an armistice. Davis knew there was a growing 'peace-party' in the North.

The Army of North Virginia secretly left Fredericksburg on 3 June and headed for Culpepper (A) and the Valley (see map, The Shenandoah Valley Region, p. 176). This movement did not long go undetected by General Hooker; a cavalry skirmish at Brandy Station on 9 June (B) warned the Union command that a major move was in progress; and on 13 June, the day before Lee reached Winchester, Hooker broke camp and marched his 100,000 men north from Falmouth, determined to keep his army (C) between Lee and Washington. The Confederate commander continued north towards Harrisburg, but he soon committed the error of detaching the dashing commander, J. E. B. Stuart, with most of his cavalry, to make a nine-day raid (D) with the intention of threatening the Federal capital and also Baltimore and thus distract Hooker. This decision deprived the main Confederate army of both news and its cavalry screen. As a result, Lee had no precise idea of where or how fast the Union forces were moving. On 27 June the Union forces reached Frederick (E), and next day, after a last dispute with Lincoln's government, 'Fighting Joe' Hooker handed over command of the army to the comparatively unknown General Meade.

On the 29th, Lee at last learnt that certain Union forces were moving to threaten his communications with Virginia. Accordingly, he halted his march on Harrisburg and ordered his corps to concentrate at Cashtown (F). Neither army, however, foresaw the great battle that was shortly to begin around the small town of Gettysburg (G). On 29 June the Union commander, Bufford, camped his reconnoitring cavalry division near that town. His presence there was spotted by a Confederate foraging brigade, which reported back to General A. P. Hill, the nearest corps commander. On his own authority, Hill ordered his men to advance against Gettysburg, and on the morning of 1 July a brisk fight broke out north-west of the town (see A on

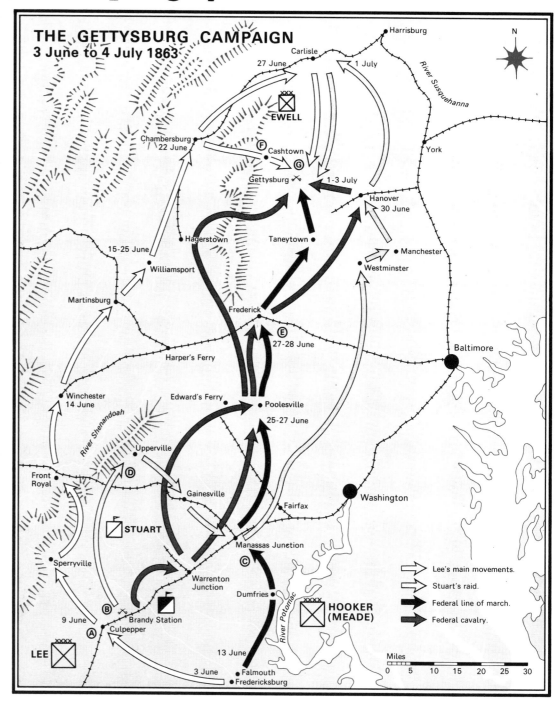

THE GETTYSBURG CAMPAIGN
3 June to 4 July 1863

Lee's main movements.
Stuart's raid.
Federal line of march.
Federal cavalry.

Miles
0 5 10 15 20 25 30

map, The Battle of Gettysburg). Although outnumbered, Bufford held on manfully and alerted General Meade, who at once sent up two small Union corps from the vicinity of Taneytown about eight miles to the south. By mid-afternoon a sizeable battle was raging. The Confederates eventually won the advantage, and by last light the Union troops had been forced to concede McPherson Ridge and Gettysburg itself, falling back to Cemetery Hill

to the south of the town (B). That evening also saw the arrival of both Lee and General Ewell with a further corps, giving the Confederates a massive superiority of force. However, Ewell disregarded a rather vague order from Lee to attack over Rock Creek towards Culp's Hill (C) before dark; a move that could have overwhelmed the Union line, but the opportunity was lost.

Overnight, General Meade arrived with more

General George Gordon Meade (1815-72) was not a particularly distinguished Union commander until he fought and won the great Battle of Gettysburg in 1863. Thereafter he fought under Grant's overall command.

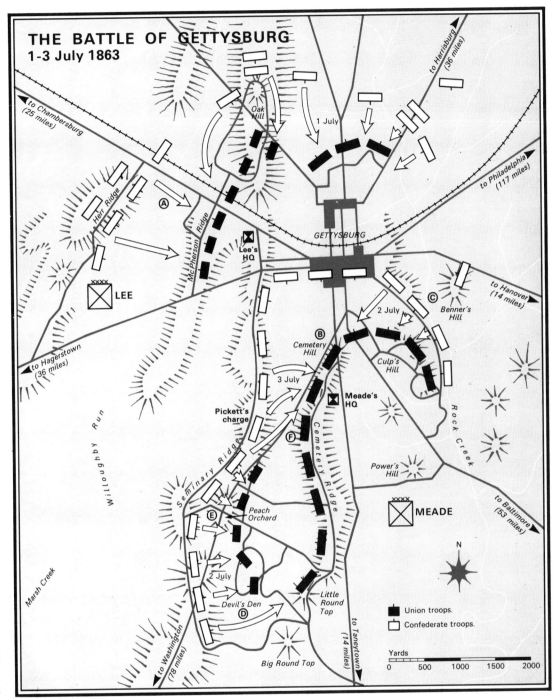

THE BATTLE OF GETTYSBURG
1-3 July 1863

to Chambersburg (25 miles)

to Hagerstown (36 miles)

to Hartisburg (36 miles)

to Philadelphia (117 miles)

to Hanover (14 miles)

to Baltimore (53 miles)

to Washington (78 miles)

to Taneytown (14 miles)

Oak Hill

1 July

GETTYSBURG

Herr Ridge

McPherson Ridge

Lee's HQ

LEE

A

Benner's Hill

C

Cemetery Hill

B

2 July

Culp's Hill

Meade's HQ

3 July

Pickett's charge

F

Willoughby Run

Seminary Ridge

Peach Orchard

E

Cemetery Ridge

Power's Hill

MEADE

Rock Creek

Marsh Creek

2 July

Devil's Den

D

Little Round Top

N

Big Round Top

Union troops.

Confederate troops.

Yards

0   500   1000   1500   2000

troops. After surveying the field, next morning he boldly decided to hold his ground, and sent messages to call up the rest of his men. But the Confederates were closer, and by noon General Longstreet had raised Lee's strength to 83,000 men; only Stuart's 5,000 horsemen were now absent, and they were riding hard for the battle-field. The odds still remained very much in Lee's favour, but they were shortening as more Union formations arrived from the south. Lee's plan was to destroy the five Union corps by means of a diversionary attack by Ewell in the north, whilst the main effort was made by Longstreet's command against the extreme left of Meade's position to the fore of Little Round Top. Longstreet, however, exploited Lee's rather vaguely-worded order, and delayed his main effort, and this allowed the Union General Sickles to extricate his men from the Peach Orchard (E), and also earned time for an engineer named Warren to extemporize an effective defence of Little Round Top. In the end, both Confederate attacks failed, although Ewell was at one stage close to success at Culp's Hill.

Overnight the Confederates at last lost their numerical superiority, as more Union troops reached the field, bringing Meade's strength up to 88,200 men and over 100 guns. Stuart (newly arrived) conferred with Lee, and they decided to mount a last major assault, this time against the centre of Cemetery Ridge, whilst the Confederate cavalry hit the Union rear up the Hanover Road. The morning passed in skirmishing, but Meade had guessed his opponent's intention, and had strengthened his centre with men and concealed artillery. Shortly after 1pm, almost 140 Confederate guns opened a massive bombardment against the Union centre, and an hour later Generals Heth, Pender and Pickett swept forward at the head of 15,000 men (F). Meade had waited for this moment, and now revealed his hidden guns in a terrible cross-fire which decimated the Confederate ranks. Only George Pickett and 150 men reached the crest of Cemetery Ridge (the 'high-water mark' of the Confederacy), and were very soon routed. In the meantime, General Gregg's Union cavalry had met and defeated Stuart's cavalry near Spangler's Spring, and by 5pm the battle was over. Lee's gamble had failed. He had lost 23,063 casualties and inflicted 23,049 over the three days of fighting, but whereas the Union could absorb these losses the Confederacy could not. Meade was so amazed by his success that he made little attempt to follow Lee's retreat back towards Virginia, but the Confederacy had lost all chance of winning the war on the Eastern front. The surrender of Vicksburg, far away on the Mississippi, on 4 July, brought with it the collapse of the Confederate front in the western theatre. The US Civil War had taken two important turns in the Union's favour, but the struggle was far from over even if the ultimate outcome was now all but decided.

The Campaign and Siege of Vicksburg, 1863

General Ulysses Simpson Grant (1822-85), Union General-in-Chief and later President of the United States. Grant emerged as the ablest overall Northern commander, and enjoyed Lincoln's full confidence. His conduct of the Vicksburg campaign in 1863 was a military masterpiece.

General Ulysses S. Grant proved both his determination and skill as a commander during this hard-fought campaign in the western, or Mississippi, theatre. The Union objective was to capture the city of Vicksburg and thus complete their overall control of the great river, effectively splitting the Confederacy in two from north to south. The climatic and physical problems of the great river valley, with its bayous or marshy tributaries and over-looking bluffs, posed daunting problems to an attacker (between late 1862 and April 1863 no less than five unsuccessful attempts were made to reach Vicksburg); but Grant's determination was unshakeable, and now he planned a bold envelopment of the position from the south and east, accepting formidable risks in the process but relying on surprise to overcome them and achieve his ultimate objective.

Employing a force of 1,200 cavalry in a raid through Mississippi and part of Louisiana to conceal his purpose, Grant moved his army southwards to the west of Vicksburg during April to the township of Hard Times (A). Covered by a naval flotilla, on 30 April two Union corps crossed the Mississippi and landed near Bruinsburg. The fact that General John Pemberton, the Confederate area commander, had placed barely 9,000 troops in this area, demonstrates the initial success of Grant's surprise move, and on 1 May the Union force defeated part of this detachment at Port Gibson (B) and compelled the evacuation of Grand Gulf by the remainder the next day. Learning that a strong Confederate force was massing near Jackson, Grant now took the bold

decision to march his force on that town to place himself between Pemberton's force and J. E. Johnston's approaching army. On 7 May, General Sherman's corps crossed the Mississippi, raising Grant's strength to 41,000 men.

The three Union corps arrived before the town of Raymond on 12 May, and there the brigade of General Gregg was repulsed (C). Pressing ahead, Grant left McClernand's corps to hold the line of the Raymond to Clinton railway against Pemberton, who was advancing eastwards from Vicksburg, and arrived outside Jackson (D) on the 14th. There, the 6,000 available Confederate troops were defeated after a sharp fight, and by late afternoon Johnston's force was retreating northwards in disarray. Sherman was left to destroy Confederate supply dumps at Jackson, whilst Grant lost no time in advancing west along the line of the Alabama and Vicksburg railroad through Clinton in the general direction of Vicksburg. The two Union corps of McClernand and McPherson approached Champion's Hill (E), where Pemberton was determined to make a stand at the head of 22,000 Confederates. Nominally, Grant had 29,000 troops available, but in fact McPherson's corps took the brunt of the action as McClernand was badly delayed. The result was a very hard-fought battle of varying fortunes before Pemberton decided to retire. Grant lost 2,441 casualties to the Confederates' 3,851, but the division that Pemberton ordered to cover his retreat could only escape southwards, and was unable to rejoin and take part in subsequent operations.

Pemberton fell back all the way to Vicksburg, leaving 5,000 men to hold the bridge at Big Black River (F). Sherman, who had now rejoined Grant's army, outflanked this outpost from the north, and on 17 May the Confederates were overwhelmed, losing 1,700 men and eighteen guns to Grant's frontal attack. By this time, ignoring J. E. Johnston's order that he was not to be bottled-up within the city, Pemberton had pulled back his remaining 20,000 men into the defences of Vicksburg. On 19 May, Grant tried an immediate assault (G), but was repulsed. However, he was now in a position to resume use of his communications running up the Mississippi. His bold eighteen-day offensive into Confederate territory—during which he had virtually abandoned all formal communications and lines of supply—had involved his troops in covering well over two hundred miles, but now he had achieved his major strategic objective: the complete isolation of Vicksburg.

The defences of Vicksburg were considerable: a series of strong redans and forts linked

by trenches and fire positions (see map, The Siege of Vicksburg). Their strength was again revealed on 22 May when Grant ordered attacks on six of the Confederate forts simultaneously. General Blair's division of Sherman's corps failed to break into the Stockade Redan in the north, and all other attacks were similarly beaten off by Pemberton's determined defenders. After losing 3,200 men in the day's fighting, Grant called off the attack and settled down to prosecuting a regular siege. Batteries were sited, approach and parallel trenches dug, and day by day the Union grip on Vicksburg grew tighter. Learning that J. E. Johnston was gathering a relief army around Jackson once more, Grant built up his strength to 71,000 men, using over 30,000 of them to conduct the siege whilst the remainder were stationed to the north and east of the siege area ready to intercept any relief attempt. But he made no more attempts at direct assault, relying on starvation and disease to overcome Vicksburg's resistance.

Pemberton had four divisions within the defences: three manning the lines as indicated, the fourth (General Bowen) constituting his reserve. Conditions rapidly deteriorated within the city; incessant Union shelling caused much damage, and soon both the civil and military populations were living in caves and holes in the ground. Under such conditions, disease was not slow in appearing, and by early July fully half of Pemberton's command had been killed, wounded or laid hors de combat by illness. Realizing that he could sustain the defence no longer, on 4 July Pemberton accepted the only terms Grant was willing to offer—unconditional surrender.

This success had cost the Union Army over 9,350 casualties, but it completed their war-effort on the western front. Johnston's relief force (31,000 men) had pushed forward as far as the Big Black River when he learnt of Pemberton's surrender. He at once fell back to Jackson. With the capture of the isolated Confederate post at Port Hudson on 9 July, Union control of the mighty Mississippi was complete. The Confederacy was now split from north to south, being deprived of the resources in terms of men and foodstuffs represented by Texas to the west of the river. Together with the loss of Gettysburg, the fall of Vicksburg doomed the Confederacy to ultimate defeat. Although there would be almost two more years of fighting before the end came, the die was cast from early July 1863. Grant, moreover, had emerged from this campaign with a greatly enhanced reputation as commander, and Abraham Lincoln would soon appoint him General-in-Chief of the Union forces.

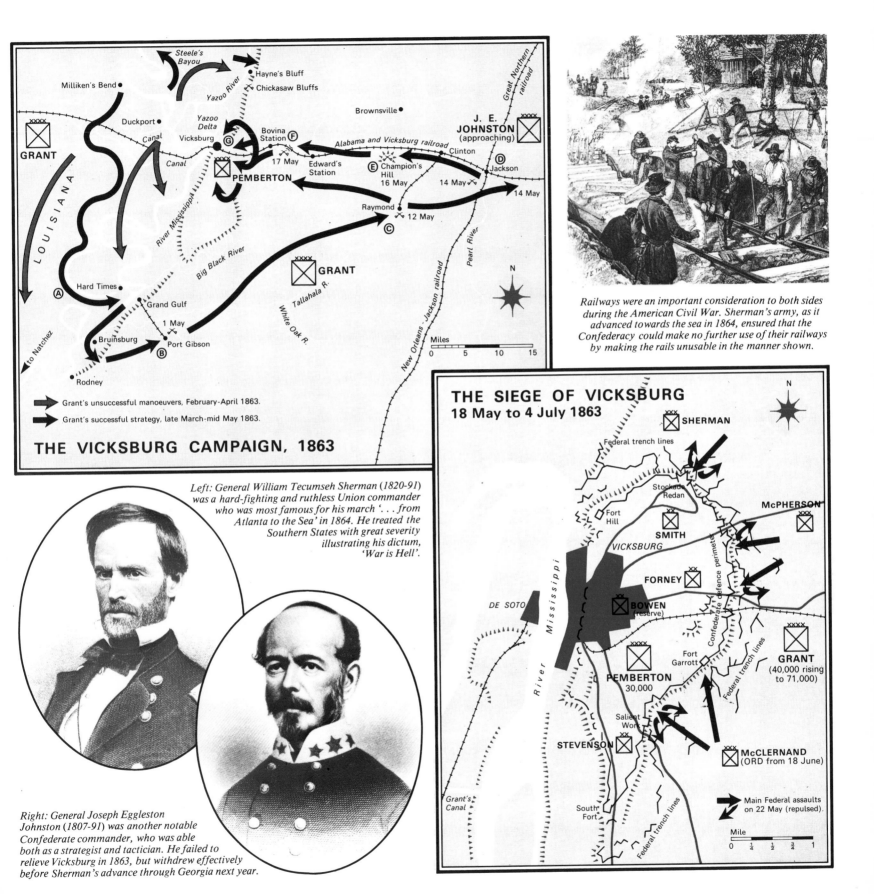

THE VICKSBURG CAMPAIGN, 1863

Steele's Bayou

Milliken's Bend

Hayne's Bluff
Chickasaw Bluffs

Duckport

Yazoo River

Brownsville

J. E. JOHNSTON
(approaching)

GRANT

Yazoo Delta Canal

Vicksburg

Bovina Station (F)

Alabama and Vicksburg railroad

Clinton

Canal

17 May

PEMBERTON

Edward's Station

(E) Champion's Hill
16 May

14 May

(D) Jackson

14 May

LOUISIANA

River Mississippi

Canal

Big Black River

Raymond

12 May

(C)

GRANT

to Natchez

Hard Times

Grand Gulf

Tallahala R.

(A)

1 May

White Oak R.

Pearl River

New Orleans - Jackson railroad

N

Bruinsburg

Port Gibson

(B)

Rodney

Miles
0 5 10 15

→ Grant's unsuccessful manoeuvers, February–April 1863.

→ Grant's successful strategy, late March–mid May 1863.

Railways were an important consideration to both sides during the American Civil War. Sherman's army, as it advanced towards the sea in 1864, ensured that the Confederacy could make no further use of their railways by making the rails unusable in the manner shown.

Left: General William Tecumseh Sherman (1820-91) was a hard-fighting and ruthless Union commander who was most famous for his march '. . . from Atlanta to the Sea' in 1864. He treated the Southern States with great severity illustrating his dictum, 'War is Hell'.

Right: General Joseph Eggleston Johnston (1807-91) was another notable Confederate commander, who was able both as a strategist and tactician. He failed to relieve Vicksburg in 1863, but withdrew effectively before Sherman's advance through Georgia next year.

THE SIEGE OF VICKSBURG
18 May to 4 July 1863

N

SHERMAN

Federal trench lines

Stockade Redan

Fort Hill

McPHERSON

SMITH

VICKSBURG

Confederate defence perimeter

DE SOTO

FORNEY

River Mississippi

BOWEN
(reserve)

Fort Garrott

GRANT
(40,000 rising to 71,000)

Federal trench lines

PEMBERTON
30,000

Salient Work

Grant's Canal

STEVENSON

McCLERNAND
(ORD from 18 June)

South Fort

Federal trench lines

→ Main Federal assaults on 22 May (repulsed).

Mile
0 ¼ ½ ¾ 1

183

The Victory of the North, 1863-65

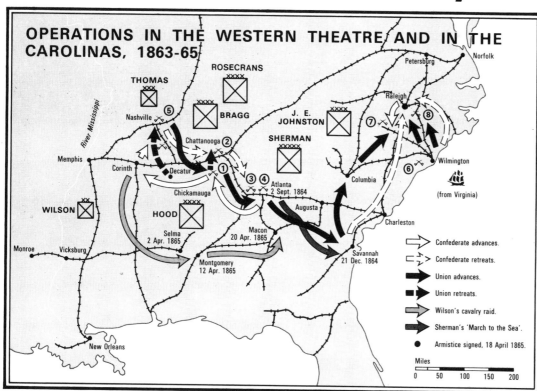

OPERATIONS IN THE WESTERN THEATRE AND IN THE CAROLINAS, 1863-65

Confederate advances.

Confederate retreats.

Union advances.

Union retreats.

Wilson's cavalry raid.

Sherman's 'March to the Sea'.

● Armistice signed, 18 April 1865.

Miles
0   50   100   150   200

MAJOR ENGAGEMENTS:

1 Battle of Chickamauga, 19-20 Sept. 1863: Rosecrans v Bragg.
2 Battle of Chattanooga, 23-25 Nov. 1863: Rosecrans defeats Bragg.
3 Battle of Kennesaw Mt., 27 June 1864: J. E. Johnston defeats Sherman.
4 Battles of Atlanta, 20,22,28 July 1864: Sherman repulses Hood three times.
5 Battles of Franklin and Nashville, 30 Nov. and 15-16 Dec. 1864: Hood defeats Union but is routed by Thomas.
6 Battle of Fort Fisher, 15 Jan. 1865: Confederates defeated.
7 Battle of Fayetteville, 11 March 1865: Confederates defeated.
8 Battle of Bentonville, 19 March 1865: Confederates defeated.

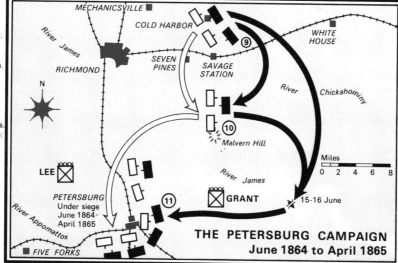

THE PETERSBURG CAMPAIGN
June 1864 to April 1865

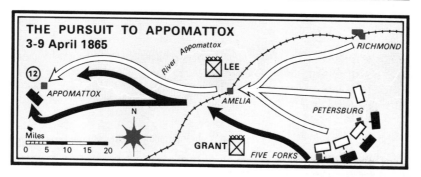

THE PURSUIT TO APPOMATTOX
3-9 April 1865

Operations in the Western Theatre and in the Carolinas, 1863-65

Shortly before Vicksburg and Gettysburg, the Federal General Rosecrans marched from Murfreesboro, Tennessee, and headed for the important road and rail junction of Chattanooga; the capture of which would effectively open the south-east of the Confederacy to the Union armies. The outnumbered Confederate General Bragg was promised the aid of Longstreet's corps from the Eastern theatre, but it was tardy arriving (only on 18 September in fact) and Rosecrans had reached the Tennessee River a month earlier. The outnumbered Bragg slipped away south, and when the Union army set off in pursuit through mountainous country, the Confederates turned in their tracks and on 17 September both sides were approaching Chickamauga Creek. After several days of skirmishing around the waterways, Longstreet made a timely appearance. In the two-day battle (1) that followed, Bragg's 62,000 men had the better of Rosecrans' 65,000 in a desperately gory engagement. After losing 16,500 men, the Union army ended up by being shut up in Chattanooga, but Bragg had lost 18,000 to achieve this.

Lincoln at once reacted in alarm as Confederate cavalry raided deep into Tennessee. General Hooker was transferred by rail with two corps from the east, and Sherman's corps arrived by river-boat from the west in a superb reinforcement move. The President also decided to replace Rosecrans with General Ulysses S. Grant, now made General-in-Chief of both east and west theatres. On 26 October, Hooker was over the Tennessee attacking westwards, and forces began to converge on Bragg who ill-advisedly detached Longstreet to besiege Knoxville. Grant, with Thomas's Army of Cumberland, Sherman's two corps of the Army of Tennessee and Hooker's pair from the Army of the Potomac in the east totalled over 60,000 men against Bragg's remaining 40,000. Between 23 and 25 November the Battle of Chattanooga (2) was fought, which, despite some big risks, ended in the collapse of Bragg's army, which retired into Georgia to regroup. Grant had won a big victory, at one stroke securing Chattanooga once and for all, winning a valuable supply area, and piercing the Confederate mountain defence line. All was now ready for a major invasion of the Deep South in 1864.

On 17 March 1864 Grant met Sherman at the important Federal base and supply depot of Nashville, which was second only to Washington DC in importance to the Federal cause. They discussed the steps needed to end the war: namely, a great sweep to the eastern seaboard by Sherman from the mid-west, followed by a northern advance to take Lee's army from the rear whilst Grant maintained a growing pressure on the Confederacy from the north. By May 1864, Sherman was ready to move from Chattanooga towards Atlanta. On 4 May he set out at the head of over 100,000 men and 254 guns. His immediate opponent was General J. E. Johnston leading two Confederate armies totalling 65,000 men. The Confederates, from the first, fought on the defensive, and the first two months passed in manoeuvre and skirmishing; Johnston's aims being to hold Atlanta and keep his army intact, in the hope that political development in Washington or overseas might even now end the war with a compromise.

Step by step, river line by river line, Sherman pushed Johnston back, outflanking one Confederate position after another. Tiring of this repetitive pattern, on 27 June Sherman attempted an assault at Kennesaw Mountain (3), but was worsted. In sum, Johnston's Fabian strategy was earning the Confederacy a little time: Sherman had barely covered one hundred miles in the last seventy days. Unfortunately, President Jefferson Davis decided that more action was required, and the cautious but effective Johnston was replaced by the fire-eater and over-bold General Hood. On 20 July, Hood left a strong position to attack Sherman at Peach Tree Creek north-east of Atlanta, but his attacks failed three times (4). Union cavalry were now ranging the area cutting railway lines, and on 1 September Sherman closed in around Atlanta. Hood abandoned the city without a fight, retiring to Alabama, and by 2 September Union forces had occupied Atlanta. The half-way point of Sherman's great wheel through the Southern States towards the distant Atlantic had been safely reached.

Sherman's bold scheme was now to march for the coast, abandoning his lines of communications behind him, relying on living off the countryside until he could reach the Atlantic, where he hoped that Admiral Farragut might already be master of Mobile. In fact, the US Navy had been in Mobile Bay since 5 August, but with no troops aboard there was no question of capturing the port. Nevertheless, Sherman was sure his strategy would work. After detaching two corps to reinforce General Thomas at Nashville, he destroyed the railroads and supply depots of Atlanta and in November the 'march to the sea' began. Moving on a sixty-mile front, Sherman systematically burned and plundered his way eastwards. With hardly any forces to oppose him, Sherman reached the coast near Savannah after marching through

Georgia on 10 December. Eleven days later, the local Confederate troops abandoned the city and port. Soon Sherman was resupplying his army from the fleet train.

Away in the interior, meanwhile, General Hood had made the anticipated countermove and attacked Nashville with 30,000 Confederates. Thomas, with 50,000 men, held his ground at Spring Hill (29 November) and Franklin (30th) counterattacked from Nashville (15/16 December) (5) and routed Hood with heavy casualties. With the Confederate Army of Tennessee in ruins, only one substantial force remained in the field—the most famous of all, the Army of North Virginia, under command of the redoubtable Robert E. Lee.

Final Operations in the Eastern Theatre: Operations around Raleigh and Petersburg

His army resupplied, Sherman remorselessly turned north, and whilst the Confederates fell back from Savannah to Charleston and thence to distant Raleigh, the Union army marched by the inland flank through Columbia towards Virginia. A Federal force was landed at Wilmington and, after winning the fight for Fort Fisher (6) on 15 January 1865, these troops pressed inland to catch the Confederates between two fires. J. E. Johnston sortied out from Raleigh only to be defeated by Sherman's army at Fayetteville (7) on 11 March, and a week later the Confederates sustained yet another defeat at Bentonville (8) to the east of Raleigh.

General Grant, meanwhile, was closing inexorably on Petersburg. In early May 1864, his army had crossed the River Rapidan west of Chancellorsville, and, after a tough fight in 'the Wilderness', by early June had pressed Lee back into the defences of Richmond. Grant now decided to turn Lee's position from the south and thus avoid costly frontal assaults against the defences of the Confederate capital. To do this, he set about capturing Petersburg (see map, The Petersburg Campaign). Fighting a series of sharp actions at Cold Harbor (9) and Malvern Hill (10), Grant skilfully broke contact with Lee and, by brilliant staff work and transportation, completely fooled his enemy by moving first over the Chickahominy River and then the far larger James River, over a 2,100ft. pontoon bridge (the largest built to date). Grant then marched upon Petersburg. After forty-four days of continuous fighting and manoeuvre, Lee found himself at last pinned down to positional warfare. The struggle now came to depend on long trench lines. Grant continuously extended his lines, seeking to outflank the Confederate position, but Lee matched every move. Indeed, growing

bold, in July Lee detached General Early's corps for one last dash towards Washington DC by way of the well-trodden Shenandoah Valley. Delayed at Frederick on 9 July, Early reached the approaches to Washington on the 11th, only to find that Grant had calmly transferred VI Corps by sea to the capital and that Lincoln's nerve was unshaken. Early escaped on the 12th, and to prevent any recurrence of this type of raid Grant sent Sheridan to devastate the Valley.

Lee's "last 100 days" from early 1865 centred around the siege of Petersburg (11). Already under blockade from April 1864, the operations were fought with great ferocity on both sides. The ten-month siege dragged on, partly due to lack of Union resolution and partly to Lee's brilliant tactical defence. Grant, by February, had 124,700 troops at his disposal against Lee's 57,000 men, but despite numbers of probes and small battles Grant could still not quite bring Lee to book. On 2 April, however, Grant at last broke through the Confederate right wing at Five Forks, and this compelled Lee to abandon both Richmond and Petersburg. Moving west in the hope that somehow he might yet make contact with J. E. Johnston's remnants near Raleigh, Lee was making his last gamble. But now that he had his wily opponent out in the open, Grant was able to make full use of his superior forces. Moving rapidly west himself, Grant continuously interposed his forces between Lee and Johnston. On 6 April, Sheridan and the Union cavalry occupied Appomattox Court-House (12), severing Lee's line of retreat.

Robert E. Lee realized that at last the game was up. On 9 April he met Grant at Appomattox, and agreed to surrender the Army of North Virginia. Magnanimous in victory, Grant allowed 28,356 Confederates to give their parole, and to keep their horses and mules, and tactfully forbade open rejoicing in the Union camps. He also provided rations for the famished Confederates. The war was won, the Union had been saved and the reassimilation of the Confederate States could begin.

Lincoln did not live to savour the fruits of victory, for on 14 April he was assassinated at the theatre. Twelve days later, J. E. Johnston surrendered the last Confederate force to Sherman at Raleigh, having accepted an armistice on 18 April. The very last Confederate force of any size laid down its arms over the Mississippi River on 26 May. The great American Civil War was at last over. The Union cause had triumphed, but in failing, the Confederacy and its armies had written some immortal pages of military history. Some have called it the first truly modern war.

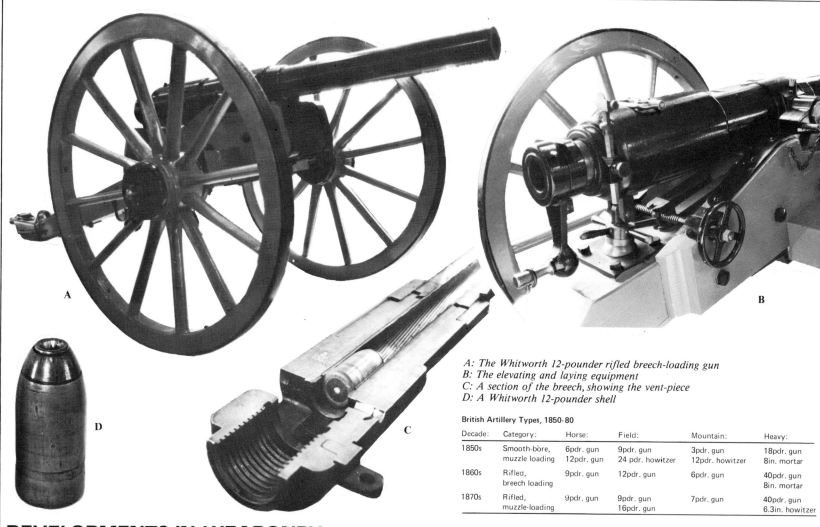

A: The Whitworth 12-pounder rifled breech-loading gun
B: The elevating and laying equipment
C: A section of the breech, showing the vent-piece
D: A Whitworth 12-pounder shell

British Artillery Types, 1850-80

Decade	Category:	Horse:	Field:	Mountain:	Heavy:
1850s	Smooth-bore, muzzle loading	6pdr. gun 12pdr. gun	9pdr. gun 24 pdr. howitzer	3pdr. gun 12pdr. howitzer	18pdr. gun 8in. mortar
1860s	Rifled, breech loading	9pdr. gun	12pdr. gun	6pdr. gun	40pdr. gun 8in. mortar
1870s	Rifled, muzzle-loading	9pdr. gun	9pdr. gun 16pdr. gun	7pdr. gun	40pdr. gun 6.3in. howitzer

DEVELOPMENTS IN WEAPONRY

Artillery

The design of artillery progressed little between 1815 and the 1850s, but thereafter considerable developments took place. These were basically the introduction of rifled artillery barrels and breech-loading. The French attempted to make rifled artillery in 1846, and during the Crimean War the British Army converted a few muzzle-loading smoothbore 68-pounders and 8-inch guns into a rudimentary form of rifled ordnance. Although ranges of up to 2,600 yards and greater accuracy than heretofore were obtained at the siege of Sebastopol, these weapons proved unpopular and were soon dropped.

The major advance was the Armstrong 3-, 9- and 12-pounder rifled breech-loading guns, which appeared between 1854 and 1859. Soon, the principles had been adapted to suit 40- and 64-pounders as well. The tangent scale was added to increase accuracy of laying and aiming (illustration B). On the smaller pieces, the breech block (or vent-piece) screwed into the rear of the reinforced barrel; in the larger cannon a side closing vent-piece was employed (illustration C). Simultaneously, the Whitworth 12-pounder rifled breech-

loading gun was introduced, with a hexagonal bore, a 2.75in calibre and a range of 3,000 yards at five degrees elevation. This piece was used, alongside the smoothbore muzzle-loading 12-pounder Napoleon and rifled Parrot guns, during the American Civil War with considerable effect (illustration A). Surprisingly, however, a reaction set in against breech-loading, and whilst the Prussian Krupp and other European artillery manufacturers and War Ministries were beginning to see the advantages of the system and ordering its adoption, the British Army reverted to rifled muzzle-loaders from 1869 to the mid-1880s believing them to be more reliable. By 1870, however, the smooth-bore cannon had virtually disappeared from all major modern armies, although its use lingered on in India.

Fortress guns remained massive pieces, often mounted upon sloping runways so as to check the effects of recoil (illustration E). Large howitzers began to appear in the early 1880s.

The French concentrated on improving their infantry weapons (see below), and neglected the modernization of their main field artillery, going to war in 1870 with the

guns they had used at Solferino in 1859. Their shells were fitted with outdated time fuzes, ranged for only 1,200 or 2,800 yards, whilst the Prussian steel-cast, breech-loading Krupp guns (introduced from 1866) fired shells loaded with cordite which exploded on impact.

Napoleon III and his advisers placed great reliance on their secret weapon—the Mitrailleuse—to redress the known inferiority of their artillery. This weapon, with its thirty barrels, could fire accurately to a range of 2,000 yards at a rate of 150 rounds a minute. Unfortunately for the French, the weapon was not employed as an infantry-support piece but as an artillery piece kept well to the rear, and it proved wholly disappointing in this role. As early as 1862 the Americans had introduced the Gatling gun (illustration F)—the ancestor of the machine-gun—with a range of 1,000 yards and a rate of fire of 350 rounds a minute, fired from at first ten and later six barrels.

The development of artillery ammunition kept pace with these inventions. During the 1860s spherical shot and shell gave way to elongated projectiles (many equipped with studs) as needed for the new rifled barrels.

An armour-piercing shell was invented by a Major Palliser, and new 'segment' shells—a cross between case and shrapnel—came into use. Time fuzes were gradually replaced by impact-fuzes, and great progress was made in fuze design after 1870. Although cordite was in use as a propellant on the Continent, the British Army clung on to gunpowder until 1891. The addition of gas-checks to shells reduced the wear and tear on the rifled barrels by preventing the forward escape of gases when a round was fired. As an improved means of achieving an efficient discharge, a friction tube (successor to the percussion tube) was introduced in 1853. Instead of relying upon the old port-fire or slow-match, a copper-tube filled with fine powder was inserted into the charge through the touchhole, and was ignited by means of a lanyard dragging a roughened bar through the detonating composition, which was sparked off by the effects of friction thus caused.

Artillery, therefore, was improving in terms of range, rate of fire and lethality. Indirect fire was still in the future, however. The chart summarizes the major types of guns on issue to the Royal Artillery between 1850 and 1880.

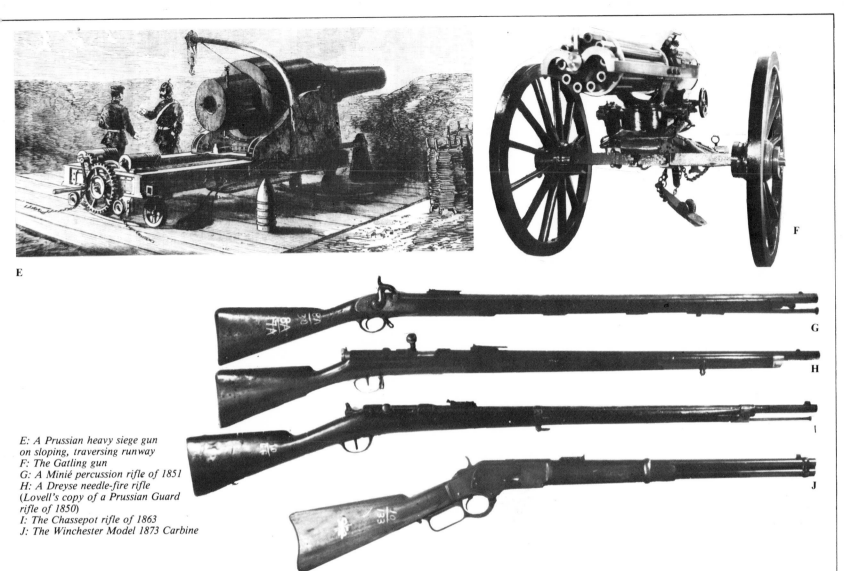

E: A Prussian heavy siege gun
on sloping, traversing runway
F: The Gatling gun
G: A Minié percussion rifle of 1851
H: A Dreyse needle-fire rifle
(Lovell's copy of a Prussian Guard
rifle of 1850)
I: The Chassepot rifle of 1863
J: The Winchester Model 1873 Carbine

Small Arms

Although there were scant changes in terms of tactical method before the American Civil War—the armies that fought at Balaclava or Inkerman might just as well have served at Waterloo—there were notable advances in the development of firearms, as with artillery, linked to the improvement of industrial techniques and general inventiveness. Great Britain made a notable series of early contributions. Joseph Manton, the gunsmith, developed an earlier idea of a Dr. Forsyth and produced a perfected version of the copper percussion cap during the 1820s. No longer was a weapon's discharge dependent upon a flint striking a spark from steel. By 1835 the percussion cap had been adopted by most continental armies, and even by the innately-conservative British Army in 1842. The American inventor, Samuel Colt, employed it in his development of the first practicable repeating revolver (1835), which also incorporated the novelty of inter-changeable parts. Its magazine cylinder held six cartridges, but the hammer had to be pulled back manually to revolve the chamber.

The most significant changes, however, related to the rifle. In 1847, Charles Minié developed a paper cartridge incorporating a hollow-based pointed and elongated bullet, which was easy to load down the rifling of a barrel when compared to the earlier Baker rifle, whose users had to have resort to a small mallet to hammer their ball home with the aid of the iron ramrod. Once fired, moreover, the Minié bullet expanded to fit the bore very tightly, thus avoiding serious gas leaks. Weighing 10lbs, and with an 18mm.calibre, the French Minié rifle (illustration G) was fired by a percussion cap being hit by the hammer of the cocking-piece. It was far more accurate, and longer-ranged, than the smoothbore muskets of 'Brown-Bess' vintage, and all major armies were equipped with it by the mid-1850s, although the British soon replaced it with the Enfield.

However, improvements now followed fast and furious. As early as 1838 the Prussian Johann von Dreyse had invented a breech-loading needle gun, adopted by the Prussian Army in 1841. This weapon (illustration H) weighed 10¾lbs had a 0.66in or 15.4mm.calibre, was sighted to 600 yards and for reloading used a breech bolt-action. It fired a self-igniting paper cartridge and a rate of ten rounds a minute could be obtained. A needle on the bolt passed through the powder-charge to explode a percussion pellet on the base of the bullet itself. There was no need for ejection, as the paper-cartridge was destroyed in the explosion. This weapon achieved a far higher rate of fire than ever obtained with muzzle-loading weapons, and implied a revolution in infantry fighting. The French realized the need to develop an equal, or superior weapon, and in 1863 came up with the 0.43in.calibre Chassepot rifle, (illustration I) in every way a superior weapon to the Prussian Zündnadelgewehr. It incorporated superior breech-sealing devices made of rubber, and an improved bolt mechanism that raised the rate of fire. A reduction in calibre raised the sighting to a theoretical 1,200 yards (although in battle anything over 300 yards was exceptional in practice), and lightened the weight of the cartridges and so increased the number a soldier could carry. Opposition to the weapon's use was overcome, and by 1870 one million were on issue to Napoleon III's army. The breech-loader was not adopted overnight. In the US Civil War, for example, the muzzle-loading firearm long remained in favour (partly for reasons of economy), but by 1865 breech-loaders were on general issue to the Union armies, including repeaters such as the Henry rifle, the famous Winchester cavalry carbine, and the Spencer carbine (illustration J). Weighing 7½lbs, and with a .56in.calibre, the Spencer held seven metallic rim-fire cartridges in a tubular magazine concealed in the butt. Reloading was achieved by manipulating a lever which revolved the rolling-block breech to eject the spent cartridge case and load a new bullet. This weapon was credited with a rate of fire of thirty rounds a minute, and greatly assisted the North's victory in the Civil War. Such weapons, of course, required copper (later brass) cased gas-tight cartridges, incorporating the percussion cap in its base. These, together with the rimmed cartridge, were French inventions that were greatly improved by American arms manufacturers. In sum, these improved weapons would mean that, by 1885 when the introduction of smokeless powder had taken place, infantry could engage at ranges of up to 500 yards. Fire-power, whether dispensed by artillery or rifle-carrying troops, finally ruled.

The Unification of Germany under Prussia

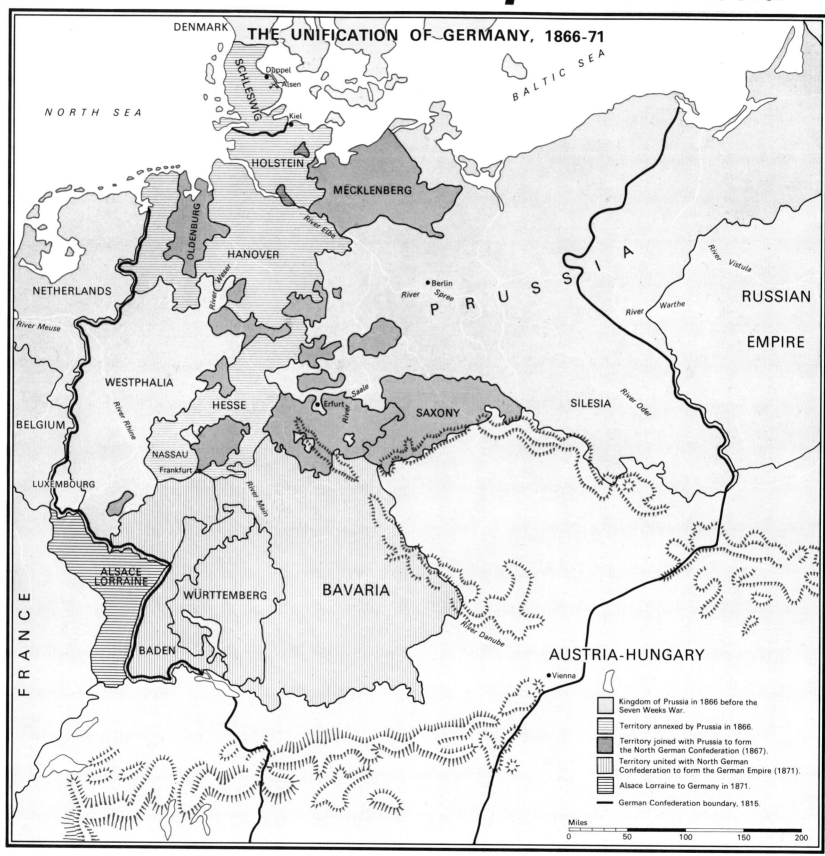

THE UNIFICATION OF GERMANY, 1866-71

DENMARK

BALTIC SEA

NORTH SEA

Düppel
Alsen
Kiel

SCHLESWIG

HOLSTEIN

MECKLENBERG

River Elbe

OLDENBURG

HANOVER

River Weser

NETHERLANDS

River Meuse

Berlin
River Spree

P R U S S I A

River Warthe

River Vistula

RUSSIAN

EMPIRE

WESTPHALIA

River Rhine

HESSE

Erfurt
River Saale

SAXONY

SILESIA

River Oder

BELGIUM

NASSAU

Frankfurt

LUXEMBOURG

River Main

F R A N C E

ALSACE
LORRAINE

WÜRTTEMBERG

BAVARIA

River Danube

BADEN

AUSTRIA-HUNGARY

Vienna

Kingdom of Prussia in 1866 before the Seven Weeks War.

Territory annexed by Prussia in 1866.

Territory joined with Prussia to form the North German Confederation (1867).

Territory united with North German Confederation to form the German Empire (1871).

Alsace Lorraine to Germany in 1871.

German Confederation boundary, 1815.

Miles
0   50   100   150   200

The kingdom of Prussia emerged from the Napoleonic Wars with greatly enhanced prestige, but the German Confederation of 1815 was in many ways unsatisfactory for Prussian aspirations to leadership. Otto von Bismarck, President of the Ministry and Minister for Foreign Affairs from 1862, was determined to defeat Austrian claims to supremacy and to replace the loose Confederation with a truly united Germany under the leadership of Prussia. He faced much opposition to his schemes from within Prussia itself, but pursued his intentions with determination.

Failing to conciliate the Prussian liberals, he high-handedly ruled without the support of a majority, and levied the money he required to expand and up-date the army. In 1863, he aided Russia by closing the Polish frontiers whilst Tsar Alexander II repressed the nationalist revolt (see p. 138) and thus earned Prussia its only ally at that time. Next, when a Congress of German Princes was summoned to Frankfurt in August that year, he boycotted the meeting and thus ensured that it would fail, and in so doing prevented Austria emerging as the leader of a remodelled German Confederation.

Danish claims to the two Duchies of Schleswig and Holstein came to a head in 1863 when Frederick VII claimed Schleswig to be part of Denmark and demanded compensation for Holstein. The Diet of the Confederation sent in Saxon and Hanoverian troops, and for his own ends (the desire to secure Kiel) Bismarck opposed the Danish claim, and thus effectively sided with Austria. The Danish War followed (February-October 1864), during which Prussian and Austrian troops occupied Schleswig and invaded Denmark. The storm of Duppel on 18 April ended the initial military phase, but attempts by a London conference to mediate in the dispute failed. Bismarck subtly used the Austrian military success to convince Prussia of the need for a larger army. Fighting recommenced, but the surrender of the Danish Army at Alsen on 29 July 1864 led to peace preliminaries on 1 August, and the Treaty of Vienna in October. Denmark renounced all claim to the Duchies, which were handed over to Austria and Prussia. Schleswig was jointly occupied, but Prussia soon forced the troops of the German Diet to withdraw from Holstein, replacing them with her own. This soon became a major source of friction between Vienna and Berlin. Negotiations at Gastein (August 1865)—whereby Austria took over Holstein and Prussia assumed responsibility for Schleswig—only papered over the cracks. More manoeuvring followed, but after concluding a treaty with Italy (promising mutual military support), Bismarck was ready for the show-down. On 14 June 1866, Prussia formally left the German Confederation over the Holstein issue and began to mobilize. Most German states rallied to Austria in the crisis, but Prussia had by now a far superior army.

The Austro-Prussian War (the events of which are summarized on pp. 190-191) was ended by the Peace of Prague, which was signed on 23 August 1866. By its terms, the German Confederation of 1815 was abolished,

Prince Otto Edward Leopold von Bismarck (1815-98), the 'Iron Chancellor' of Prussia, was the dominant statesman of the mid-nineteenth century. He achieved the unification of Germany by victorious wars against Austria and then France. The result was the Second German Empire.

and Austria withdrew from direct participation in German affairs, agreeing to pay a war indemnity to Prussia, and ceding Venetia to Italy (see p. 159). The new North German Confederation—consisting of states north of the River Main—was brought into existence, very much subject to Prussian influence. Prussia also gained substantial territory, including the Danish Duchies, Hesse-Cassel, Nassau, Hanover and part of Hesse-Darmstadt. These acquisitions brought Prussia four million new citizens, and made her the predominant state in Germany. Yet it was a reasonable peace treaty, for Bismarck was anxious to avoid a prolonged alienation of Austria. But Vienna's interests would soon be absorbed by Hungarian and Balkan developments. The South German states, however, remained subordinate to neither Austria nor Prussia.

The consolidation of North Germany aroused the alarm of Napoleon III, and relations between Paris and Berlin rapidly worsened. Bismarck encouraged the development of a crisis, for he was convinced that only a struggle with France would induce the South German states to seek unification under Prussian hegemony. Using the Spanish Succession crisis (see p. 192) to further his ends, Bismarck manoeuvred Napoleon III into taking the initiative. The Franco-Prussian War followed (1870-71), and the triumph of Prussia's armies (see maps on pp. 192, 194 and 195) led to the declaration of the German Empire at Versailles in January 1871, by which the South German states joined the North German states. By the Treaty of Frankfurt (10 May 1871) Prussia gained Alsace and much of Lorraine, including valuable iron-ore resources. France was also required to pay a large indemnity. These severe terms, which moved France's eastern frontier back from the Rhine to the line of the Vosges mountains, were harbingers of trouble and would lead to ever-increasing resentment within France and to a growing demand for revenge. The war also left France a Republic once again, whilst in Italy King Victor Emmanuel took advantage of the withdrawal of French troops from the Holy See to enter Rome. Farther east, the Tsar (encouraged by Bismarck) repudiated the Black Sea clauses of the Treaty of Paris (see p. 142). In 1878 the German Empire displayed its prestige by mediating at the Congress of Berlin in the perennial Balkan dispute (see p. 196). Next year, the development of the Dual Alliance between Germany and Austria-Hungary created a strong Central European Power base.

Thus the creation of the German Empire, although achieving Bismarck's ambitions, left many seeds of future discord. Large conscript armies and massive armaments were retained by the major continental powers as the sole means to secure their self-protection, and stage by stage the lines were being drawn for a shattering war. By the end of the century, colonial rivalry, European tension and the naval arms race—together with the traditional instability of the Balkan region—would be threatening catastrophe.

The Six Weeks (Austro-Prussian) War of 1866

Determined to gain Prussian paramountcy in Germany, Bismarck prepared to fight Austria in Bohemia and (with the aid of his allies) in North Italy. On 14 June 1866, hostilities opened on both fronts as Prussian troops advanced into Saxony and their Italian allies invaded Venetia. As the Austrian commander, von Benedek, began to assemble the main Austrian army in Bohemia (see map, Campaign in Bohemia, opposite) it became obvious that this was likely to prove the major theatre of war.

The Prussian mastermind, Helmuth, Graf von Moltke, accordingly started to assemble three armies around the Bohemian and Saxon frontiers—on the Elbe, in Northern Silesia and Saxony respectively—using five railways to expedite the concentration of his forces. The Army of the Elbe gathered around Dresden, the First Army around Gorlitz, and the Second Army near Schweidnitz. Moltke hoped to be able to carry out these moves fast enough to forestall Benedek's concentration, and then to advance deep into Bohemia from all three directions.

In fact, Benedek moved with unanticipated speed, massing 250,000 men near Josephstadt by 26 June, days before the Prussian Second Army had entered the mountain passes. Their First and Elbe armies were in touch to the north-west, but were at least a week's march away from the Crown Prince advancing from Schweidnitz. An opportunity thus existed for the Austrians to fall upon the Prussian Second Army as it emerged from the difficult mountain passes, but the Austrian high command was excessively defensive-minded, and the chance was allowed to pass. Even worse, by 2 July they had allowed their troops to be shepherded into a central position on the high ground between the Elbe and the River Bistritz, and along the main road running south-east from Königgrätz in the direction of Vienna. On that date the Prussian First Army's advance forces were in contact with Austrians near Sadowa, with the Second Army coming up from the north and the Army of the Elbe converging from the west. Moltke was thus presented with the opportunity of enveloping his enemies after the fashion of Napoleon. He made full use of the chance, and the great and decisive Battle of Königgrätz was the outcome.

Although the Austrian commander, the Archduke Albert, had more than held his own in North Italy, defeating General La Marmora's Italian army at Custozza on 24 June and driving back the remnants over the River Mincio, the war was effectively over after Königgrätz. Bismarck, aware that France might enter the struggle, advised the King of Prussia to grant Austria generous terms in the Peace of Prague. Nevertheless, Prussia was now undisputed leader of the new North German Confederacy.

The Battle of Königgrätz, 1866

Field-Marshal-Lieutenant von Benedek, at the head of seven Austrian and one Saxon army corps (or some 270,000 men), placed his army in a defensive position west of the Elbe near Königgrätz, and awaited the arrival of the converging Prussian forces. He massed his guns along the crests of the heights of Chlum and Lipa (A), and placed four corps in two lines to form his centre. Benedek's left comprised the Saxon Corps and part of the Austrian 8th, with a cavalry division in rear (B). He placed his right wing (the 4th and 2nd Corps) in a refused position running from near Chlum to Lochenitz through Nederlist (C). Benedek was relying on his superior artillery to wear down any attack.

In mid-morning of 2 July, von Moltke learnt for the first time that the whole Austrian Army was west of the Elbe. He at once ordered Prince Frederick Charles to advance with his First Army towards the River Bistritz and Sadowa and launch a frontal attack to pin Benedek, whilst the Crown Prince of Prussia led the Second Army to envelop the Austrian right, and the Austrian Army of the Elbe swung south to attack the other flank. He thus planned a double-envelopment, employing all his 278,000 troops.

At 3am on 3 July the Prussian 8th Division attacked, and took, Sadowa, but was halted by the Austrian gun-line in the woods along the ridge behind (D). All morning the action grew, until all six divisions of the Prussian First Army were engaged and tied down, after suffering 5,000 casualties. One (the 7th Division) was counter-attacked by the Austrian 3rd Corps (E) and almost wiped out. The Prussian Army of the Elbe was only becoming involved against the Saxons by 11am (F), and its effect was not immediate. The First Prussian Army was on the verge of defeat.

Benedek was about to order an all-out attack by his reserves when he discovered that the commanders of the 4th and 2nd Corps had advanced to the front without orders, and taken post north of Chlum (G). Furious at this insubordination, he ordered them to resume their assigned stations, but it took three sets of orders before Generals Festetics and Thun obeyed. They were in the act of falling back when the opportune arrival of the Prussian Crown Prince's leading formations caught them in flank (H). The Prussians routed the troops in their path and stormed into Chlum (I) itself, reversing the fortunes of the day. However, Benedek's men rallied to the crisis, and Second Army's advance was held. A fine cavalry charge and heroic fighting by the Austrian gunners allowed Benedek to withdraw his men over the Elbe, but he had lost 44,000 casualties and could not renew the fight. Negotiations led to a ceasefire, and the end of the Six Weeks' War. Moltke had lost 9,000 casualties in achieving this great, though hard-fought, victory.

The Prussian Army in the field: they learned the hard way the need to adopt fluid infantry tactics in replacement of close-order battle formations. (John D. Walter)

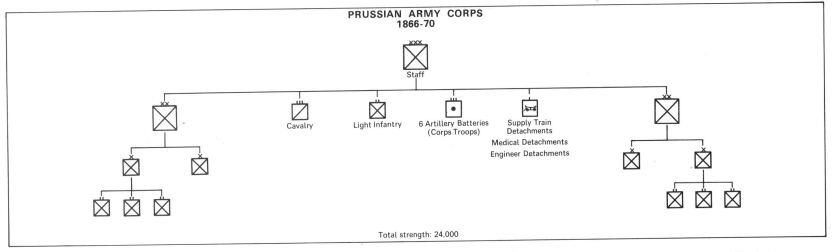

PRUSSIAN ARMY CORPS
1866-70

Staff

Cavalry

Light Infantry

6 Artillery Batteries (Corps Troops)

Supply Train Detachments
Medical Detachments
Engineer Detachments

Total strength: 24,000

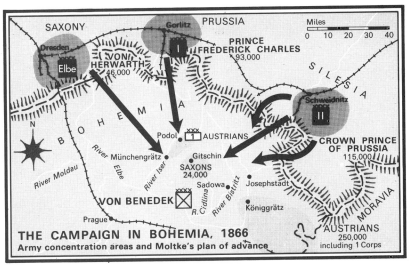

THE CAMPAIGN IN BOHEMIA, 1866
Army concentration areas and Moltke's plan of advance

SAXONY   Görlitz   PRUSSIA

PRINCE FREDERICK CHARLES
93,000

Dresden   Elbe   VON HERWARTH 46,000

SILESIA

Schweidnitz

CROWN PRINCE OF PRUSSIA
115,000

Podol   AUSTRIANS

River Münchengrätz   Gitschin

SAXONS 24,000

River Moldau   River Elbe   River Iser   Sadowa   Sadlina   Josephstadt

VON BENEDEK

R. Cidlina   River Bistritz   Königgrätz

Prague   MORAVIA

AUSTRIANS 250,000 including 1 Corps

Miles
0 10 20 30 40

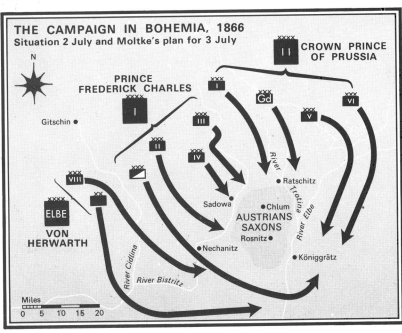

THE CAMPAIGN IN BOHEMIA, 1866
Situation 2 July and Moltke's plan for 3 July

CROWN PRINCE OF PRUSSIA

PRINCE FREDERICK CHARLES

Gitschin

I   Gd   V   VI

III   II   IV

VIII

ELBE   VON HERWARTH

River Cidlina   Sadowa   Chlum   River Trotina   River Elbe

AUSTRIANS SAXONS

Ratschitz

Rosnitz

River Bistritz   Nechanitz   Königgrätz

Miles
0 5 10 15 20

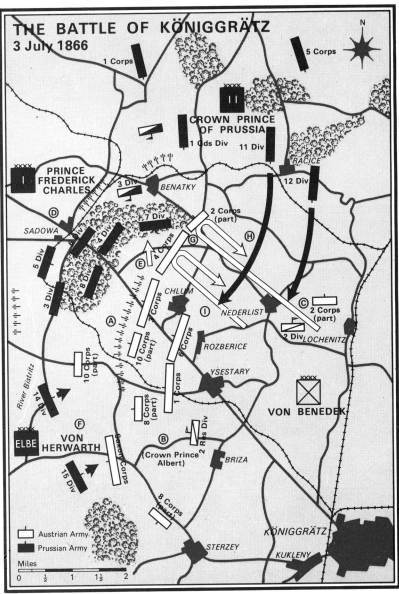

THE BATTLE OF KÖNIGGRÄTZ
3 July 1866

N

1 Corps   5 Corps

II   CROWN PRINCE OF PRUSSIA

1 Gds Div   11 Div

RACICE

I   PRINCE FREDERICK CHARLES

3 Div   BENATKY   12 Div

2 Corps (part)

SADOWA   7 Div   G   H

5 Div   4 Corps   E   NEDERLIST   C   2 Corps (part)

3 Div   8 Div   3 Corps   CHLUM   I   2 Div LOCHENITZ

A   ROZBERICE

10 Corps (part)   6 Corps

10 Corps (part)   1 Corps   YSESTARY

River Bistritz   14 Div   8 Corps (part)   VON BENEDEK

F   B   2 Res Div

ELBE   VON HERWARTH   Saxony Corps   BRIZA

15 Div   (Crown Prince Albert)

8 Corps (res)

Austrian Army
Prussian Army

STERZEY   KÖNIGGRÄTZ

KUKLENY

Miles
0 ½ 1 1½ 2

191

The Franco-Prussian War, 1870-71

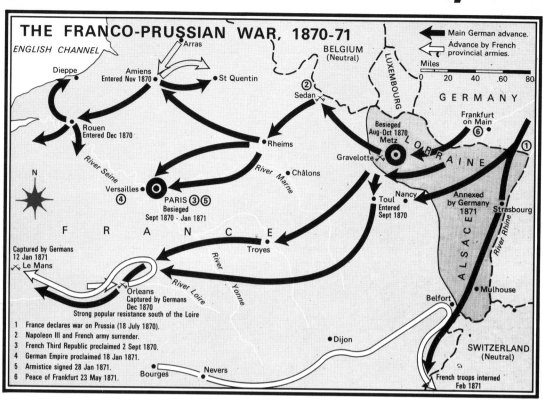

THE FRANCO-PRUSSIAN WAR, 1870-71

Main German advance.
Advance by French provincial armies.

1. France declares war on Prussia (18 July 1870).
2. Napoleon III and French army surrender.
3. French Third Republic proclaimed 2 Sept 1870.
4. German Empire proclaimed 18 Jan 1871.
5. Armistice signed 28 Jan 1871.
6. Peace of Frankfurt 23 May 1871.

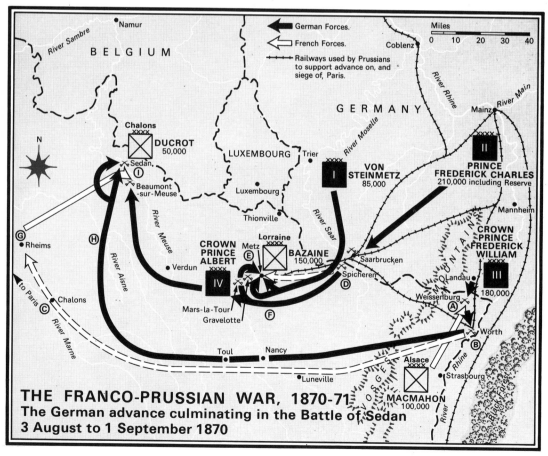

THE FRANCO-PRUSSIAN WAR, 1870-71
The German advance culminating in the Battle of Sedan
3 August to 1 September 1870

German Forces.
French Forces.
Railways used by Prussians to support advance on, and siege of, Paris.

The Franco-Prussian War was ostensibly caused by Prince Leopold of Hohenzollern's claim to the Spanish throne, but it also served Otto von Bismarck's desire to cement German unification under Prussia. In fact, Prince Leopold—aware that France was strongly opposed to his candidature—withdrew, but by wily and unscrupulous diplomatic moves the Prussian Chancellor manoeuvred Napoleon III into declaring war on 19 July 1870. This was a rash decision, for it was only with the utmost difficulty that seven army corps and the Imperial Guard could be mobilized and moved to the Rhine frontier; three French armies were eventually assembled at wide intervals. Some 100,000 men gathered around Strasbourg, another 150,000 massed near Metz, and a 50,000-strong Reserve Army collected at Châlons on the River Marne.

The Prussian mobilization went far more smoothly, thanks to the genius of Helmuth von Moltke and the efficiency of his General Staff, which made full use of the railways. Some thirteen corps (475,000 men) were mobilized, and subdivided into four forces in the Palatinate area: two corps forming the Prussian First Army at Trier, four corps making up the Second Army at Mainz, as many more constituting the Third Army around Landau and Maxau, and finally a reserve of two more corps collecting behind Mainz. These initial moves were completed by 1 August (see map, The German Advance culminating in the Battle of Sedan).

The French plan conceived of a major advance by their two main armies acting in conjunction to the River Main, thus splitting Germany and then On to Berlin! However, so unprepared were they that no serious forward movement could be undertaken and, accordingly, Prussia was able to take the offensive on 3 August against a largely passive opponent. Von Moltke ordered his three armies to close up towards the inner flank, and to attack between the River Saar and the River Lauter. His intention was to execute a gigantic wheel, pivotting on the right, to drive the French against the Belgian frontier, whilst masking Strasbourg and keeping Macmahon and Bazaine apart.

On 4 August the German Third Army met the 1st Corps of Macmahon's army at Weissenburg (A) and drove it back. On the 6th, the Battle of Worth (B) again ended in a French defeat, and Macmahon, having lost the barrier of the Vosges, decided to abandon Alsace and retire to Châlons (C); a march that he conducted (without interference) between the 7th and 14th, whilst the Prussian Third Army advanced through Nancy and Toul to the Meuse.

Meanwhile, the Prussian First and Second Armies were also on the move. On 2 August there was a skirmish at Saarbrucken, followed four days later by the Battle of Spicheren (D), where the Second Army defeated two French corps, and induced Bazaine to pull back his forces to Metz (E) and abandon all idea of direct co-operation with Macmahon. Thus Moltke's intention of holding the French armies apart had been achieved. The First and Second Armies next crossed the Moselle south of Metz in an enveloping move (F) and, after a number of small battles south and west of the city, including Mars-la-Tour, Vionville and Rézonville on the 16th, Bazaine withdrew nearer to Metz. Two days later, the important engagement of Gravelotte-St. Privat (see map, p. 194) was fought, resulting in the bottling up of the Army of Lorraine within Metz.

Moltke ordered his Third and the newly-created Fourth Armies (the latter, sometimes called the Army of the Meuse, comprising three corps under Crown Prince Albert of Saxony) to press ahead with their joint 220,000 men towards Châlons and Paris. Macmahon, now at Châlons with four corps totalling 130,000 troops, was induced by popular pressure to undertake an attempt at the relief of Bazaine in Metz and, as a preliminary move, transferred his command to Rheims (G), planning to make a northerly detour to reach the besieged city. This invited a Prussian turning movement by Third and Fourth Armies to cut him off, and this was duly carried out (H). Abandoning all hope of reaching Metz after a repulse at Beaumont (I), Macmahon entered Sedan. On 1 September, General Ducrot, succeeding the wounded Macmahon in command, and with Napoleon III at his side, fought and lost the Battle of Sedan as the Prussians and Germans encircled the city. The next day, Napoleon III surrendered at the head of 83,000 men and 449 guns. The battle itself had cost the French 17,000 casualties to the German 9,000.

On 4 September the German advance on Paris recommenced, and the siege of the French capital began on 20 September (see p. 195). The Third Republic had been proclaimed on 4 September, and a government of National Defence set up with a 'delegation' operating from Tours. An Army of the Loire (six corps or possibly 750,000 men in all) was raised, under General de Paladines, and soon resistance was being offered to the invaders around Lille, in Brittany, at Lyons and Bourges.

As the siege of Paris continued, on 27 October Metz capitulated, releasing two Prussian armies. The French Army of the Loire recaptured Orléans by 4 December, but its intention of marching to relieve Paris was thwarted by the arrival of the Second Army from Metz, and the city was lost again on the 5th. Other regional armies continued a series of operations, but they were badly co-ordinated, and 90,000 troops sought internment in Switzerland in late January.

THE GENERAL STAFFS

Prussia, under a series of great military men ranging from Scharnhorst after Jena through von Müffling and Schellendorf to the great Moltke, developed the most refined and professional General Staff system of the nineteenth century. Its major features are summarized in diagrammatic form (right). A major step for the Prussian General Staff was its ability (through royal patronage) to break away from the control of the Berlin Ministry of War. The argument gained ground that when a state of actual war existed, the chief of the general staff should receive full powers to direct it and, conversely, that in times of peace he should have complete liberty to make his plans and preparations against all likely contingencies.

Of equal importance was the development of closely-integrated field staffs, which linked the Great General Staff at Berlin with the fighting formations in the field. The army corps—inherited from the Napoleonic period (see p. 98) and further refined by the late 1860s—was the key formation, its staff being divided into general staff (operations), routine staff (administration), legal branch and intendance (quarter-master) sections. This division of responsiblities mirrored the major sections of the Great General Staff from as early as 1828, and great pains were taken to train promising officers in the requirements and refinements of staff work. However, they soon became a caste set apart from the officers of the regiments; a development that had both its advantages and disadvantages. The staff system did not, in fact, alter a great deal after 1828, but in 1867 at Moltke's insistence the Great General Staff was subdivided (until 1898) into the Haup-État, concerned with preparation and training for war, and the Neben-État, the junior branch, concerned with scientific assignments.

Britain did not possess a regular staff system until 1906, when the Imperial General Staff belatedly appeared. Other countries (Austria, Hungary and Russia in particular) modelled their General Staffs on adaptations of the Prussian system. The French État Major still owed much to the rather top-heavy Napoleonic system. After 1815, a major development was the establishment of staff training schools by Gouvion St. Cyr and Baron Jomini (a step that Napoleon had deliberately avoided), and eventually the Corps d'État Major evolved, starting in 1818. Staff officers were selected by competitive examination and, once chosen, members became part of a closed service (from 1833 until 1870). The disasters of 1870-71 showed up the inadequacies of the system (as, indeed, had the Crimean War in the 1850s); the prophetic warnings of such soldiers as Colonel Stoffel and Marshal Niel (Minister of War in 1867) having gone unheeded. Thereafter, reforms and improvements began to be implemented. By 1900, different staff functions were allotted to distinct bureaus: the First, Second and Third, dealing with personnel, intelligence and administration respectively.

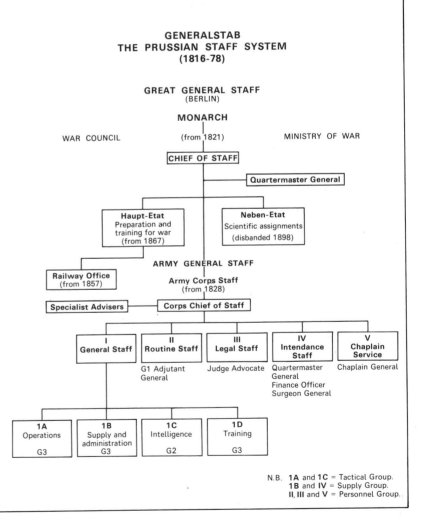

GENERALSTAB
THE PRUSSIAN STAFF SYSTEM
(1816-78)

N.B. **1A** and **1C** = Tactical Group.
1B and **IV** = Supply Group.
II, III and **V** = Personnel Group.

The Franco-Prussian War, 1870-71

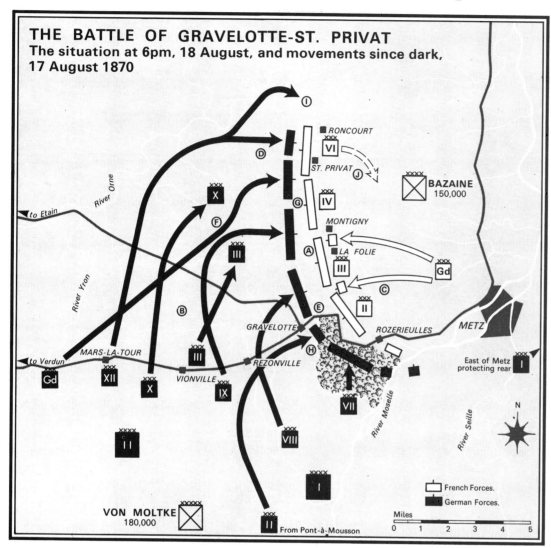

THE BATTLE OF GRAVELOTTE-ST. PRIVAT
The situation at 6pm, 18 August, and movements since dark, 17 August 1870

RONCOURT
VI
ST. PRIVAT
J
BAZAINE 150,000
D
River Orne
X
to Etain
F
IV
G
MONTIGNY
III
A
LA FOLIE
III
Gd
B
C
II
River Yron
E
GRAVELOTTE
ROZERIEULLES
METZ
MARS-LA-TOUR
to Verdun
III
REZONVILLE
H
East of Metz protecting rear
I
Gd
XII
X
VIONVILLE
IX
VII
River Moselle
N
II
VIII
River Seille
I
French Forces.
German Forces.
VON MOLTKE 180,000
II
From Pont-à-Mousson
Miles
0 1 2 3 4 5

The fate of Paris would determine the end of the war. On 28 January the capital surrendered, and only the garrison of Belfort held out for a further nineteen days. On 10 May a peace treaty was signed at Frankfurt-on-Main, whereby France agreed to cede Alsace and north-east Lorraine to the newly-proclaimed Empire of Germany, and also to pay a war indemnity of five billion francs.

The Battle of Gravelotte-St. Privat, 1870
From the outset of clashes on the frontier, the German General Staff set itself the task of separating the two main French field armies: Macmahon's near Strasbourg and Bazaine's around Metz. The latter commander, after some stiff fighting near Mars-la-Tour on 16 August, took the unwise decision (as it proved) to swing back towards Metz in order to take up a strong position (A) between the Rivers Orne and Moselle, with the great fortress in rear to

the east. General Helmuth von Moltke lost no time in executing a daring swing to the north and east during the night of the 17th (B) so as to come face to face with the new French line. A reversed-front battle was in the making, with each army facing towards its base.

Bazaine anticipated that the Prussians might attempt to force his southern flank, so, after deploying four of his corps between Roncourt and Rozerieulles, he moved the Guard Corps to a position from which it could strengthen his left flank (C). Then, with his 150,000 and 520 guns, he awaited the onset of the anticipated Prussian attack, and set about entrenching his position.

In fact the Prussians had made a dangerous miscalculation. Believing that the French position only extended as far north as Montigny, the Second Army sent the XII Corps and the Prussian Guards from Mars-la-Tour to the vicinity of the Orne (D) with orders to

envelop the French right wing, which incomplete reconnaissances had reported to lie invitingly open. Meanwhile, the VII and VIII Corps were to attack the French between Montigny and the Moselle (E) aided by the IX Corps; the remaining pair of formations (which brought Prussian strength up to 180,000 men and 730 guns) were held in generally central reserve (F) ready to exploit the anticipated collapse of the French right wing.

The morning's fighting on the 18th soon revealed his error to von Moltke. Facing heavy fighting along a complete ten-mile front instead of one of only six miles, the Prussian attack lost momentum and made scant progress. A series of rather badly co-ordinated Prussian attacks against the various sectors of the French position continued during the afternoon, and the possibility of Prussian defeat became increasingly strong. In the north the Prussian Guard Corps and its neighbour to the south (the IXth) suffered heavy casualties when they advanced in mass formation (G); much the same was true in the southern sector, where only the fortunate arrival of the fresh II Corps from Pont-à-Mousson at the height of the fighting saved the Prussian right from disaster amidst the forest (H) and the area of Gravelotte, as Bazaine launched the French Guard Corps forward in a heavy counterattack. The reinforcements, however, enabled the Prussian right to withstand this onslaught.

In fact, Bazaine had sent in his reserves rather too early, for he was soon to need the Guards elsewhere. At about 7pm, part of the Prussian XII Corps, after making a considerable detour, at last managed to envelop the French VI Corps to the north of Roncourt and St. Privat (I). This disturbing development induced the French corps commander to abandon his position and fall back southwards (J). The remainder of Bazaine's line was forced to comply, and by 9pm the whole French army was in retreat. Overnight, Bazaine withdrew all his men within the fortress of Metz and its out-lying defences. He had lost 13,000 casualties over the day, and inflicted an estimated 20,000 on von Moltke, but the outcome of the day's fortunes was decisive. Strategically, the main campaign was almost decided. The obsession of keeping all his troops to hold Metz (a single corps would have sufficed for the task) deprived Bazaine of all will-power to attempt the desired junction with Macmahon around Sedan. Thus the exploitation of a small success in the late evening of a hard-fought day, enabled the Prussians to carry out von Moltke's master-strategy; the twin disasters of Metz and Sedan would set the seal on the fate of Napoleon III's French Second Empire.

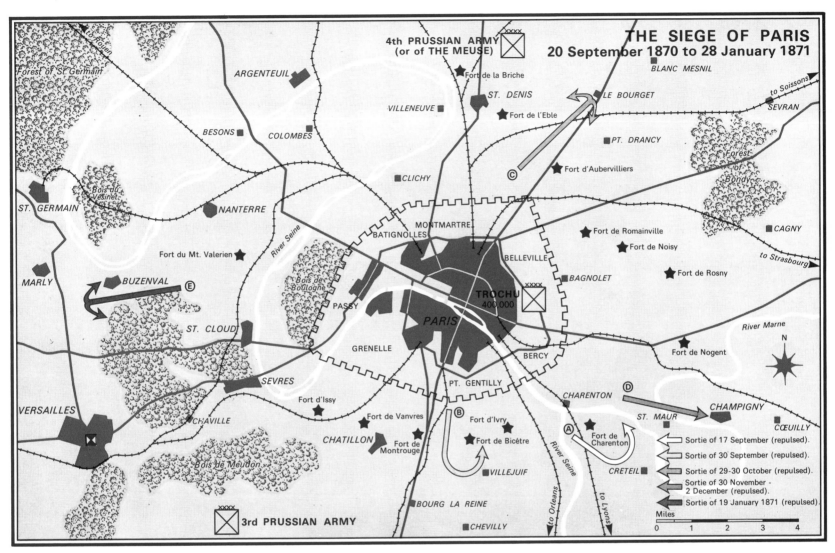

The Siege of Paris, 1870-71

After shutting up Bazaine's army in Metz and eliminating Macmahon's army at Sedan, the Prussians were free to advance on Paris. By 20 September the Army of the Meuse had sealed off the French capital from the north of the Seine, and the Prussian Third Army closed in from the south; 240,000 men with 900 field- and 240 siege-guns, with headquarters at Versailles, formed a siege line of more than 50 miles; within Paris (where the Third Republic had been proclaimed) General Trochu commanded almost 400,000 troops, to include 18,000 sailors and many auxiliaries, backed by almost 2,200 guns. Besides such numerical strength, Paris had the advantage of a series of 14 modern detached forts and other strong fortifications (totalling 61 positions) erected a generation earlier.

So daunting were these outlying positions that von Moltke ruled out a direct assault on the city and ordered his men to settle down to starve the garrison and population into surrender. Bismarck, the German Chancellor, demanded more action, but von Moltke held his ground, and only on 5 January 1871 did the Prussian heavy guns open a real bombardment against selected forts; with some success, but negligible effect on the overall population, or on the resolve of the Government of National Defence. In all, 10,000 shells were fired into the city.

For their part, the French commanders regarded many of their National Guardsmen as too unreliable to commit to a full-scale sortie, and such minor attacks as were ordered—on 17 and 30 September, and 29 October (A, B and C), the largest being between 30 November and 2 December—were designed as morale-raisers rather than serious break-out attempts. On the latter occasion Ducrot led 80,000 men through Champigny to the south-east of the city (D), with the hope of linking up with the French Army of the Loire which was operating towards the city, but the Prussians drove back the assaults with little difficulty.

The siege dragged on and conditions within Paris rapidly worsened. Some contact with the outer world was maintained by intrepid balloonists, but no food could be brought in. The situation could not be maintained indefinitely and, after a final attempt at a break-out through Buzenval to the west on 19 January (E), the end was in sight. With growing starvation and riots within the city, the new Military Governor Vinoy opened negotiations, and on 28 January an armistice was agreed which ended both the siege and the Franco-Prussian War. In terms of military casualties the losses sustained were fairly moderate, the Prussians losing some 12,000 men and inflicting perhaps 24,000 killed, wounded and missing on the French.

The Balkan Crisis of 1877-78 and the Europea

The Russo-Turkish War, 1877-78.

As a result of Turkish atrocities committed in Bosnia and Herzegovina in 1875, Serbia and Montenegro declared war on Turkey in 1876. Russia soon intervened in the struggle and this led to the Russo-Turkish War of 1877. In April of that year, Russian forces under Grand Duke Nicholas invaded Rumania, and were faced by Abdul Krim. Gradually the Russians cleared the country, despite grave logistical problems, and achieved a crossing over the Danube. In late June a second Russian army, under General Krudener, invaded Bulgaria with the aid of Rumanian patriots, and soon the siege of Plevna was under way. It lasted from July until December, when Osman Pasha surrendered after the failure of a desperate Turkish attempt to break out. But the Russian advance had been held up for five months. Fighting continued in Central Bulgaria and in the Caucasus and, after the great success of Senova (8-9 January 1878), the Russians advanced on Adrianople, and thence to within sight of Constantinople. On 31 January an armistice was signed which led, on 3 March, to the conclusion of a treaty.

The Treaty of San Stefano brought the Russo-Turkish War to a close. Turkey agreed to pay a large indemnity, and conceded the independence of Montenegro, Serbia and Rumania, whilst a greatly enlarged Bulgaria would become an independent state under Russian supervizion. Russian gains were mainly on the Caucasus front, including Ardahan, Kars, Batum and Bayazid in Armenia, and part of Bessarabia lost after the Crimean War. By these terms, Turkey would have lost practically all her territories in Europe. However, Britain and Austria were strongly opposed to these terms, which they felt favoured Russia excessively. For a time there was danger of a wider conflict breaking out, but to avert this Bismarck invited all parties to the Conference of Berlin, there to consider revising the terms.

The Treaty of Berlin. After lengthy negotiations, the powers agreed to certain important modifications. The proposed, enlarged Bulgaria was divided into three parts: Bulgaria itself, with self-government under Turkey; Eastern Rumelia, also to be governed by Turkey but under a Christian governor; and Macedonia, restored completely to Turkey. In this way Turkish territory in Europe remained in a cohesive form. Montenegro and Serbia remained independent, but lost some proposed territorial adjustments. Russia's gains in the Caucasus and Bessarabia were confirmed, as was the cession of the Dobrudja to Rumania. It was further arranged that Bosnia and Herzegovina would be administered by Austria, whilst Britain occupied Cyprus. Thus, once

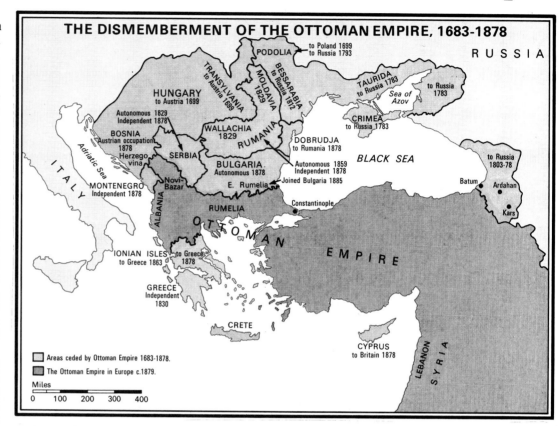

THE DISMEMBERMENT OF THE OTTOMAN EMPIRE, 1683-1878

again, Turkey was restored to a semblance of authority, but the treaty also held the seeds of future discord between Austria and Russia.

The Siege of Plevna, 1877

As strong Russian forces invaded Bulgaria, Osman Pasha marched at the head of 40,000 troops and 77 guns from Widdin to protect the threatened territories. Before he could arrive, however, the Russians had captured Nicopolis on the Danube, so Osman Pasha fell back to Plevna, and set about fortifying the area with 25 miles of defences incorporating 20 redoubts. The position threatened the right flank of the invading armies, and before they could press forward with their advance it was necessary to reduce it by siege.

The first major Russian attack was made on 30 July, but it was repelled with the loss of 8,000 casualties. The next month Osman Pasha launched his counter-offensive from Plevna, with supporting actions on the River Lom and in the Shipka Pass by other Turkish commanders, but he was repulsed from the town of Zgalevitsa and compelled to fall back within his defences. The Russians were constrained to call on Rumania for aid, and Prince Charles of Rumania joined General Krudener's forces before Plevna, bringing his total strength up to 110,000 men and almost 500 guns.

On 11 September, the combined Russo-Rumanian forces attempted a major storming of the defences of Plevna, but the Turks held their ground manfully and only lost a single position, the Grivitza No. 1 redoubt. To achieve this cost the Allies the grim figure of 18,000 casualties.

As the blockade continued, Osman Pasha continued to improve his fortifications, being particularly concerned to keep open his communications with Sofia, 100 miles to the south-west. However, on 24 October a force of 35,000 Russians managed to cross the River Vid, and took possession of Gorni-Dubnik, thus completing the isolation of Plevna. Osman continued to offer a staunch resistance, but with growing shortages of supplies and ammunition to contend with, on the night of 9-10 December he attempted a major sortie, hoping to break-out through the Russian lines. These hopes were dashed after a bitter struggle of several hours duration, and later on the 10th he agreed to an unconditional surrender. To capture Plevna had cost the Russians and Rumanians about 40,000 casualties; Osman Pasha's losses were in the region of 30,000. But Osman Pasha had at least imposed a five months' delay on the Russian invasion of Bulgaria, virtually paralysing all the forces of Grand Duke Nicholas for that period.

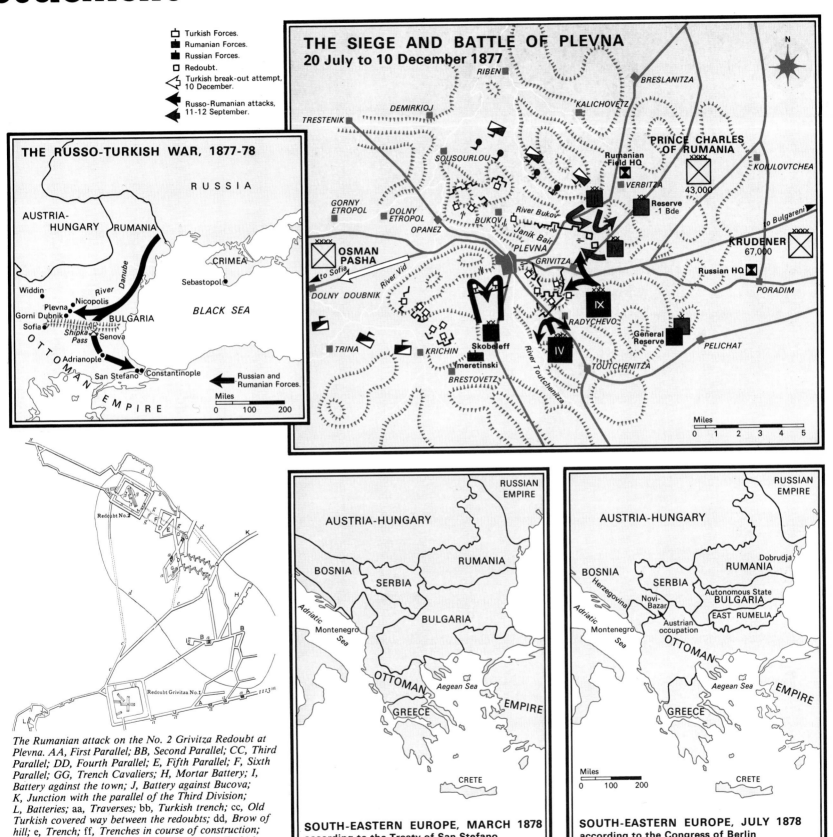

Turkish Forces.
Rumanian Forces.
Russian Forces.
Redoubt.
Turkish break-out attempt, 10 December.
Russo-Rumanian attacks, 11-12 September.

THE SIEGE AND BATTLE OF PLEVNA
20 July to 10 December 1877

N

RIBEN
BRESLANITZA
KALICHOVETZ
DEMIRKIOJ
TRESTENIK
SOUSOURLOU
Rumanian Field HQ
PRINCE CHARLES OF RUMANIA
43,000
VERBITZA
KOIULOVTCHEA
GORNY ETROPOL
DOLNY ETROPOL
OPANEZ
River Bukov
BUKOV
Janik Bair
Reserve -1 Bde
to Bulgareni
OSMAN PASHA
to Sofia
River Vid
PLEVNA
GRIVITZA
KRUDENER
67,000
Russian HQ
DOLNY DOUBNIK
IX
PORADIM
RADYCHEVO
General Reserve
Skobeleff
IV
PELICHAT
TRINA
KRICHIN
Imeretinski
River Toutchenitza
TOUTCHENITZA
BRESTOVETZ

Miles
0 1 2 3 4 5

THE RUSSO-TURKISH WAR, 1877-78

RUSSIA
AUSTRIA-HUNGARY
RUMANIA
River Danube
CRIMEA
Sebastopol
Widdin
Nicopolis
Plevna
Gorni Dubnik
BULGARIA
BLACK SEA
Sofia
Shipka Pass
Senova
Adrianople
San Stefano
Constantinople
OTTOMAN EMPIRE

Miles
0 100 200

Russian and Rumanian Forces.

The Rumanian attack on the No. 2 Grivitza Redoubt at Plevna. AA, First Parallel; BB, Second Parallel; CC, Third Parallel; DD, Fourth Parallel; E, Fifth Parallel; F, Sixth Parallel; GG, Trench Cavaliers; H, Mortar Battery; I, Battery against the town; J, Battery against Bucova; K, Junction with the parallel of the Third Division; L, Batteries; aa, Traverses; bb, Turkish trench; cc, Old Turkish covered way between the redoubts; dd, Brow of hill; e, Trench; ff, Trenches in course of construction; gg, Mine galleries; x, Covered way leading to Turkish works of Bucova; BB, Fougasses.

Redoubt No. 2
Redoubt Grivitza No. 1
1113ᵐ

RUSSIAN EMPIRE
AUSTRIA-HUNGARY
RUMANIA
BOSNIA
SERBIA
BULGARIA
Adriatic Sea
Montenegro
OTTOMAN EMPIRE
Aegean Sea
GREECE
CRETE

SOUTH-EASTERN EUROPE, MARCH 1878
according to the Treaty of San Stefano

RUSSIAN EMPIRE
AUSTRIA-HUNGARY
Dobrudja
RUMANIA
BOSNIA
Herzegovina
SERBIA
Novi-Bazar
Autonomous State BULGARIA
Austrian occupation
EAST RUMELIA
Adriatic Sea
Montenegro
OTTOMAN EMPIRE
Aegean Sea
GREECE
CRETE

Miles
0 100 200

SOUTH-EASTERN EUROPE, JULY 1878
according to the Congress of Berlin

The Genius of Helmuth von Moltke

Field Marshal Helmuth Karl Bernhard, Graf von Moltke (1800-91), the great Prussian commander and mastermind who developed the Great General Staff and planned, as Chief of Staff, the Prussian Army's strategies that led to great victories in 1866 and 1870-71.

Probably the ablest military mind since Napoleon, Field Marshal Helmuth von Moltke (1800-91) created the modern General Staff System and led his country's forces to amazing successes against Austria in 1866 (see pp. 190-191) and France in 1870-71 (see pp. 192-195). His intellectual brilliance set him apart from his contemporaries.

His first military service was in the Danish Army, but as a lieutenant he transferred to the Prussian Army in 1821. He became a 2nd Lieutenant in the 8th Infantry Regiment. From 1823-26 Moltke studied at the Kriegsakademie, achieving good results and, to add a little to his meagre pay (he had lost seniority on account of the transfer), he began to write on military matters. From 1834 to 1839 he was permitted to travel in the Turkish Empire, where he shared in part of the second war against Mohammed Ali and learnt much of value. Turkish incompetence inspired him to seek perfect efficiency in staff matters, but a superior could still claim "That man will never make a soldier".

Long years on the Prussian General Staff followed, during which he took an English wife and continued to write, above all concerning the military use of railways (see p. 166). In 1857 he was appointed head of the Great General Staff, and in 1864 served as chief of staff in the short Austro-Prussian War with Denmark over Schleswig-Holstein (see p. 189). His performance impressed King William I of Prussia,

and in 1866 he ensured that Moltke became commander of the Prussian forces unleashed against Austria in the famous Six Weeks War. His methods of using railways to achieve rapid mobilization of forces, albeit at widely separated points, his equally rapid concentration of force by employing converging lines of march, and the use of railways for bringing up reserves and supplies (despite a number of considerable hitches) outclassed his opponents. The von Moltke form of war was then tried against supposedly the greatest military power in the world: the French Empire of Napoleon III. Rapid victory was achieved against the regular French armies, but the popular war that followed extended well into 1871 before the triumph of the new German Empire was consolidated. A grateful monarch and Chancellor (Bismarck) promoted him to Field Marshal, and the title of Count was awarded.

Moltke remained chief of staff until 1888, but so complete had been his victories and so effective was the diplomacy of Bismarck that he was never required to employ his skills again. Moltke's successor was von Schlieffen, and, after him, his less able nephew (the younger Moltke) was favoured by the Kaiser. Neither was of the calibre of the elder von Moltke, whose austere dedication to the profession of arms and military study became a byword in his own and succeeding generations. He always stressed the need for flexibility.

GK BERLIN
MINISTRY OF WAR
GREAT GENERAL STAFF
GARDEKORPS comprising:
1 Garde-Division (2 Garde-inf. Bdes.=7 Regts.,
1 Garde Fd. Army Bde.=2 Regts.)
2 Garde-Division (3 Garde-inf. Bdes.=7 Regts.,
1 Garde Fd. Army Bde.=2 Regts.)
Garde-kavallerie Division (4 Garde-kav. Bdes.
=8 Regts.)
— also numerous support and training units —
Berlin, Spandau, Potsdam.

I KÖNIGSBERG
I ARMEEKORPS comprising:
1 Division (Königsberg) (2 Inf. Bdes.=4 Regts.,
1 Cav. Bde.=2 Regts., 1 Fd. Arty. Bde.=2 Regts.)
2 Division (Insterburg) (2 Inf. Bdes.=4 Regts.,
2 Cav. Bdes.=4 Regts., 1 Fd. Arty. Bde.=2 Regts.)
Formations stationed at Königsberg, Insterburg,
Tilsit, Rastenburg, Gumbinnen,—also supporting
units (not shown).

II STETTIN
II ARMEEKORPS comprising:
3 Division (Stettin) (2 Inf. Bdes.=4 Regts.,
2 Cav. Bdes.=4 Regts., 1 Fd. Arty. Bde.=2 Regts.)
4 Division (Bromberg) (2 Inf. Bdes.=4 Regts.,
1 Cav. Bde.=2 Regts., 1 Fd. Arty. Bde.=2 Regts.)
Formations stationed at Stettin and Bromberg
— also supporting units (not shown).

III BERLIN
III ARMEEKORPS comprising:
5 Division (Frankfurt-on-Oder) (2 Inf. Bdes.=
4 Regts., 1 Cav. Bde.=2 Regts., 1 Fd. Arty. Bde.=
2 Regts.)
6 Division (Brandenburg) (1 Inf. Bde.=4 Regts.,
1 Cav. Bde.=2 Regts., 1 Fd. Arty. Bde.=2 Regts.)
Formations stationed at Frankfurt-on-Oder and
Brandenburg, — also supporting units (not shown).

IV MAGDEBURG
IV ARMEEKORPS comprising:
7 Division (Magdeburg) (2 Inf. Bdes.=4 Regts.,
1 Cav. Bde.=2 Regts., 1 Fd. Arty. Bde.=2 Regts.)
8 Division (Halle) (2 Inf. Bdes.=4 Regts.,
1 Cav. Bde.=2 Regts., 1 Fd. Arty. Bde.=2 Regts.)
Formations stationed at Magdeburg, Halberstadt,
Halle and Torgau, — also supporting units (not
shown).

V POSEN
V ARMEEKORPS comprising:
9 Division (Glogau) (2 Inf. Bdes.=4 Regts.,
1 Cav. Bde.=2 Regts., 1 Fd. Arty. Bde.=2 Regts.)
10 Division (Posen) (2 Inf. Bdes.=4 Regts.,
1 Cav. Bde.=2 Regts., 1 Fd. Arty. Bde.=2 Regts.)
Formations stationed at Posen, Glogau,
Liegnitz and Ostrowo. — also supporting units
(not shown).

VI BRESLAU
VI ARMEEKORPS comprising:
11 Division (Breslau) (2 Inf. Bdes.=4 Regts.,
1 Cav. Bde.=2 Regts., 1 Fd. Arty. Bde.=2 Regts.)
12 Division (Neisse) (2 Inf. Bdes.=4 Regts.,
1 Cav. Bde.=2 Regts., 1 Fd. Arty. Bde.=2 Regts.)
Formations stationed at Breslau, Neisse,
Schweidnitz, Gleiwitz and Brieg. — also
supporting units (not shown).

VII MÜNSTER
VII ARMEEKORPS comprising:
13 Division (Münster) (2 Inf. Bdes.=4 Regts.,
1 Cav. Bde.=2 Regts., 1 Fd. Arty. Bde.=2 Regts.)
14 Division (Düsseldorf) 2 Inf. Bdes.=4 Regts.,
1 Cav. Bde.=2 Regts., 1 Fd. Arty. Bde.=2 Regts.).
Formations stationed at Münster, Koln, Wesel,
Minden and Düsseldorf, — also supporting units
(not shown).

VIII COBLENZ
VIII ARMEEKORPS comprising:
15 Division (Koln) (2 Inf. Bdes.=4 Regts.,
1 Cav. Bde.=2 Regts., 1 Fd. Arty. Bde.=2 Regts.)
16 Division (Trier) (2 Inf. Bdes.=4 Regts.,
1 Cav. Bde.=2 Regts., 1 Fd. Arty. Bde.=2 Regts)
Formations stationed at Coblenz, Cologne,
Aachen, Bonn and Trier, — also supporting units
(not shown).

IX ALTONA
IX ARMEEKORPS comprising:
17 Division (Schwerin) (3 Inf. Bdes.=6 Re
1 Cav. Bde.=2 Regts., 1 Fd. Arty. Bde.=2
18 Division (Flensburg) (2 Inf. Bdes.=4 Re
1 Cav. Bde.=2 Regts., 1 Fd. Arty. Bde.=2
Formations stationed at Altona, Schwerin,
Lübeck, Flensburg and Rendsburg, — also
supporting units (not shown).

X HANOVER
X ARMEEKORPS comprising:
19 Division (Hanover) (2 Inf. Bdes.=4 Reg
1 Cav. Bde.=2 Regts., 1 Fd. Arty. Bde.=2
20 Division (Hanover) (2 Inf. Bdes.=4 Reg
1 Cav. Bde.=2 Regts., 1 Fd. Arty. Bde.=2
Formations stationed at Hanover, Oldenbu
and Braunschweig, — also supporting unit
(not shown).

XI KASSEL
XI ARMEEKORPS comprising:
22 Division (Kassel) (2 Inf. Bdes.=4 Regts
1 Cav. Bde.=2 Regts., 1 Fd. Arty. Bde.=2
38 Division (Erfurt) (2 Inf. Bdes.=4 Regts.,
1 Cav. Bde.=2 Regts., 1 Fd. Arty. Bde.=2
Formations stationed at Kassel and Erfurt,
also supporting units (not shown).

XIV KARLSRUHE
XIV ARMEEKORPS comprising:
28 Division (Karlsruhe) (2 Inf. Bdes.=4 Re
1 Cav. Bde.=2 Regts., 1 Fd. Arty. Bde.=2
29 Division (Freiburg-im-Breslau) (3 Inf. B
6 Regts., 1 Cav. Bde.=2 Regts., 1 Fd. Arty
2 Regts.)
Formations stationed at Karlsruhe, Rastatt
Freiburg, Mülhausen and Lahr, — also sup
units (not shown).

XV STRASBURG
XV ARMEEKORPS comprising:
30 Division (Strasburg) (2 Inf. Bdes.=4 Re
1 Cav. Bde.=2 Regts., 1 Fd. Arty. Bde.=2
39 Division (Colmar) (1 Inf. Bde.=3 Regts.,
1 Cav. Bde.=2 Regts., 1 Fd. Arty. Bde.=2
Formations stationed at Strasburg and Coln
also supporting units (not shown).

XVI METZ
XVI ARMEEKORPS comprising:
33 Division (Metz) (2 Inf. Bdes.=4 Regts.,
1 Cav. Bde.=2 Regts., 1 Fd. Arty. Bde.=2
34 Division (Metz) (2 Inf. Bdes.=4 Regts.,
2 Cav. Bdes.=4 Regts., 1 Fd. Arty. Bde.=2
Formations stationed at Metz, Saarlouis an
St. Avold, — also supporting units (not sho

XVII DANZIG
XVII ARMEEKORPS comprising:
35 Division (Thorn) (2 Inf. Bdes.=4 Regts.,
1 Cav. Bde.=2 Regts., 1 Fd. Arty. Bde.=2 R
36 Division (Danzig) (2 Inf. Bdes.=4 Regts.,
1 Cav. Bde.=2 Regts., 1 Fd. Arty. Bde.=2 F
Formations stationed at Thorn, Danzig and
Graudenz, — also supporting units (not sho

XVIII FRANKFURT-ON-MAIN
XVIII ARMEEKORPS comprising:
21 Division (Frankfurt) (2 Inf. Bdes.=4 Regt
1 Cav. Bde.=2 Regts., 1 Fd. Arty. Bde.=2 R
25 Division Grossherzoglich Hessische
(Darmstädt) (2 Inf. Bdes.=3 Regts.,
1 Cav. Bde.=2 Regts., 2 Fd. Arty. Bdes.=4 R
Formations stationed at Frankfurt, Mainz
and Darmstädt, — also supporting units (no
shown).

XX ALLENSTEIN
XX ARMEEKORPS comprising:
37 Division (Allenstein) (2 Inf. Bdes.=4 Reg
1 Cav. Bde.=2 Regts., 1 Fd. Arty. Bde.=2 R
41 Division (Deutsch-Eylau) (2 Inf. Bdes.=
4 Regts., 1 Cav. Bde.=2 Regts., 1 Fd. Arty.
2 Regts.)
Formations stationed at Allenstein, Lyck,
Osterode, Marienburg and Deutsch-Eylau, —
supporting units (not shown).

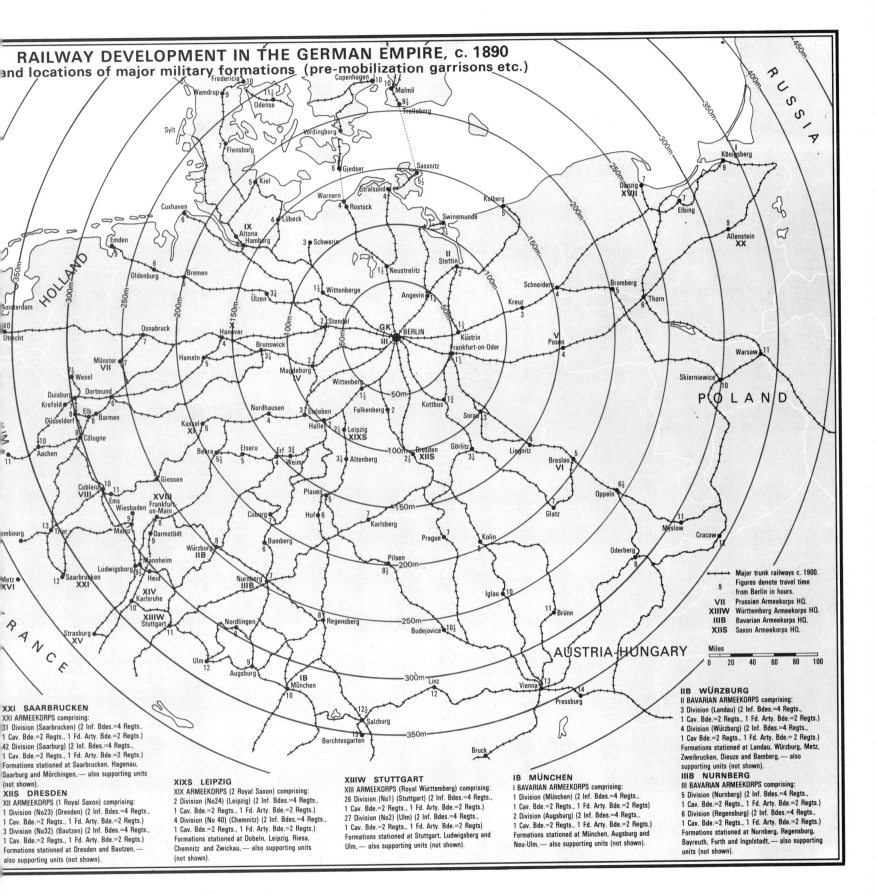

RAILWAY DEVELOPMENT IN THE GERMAN EMPIRE, c. 1890
and locations of major military formations (pre-mobilization garrisons etc.)

RUSSIA

HOLLAND

POLAND

FRANCE

AUSTRIA-HUNGARY

Major trunk railways c. 1900.
Figures denote travel time from Berlin in hours.

VII Prussian Armeekorps HQ.
XIIIW Württemberg Armeekorps HQ.
IIIB Bavarian Armeekorps HQ.
XIIS Saxon Armeekorps HQ.

Miles
0 20 40 60 80 100

XXI SAARBRUCKEN
XXI ARMEEKORPS comprising:
31 Division (Saarbrucken) (2 Inf. Bdes.=4 Regts.,
1 Cav. Bde.=2 Regts., 1 Fd. Arty. Bde.=2 Regts.)
42 Division (Saarburg) (2 Inf. Bdes.=4 Regts.,
1 Cav. Bde.=2 Regts., 1 Fd. Arty. Bde.=2 Regts.)
Formations stationed at Saarbrucken, Hagenau,
Saarburg and Mörchingen, — also supporting units
(not shown).

XIIS DRESDEN
XII ARMEEKORPS (1 Royal Saxon) comprising:
1 Division (No23) (Dresden) (2 Inf. Bdes.=4 Regts.,
1 Cav. Bde.=2 Regts., 1 Fd. Arty. Bde.=2 Regts.)
3 Division (No32) (Bautzen) (2 Inf. Bdes.=4 Regts.,
1 Cav. Bde.=2 Regts., 1 Fd. Arty. Bde.=2 Regts.)
Formations stationed at Dresden and Bautzen, —
also supporting units (not shown).

XIXS LEIPZIG
XIX ARMEEKORPS (2 Royal Saxon) comprising:
2 Division (No24) (Leipzig) (2 Inf. Bdes.=4 Regts.,
1 Cav. Bde.=2 Regts., 1 Fd. Arty. Bde.=2 Regts.)
4 Division (No 40) (Chemnitz) (2 Inf. Bdes.=4 Regts.,
1 Cav. Bde.=2 Regts., 1 Fd. Arty. Bde.=2 Regts.)
Formations stationed at Dobeln, Leipzig, Riesa,
Chemnitz and Zwickau, — also supporting units
(not shown).

XIIIW STUTTGART
XIII ARMEEKORPS (Royal Württemberg) comprising:
26 Division (No1) (Stuttgart) (2 Inf. Bdes.=4 Regts.,
1 Cav. Bde.=2 Regts., 1 Fd. Arty. Bde.=2 Regts.)
27 Division (No2) (Ulm) (2 Inf. Bdes.=4 Regts.)
1 Cav. Bde.=2 Regts., 1 Fd. Arty. Bde.=2 Regts.)
Formations stationed at Stuttgart, Ludwigsberg and
Ulm, — also supporting units (not shown).

IB MÜNCHEN
I BAVARIAN ARMEEKORPS comprising:
1 Division (München) (2 Inf. Bdes.=4 Regts.,
1 Cav. Bde.=2 Regts., 1 Fd. Arty. Bde.=2 Regts)
2 Division (Augsburg) (2 Inf. Bdes.=4 Regts.,
1 Cav. Bde.=2 Regts., 1 Fd. Arty. Bde.=2 Regts.)
Formations stationed at München, Augsburg and
Neu-Ulm, — also supporting units (not shown).

IIB WÜRZBURG
II BAVARIAN ARMEEKORPS comprising:
3 Division (Landau) (2 Inf. Bdes.=4 Regts.,
1 Cav. Bde.=2 Regts., 1 Fd. Arty. Bde.=2 Regts.)
4 Division (Würzburg) (2 Inf. Bdes.=4 Regts.,
1 Cav. Bde.=2 Regts., 1 Fd. Arty. Bde. = 2 Regts.)
Formations stationed at Landau, Würzburg, Metz,
Zweibrucken, Dieuze and Bamberg, — also
supporting units (not shown).

IIIB NURNBERG
III BAVARIAN ARMEEKORPS comprising:
5 Division (Nurnberg) (2 Inf. Bdes.=4 Regts.,
1 Cav. Bde.=2 Regts., 1 Fd. Arty. Bde.=2 Regts.)
6 Division (Regensburg) (2 Inf. Bdes.=4 Regts.,
1 Cav. Bde.=2 Regts., 1 Fd. Arty. Bde.=2 Regts.)
Formations stationed at Nurnberg, Regensburg,
Bayreuth, Furth and Ingolstadt, — also supporting
units (not shown).

Towards Armageddon

This Atlas has examined certain, selected aspects of the development of the art and science of land warfare between the years 1618 and 1878. Two hundred and sixty years of world history had contained all of 375 years of warfare (including overlapping conflicts but exclusive of local minor struggles), and but for the relative quiescence of the 1815-45 period the situation would have been even darker. It is part of the tragedy of mankind that his recorded history has been so dominated by wars, but it is also part of the fascination of the human story, for conflicts have been known to bring out the best as well as the worst in both nations and individuals, and they have also provided invaluable stimuli for scientific and medical advances. Many of these developments have been of a destructive nature: improved weaponry, better ammunition, more sophisticated instruments of war of every kind; but equally it is true that many inventions have been to the positive advantage of mankind, particularly in the fields of medicine and communications.

It has been estimated that, since the start of recorded history (in about 7,000 BC), this embattled planet has known barely 250 years of complete peace. This appalling indication of the competitive and destructive drive that has inspired both men and states is not particularly reassuring when we turn to consider the future; and, indeed, the 100 years that have passed since 1878 have held even larger and more destructive wars than previous millenia. It would seem, therefore, that the instinct to survive by fighting is indeed an ingrained aspect of the human condition, but we may also express the hope that succeeding generations—aware of the appalling effect of new weapons systems—will develop the sense to avoid direct confrontation as the ultimate means to settle serious differences of interest or policy. However, it is also probable that the newly-emergent Third World nations may need to pass through the fire that First and Second World nations have undergone over the last three centuries.

The important part military struggle has played in world history makes its study inevitable and indeed desirable by all with a genuine interest in the past, both for its own sake and for its possible use in casting light on the present and future. The period covered by this book saw a number of great and significant changes, and these can now be briefly summarized.

First, the attitude of peoples and governments towards warfare underwent considerable change. In the first half of the seventeenth century, where out story began, military means in terms of weaponry and support systems were relatively crude and ineffective by modern standards. Yet, the bitter ideological factors, particularly those associated with religion, caused the struggles in Europe to be waged with a ferocity and a near-totality (in certain key areas) that would not be surpassed or even approached for the next 150 years. The revulsion that the excesses and suffering caused led to a long period of so-called limited war, from the 1660s to the 1790s, during which wars were fought in the main by small professional armies, for restricted dynastic, territorial and later national and colonial objectives and, although weapons slowly improved over the period, there was a degree of restraint in the conduct of war suitable to the Age of Reason. Naturally, this should not be overstressed, for some struggles were as bitterly waged as ever, but in general terms restraint was in evidence. The coming of the French Revolution, itself in part foreshadowed by aspects of the American Revolution that preceded it, created the concept of the modern national state and the organization of the 'nation-at-arms', and returned an ideological content to warfare. And so the pendulum swung back towards totality, and the long Revolutionary and Napoleonic Wars were fought to a conclusion on an unprecedented scale both in terms of national involvement and destructiveness. Little wonder, then, if in 1815 another period of strong reaction set in, and that major wars receded almost from sight (although many 'little wars' in South America, the Balkans and elsewhere still continued). But then the relative peace was shattered by the Crimean War, followed by the American Civil War, and once again the march towards totality proceeded, through the struggles in Italy and Austria to the Franco-Prussian War, after which mankind was on the slippery slope once more. After a deceptive period of relative peacefulness in the late-nineteenth century (colonial and Balkan struggles apart), the armed camps and arms races of Europe began to form and gather momentum, and in 1914 the world reached a cataclysmic moment.

The period of our Atlas witnessed some startling swings in the fortunes of individual countries. The early predominance of France gave place to the rise of England to a position of paramountcy, whilst in Eastern Europe the short-lived blaze of Swedish success gave way before the emergence of Russia to great-power status. The development of British power continued through the first and second thirds of the eighteenth century, only to receive a major setback with the loss of the Old Empire. However, the sudden resurgence of France to the centre of the European stage from 1789 still further enhanced British prestige in the long run, and led to the rapid development of the 'Second' British Empire. New nation states—Prussia, Austria and Tsarist Russia—also benefited from the ultimate French cataclysm but to a lesser degree. Then, the nineteenth century saw long periods of British 'splendid isolation' from the European scene, with occasional exceptions linked with the problems of Balkan and Turkish affairs. However, step by step British power overseas grew, fitfully perhaps until 1870, and thereafter deliberately and dramatically. By that time, however, the growing power of Prussia presented a new focus within Europe, and the emergence of the German Empire from the humiliation of France in 1870-71 brought in a new era of German predominance. This held the seeds of two devastating World Wars.

One continuing theme, from 1660 to 1878, was the steady and continuous decline of Turkey. The effects of this, in the problems it caused by creating a power vacuum in the Balkans and Middle Eastern regions—and the rivalries it engendered between Russia and Austria, as each manoeuvred to exploit the situation to its own advantage—led to much international instability and conflict, drawing in many other countries in attempts to preserve an acceptable balance of power. Similarly, the earlier decline of Poland had led to fierce rivalries in mid-Eastern Europe, involving Russia, Prussia and Austria. So the fortunes of states waxed and waned as armies marched and fought and alliances and treaties were drawn up and dissolved.

Our period saw an amazing selection of notable commanders. In Gustavus Adolphus, called 'the Father of Modern Warfare', Sweden produced a genius who left an indelible stamp on the armies of the seventeenth century. Then France, in Condé, Turenne and Luxembourg, produced a trio of pre-eminent commanders, whose superiority underpinned the mounting power and 'la gloire' of France under Louis XIV until a generation of lesser generals fell victim to the genius of the 'Twin-Captains', Marlborough and Prince Eugene of Savoy. At the same time the restless demon sent Charles XII into his dynamic but ultimately disastrous military career; whilst Eugene, continuing the work of earlier Habsburg commanders, added fresh laurels to his crown in a series of devastating campaigns that initiated the military decline of Turkey, whose power, checked before Vienna in 1680, rapidly receded after Belgrade thirty-seven years later.

The mid-eighteenth century military scene was dominated first by Marshal de Saxe and then Frederick the Great of Prussia, a commander of unlikely gifts who emerged as one of the Great Captains of all time through the tempering fires of military disasters and threatened catastrophe. But even Frederick's reputation paled before the genius and power of the next pre-eminent commander: Napoleon Bonaparte, who left a stamp on the conduct of war that remained dominant to 1914 and in some ways to the present day. However, the nineteenth century also produced notable, if less dynamic commanders: Wellington and the Archduke Charles, Bolivar and Garibaldi, Grant and Lee, and then the great Moltke the Elder. Their achievements and concepts have figured large in the preceding pages.

Weaponry was also transformed, although only slowly, over the period under review. The clumsy matchlocks and pikes of the 1630s, gave way before the challenge of the smoothbore flintlock musket and socket bayonet of the early 1700s, and this combination in turn was superceded, in the mid-nineteenth century, by the breech-loading rifle and then the machine-gun in its earliest manifestations. Similar developments in artillery also took place. From the monsters of cast-iron and bronze of Marlborough's day, guns developed to the refined and standardized pieces of the late eighteenth century, and these in turn to rifled, breech-loading cannon of immeasurably greater power and range by the later years of the nineteenth century.

Regular armies gave way to large conscript citizen-forces, save in the case of Great Britain and the post-Civil War United States. Means of transportation improved: mud tracks gave way to metalled roads, rivers to canals, and both eventually ceded primacy to railways, which revolutionized ideas of military mobilization and concentration of force to those able to appreciate the opportunities and exploit them, although they proved less adept at using the iron-roads for the logistical support of armies already committed to the field. Similarly, sailing ships slowly gave way to steam as the main means of propulsion over water, and new weapons (revolving turrets, mines and then the early submarines and torpedoes) added new dimensions to the prosecution of naval wars. New industries arose to supply the new weapons and equipment, and concepts of mass-production supplanted cottage-industry, and machines in large measure replaced the skill of the individual craftsman. As populations grew, results of the growing industrialization and the beginnings of effective medicine, so did national budgets and the ability to build and maintain and equip ever larger and on the whole more effective armed forces, on land and at sea.

Such has been the main pattern of events unfolded in this Atlas of military strategy.

The lasting legacy of the Franco-Prussian War would be an increased French desire for 'la revanche' and an inveterate Prussian suspicion of France's intentions. These would be two factors in unleashing the First World War in 1914.

Select Bibliography

Atlases

A Military History and Atlas of the Napoleonic Wars. Edited by V. J. Esposito and J. R. Elting. Praeger, New York, 1964

Atlas of the American Revolution. Edited by K. Nebenzahl. Rand McNally, Chicago, 1974

Bayerischer Geschictatlas. Edited by J. Engel. Vol. 3. Bayerischer Schulbuch-Verlag, Munich, 1969

Diercke Schulatlas für Hohere Lehranstalten. George Westermann, Berlin, 1914

DTV Atlas zur Weltgeschichte. Edited by H. Kinder and W. Hilgemann. 2 vols. Deutscher Taschenbuch-Verlag, Munich, 1964, 1966

Gilbert, M. *Recent History Atlas, 1870 to the present day.* Weidenfeld & Nicolson, London, 1966

Meer, F. van der. *Atlas of Western Civilization.* English version by T. A. Birrell. Cleaver-Hume Press, London, 1955

The New Cambridge Modern History Atlas. Cambridge Univ. Press, 1970

Philips' New Historical Atlas for Students. Edited by R. Muir. Philips, London, 1914

The Times Atlas of World History. Edited by G. Barraclough. Times Books, London, 1978; Hammond, Maplewood, N.J., 1978

The West Point Atlas of American Wars. Edited by V. J. Esposito. 2 vols. Praeger, New York, 1959

General Works

Boudet, J. *The Ancient Art of Warfare.* 2 vols. Barrie & Cresset, London, 1969

Burne, A. H. *The Art of War on Land.* Methuen, London, 1944; Military Service, Harrisburg, Pa., 1947

'Captain J.S.' *Military Discipline—or the Art of War.* London, 1689

Carnot, L. *De la Défence des places fortes.* Paris, 1810

Chandler, D. G. *A Traveller's Guide to the Battlefields of Europe.* 2 vols. Hugh Evelyn, London, 1965; USA title: *A Guide to the Battle-fields of Europe.* Chilton, Philadelphia, 1966.
The Art of Warfare on Land. Hamlyn, London, 1974

Colin, J. *The Transformations of War.* Hugh Rees, London, 1912

Creasey, E. S. *The Fifteen Decisive Battles of the Western World.* Simkin, Marshall, Hamilton & Kent, London, 1851

Creveld, M. van. *Supplying War.* Cambridge Univ. Press, 1977; Cambridge Univ. Press, New York, 1977

Delbruck, H. *Geschichte der Kriegskunst.* 4 vols. Dresden, 1900-20

Duffy, C. J. *Fire and Stone: the Science of Fortress Warfare, 1660-1860.* David & Charles, Newton Abbot, 1975; Hippocrene, New York, 1975

Dupuy, R. E. and T. N. *Encyclopedia of Military History.* Macdonald, London, 1970; Harper & Rowe, New York, 1970

Eggenberger, D. *A Dictionary of Battles.* Allen & Unwin, London, 1967; Crowell, Scranton, Pa., 1967

Falls, C. *The Art of War from the Age of Napoleon to the Present Day.* Oxford Univ. Press, 1961; Oxford Univ. Press, New York, 1961
(ed.) *Great Military Battles.* Weidenfeld & Nicolson, London, 1964; MacMillan, New York, 1964

Fuller, J. F. C. *The Decisive Battles of the Western World and their Influence on History.* 3 vols. Eyre & Spottiswoode, London, 1954-56; Funk, New York, 1954-56; *The Conduct of War, 1789-1961: A Study of the Impact of the French, Industrial, and Russian Revolutions on war and its Conduct.* Eyre & Spottiswoode, London, 1961; Rutgers Univ. Press, New Brunswick, 1961

Howard, M. E. *War in European History.* Oxford Univ. Press, 1976; Oxford Univ. Press, New York, 1976

Hughes, Q. *Miliary Architecture.* Hugh Evelyn, London, 1974; St. Martin's Press, New York, 1975

Lawford, J. *Cavalry, Techniques and Triumphs of the Military Horsemen.* Sampson Low, London, 1976; Bobbs Merril, New York, 1976

Liddell Hart, B. H. *Strategy, the Indirect Approach.* Faber, London, 1954; Praeger, New York, 1954

Maguire, T. M. *Military Geography.* Cambridge Univ. Press, 1899

Montgomery, B. L. *A History of Warfare.* Collins, London, 1968; World Publishing, Cleveland, 1968

Preston, R. A., Wise, S., and Werner, H. O. *Men in Arms: A History of Warfare and Its Interrelationships with Western Society.* Atlantic Press, London, 1957; Praeger, New York, 1970

Spaulding, O. L. et al. *Warfare: A Study of Military Methods from the Earliest Times.* Harrap, London, 1925; Arno, New York, 1972

Toynbee, A. *A Study of History.* Thames & Hudson, London, 1976

Wanty, E. *L'Art de la Guerre.* 3 vols. Marabout Université, Paris, 1967

Warner, O. *Great Sea Battles.* Weidenfeld & Nicolson, London, 1963; MacMillan, New York, 1963

Weller, J. *Weapons and Tactics.* Vane, London, 1966; St. Martin's Press, New York, 1966

Part I: 1618-1721

Atkinson, C. T. *Marlborough and the Rise of the British Army.* Putnam, London, 1921

Barnett, C. *Marlborough.* Eyre Methuen, London, 1974

Bloomfield, R. *Sebastien le Prestre de Vauban.* London, 1938

Burne, A. H. and Young, P. *The Great Civil War: a military history of the first civil war.* Eyre & Spottiswoode, London, 1959

Chandler, D. G. *Marlborough as Military Commander.* Batsford, London. 1973; Scribner, New York, 1973
The Art of War in the Age of Marlborough. Batsford, London, 1976; Hippocrene, New York, 1976

Churchill, W. S. *Marlborough, His Life and Times.* 6 vols. Harrap, London, 1933-38; Scribner, New York, 1933-38

Craig, G. A. *The Politics of the Prussian Army, 1640-1945.* Oxford Univ. Press, 1955; Oxford Univ. Press, New York, 1964

Firth, C. H. *Cromwell's Army.* Methuen, London, 1902

Francis, D. *The First Peninsular War, 1702-13.* Benn, London, 1975; St. Martin's Press, New York, 1975

Green, D. *Blenheim.* Collins, London, 1974; Scribner, New York, 1975

Hatton, R. *Charles XII of Sweden.* Weidenfeld & Nicolson, London, 1968

Henderson, N. *Prince Eugen of Savoy.* Weidenfeld & Nicolson, London, 1964; Praeger, New York, 1965

Longworth, P. *The Art of Victory.* Thames & Hudson, London, 1965; Holt, New York, 1966

McKay, D. *Prince Eugene of Savoy.* Thames & Hudson, London, 1977; Thames & Hudson, New York, 1978

Mitford, N. *Frederick the Great.* Hamish Hamilton, London, 1970; Harper Row, New York, 1970

Roberts, M. *Gustavus Adolphus: A history of Sweden, 1611-1632.* 2 vols. Longman, Harlow, 1953
'The Military Revolution, 1560-1660', in *Essays in Swedish History.* Weidenfeld & Nicolson, London, 1967; Univ. of Minnesota Press, Minneapolis, 1968

Scouller, R. E. *The Armies of Queen Anne.* Oxford Univ. Press, 1966; Oxford Univ. Press, New York, 1966

Shy, J. *Towards Lexington.* Princeton Univ. Press, 1965

Trevelyan, G. M. *England Under Queen Anne.* 3 vols. Longman, London, 1930-34

Vaughan, D. M. *Europe and the Turk, 1350-1700.* Liverpool Univ. Press, 1954; AMS Press, New York, 1976

Wedgwood, C. V. *The Thirty Years War.* Cape, London, 1938; Yale Univ. Press, 1939

Young, P. and Holmes, E. R. *The English Civil Wars.* Eyre Methuen, London, 1974

Part II: 1722-1815

Aubry, O. *Napoleon.* Hamlyn, London, 1964; Crown, New York, 1964

Barnett, C. *Bonaparte.* Allen & Unwin, London, 1978

Brett-James, A. *Wellington at War, 1794-1815.* Macmillan, London, 1961; St. Martin's Press, New York, 1961

Life in Wellington's Army. Allen & Unwin, London, 1972; Allen & Unwin, Winchester, Ma., 1972

Chandler, D. G. *The Campaigns of Napoleon.* Weidenfeld & Nicolson, London, 1978; Macmillan, New York, 1978

Dictionary of the Napoleonic Wars. Arms and Armour Press, London and Melbourne, 1979; Macmillan, New York, 1979.

Napoleon. Weidenfeld & Nicolson, London, 1974

Colin, J. *Les Campagnes de Maurice de Saxe.* 5 vols. Paris, 1901-6

Corbett, J. S. *England in the Seven Years' War.* 2 vols. Longman, Harlow, 1907; Reprint of 1918 edition by AMS Press, New York

Duffy, C. J. *The Army of Frederick the Great.* David & Charles, Newton Abbot 1974; Hippocrene, New York, 1977

The Army of Maria Theresa. David & Charles, Newton Abbot, 1977; Hippocrene, New York, 1977

Austerlitz, 1805. Seeley Service, London, 1976; Shoe String Press, Hamden, Conn., 1977

Borodino and the war of 1812. Seeley Service, London, 1972; Scribner, New York, 1973

Glover, M. *The Peninsular War, 1807-14: A Concise Military History.* David & Charles, Newton Abbot, 1974; Shoe String Press, Hamden, Conn., 1974

Wellington's Army: In the Peninsula, 1808-14. David & Charles, Newton Abbot, 1976; Hippocrene, New York, 1977

Glover, R. *Peninsular Preparation; the reform of the British Army, 1795-1809.* Cambridge Univ. Press, 1963; Cambridge Univ. Press, New York, 1963

Lachouque, H. *The Anatomy of Glory.* Tr. by A. S. Brown. 3rd edition. Arms & Armour Press, London, 1978; Hippocrene, New York, 1978

Waterloo. Arms & Armour Press, London, 1975; Hippocrene, New York, 1978

Liddell-Hart, B. H. *The Ghost of Napoleon.* Faber, London, 1933

Longford, E. *Wellington, the Years of the Sword.* Weidenfeld & Nicolson, London, 1968; Harper, New York, 1969

Macksey, P. *The War for America, 1775-83.* Longman, London, 1964

Marshall-Cornwall, J. *Napoleon as Military Commander.* Batsford, London, 1967; Van Nostrand, New York, 1967

McGuffie, T. H. *The Siege of Gibraltar.* 7 vols. Batsford, London, 1965; Dufour, Chester Springs, Pa., 1965

Morvan, J. *Le soldat imperial.* Paris, 1904

Oman, C. *A History of the Peninsular War.* 7 vols. Oxford Univ. Press, 1902-20; Reprint of 1930 edition by AMS Press, New York

Sir John Moore. London, 1943

Wellington's Army. London, 1912

Paret, P. *Yorck and the Era of Prussian Reform.* Princeton Univ. Press, 1966

Parkinson, R. *The Hussar General: the life of Blucher.* Peter Davies, London, 1975

Phipps, R. *The Armies of the First French Republic and the rise of the Marshals of Napoleon.* 5 vols. Oxford Univ. Press, 1926-39; Oxford Univ. Press, New York, 1926-39

Quimby, R. *Background of Napoleonic Warfare.* Oxford Univ. Press, 1958; Columbia Univ. Press, 1957

Reichel, D. *Davout et l'Art de la Guerre.* Delachaux et Nestle, Neuchatel, 1975

Rogers, H. C. B. *The British Army in the 18th Century.* Allen & Unwin, London, 1976; Hippocrene, New York, 1976

Rothenburg, G. E. *The Art of Warfare in the Age of Napoleon.* Batsford, London, 1978; Indiana Univ. Press, Bloomington, 1978

Savory, R. G. *His Brittanic Majesty's Army in Germany during the Seven Years War.* Oxford Univ. Press, 1965; Oxford Univ. Press, New York, 1965

Vachée, A. *Napoleon at Work.* A. & C. Black, London, 1914

Ward, S. G. P. *Wellington's Headquarters.* Oxford Univ. Press, 1957; Oxford Univ. Press, New York, 1957

Weller, J. *Wellington at Waterloo.* Longman, Harlow, 1967; Crowell, New York, 1967

Wellington in India. Longman, Harlow, 1972

Wellington in the Peninsula, 1808-1814. Vane, London, 1962

White, J. M. *Marshal of France: Maréchal de Saxe.* Hamish Hamilton, London, 1962

Wilkinson, S. *The Coming of General Bonaparte.* Oxford Univ. Press, 1921

The French Army Before Napoleon. Oxford Univ. Press, 1915

Winsor, S. *The American Revolution.* Sons of Liberty Chronicle, New York, 1974

Young, P. *Napoleon's Marshals.* Hippocrene, New York, 1973

Part III: 1816-1878

Bond, B. J. *Victorian Military Campaigns.* Hutchinson, London, 1967; Praeger, New York, 1967

Craig, G. A. *The Battle of Königgrätz.* Weidenfeld & Nicolson, London, 1965; Lippincott, Philadelphia, 1964

Catton, B. *The Centennial History of the Civil War.* 3 vols. Gollancz, London, 1962-66; Doubleday, New York, 1961-65

Clausewitz, C. von. *On War.* Edited by M. E. Howard and P. Paret. Princeton Univ. Press, 1976

Earle, E. M. (ed.) *Makers of Modern Strategy: Military Thought from Machiavelli to Hitler.* Princeton Univ. Press, 1943

Edwardes, M. *The Red Year.* (The Indian Mutiny) Cardinal, London, 1975

Falls, C. *Hundred Years of War.* Duckworth, London, 1953; Macmillan, New York, 1954

Featherstone, D. *Colonial Small Wars, 1837-1901.* David & Charles, Newton Abbot, 1973; David & Charles, Pomfret, 1974

Fuller, J. F. C. *War and Western Civilisation, 1832-1932.* Duckworth, London, 1932; Arno, New York

Furneaux, R. *The Siege of Plevna.* A. Blond, London, 1958; US title: *Breakfast War.* Crowell, New York, 1960

Gorlitz, W. *History of the German General Staff, 1675-1945.* Tr. by B. Battershaw. Hollis, London, 1953; Praeger, New York, 1953

Henderson, G. F. R. *Stonewall Jackson.* 2 vols. Longman, London, 1904

Howard, M. E. *The Franco-Prussian War.* Hart-Davis, London, 1961; Macmillan, New York, 1961

Hozier, H. M. *The Seven Weeks' War.* Macmillan, London, 1871

Jomini, H. *Summary of the Art of War.* Lippincott, Philadelphia, 1871

Judd, D. *The Crimean War.* Hart-Davis MacGibbon, London, 1975

Kinglake, A. W. *The Invasion of the Crimea.* 9 vols. London, 1888; AMS Press, New York, 1972

McElwee, W. E. *The Art of War—Waterloo to Mons.* Weidenfeld & Nicolson, London, 1974; Indiana Univ. Press, Bloomington, 1975

Nickerson, H. *The Armed Horde, 1793-1939.* Putnam, London, 1941; Putnam, New York, 1940

Pratt, E. A. *The Rise of Railway Power in War and Conquest, 1833-1914.* R. S. King, London, 1915

Selby, J. M. *The Paper Dragon; an account of the China Wars, 1840-1900.* Barker, London, 1968; Praeger, New York, 1968

Whitton, F. E. *Moltke.* Constable, London, 1971; Arno, New York, 1972

Index

Abercromby, Sir Ralph, British general, 82, 97
Abdul Krim, Turkish commander, 196
Aberdeen, George Gordon, 4th Earl of, 146
Aboukir (Nile), Battle of, 96, 97
Aboukir Bay, 82
Abraham, Heights of, 82
Abyssinia, British expedition to, 154
Acre, Austro-British bombardment of, 137; siege of, 97
Acropolis, Egyptian capture of the, 137
Adige, River, 41, 92
Adrianople, 57
Afghanistan, Persian invasion of, 149
Afghan War, First, 149
Alabama, 185
Alamo, siege of, 162
Alba de Tormes, 112
Albany, 85, 86
Albert, Duke of Teschen, Archduke of Austria, 159, 190
Albert, Crown Prince of Saxony, 193
Albuera, Battle of, 111
Albuquerque, Mathias d', Portuguese general, 28
Aldea Tejada, 112
Alexander the Great, 14
Alexander I, Tsar of Russia, 119, 120, 138
Alexander II, Tsar of Russia, 138, 147
Alexander VII, Pope, 54
Alexandria, Convention of, 137; siege of, 97; surrender of Turkish fleet at, 137
Aliwal, Battle of, 152
Al-Jazzar, Pasha of Damascus, 97
Allahabad, 156
Allegheny, River, 77
Allenstein, 106
Alma, Battle of the, 142, 144, 145
Almanza, Battle of, 49
Almeida, fortress of, 111
Alsace, evacuation of, 27
Alsen, Battle of, 189
Altranstadt, 58
Alvarez, Don Martin Alvarez de Sota Mayor, Spanish general, 82
Alvintzi, Joseph d', Baron, Austrian general, 92, 93
Amboyna, massacre of, 62
Amerdingen, 44
American Civil War, 7, 129, 168-185; defences in, 141; interference by politicians in conduct of, 130, 131; military lessons learnt from, 131; tactics used in, 165; use of railways in, 166; use of telegraph in, 167; weaponry used in, 186, 187
American-Mexican War, 164
American War of 1812, 116-117
American War of Independence, 67, 82, 84-87, 200; affects stability in South America, 132
Amherst of Arracan, William Pitt Amherst, 149
Amiens, Preliminaries of Peace signed at, 97
Ammunition, 186, 187
Ampudia, Pedro de, Mexican general, 164
Anaconda Plan, 169, 170, 173
Anderson, Major-General Richard H., Confederate soldier, 179
Anhalt-Dessau, Prince of, see: Moritz
Anna Ivanovna, Empress of Russia, 70, 138
Annapolis, 63
Anne, Queen of England, 20, 32, 42, 65; military expenditure during the reign of, 22
Annesley Bay, 154, 167
Anse de Foulon, 82
Anson, George, British general, 156
Antalo, 154
Antietam, Battle of, 170
Antwerp, 38, 141
Appomattox, 171, 185
Arcis-sur-Aube, Battle of, 122
Arcola, Battle of, 9, 92
Arcot, 140; British seizure of, 78
Arinez, action near, 115
Armstrong gun, introduction of the, 186
Army reforms, British, 128, 131
Army strengths, 22
Arnold, Benedict, American turncoat, 85, 86
Artillery, 37; development of, 186-187; of French Army, 1770-1815, 95; revolutionized by Gustavus Adolphus, 27
Artois, 34
Ashburnham, Lieutenant-Colonel the Hon. Thomas, British soldier, 152
Ashby, Turner, Confederate general, 176
Aspern-Essling, Battle of, 107
Aspromonte, Battle of, 159
Assaye, Battle of, 78, 80; envelopment tactic used at, 12
Astley, Major-General Sir Jacob, (later Baron Astley), British soldier, 32
Athens, Venetian capture of, 54
Attack: in oblique order, 12; from a defensive position, 13
Atlanta, 170, 171, 185
Aube, River, 122
Auckland, George Eden, Earl of, Governor-General of India, 149
Auerstadt, Battle of, 98, 103, 104
Augereau, Pierre Francois Charles, Duc de Castiglione, French marshal, 92, 104, 106
Augsburg, 44, 45
Augsburg (Nine Years War), War of the League of, 19, 39
Augustin I (Augustin de Yturbide), Emperor of Mexico, 133
Auma, 103
Austerlitz, Battle of, 10, 98, 100-101; Allied strategic error at, 12
Austrian Succession, War of the, 66, 69, 70, 77
Austro-Prussian (Six Weeks) War, 159, 189, 190, 198
Ayacucho, Battle of, 132
Azov, fortress of, 55, 61

Badajoz, 111; sack of, 114; siege of, 112
Baddowal, action at, 152
Baden, Margrave of, see: Louis, Prince
Badli-ke-Serai, Battle of, 156
Bagration, Prince Peter Ivanovich, Russian general, 101, 119, 120
Bahadur Shah II, Moghul Emperor, 156
Bailen, action at, 111
Baird, Sir David, British general, 97
Balaclava, Battle of, 142, 145, 146
Ball's Bluff, Battle of, 170
Baner, Johann, Swedish general, 27
Banks, Nathaniel Prentiss, US general and politician, 174, 176, 177, 178
Bara Kalra, 153
Barbados, 63
Barcelona, struggle for, 49
Barclay de Tolly (Prince Michael Bogdanovich), Russian general, 119
Barrackput, 156
Bashillo, River, 154
Bassano, Battle of, 92
Baton Rouge, Union raid on, 171
Bautzen, Battle of, 12, 98, 122
Bavaria, Elector of, see: Maximilian II Emmanuel
Bayonets, 37
Bazaine, Achille, French general, 167, 192, 193, 194, 195
Beaconsfield, Benjamin Disraeli, Earl of, British politician, 151
Beauharnais, Eugène de, Viceroy of Italy, 119, 121
Beaulieu, Jean Pierre, Austrian general, 90
Beaumont, action at, 193
Beauport, action at, 82
Beauregard, Pierre Gustave Toutant, Confederate general, 171, 172, 173
Beirut, Austro-British bombardment of, 137
Belfort Gap, 40
Belgrade, Battle of, 56-57, 137, 201; siege of, 12, 13, 19, 55-57; Turkish capture of, 54
Belidor, Bernard Forest de, French engineer, 140
Belle-Isle, Charles L. A. F., French marshal, 70
HMS *Bellerophon*, 126
Belliard, Auguste-Daniel, Comte, French general, 97
Belmont, action at, 170
Bemis Heights, Battle of, 86
Bender, 54, 59, 61
Benedek, Ludwig August, Ritter von, Austrian general, 161, 166, 190
Bennigsen, Theophil, Count, Russian general, 106
Bentinck, Lord William Cavendish, Governor-General of India, 149
Bentonville, Battle of, 185
Berar, Rajah of, 80
Beresford, William Carr, British general, 111
Beresina, Battle of, 120, 121
Berlin, 74; Congress of, 131, 189; French occupation of, 103
Bernadotte, Jean-Baptiste-Jules (Charles XIV of Sweden), French marshal, 100, 101, 103, 104, 106
Bernard, Duke of Saxe-Weimar, German general, 26
Berthier, Louis-Alexandre, French marshal, 93
Berwick, James Fitzjames, Duke of, French marshal, 49, 53, 70
Bessarabia, Turks cede, 137
Bessemer, Sir Henry, British inventor, 129
Bessières, Jean-Baptiste, Duc d'Istrie, French marshal, 107
Big Black River, 182
Bismarck, Otto von, Count, Chancellor of Prussia, 7, 131, 159, 189, 190, 192, 195, 198
Black Sea Clauses, 138
Bladensburg, action at, 116
Blair, Major-General Frank P., Union soldier, 182
Blaswitz, action at, 101
Blenheim, Battle of, 12, 42, 45, 46, 47; Marlborough's concentration of force at, 10
Blücher, Gebhard Leberecht von, Prince of Wahlstadt, Prussian general, 10, 103, 122, 125
Boer War, First, 151
Bohemia, rebellion in, 24-25
Bokhara, khanate of, 138
Bolivar, Simon, South American pàtriot, 132, 133, 201
Bomarsund, Allied naval expedition to, 142
Bombay, 77, 78
Bonaparte, Jérome, King of Westphalia, 109, 119
Bonaparte, Joseph, King of Naples and Spain, 109, 111, 114, 115
Bonaparte, Louis, King of Holland, 103, 109
Bonaparte, Louis-Napoleon (Napoleon III), Emperor of the French, 129, 142, 146, 158, 159, 161; his conduct of Franco-Prussian War, 131
Borodino, Battle of, 10, 98, 120, 121
Bosnia, Turkish atrocities in, 196
Bosquet, Pierre Jean François, French general, 144, 145, 147
Boston, British evacuation of, 85
Botany Bay, 151
Bouchain, Battle of, 38; capture of, 41
Boudet, Jean, French general, 94, 98
Boufflers, Louis-Francois, French marshal, 43, 52, 53
Boulogne, camp of, 100
Bournonville, Alexandre Hippolyte of, Prince, Austrian general, 40
Boves, José Tomas, Spanish soldier, 132
Bowen, Brigadier-General John S., Confederate soldier, 182
Boxer Revolt, 136
Boyer, Pierre François Joseph, French general, 112
Brabant, Forcing of Lines of, 40-41
Braddock, Edward, British general, 77, 84
Bragg, Braxton, Confederate general, 171, 184
Brandy Station, skirmish at, 180
Brandywine, Battle of, 67, 86
Brazil, independence movement in Brazil, 135
Breisach, 34, 37
Breitenfeld, Battle of, 19, 26, 27
Brenner Pass, 166
Brienne, action at, 122
British Army: reforms in, 128, 131; rôle in nineteenth century, 148
Brock, Brigadier-General Isaac, American soldier, 116
Broglie, François-Marie, Duc de, French marshal, 70
Broglie, Victor, Duc de, French general, 71, 76
Brooklyn, Battle of, 85
Brown, John, American abolitionist, 169
Brown, Jacob, American general, 116
Brown's Gap, 177
Brueys, François Paul, French admiral, 96
Bruges, 53
Brunswick, Ferdinand, Duke of, Prussian general, 67, 71, 103
Brussels, 52, 53, 125; French capture of, 70
Buda, siege of, 54
Buena Vista, Battle of, 164
Bufford, Major-General John, Union soldier, 180
Bull Run: Battle of First (Manassas), 170, 172-173; Battle of Second, 130, 170
Bultaghi, Turkish Grand Vizier, 61
Burgos, 111
Burgoyne, John, British general, 67, 85, 86, 87
Burgoyne, Sir John, British engineer, 146
Burgundy, Louis, Duke of, 53
Burma War: First, 149; Second, 151
Burne, Colonel Alfred, 10
Burnside, Ambrose Everett, Union general, 170, 179
Busaco, Battle of, 111
Butler, Captain James Armar, British soldier, 142
Butler, Benjamin Franklin, Union genéral, 170
Buxar, Battle of, 78
Buxhowden, Friedrich Wilhelm von, Count, Russian general, 100, 101
Buzenval, 195
Byng, John, British admiral, 11

Cadiz, 111; British raid on, 82
Cairo, 169; British capture of, 97; French capture of, 96
Calamita Bay, 142
Calatafini, Battle of, 159
Calcutta, 77, 78
California, 169; pacification of, 164
Callao, 132
Cambrai, 37
Cambridge, George William Frederick Charles, Duke of, 128, 145
Camon, Henri, French general and historian, 98
Camp, armies in, 33
Campaigning: in Europe, 37; difficulties of, 49; seasonal factors and, 18-19
Campbell, Sir Colin, British general, 145, 157
Camp followers, 22
Canada Act, 151
Candia, 19; Turkish capture of, 54
Canning, Charles John Canning, 1st Earl, British Governor-General of India, 156
Canrobert, François, French general, 142, 146
Cape Coast Castle, English occupation of, 63
Capes, Battle of the, see: Chesapeake Bay
Cap la Hogue, Battle of, 20
Caprera, Isle of, 159
Caraman, Philippe, French general, 41
Cardigan, James Thomas Brudenell, Lord, British general, 145
Cardwell, Edward, British Secretary of State for War, 131
Carleton, Colonel Guy, British soldier, 85
Carnot, Lazare Nicolas, French engineer, 89, 140
Castel Ceriolo, action at, 94
Castiglione, 161; Battle of, 12, 92, 98
Catalan revolt, 48-49
Cathcart, Sir George, British general, 145
Catherine II, Empress of Russia, 138
Catholic League, 24, 25
Catinat, Nicolas de, French general, 19
Cavour, Camillo di, Count, Piedmontese parliamentarian, 142, 158, 159, 166
Cawnpore, siege of, 156, 157
Centreville, 172, 173
Cerro Gordo, 164
Ceva, Battle of, 90
Chacabuco, Battle of, 132
Châlons, 192, 193
Champaubert, action at, 122
Chancellorsville, 167; campaign and battle of, 130, 171, 179
Chandernagore, French trading post at, 78
Chapultepec, action at, 165
Charleroi, 37
Charles of Rumania, Prince, 196
Charles I, King of England, 30, 32
Charles II, King of England, 30, 62
Charles II, King of Spain, 19, 48, 49
Charles V, Duke of Lorraine, 54, 55, 56
Charles V, King of Spain, 19, 48
Charles VI, Holy Roman Emperor, 19, 66
Charles XII, King of Sweden, 8, 20, 42, 54, 55, 58, 65, 76, 201; and 'principles of war', 9, 10; at Narva, 59-61; at Poltava, 61; relationship with his men, 11
Charles, Archduke (later Charles VI, Holy Roman Emperor), 19
Charles of Lorraine, Prince, 75
Charles Albert, King of Sardinia-Piedmont, 158
Charles-Louis, Archduke, 92, 100, 107, 201
Charleston, 185
Chassepot rifle, 131; invention of, 187
Chatham, William Pitt, 1st Earl of, British politician, 66, 67, 69, 77
Chattanooga, 166, 169, 170, 185; Battle of, 184; Union capture of, 171
Chattar Singh, Sikh soldier, 153
Cherasco, Armistice of, 90
Chesapeake, River, 116
Chesapeake Bay (Capes), Battle of, 87
Chickamauga, Battle of, 171, 184
Chihuahua, 164
Chile, independence movement in, 132
Chilianwala, Battle of, 153
China Wars, 136
Chinat, action at, 156
Chota Kalra, 153
Christian IV, King of Denmark, 25
Churchill, Sir Winston, British politician, 8
Ciudad Rodrigo, 13, 111, 112, 114
Civitavecchia, 158, 159
Clam-Gallas, Edward von, Count, Austrian general, 159
Clausel, Baron Bertrand, French general, 112
Clausewitz, Karl von, Prussian general and military theorist: definition of strategy and tactics, 9; on warfare, 129
Clay, Henry, American politician, 169
Clinton, Sir Henry, British general, 86
Clive of Plassey, Robert Clive, 1st Baron, British general and statesman, 67, 77, 78, 80, 156
Coblenz, 45
Cochrane, Sir Alexander Inglis, British admiral, 116
Cochrane, Thomas, 10th Earl of Dundonald, British admiral, 82, 132, 133
Cockburn, Sir George, British admiral, 116
Coehoorn, Menno, Baron, Dutch military engineer, 50
Colbert, Jean-Baptiste, Marquis de Seignelay, French statesman, 34
Cold Harbor, action at, 185
Colli, Michael von, Baron, Piedmontese general, 90
Colmar, Battle of, 27
Colonies, 62-63; in nineteenth century, 134-136, 148; rivalry between Gt. Britain, France and Spain in, 77
Colt, Samuel, American inventor, 186
Columbia, independence movement in, 132
Communications, 166-167
Condé, see: Enghien
'Congress System', 128
Conscription, 89, 128, 129; in nineteenth century, 89, 128, 129
Constantinople, 136
Contades, Louis, French marshal, 71
Continental Army, Washington's, 22, 67, 84, 86
Continental Congresses, 85
Continental System, 109, 111
Coote, Sir Eyre (the Younger), British general, 97
Coote, Sir Eyre (the Elder), British general, 78
Corinth, 166, 169
Cornwallis, Charles, 1st Marquis, British general, 78, 86, 87
Corps d'armee, Napoleonic, 98

Corunna, Battle of, 111
Courtrai, 37
Covenanters, 30
Cracow, Polish occupation of, 54
Craonne, Battle of, 122
Cretan insurrection, 137
Creuz, General Carl Gustav, Swedish soldier, 61
Crillon, Louis des Balbes de Berton, Duc de, French general, 82
Crimean War, 129, 137, 138, 141, 142-147, 148, 156; artillery used in, 186
Cromwell, Oliver, English general and dictator, 20, 27, 30, 32
Cross Keys, Battle of, 178
Crouch, General Darius, US soldier, 179
Crown Point, British occupation of, 86
Croy, Charles de, French general, 56
Crusades, 14
Crysler's Farm, Battle of, 116, 117
Culloden, Battle of, 69
Culp's Hill, 180, 181
Cumberland, William Augustus, Duke of, British general, 70, 71
Curtis, Captain Roger, British naval officer, 82
Custozza, Battle of, 158, 159, 190

Dalhousie, George Ramsay, 9th Earl of, British soldier, 115
Dalhousie of Dalhousie Castle and the Punjab, James Andrew Brown-Ramsay, Marquis of, British statesman, 151, 153, 154, 156
Damascus, Pasha of, see: Al-Jazzar
Danish War, 165, 189
Dannenberg, P. A., Russian general, 145
Danube, Marlborough's march to the, 13, 44
Dardanelles, naval action in, 142
Daun, Leopold Joseph Maria, Count, Austrian soldier, 74, 75
Davidovitch, Paul, Austrian general, 92
Davies, Major-General Thomas A., Union soldier, 172
Davis, Jefferson, President of the Confederate States, 131, 164, 169, 171, 174, 180, 185
Davout, Louis Nicolas, Duc d'Auerstadt, French marshal, 100, 101, 103, 104, 106, 107, 119, 121
Dearborn, Henry, American general, 116
Dego, First and Second Battles of, 90
Delaware, River, 62, 86
Delhi, Indian mutineers' march on, 156
Demotika, 59
Denain, Battle of, 55
Desaix de Veygoux, Louis, French general, 94
Descartes, René, French philosopher, 29
Dessau, Battle of, 25
Detached forts, system of, 140
Detroit, Battle of, 116
Dettingen, Battle of, 69
Dick, Major-General Sir Robert, British soldier, 153
Dijon, 93
'Diplomatic Revolution', 66, 77
Disraeli, see: Beaconsfield
Dnieper, River, 120
Dönauwörth, Allied capture of, 44
Doniphan, Colonel Alexander W., American soldier, 164
Dost Mohammed, Amir of Kabul, 149
Douai, siege of, 37, 53
Doulut Rao, Prince of Scindia, Maratha leader, 80
Dreikaiserbund (Three Emperors' League), 138
Dresden, 122, 190
Dreyse, Johann von, Prussian inventor, 186
Driesen, G. W., Prussian general, 75
Drissa, manoeuvre of, 119
Drogheda, massacre of, 30
Drummond, Gordon, British general, 116
Ducrot, Auguste, French general, 193, 195
Dum-Dum, 156
Dumouriez, Charles du Perier, French general, 89
Dunaburg, 119
Dunbar, Battle of, 30
Dunes, Battle of the, 27, 30
Dupleix, Joseph François, Marquis de, Governor of French India, 78
Dupont, Pierre, Comte de l'Etung, French general, 111
Duppel, storm of, 189
Dürer, Albrecht, German painter, 29
Durham, 171
Durham, John George Lambton, 1st Earl of, British statesman, 151

Early, Jubal Anderson, Confederate general, 172, 179, 185
East India Company, 62, 77, 78, 80, 148, 151, 152; their rule of India transferred to British Crown, 157
Edgehill, Battle of, 30
Egypt, British invasion of, 82; British reconquest of, 97
Egyptian Wars, First and Second, 137
Eighty Years War, 24
Elba, Isle of, 122, 125
El Brazito, skirmish at, 164
Elchingen, action at, 100
Eliott, Sir George Augustus, British general, 82, 86
Elizabeth Petrovna, Empress of Russia, 67, 69, 75, 138
Ellenborough, Edward Law, Earl of, British Governor-General of India, 149
Elvas, fortress of, 111
Emancipation Proclamation, 170
Enfield rifle, 186
Enghien, Prince Louis II of Bourbon and Duc d' (the Great Condé), 12, 19, 26, 37, 39, 201; at Battle of Rocroi, 27, 28
English Civil Wars, 20, 24, 30-32
Entzheim, Battle of, 40
Envelopment: of single flank, 12; of both flanks, 12
Eprenan, M. d', French general, 28
Erfurt, capture of, 103
Erfurt Agreements, 118
Erlon, Jean-Baptiste Drouet, Comte d', French general, 115, 125
Erzerum, Battle of, 137
Essex, Robert Devereux, 20th Earl of, 30
Eugene, Prince of Savoy, 12, 13, 19, 27, 40, 42, 52, 53, 55, 65, 70, 76, 201; and campaign of 1704, 44, 45, 46; and grand tactics, 9; at siege and battle of Belgrade, 56-57; cooperation with Marlborough, 10; marches to Turin, 41; relationship with his men, 11
Eupatoria, 146; action at, 142
Eureka Stockade, storming of, 151
Evans, Captain N. G., Confederate soldier, 173
Ewell, Richard S., Confederate general, 173, 176, 180, 181
Eylau, Battle of, 106

Fairfax of Cameron, Thomas Fairfax, 3rd Baron, English parliamentary military commander, 27, 30, 32
Fair Oaks, Battle of, 175
Falmouth, 180
Farragut, David Glasgow, American admiral, 171, 185
Fayetteville, Battle of, 185
Fehrbellin, Battle of, 20
Fenian raids, 151, 162
Ferdinand, Archduke of Austria, 100
Ferdinand II, Holy Roman Emperor, 24, 25
Ferdinand III, Holy Roman Emperor, 26
Ferdinand VII, King of Spain, 132
Ferey, Claude Francois, Baron, French general, 112
Ferozepore, 152
Ferozeshah, Battle of, 151, 152
Ferrol, British raid on, 82
Festetics de Tolna, Field Marshal Tassilo, Count, 190
Finland, Russian annexation of, 138
Firearms: 37, 65; revolutionized by Gustavus Adolphus, 27
First Coalition, War of the, 89
First World War, 131
Five Forks, 171, 185
Fletcher, Colonel Richard, British soldier, 112
Fleurus, Battle of, 39
Florida, 62, 136
Folard, Le Chevalier Jean-Charles de, French military writer, 70; on tactics, 76
Fontaine, M. de, French general, 28
Fontainebleau, 122
Fontenoy, Battle of, 14, 70-71
Forrest, General Nathan Bedford, Confederate soldier, 166, 171
Forsyth, Rev. Alexander John, British inventor, 186
Fort Bard, 93
Fort Donelson, Battle for, 171
Fort Duquesne (renamed Fort Pitt), 77
Fort Erie, 116
Fort Fisher, action at, 185
Fort George, 116, 117, 140
Fort Henry, Battle for, 171
Fortifications, 50; development of, 140-141
Fort Leavenworth, 164
Fort Malden, 116
Fort Monroe, 170
Fort Niagara, 116
Fort Picurina, 112
Fort Pitt (formerly Fort Duquesne), 77
Fort San Christoval, 114
Fort Sumter, Confederate capture of, 169
Foy, Maximilien S., French general, 112
Francis Joseph, Emperor of Austria, 161
Franco-Prussian War, 131, 159, 189, 192-195; artillery used in, 186; tactics used in, 187
Franco-Spanish War, 48
Franklin, 177; action at, 185
Frederick, 180, 185
Frederick II (the Great), King of Prussia, 16, 66, 67, 69, 89, 98, 201; and indirect approach, 13; and military expenditure, 22; and principles of war, 9; at Kolin, 73, 74; at Leuthen, 75; at Rossbach, 74; employment of advance guard at Leuthen, 10; his contribution to art of warfare, 72-75; on tactics, 76; relationship with his men, 11; use of oblique order, 10, 12
Frederick V, King of Bohemia, 24, 25
Frederick VII, King of Denmark, 189
Frederick Augustus II, Elector of Saxony, 70
Frederick Charles, Prince of Prussia, 39, 190
Frederick William I, King of Prussia, 66, 72
Frederick William III, King of Prussia, 103
Fredericksburg, 167, 170, 177, 179, 180
Frederickshal, Battle of, 59; siege of, 9
Freiburg, 37
Fréjus, 97
Fremont, John Charles, Union general, 176, 177, 178
French Revolution, 16, 67, 72, 132, 200
Friuli, French invasion of, 92
Front Royal, 177
Fuentes de Oñoro, Battle of, 111
Fürstenberg, Egon, German general, 27
Futak, camp of, 56

Gadsden Purchase, 165
Gaeta, 159
Gage, Thomas, British general, 85
Gaines Mill, action at, 175
Gallipoli, 142
Galway, Henry de Massue, Earl of, British general, 49
Garibaldi, Giuseppe, Italian patriot, 159, 160, 161, 201
Gassion, Jean de, French marshal, 28
Gates, Horatio, American general, 86
Gatling gun, introduction of, 186
Gavre, Allied capture of, 53
Gazan, Honoré, French general, 115
General Staff organizations, evolution of, 127
Geneva Convention, 129
Genoa, siege of, 93
George I, King of Greece, 138
George II, King of Britain, 69
George III, King of Britain, 67, 85, 103
George William, Elector of Brandenburg, 26
German Confederation, 189
German Princes, Congress of, 189
Germantown, Battle of, 85, 86
Gettysburg, Battle of, 130, 171, 179, 180-181, 182
Ghent, 37, 38
Ghulab Singh, Sikh commander, 153
Gibraltar, 49; siege of, 67, 82, 86
Gilbert, Sir Walter, British general, 153
Gilchrist, Peter Carlyle, British inventor, 129
Glorious Revolution, 30
Gneisenau, August Wilhelm Anton von, Prussian field marshal, 125
Good Hope, Cape of, 151
Gorchakov, Prince Mikhail, Russian general, 145, 147
Goring, Lieutenant-General George, Baron, English Royalist general, 30
Gorlitz, 190
Gorni-Dubnik, 196
Gough, Sir Hugh, British general, 152, 153, 154
Graham, Thomas, Lord Lynedoch, British general, 114, 115
Grand strategy, definition of, 8
Grand tactics, definition of, 8, 9
Grant, Ulysses Simpson, Union general, 130, 171, 182, 184, 185, 201
Grasse, Francois J. P. de, French admiral, 86, 87
Grussin, Pierre, French general, 70
Grassin Legion, 70
Gravelotte-St. Privat, Battle of, 131, 193, 194; casualties at, 165
Great Northern War, 20, 58, 59, 66
Greek War of Independence, 137
Gregg, David McMurtrie, Union general, 181
Gregg, William, Confederate general, 182
Grey, Sir George, Governor of New Zealand, 151
Gribeauval, Jean-Baptiste Vacquette de, Comte, French general, 89
Grierson, Benjamin H., Union general, 171
Grodno, Swedish occupation of, 58
Grossenzersdorf, 107
Gross Heppach, 45
Grouchy, Emmanual, Marquis de, French marshal, 125, 126
Guatemala, independence of, 133
Guayaquil, 132
Guilford, Frederick North, 4th Earl of, 85
Guibert, Jacques, Comte, military theorist, 76
Gujerat, Battle of, 151, 153
Gurkha War, 149
Gustavus II Adolphus, King of Sweden, 19, 26, 28, 58, 73, 201; and principles of war, 8, 9; at Breitenfeld, 27; his contribution to warfare, 27; relationship with his men, 11; revolutionizes tactical concepts, 65
Gwalior, British capture of, 157; revolt of, 151

Habsburg-Turkish Wars, 55
Hague, strategic importance of the, 38
Haguenau, 40
Haldane, Richard Burdon Haldane, Viscount, British statesman, 131
Halifax Court-House, 171
Halle, action off, 103
Hamburg, 103
Hamley, Sir Edward, British general, 8
Hampton, 116
Hanau, Battle of, 122
Hannibal, Carthaginian general, 14
Hanover, 71
Hardinge of Lahore and Kings Newton, Henry Hardinge, 1st Viscount, British Governor-General of India, 152
Hard Times, 182
Harper's Ferry, action at, 169, 170
Harris, George, Lord, British general, 78
Harrison's Landing, 175
Harrison, William Henry, American general, 116
Harrisonburg, 176, 177
Haslach, action at, 100
Hastings, Francis Rawdon-Hastings, Earl of Moira, 1st Marquis of, British general, 78, 149
Hastings, Warren, British Governor-General of India, 78
Hauterive, capture of, 53
Havelock, Sir Henry, British general, 156
Heidelberg, 44
Heintzelman, Samuel P., Union general, 172
Hely-Hutchinson, John, British general, 97
Henry IV, King of France, 19, 34
Henry House Hill, 173
Henry rifle, 187
Herzegovina, Turkish atrocities in, 196
Hess, Henry Hermann Joseph, Baron, Austrian general, 161
Heth, Major-General Henry, Confederate soldier, 181
Hidalgo y Costilla, Miguel, Mexican revolutionary priest, 132
Hill, Ambrose Powell, Confederate general, 179, 180
Hill, Roland Hill, 1st Viscount, British general, 114
Höchstädt, Battle of, 43
Hohenfriedberg, Battle of, 69
Hohenlinden, Battle of, 93, 97
Hohenlohe-Ingelfingen, Prince Frederick Louis von, 103, 104
Hohenzollern-Sigmaringen, Leopold von, 192
Holmes, Charles, British admiral, 82
Holmes, Lieutenant-General Theophilus, Confederate soldier, 173
Holstein, General G. L., Duke, 71
Hood, John Bell, Confederate general, 171, 185
Hooghly, River, 80
Hooker, Joseph ('Fighting Joe'), Union general, 170, 179, 180, 184
Hôpital, François de L', Comte de Rosnay, French marshal, 28
Hopton of Stratton, Ralph Hopton, Baron, English Royalist commander, 30
Horn, Gustav, Swedish general, 26, 27
Hougoumont, 125
Housatanic, 129
Howard, Oliver O., Union general, 179
Howe, Richard, British admiral, 82
Howe, William, 5th Viscount, British general, 85, 86
Hudson Bay, 37; French seize posts in, 63
Hudson's Bay Company, 62
Hull, William, American general, 116
Hülsen, General J. D., 74
Hunter, David, Union general, 172
Hyder Ali, Muslim military adventurer, 78

Iberian peninsula: difficulties of campaigning in, 49
Ibrahim Pasha, Egyptian general, 137
Île d'Orleans, 82
Immigrants in New World, European, 136
India, British acquisition of, 78-80
Indian Mutiny, 156-157
Ingoldsby, Brigadier James, British soldier, 71
Ingolstadt, 45
Inkerman, Battle of, 142, 145, 146
Inkovo, 106
Intelligence, military: in Peninsular War, 11
Interregnum, 30
Ireton, Lieutenant-General Henry, English soldier and politician, 32
Irgai, Battle of, 138
Isembourg, M. de, Count, Flemish cavalry commander, 28
Italian War of Independence, 129, 158-161; use of railways in, 166

Jackson, 182
Jackson, Andrew, US general and president, 116
Jackson, J. F. G., Confederate soldier, 175, 176, 177, 178
Jackson, Thomas Jonathan ('Stonewall'), Confederate general, 130, 170, 173, 174, 179
Jacobite rebellion, 69
Jamaica, British capture of, 30, 63
James I and VI, King of Scotland and England, 25
James II, King of England, 42
James, River, 175
Jamestown, 87
Janissaries, 22
Jellacic, Josef, Count, Austrian general, 161
Jena, Battle of, 98, 103, 104
Jenkins's Ear, War of, 77

Index

Jhansi, Rani of, 157
Jhelum, River, 153
John, Archduke, Austrian general, 100
John VI, King of Portugal, 133
John George I, Elector of Saxony, 26
John George III, Elector of Saxony, 55
Johnston, Albert Sidney, Confederate general, 171
Johnston, Joseph Egglestone, Confederate general, 171, 172, 174, 175, 176, 182, 185
Jomini, Henri A., Baron, Swiss general and military theorist, 129
Joubert, Barthélemy C., French general, 92
Jourdan, Jean-Baptiste, Comte, French marshal, 89, 114
Junot, Andoche, French general, 111

Kaffir Wars, 151
Kara Mustafa, Grand Vizier of Turkey, 55, 56
Kars, Russian capture of, 142
Kassai of Shoa, Prince, 154
Katahia, Convention of, 137
Kearny, Colonel Stephen Watts, American soldier, 164
Keith, Sir William, British admiral, 82, 83
Kellerman, François-Etienne Christophe, Duc de Valmy, French general, 94
Kelly's Ford, 179
Kempt, Sir James, British soldier, 115
Kennesaw Mountain, Union assault at, 185
Kerkhoff, Allied capture of, 53
Kernstown, Battle of, 176, 178
Khalil Pasha, Grand Vizier of Turkey, 57
Khiva, khanate of, 138
Khoczim, Battle of, 54
Khokand, khanate of, 138
Kiel, 189
'King William's War', 62
Kléber, Jean Baptiste, French general, 97
Knoxville, 184
Knyphausen, Wilhelm von, Prussian general, 86
Kolin, Battle of, 73-74
Königgrätz (Sadowa), Battle of, 131, 190
Kornilov, V. A., Russian admiral, 145, 146
Krasnöe, action at, 120
Krudener, Nicolas, Russian general, 196
Krupp guns, 186
Kulm, Battle of, 122
Kunersdorf, Battle of, 69, 75
Kutusov, Mikhail Illarionovich, Prince, Russian field marshal, 100, 101, 120, 121

La Belle Alliance, 125
Lafayette, Marie Joseph Paul Roch Yves Gilbert Motier, Marquis de, French soldier and statesman, 22, 85
La Ferté-Senneterre, Henri de, French general, 28
Laffeldt, Battle of, 70
La Haie Sainte, 126
Lake Erie, action on, 116
Lake, Sir Gerard, British general, 78
Lake Ontario, 116
Lake Ticonderoga, 116
Lake Van, 137
Lally, Comte de, Governor-General and Commander-in-Chief of French India, 78
La Marmora, Alfonso Ferrero di, Sardinian general, 146, 159, 190
Landau, 42; Allied capture of, 45
Landen, (Neerwinden), Battle of, 39
Landsberg the Younger, Prussian engineer, 140
Langdale, Sir Marmaduke, (later Baron Langdale), 32
Lannes, Jean, Duc de Montebello, French marshal, 93, 94, 101, 103
Laon, Battle of, 122
Lapoype, French general, 94
La Rothière, action at, 122
Las Lineas, British assault on, 82
Lasalle, Antoine Charles, French general, 103
Lasta, Prince of, 154
Latin-American Wars of Independence, 132
Laubat, Chasseloup de, French engineer, 140
Laud, William, Archbishop of Canterbury, 30
Launsheim, 44
Lauriston, Jacques Alexandre Bernard Law, Comte, 11
Lawrence, Henry, English colonial administrator in India, 78
Lawrence, Sir Henry Montgomery, British soldier, 151, 156
Lawrence of the Punjab and of Grately, John Lawrence, 1st Baron, Governor-General of India, 151, 156
League of Armed Neutrality, 67
Le Caillou, 125
Lee, Robert Edward, Confederate general, 130, 170, 171, 174, 175, 179, 180, 181, 185, 201
Leeward Islands, 63
Leffinghe, 53
Legnago, fortress of, 159
Leicester, 32
Leipzig, 122
Leith, Sir James, British general, 112
Le Marchant, John G., British general, 112
Lemberg, 54
Leoben, 92, 93
Leopold I, Holy Roman Emperor, 10, 19, 55
Leopold II, Prince of Anhalt-Dessau, 73, 76
Lermanda, action near, 115
Lesczynski, Stanislas, King of Poland, 70
Leslie, David, (later Baron Newark), English-Scottish Parliamentary general, 30
Lestocq, C. Anton Wilhelm, Prussian general, 106
Le Tellier, Michel de Louvois, French Secretary of State for War, 27, 34
Leuthen, Battle of, 10, 12, 69, 75
Leven, Alexander Leslie, 1st Earl of, Swedish-Scottish general, 30
Lewenhaupt, Adam, Swedish general, 58, 61
Lexington Green, action at, 85
Lichtenstein, Wenzel, Prince, Austrian field marshal, 101
Lienbenberg, Johann von, Burgomaster of Vienna, 55
Light Brigade, charge of the, 145, 165
Ligny, 125
Ligonier, John Louis, Earl, British soldier, 71
Lille, 37; siege of, 33, 38, 42, 52-53
Lima, patriots' occupation of, 132
Lincoln, Abraham, President of the United States of America, 130, 166, 167, 169, 170, 172, 174, 175, 177, 179, 180, 182, 185
Lincoln, Benjamin, American general, 86
Lisbon, 10, 49, 112; French occupation of, 111
Littler, Sir John, British general, 152
Liverpool, Robert Banks Jenkinson, 2nd Earl of, British politician, 67
Lobau, Georges Mouton, Comte de, French general, 125
Lobau, Isle of, 107
Lodi, Battle of, 9, 90
Logistics: definition of, 9; difficulty of in Iberian peninsula, 49; during French Revolutionary Wars, 89; in seventeenth and early eighteenth century, 33
Lonato, Battle of First, 92
London, 38
Long Parliament, 30
Longa (alias Papel), Brigadier-General Juan, Spanish guerrilla leader, 115
Longstreet, James, Confederate general, 175, 181, 184
L'ordre mixte, 98; used at Marengo, 94
Lorraine, Duke of, see: Charles
Louis, Prince, Margrave of Baden, 44, 45, 54
Louis XIV, King of France, 19, 20, 27, 28, 29, 34, 37, 39, 41, 49, 53, 201
Louis XV, King of France, 66, 70
Louis XVIII, King of France, 122
Louis-Ferdinand, Prince of Prussia, 103
Louisbourg, fortress of, 77, 82, 140
Louisiana purchase, 162
Louvois, François Michel le Tellier, Marquis de, French statesman, 27, 34
Lübeck, 103
Lucan, George C. Bingham, Earl of, British soldier, 145
Lucchessi d'Averna, General Joseph, 75
Lucknow, siege of, 156, 157
Lundy's Lane, action at, 116
Lutter, Battle of, 25
Lützen, Battle of, 9, 26, 122
Luxembourg, 37
Luxembourg, Major-General le Chevalier de, 53
Luxembourg, François Henri de Montmorency-Bouteville, Duc de, French general, 19, 27, 201; campaigns of, 39

Maastricht, 37; French capture of, 70
Macdonald, Jacques Etienne Joseph Alexandre, Duc de Tarente, French marshal, 122
Macdonough, Commodore Thomas, American naval officer, 116
Mack, Karl Leiberich, Baron, Austrian general, 100, 101
Mackenzie's Farm, Battle of, 146
MacMahon, Marie Edmé Patrice de, Duc de Magenta, French marshal, 159, 161, 192, 193, 194, 195
Madras, 77, 78
Madrid, 111, 112
Magdala, Napier's march to, 154
Magdeburg, razing of, 26; siege of, 103
Magenta, Battle of, 161
Magruder, Major-General John B., Confederate soldier, 175, 177
Mahan, Alfred, American admiral and theorist, 9
Maharajpur campaign, 151
Mahratta Wars, 78, 149
Maitland, Captain Frederick Lewis, British naval officer, 126
Malaga, Battle of, 49
Malakov Fort, 142, 145, 146
Malo-Jaroslavets, Battle of, 120
Malplaquet, Battle of, 12, 14, 42, 43, 47
Malta, British conquest of, 96
Malvern Hill, action at, 175, 185
Manassas (First Bull Run), Battle of, 170, 172-173
Manassas Junction, 172
Manoeuvres: attack from a defensive position, 13; attack in oblique order, 12; envelopment (superiority strategy), 12, 90, 93, 98; feigned withdrawal, 12-13; indirect approach, 13; Napoleonic strategic battle, 98; of the central position (inferiority strategy), 90, 98
Mansfeld, Ernest von, Count, German general, 25
Manstein, C. H. von, Prussian general, 74
Manton, Joseph, British gunsmith, 186
Mantua, Battle of, 159; siege of, 92
Maori War, First, 151
Maracaibo, Second Battle of, 132
Marching: by flanks, 33; by lines, 33
Marengo, Battle of, 93-94; Napoleon's employment of l'ordre mixte at, 98
Maria Theresa, Archduchess of Austria, 66, 69
Marlborough, John Churchill, 1st Duke of, 19, 27, 33, 39, 42, 51, 52, 53, 55, 58, 65, 73, 76, 201; and 1704 campaign, 43, 45, 46, 47; forces lines of Brabant, 40-41; generalship of, 42; humanity of, 11; march to Danube, 13; penetration of centre at Malplaquet, 12; politics and, 42; principles of war and, 9, 10
Marmont, August Frederic Louis Viesse de, Duc de Raguse, French general, 12, 13, 94, 111, 114, 122
Marmorice Bay, 82
Marsala, 159
Marsin, Ferdinand, Comte de, French marshal, 10
Mars-la-Tour, action around, 194; Battle of, 193
Marston Moor, Battle of, 30
Marx, Karl Heinrich, German philosopher, 128
Massawa, 154
Masséna, André, French general, 90, 92, 107, 111, 112, 141
Matthias, Holy Roman Emperor, 24, 25
Maubeuge, 37
Maucune, Antoine-Louis, Baron, French general, 112
Maurice, Count of Nassau and Prince of Orange, 27, 65
Maximilian, Elector of Bavaria, 24
Maximilian II Emmanuel, Elector of Bavaria, 44, 45, 53, 55, 56
Maxwell, Colonel Patrick, British soldier, 80
Mazeppa, Ivan Stepanovich, Russian soldier, 59, 61
Mazzini, Giuseppe, Italian patriot, 158, 159, 161
McClellan, George Brinton, Union general, 170, 173, 174, 175, 176, 177
McClernand, John A., Union general, 182
McDowell, 177
McDowell, Irvin, Union soldier, 170, 172, 173, 174, 176, 178
McLaws, William B., Confederate general, 179
McPherson, James B., Union general, 182
Meade, George Gordon, Union general, 130, 171, 180, 181
Mechanicsville, action at, 175
Meerut, mutiny at, 156
Melas, Michael, Baron, Austrian general, 93, 94
Menin, 37, 52
Menou, Jacques F., Baron de, French general, 82, 97
Menshikov, Alexander Sergeievich, Prince, Russian general, 142, 144, 145, 146, 147
Mentana, Battle of, 159
Méry-sur-Seine, 122
Metz, 28, 34, 141, 193, 194, 195
Mexico, independence movement in, 132, 133
Mexico City, American occupation of, 164, 165
Milan, French capture, 92
Millesimo, Battle of, 90
Milroy, Robert, Union general, 176, 177
Minas, Antonio Luis de Sousa, Marquis das, Spanish commander, 49
Minden, Battle of, 67, 71
Minié, Charles, French inventor, 186
Minorca, 82; capture of, 49
Minor tactics, definition of, 9
Minsk, 120
Miranda de Azan, 112
Miranda, General Sebastian Francisco de, Venezuelan revolutionary, 132
Mir Jafar, Nawab of Bengal, 80
Missalonghi, siege of, 137
Missouri Compromise, 169
Mitrailleuse machine-gun, 131, 186
Mobile, 116, 185
Mohacs, Battle of, 54
Mohammed Ali, Pasha of Egypt, 137, 198
Molino del Rey, action at, 165
Mollwitz, Battle of, 69
Moltke, Helmuth Karl Bernhard von, Prussian marshal, 7, 129, 173, 190, 192, 193, 194, 195, 198, 201; his utilization of railways, 166; military methods are resisted, 165; perfects Staff System, 131
Moltke, Helmuth von, German general, 198
Monckton, Brigadier-General Robert, British Army officer, 82
Mondovi, Battle of, 90
Mongols, 14, 17
USS *Monitor*, 129
Monk, Lieutenant-General George (later Duke of Albemarle), 30
Monnier, Jean-Charles, French general, 94
Monongahela, River, 77
'Monroe Doctrine', 128
Monroe, James, President of United States, 128, 132
Mons, 38, 53; capture of, 42
Mons-en-Pêvelle, 53
Montalembert, Marc-René, French engineer, 140
Montcalm de Saint-Servan, Louis-Joseph de Montcalm-Gazon, Marquis de, French marshal, 82
Mont Cenis, 166
Montecuccoli, Count Raimondo, Italian soldier, 27, 54
Montenegro, independence of, 137
Montenotte, Battle of, 90
Monterrey, action at, 164
Montgomery, Field Marshal Bernard, British soldier, 8
Montgomery, Confederate capital established at, 169
Montgomery, Richard, American general, 85
Montmirail, action at, 122
Montmorency, River, 82
Montrose, James Graham, 1st Marquis of, Scottish general, 30
Mont Royal, 62
Mont St. Jean, 125
Moore, Sir John, British general, 82, 83, 111
Morale, maintenance of, 9
Moravian Town, 116
Morea, Greek massacre in the, 137; Venetian capture of, 54
Moreau, Jean Victor, French general, 89, 92, 93, 97
Morgan, Daniel, American soldier, 86
Morillo, Pablo, Spanish general, 114, 132
Moriscos, expulsion of the, 48
Moritz, Prince of Anhalt-Dessau, Prussian general, 74
Morosini, Francesco, Venetian general, 54
Morrison, Colonel Joseph, British soldier, 116
Mortier, Edouard Adolphe Casmir Joseph, Duc de Trévise, French marshal, 103, 122
Moscow, occupation of and retreat from, 120-121
Motte, Louis-Jacques du Fosse, Count, French general, 53
Mount Jackson, 176
Mouton: see Lobau
Mudki, action at, 152
Muhammad IV, Sultan of Turkey, 54
Mulhouse, Battle of, 27, 40
Multan, British capture of, 153
Munich, 69
Münnich, Burchard Christophe, Prussian engineer, 140
Murat, Joachim, Prince, King of Naples, 100, 101, 104, 106, 109, 111, 119, 120, 121
Murfreesboro, 184; Battle of, 171
Murray, Brigadier-General the Hon. James, 82
Mustafa Kuprili, Grand Vizier of Turkey, 54, 56
Mustafa Pasha, Turkish general, 56

Nadasti, General F. L., 75
Naismith, Lieutenant, British soldier, 142
Nana Sahib (Dandu Panth), Rajah of Bitpur, 156
Namur, fortress of, 38, 39
Napier, Sir Charles, British admiral, 142
Napier, Sir Charles, British general, 151, 153
Napier, Sir Robert, British general, 134, 154
Napoleon (I) Bonaparte, Emperor of the French, 14, 17, 27, 67, 68, 69, 72, 73, 75, 89, 129, 201; and Austerlitz campaign, 100-101; and 1814 Campaign, 122-123; and principles of war, 9; and Waterloo campaign, 125-126; destroys Prussian Army, 103-105; employment of the indirect approach, 13; employs attack in oblique order at Castiglione, 12; his campaigns in North Italy, 90-94; his contribution to warfare, 98-99; his effect on classical warfare, 16; his expansionist ambitions, 109; his prosecution of warfare, 16-17; his regard for military history, 8; imposes Continental System on Britain, 109; in the Levant, 96; invades Russia, 118-121; on generalship, 11; Peninsular War and, 110-115; setbacks at Eylau and Aspern-Essling, 106-107; use of strategic systems, 10
Napoleon III, Louis-Napoleon Bonaparte, Emperor of the French, 129, 131, 142, 144, 146, 158, 161, 186, 189, 192, 194, 198
Narva, Battle of, 10, 14, 58, 59-61
Naseby, Battle of, 30, 32
Nashville, 184, 185
Natal, 151
Nations', 'Battle of the, 122
Naumburg, 103, 104
Navarino, Battle of, 137
Nebel, River, 33, 46
Neerwinden, see: Landen, Battle of
Nelson of the Nile and Burnham Thorpe, Horatio Nelson, Viscount, British admiral, 96
Ne Plus Ultra, Lines of, 41
New Amsterdam, 62

Newfoundland, 37, 62
New Market, 176, 177
New Model Army, 27, 30, 32
New Orleans, 170; Battle of, 116; Union capture of, 171
New South Wales, development of, 151
New York, 62, 86
New Zealand, annexation of, 151
Ney, Michel, French marshal, 104, 106, 120, 121, 122, 125
Nezib, Battle of, 137
Niagara, River, 116
Nice, 37
Nicholas I, Tsar of Russia, 138, 142, 147
Nicholas Nikolaevich, Grand Duke of Russia, 196
Nicholson, Brigadier-General John, British soldier, 156
Nightingale, Florence, British nurse, 129, 146
Nile, see: Aboukir, Battle of
Nine Years War (War of the League of Augsburg), 39, 62
Nish, Turkish capture of, 54
Noailles, Adrien-Maurice, Duc de, French marshal, 70
Nolan, Captain Lewis, British soldier, 145
Nördlingen, Battle of, 26
North, Lord, see: Guilford
Nottingham, 30
Novara, Battle of, 158

Oberglau, 46
Oblique order, 12, 73; used at Assaye, 80
O'Higgins y Riquelme, Bernardo, Chilean general and president, 132
Ohio, River, 77
Okey, Colonel John, Parliamentary soldier, 32
Olmutz, 100
Oltenitza, Battle of, 142
Omar Pasha, Turkish soldier, 142, 146
Orange Free State, 151
'Oregon Claim', 162
O'Reilly, Austrian general of Irish extraction, 94
Orford, Robert Walpole, 2nd Earl of, British politician, 66, 69
Orléans, French recapture of, 193
Orléans, Philip II of France, Duke of Chartres and, 41, 49
Osman Pasha, Turkish general, 196
Ostend, 53
Ott, Karl, Austrian general, 94
Oudenarde, 37; Allied capture of, 53; Battle of, 42, 43, 46, 47
Oudenburg, action near, 53
Oudinot, Nicolas, French general, 158, 159, 160, 161
Oudinot, Nicolas-Charles, French marshal, 107, 121, 122
Outram, Sir James, British general, 156
Overkirk, Hendrik, Dutch field marshal, 42
Oxford, 30

Pagan, Blaise François, Comte de, 50
Pakenham, Sir Edward, British general, 116
Palermo, fall of, 159
Palestrina, Battle of, 160
Palliser, Major Sir W., British inventor, 186
Palmerston, Henry John Temple, 3rd Viscount, British politician, 146
Papineau, Louis-Joseph, Canadian politician, 151
Pappenheim, Gottfried Heinrich Zu, Count, German general, 26, 27
Paris, 122, 141; siege of, 131, 167, 193, 194, 195
Parrot gun, 186
Patterson, Major-General Robert, Union soldier, 172
Paulov, P. I., Russian general, 145
Peach Tree Creek, action at, 185
Pedro 1, Emperor of Brazil, 133
Peking, march on, 136
Pelissier, Jean Jacques, French general, 142, 146, 147
Pemberton, John C., Confederate general, 171, 182
Penal settlements, British, 136, 151
Pender, Major-General William D., Confederate soldier, 181
Peninsular War, 110-115; logistics in, 9, 11; military intelligence in, 11
Penn, William, English Quaker and colonist, 63
Pensacola, 116
Pernambuco, revolt around, 133
Perry, Commodore Oliver Hazard, American naval officer, 116
Persia, Turks' war with, 137
Peschiera, fortress of, 159
Peshawar, 156; Indian mutiny in, 153
Peter (I) the Great, Tsar of Russia, 10, 12, 20, 54, 58, 59, 60, 70; at Poltava, 61; at the Pruth, 61
Peter III, Tsar of Russia, 67, 68
Petersburg, 141, 171; siege of, 185
Peterwardein, Battle of, 56; siege of, 55
Pickett, George, Confederate general, 181
Philadelphia, British occupation of, 86; British offensive against, 85
Philip, Duke of Anjou (later Philip V, King of Spain), 19
Philip II, King of Spain, 48
Philip III, King of Spain, 48
Philip IV, King of Spain, 48
Philip V, King of Spain, 37, 49
Philiphaugh, Battle of, 30
Phillipon, Armand, French general, 112, 114
Philippsburg, 34, 37
Piacenza, French capture of, 93
Picton, Sir Thomas, British soldier, 114, 115
Pindari War, 78, 149
Pinerolo, 34, 37
Pitt, see: Chatham
Pius IX, Pope, 159
Plancenoit, 126
Plassey, Battle of, 77, 78, 80, 156
Plattsburgh, 116
Plevna, siege of, 141, 196
Po, River, 90
Point Levis, 82
Polish Succession, War of the, 70
Polk, James Knox, President of the United States, 164, 165
Poltava, Battle of, 10, 20, 59, 61
Pomerania, Swedish occupation of, 26
Pondicherry, French trading post at, 78
Pont-à-Marque, 53
Pontiac, Indian Chief, 77
Poniatowski, Josef Anton, Prince, French marshal, 121
Poor, Brigadier-General Enoch, American soldier, 86
Pope, John, Union general, 170
Populations: affect of Thirty Years war on, 19; between 1600-1800, 18, 23; effect of warfare on, 29; in nineteenth century, 129
Porcile, 92
Port Arthur, 141
Port Gibson, action at, 182
Port Hudson, 182
Port Republic, Battle of, 178
Port Royal, 170; attacks on, 63; Battle of, 62; renamed Annapolis, 63
Portsmouth, 141
Prague, 24, 69, 74, 75
Pratzen Heights, 12, 101
Prenzlau, 103
Preston, Battle of, 30
Prevost, Sir George, British Governor of Canada, 116
Princeton, Battle of, 85
Proctor, Colonel Henry, British officer, 116
Provera, Johann, Marquis of, Austrian general, 92
Pruth, Battle of the, 54, 59, 61; Turkish envelopment of both flanks at, 12
Puebla, 114; American occupation of, 165; siege of, 165
Pym, John, English politician, 30
Pyramids, Battle of the, 96

Quadrilateral fortresses, 158, 159
Quatre Bras, 125
Quebec, capture of, 77; siege of, 82, 85
'Queen Anne's War', 63
Queensland, expansion of, 151
Queenstown Heights, Battle of, 116

Radetsky, Joseph, Austrian field marshal, 165
Raglan, Fitzroy James Henry Somerset, 1st Baron, British general, 142, 144, 145, 146, 147, 167
Railways, advent of, 129; development and strategic effects of, 166-167
Raleigh, 185
Ramillies, Battle of, 9, 10, 12, 33, 41, 42, 46
Ramnagar, actions around, 153
Ranjit Singh, ruler of the Punjab, 151
Ranjodl Singh, Sikh soldier, 152
Raymond, 182
Red Cross Society, founding of, 129
Red River Expedition, 151
Reed, Sir Thomas, British general, 152
Rehnskold, Karl Gustaf, Swedish general, 61
Reid, Sir Charles, British general, 156
Reille, Honoré, Comte, French general, 115
Rensel, Russian general, 61
Reshid Pasha, Turkish soldier, 137
Restoration, Stuart, 30
Reunion Act, 151
Réunions, The, 19
Revolutionary Wars, French, 89, 200
Rézonville, 165; Battle of, 193
Rhé, Isle of, 82
Richardson, Major-General Israel B., Union soldier, 172
Richelieu, Armand Jean du Plessis, Cardinal and Duc de, 19, 26, 34
Richmond, 170, 173, 174, 176; Confederates move capital to, 169
Riedesel, Major-General Friedrich Adolph, Baron, 86
Riell, Louis, Canadian revolutionary, 151
Riga, 61, 120
'Right of Search', abolition of, 129
Rimpler, George, Prussian military engineer, 50, 55
Risorgimento, 158-161
Rivoli, Battle of, 92
Roads, 17
Rochambeau, de, Jean Baptiste Donatien de Vimeur, French general, 86
Rocoux, Battle of, 70
Rocroi, Battle of, 12, 19, 26, 27, 28
Rodney, Sir George B., British admiral, 86
Rome, siege of, 159-161
Romney, Confederate raid on, 175
Roos, Carl Gustav, Swedish general, 61
Root, Elihu, US Secretary of State, 131
Rose, Sir Hugh, British general, 157
Rosecrans, William Starke, Union general, 171, 175, 184
Ross, General Robert, 116
Rossbach, Battle of, 69, 75
Ruchel, Ernest Philip von, Prussian general, 104
Rudolstadt, action at, 103
Rupert, Prince, English soldier, 30, 32
Russell, William Howard, British journalist, 129, 142, 146
Russo-Turkish War (1710), 54; (1828-29), 137; (1877-78), 196

Saabrucken, skirmish at, 193
Sachsen-Hildburghausen, Joseph Friedrich, Prince, 74
Sackville of Drayton, Lord George Sackville, 1st Viscount, British general, 71
Sacramento, action at, 164
Sadowa, see: Königgrätz, Battle of
St. Arnaud, Armand Jacques Leroy, French marshal, 142, 144, 146, 167
St. Augustine, siege of, 77
St. Avold, 167
St. Bernard Pass, 93
St. Dizier, French occupation of, 122
St. Fraise, M., French soldier, 80
St. Gotthard-on-the-Raab, Battle of, 54
St. Gotthard pass, 93
St. Kitts, 37
St. Omer, 37
St. Petersburg, building of, 138
St. Privat, see: Gravelotte—St. Privat
Salamanca, Battle of, 9, 12, 13, 111, 112, 114
Salem Church, Battle of, 179
Samarkand, fall of, 138
San Francisco, revolt by colonists near, 162
San Gabriel, action at, 164
San Jacinto, Battle of, 162
San Luis Potosi, 164
San Martin, Jose de, Argentine general, 132
San Pasqual, action at, 164
San Sebastian, siege of, 111
Santa Anna, Antonio Lopez de, Mexican general and president, 133, 164, 165
Santa Cruz system, 76
Santa Fe, 164
Saratoga, Battle of, 67, 85, 86, 87
Sardar Lal Singh, Sikh soldier, 152
Sasbach, Battle of, 27
Saugar, relief of, 157
Saunders, Charles, British admiral, 82
Savage Station, action at, 175
Savannah, 185; building of, 77; Union capture of, 171
Saxe, Hermann-Maurice, Comte de, French marshal, 14, 66, 69-70, 72, 73, 201; on tactics, 76
Scarlett, Sir James, British general, 145
Schellenberg, storming of the, 43, 44, 45
Scheldt, River, recapture of, crossings on the, 52, 141, 53; strategic significance of, 38
Schenk, Major-General R. C., Union soldier, 172
Schleiz, action at, 103
Schleswig, occupation of, 189
Schlieffen, Alfred von, Count, German general, 166, 198
Schwarzenberg, Karl Philip, Prince, Austrian marshal, 122
Scindia, Prince of, see: Doulut Rao, Maratha leader, 80
Scotland, 'Young Pretender's' invasion of, 66
Scott, Winfield, Union general, 164, 165, 169, 170, 173
Scutari, 142, 146
Sebastopol, siege of, 141, 142, 145-147
Second Coalition, War of the, 93
Sedan, Battle of, 131, 165, 193, 194
Sedgwick, John, Union general, 179
Senafe, 154
Senova, Battle of, 196
Serbia, revolt in, 137
Seringapatam, siege of, 78
Sérurier, Jean Mathieu Philibert, Comte, French marshal, 90
Seven Years War, 66, 69, 73, 77, 78
Seydlitz, Friederick Willhelm von, Prussian general, 74
Sharpsburg, 170
Shenandoah Valley, campaign in, 174, 175
Sheridan, Philip Henry, Union general, 171, 185
Sherman, William Tecumseh, Union general, 7, 130, 171, 173, 182, 184, 185
Sher Singh, Sikh soldier, 153
Shields, James, Union general, 176, 177, 178
Shiloh, Battle of, 171; casualties at, 165
Ship-money, 30
Shuja, Shah of Afghanistan, 149
Sibertoft, 32
Sickles, Daniel E., Union general, 179, 181
Siege warfare, 14, 50-53
Sikh Wars, 151-153
Silesia, 66, 69, 72, 75
Silistria, Russian attack on, 142
Silva, José e, 133
Simpson, Sir James, British general, 142, 147
Sind, British annexation of, 151
Sinope, action at, 142
Siraj-ud-Daula, Nawab of Bengal, 78, 80
Sirot, Claude de Létouf, Baron, French general, 28
Six Weeks (Austro-Prussia) War, 190, 198
Slavery, 169
Slocum, Major-General Henry W., Union soldier, 179
Small arms, 186-187
Smith, Sir Harry, British general, 152, 153
Smith, Kirby, Confederate general, 173
Smolensk, manoeuvre of, 120
Sobieski, John III, King of Poland, 54-56
Sobraon, Battle of, 151, 152, 153
Socrates, Athenian philosopher: his definition of the qualities of generalship, 11
Sohr, Battle of, 69
Soimonov, F. I., Russian general, 145
Solferino, Battle of, 159, 161
Sombreffe, 125
Soubise, Charles de Rohan, Prince of, French marshal, 74
Soult, Nicolas-Jean de Dieu, Duc de Dalmatie, French marshal, 101, 104, 106, 111, 112, 114
Spain, revolt in, 132
Spanish Succession, War of the, 19, 20, 38, 40, 42, 46, 50, 58, 66, 76
Spencer carbine, 187
Spicheren, Battle of, 131, 167, 193
Spörcke, Lieutenant-General Freiherr von, 71
Spring Hill, action at, 185
Staff System, Prussian, 131
Stahremberg, Ernst Rüdiger von, Count, Austrian general, 55
Staunton, 177
Steinheil, Russian general, 120
Steinkirk, Battle of, 38, 39
Stockton, Commander Robert B., American naval officer, 164
Stollhofen, lines of, 44
Stoneman, George, Union general, 171
Stopford, George, Union general, 179
Stowe, Harriet Beecher, American writer, 169
Strafford, Thomas Wentworth, 1st Earl of, Lord-Lieutenant of Ireland, 30
Straits Convention, 137
Strasbourg, 37, 45
Strategy: definition of, 8, 9; grand, 8, 9
Stuart, Charles Edward (The Young Pretender), 66, 69
Stuart, James Ewell Brown, Confederate general, 173, 175, 179, 180, 181
Studienka, 120, 121
Sucre, Antonio José de, Venezuelan general, 132
Suez Canal, 148, 151
Superiority strategy (manoeuvre of envelopment), 90
Sutlej, River, 151, 152, 153
Suttee practices, suppression of, 149
Sydney, 151
Sykes, Major-General George, Union soldier, 173
Szlankamen, Battle of, 54

Tabor, Mount, 97
Tactics, 165; affected by railways, 166; cavalry, 65; experiments in mid-eighteenth century, 76; grand, 9; minor, 9; post-Napoleonic, 187; revolutionized by Gustavus Adolphus, 27; sequence of French early nineteenth century engagements, 95; use of in Franco-Prussian War, 131
Taku Forts, 136
Talavera, Battle of, 111
Tallard, Camille d'Hostan, Comte de, French marshal, 10, 43, 45, 46, 47
Tanfore, Indian attempted seizure of, 78
Tantia Topi, Maratha leader, 157
Tarvis, Battle of, 92
Tasmania, 151
Tauenzien, Bolesas Friedrich Emanuel, Count von Wittenberg, Prussian general, 103
Tauroggen, 122
Taylor, Zachary, American general and president, 164
T'Chernaya, action at, 142
Teano, 159
Tej Singh, Sikh soldier, 152, 153
Telegraph, electric, 129, 130, 167
Temesvar, Banat of, 54, 55, 56
Texan Independence, War of, 162
Texas, annexation of, 162
Thames estuary, security of, 38
Theodore III, Emperor of Abyssinia, 154
'Theodorus', 154
Thirty Years War, 7, 8, 14, 19, 24, 30, 33, 34, 37, 51
Thomas, George Henry, Union general, 171, 184
Thomières, 113
Three Emperors' League (Dreikaiserbund), 138
Thuggee practices, suppression of, 149

Index

Thun-Hohenstadt, Field Marshal Karl von, Count, Austrian soldier, 190
Thuringer Wald (Thuringian Forest), 103
Ticonderoga, 85, 86
Tilly, Johan Tzerclaes, Count of, Imperial marshal, 19, 25, 26, 27
Times, The, 129, 146
Tippoo Sahib, ruler of Mysore, 78
Todleben, Count Franz, Russian engineer, 141, 142, 145, 146
Torres Vedras, lines of, 10, 13, 111, 112, 141
Toul, 28, 34
Toulon, 96, 141; Battle of, 55; siege of, 89
Toulouse, Battle of, 112
Tournai, 37, 53; siege of, 70
Tours, 193
Townsend, Brigadier-General the Hon. George, British soldier, 82
Trans-Siberian railway, 138
Transvaal, 151
Treaties: Abo, 138; Adrianople, 137; Aix-la-Chapelle (1668), 19, 34; (1748), 66, 69, 77; Berlin, 7, 138, 196; Bucharest, 137; Buczacs, 54; Campo Formio, 92; Eisenburg (Vasvar), 19, 55; Erzerum, 137; Frankfurt, 189; Ghent, 116; Guadalupe, 165; Hubertusburg, 69; Jassy, 138; Karlowitz, 54, 55; Kasure, 153; Lübeck, 25; London, 137; Munster, 28; Nijmegen, 19, 37; Nystadt, 20, 59; Osnabruck, 28; Paris (1763), 67, 77, 86; (1814), 125; (1856), 138, 142, 189; Passarowitz, 55; Prague, 138, 161, 189, 190; Pruth, 54, 59, 61; Pyrenees, 19, 34; Rastadt, 19, 37, 49, 64; Ratisbon, 37; Ryswick, 19, 37, 63; San Stefano, 138, 196; Turin, 37; Utrecht, 19, 20, 37, 63, 64, 77; Vasvar, see Eisenburg; Versailles, 67, 86, 87; Vienna, 189; Villafranca, 159; Waitingi, 151; Westphalia, 19, 20, 28, 34
Trent, 92
Trenton, Battle of, 85
Treskow, General, 74
Tres Puentes, 115
Trinchinopoly, 78
Trochu, French general, 195
Troyes, French occupation of, 122
Tshitsagov, Admiral Pavel, Russian commander, 120, 121
Turckheim, Battle of, 27
Turenne, Henri de la Tour d'Auvergne, Vicomte de, French marshal, 19, 39, 42, 201; and grand tactics, 9; campaigns in the Vosges, 40; his skill as a commander, 27; relationship with his men, 11
Turin, Battle of, 41
Turkenschanz fort, 56
Turnham Green, Battle of, 30

Ucholodi, 121
Ulm: Allied capture of, 44, 45; as example of 'the indirect approach', 13; manoeuvre at, 100
Urban, Lieutenant-General Sir Benjamin d', British soldier, 112
US Ford, 179
Utitsa, 121

Vaillant, Jean-Baptiste, French marshal, 161
Valenciennes, 37
Vallière, Jean Florent de, French general, 28
Valmy, Battle of, 98
Vandamme, Dominique Joseph René, Comte d'Unsebourg, French general, 122
Van Rensselaer, Stephen, American general, 116

Varna, 142
Vauban, Sebastien le Prestre, French marshal and military engineer, 34, 37, 38, 50, 52, 70, 140, 141; his timetable for a standard siege, 51
Vaubois, C. H., Count, French general, 92
Vauchamps, action at, 122
Vaudreuil-Cavagnal, Pierre de Riguad, Marquis de, French Governor of Canada, 82
Velletri, Battle of, 160
Venables, Robert, British general, 63
Vendôme, Louis-Joseph, Duc de, French marshal, 9, 41, 49, 53
Venetia, ceded to Italy, 189
Venezuela, independence movement in, 132
Vera Cruz, 164; US evacuation of, 165
Verdun, 28, 34
Verona, fortress of, 159
Vial, Honoré, Baron, French general, 92
Vicksburg, 141, 169, 170; campaign and siege of, 130, 171, 181, 182
Victor, Claude Victor-Perrin, Duc de Bellune, French marshal, 94, 120
Victor Amadeus II, King of Savoy, 37, 41
Victor Emmanuel II, King of Italy, 159, 161, 189
Victoria, Queen of Britain, 151
Vienna: Congress of, 138, 158; French occupation of, 107; siege of, 19, 42, 54, 55-56, 201
Villars, Claude-Louis Hector de, Duc de, French marshal, 12, 43, 69
Villeroi, François de Neufville, Duc de, French general, 10, 41
Vilna, 119, 120
Vimeiro, Battle at, 111
Vinoy, General, Governor of Paris, 195
Vionville, Battle of, 193
Vitoria, Campaign and Battle of, 111, 114, 115
Vladivostok, 138
Voltaire, François Marie Arouet, French writer and thinker, 11
Volturno, River, 159
Vorskla, River, 61

Wagram, Battle of, 107
Waldeck, George Frederick, Prince of, Prussian general, 42, 55
Wallachian Revolt, 137
Wallenstein, Albrecht Eusebius Wenzel von, Duke of Friedland and Mecklenburg, Imperial general, 25, 26
Walpole, see: Orford
Wandewash, Battle of, 78
Wangenheim, George August, German general, 71
Warfare: and its effect on economies, 22; art of in seventeenth and early eighteenth century, 33; attitudes in nineteenth century towards, 128, 129; classical, 14-17; effects on populations, 19, 22, 29; generalship, 11; principles of, 8-11; railways and, 166-167; siege, 14, 50-53; variable factors in, 10
Warren, G. K., Union general and engineer, 181
Warsaw, 55; French occupation of, 106
Washington, 116
Washington, George, 1st President of the United States of America, 22, 67, 84, 85, 86
Waterloo, Battle of, 9, 10, 67, 76, 124-126
Watson, Charles, British admiral, 78
Wavell, Sir Archibald Percival, British field marshal, 8; definition of tactics, 9; views on

essentials of generalship, 11
Wavre, 125, 126
Waynesboro, 177
Weapons: 37, 131; artillery, 27, 37, 95; bayonets, 37; development of, 186-187; firearms, 27, 37, 65
Webb, John Richmond, British general, 53
Webster, Daniel, American politician, 169
Weimar, 103
Weissenburg, action at, 192
Wellington, Arthur Wellesley, 1st Duke of, 10, 12, 67, 78, 80, 111, 114, 115, 141, 201; administrative skill of, 11; and the campaign of 1814, 122; at Waterloo, 125-126; his use of classical methods of warfare, 16-17; in Peninsular War, 112-113; interpretation of military intelligence in Peninsular War, 11; lines of communication in the Peninsula, 9, 10; opposes attempts at modernizing British Army, 128; relationship with his men, 11; uses feigned withdrawal in Peninsula, 13
West Chester, 86
Wheeler, Sir Hugh, British general, 156
White Mountain, Battle of the, 25
Whitworth gun, introduction of, 186
Wilderness', 'the, 171, 179, 185
Wilkinson, Major-General J., American soldier, 116
William I, King of Prussia, 190, 198
William III, Prince of Orange and King of England, 19, 20, 32, 34, 39, 42, 65
Williams, Sir William Fenwick (William Pasha), British general, 142
Wilmington, 185
Wilson, Sir Archdale, British general, 156
Wimpffen, Emmanuel Felix de, French general, 161
Winchester, 175, 176; Battle of, 177
Winchester carbine, 187
Withdrawal, feigned, 12
Wittgenstein, Ludwig Adolf Peter, Russian general, 120, 122
Wolfe, General James, British soldier, 67, 77, 82
Woodstock, 176
Wool, John E., American general, 164
Worcester, Battle of, 30
Worth, Battle of, 192
Wrangel, Friedrich Heinrich von, Prussian general, 165
Wurmser, Dagobert Siegmund, Count, Austrian general, 92
Württemberg, Crown Prince William of, 103
Würzburg, 103
Wynendael, action at, 53

'Year of Revolutions', 128, 142, 158
Yorck, Hans David Ludwig, Count, Prussian general, 122
Yorktown, Battle of, 67, 86, 170
Ypres, 37
Yturbide, see: Augustin I

Zach, Austrian general, 94
Zenta, Battle of, 19, 54
Zgalevitsa, action at, 196
Ziethen, Hans Joachim von, Prussian soldier, 74
Ziethen, Wieprecht Hans von, Prussian general, 126
Zorndorf, Battle of, 69, 75
Zula, 154
Zulu War, 151
Zündnadelgewehr, introduction of the, 186, 187

Above right: The penalty of defeat: Russian gun-teams crash through the ice on the Satschanmere at Austerlitz, as the defeat of the Allies by Napoleon's Grande Armée turns into a rout (2 December 1805). (DAG)

Right: Bull Run was the scene of a second tough battle—on 29/30 August 1862—when Lee and Jackson inflicted a heavy defeat on General Pope and destroyed Union stores in the rail depot.

Overleaf: The celebrated 'Charge of the Light Brigade' at the Battle of Balaclava—immortalized by Lord Tennyson's poem—was in fact a perfect example of the futility of warfare, as fought in the mid-nineteenth century. (DAG)